FLORA OF THE ARABIAN PENINSULA AND SOCOTRA

VOLUME 1

FLORA OF THE ARABIAN PENINSULA AND SOCOTRA

VOLUME 1

A. G. MILLER AND T. A. COPE
WITH THE ASSISTANCE OF
J. A. NYBERG

EDINBURGH UNIVERSITY PRESS
IN ASSOCIATION WITH
ROYAL BOTANIC GARDEN EDINBURGH
ROYAL BOTANIC GARDENS, KEW

© A. G. Miller and T. A. Cope 1996

Edinburgh University Press
22 George Square, Edinburgh

Designed by Geoff Green

British Library Cataloguing in Publication Data
is available for this book

ISBN 0 7486 0475 8

Set in Monotype Times by BPC Digital Data Ltd, Glasgow
and printed and bound in Great Britain

CONTENTS

PREFACE	vii
ACKNOWLEDGEMENTS	ix
ABBREVIATIONS	xi
BIBLIOGRAPHY	xiii
LIST OF FAMILIES IN VOLUME ONE	xvii
LIST OF FIGURES	xix
INTRODUCTION	1
Geology and Topography	1
Climate	8
Vegetation	13
Floristics	22
Phytogeography	25
History of Botanical Exploration	27
Conservation and Threats	28
PLAN OF THE FLORA	30
THE FLORA	32
Pteridophyta	33
Spermatophyta	71
Gymnospermae	71
Angiospermae	80
MAPS	495
INDEX	574

PREFACE

Scientific studies of the Arabian flora began over 200 years ago with Pehr Forsskål's ill-fated expedition to 'Arabia Felix' and the posthumous publication in 1775 of his *Flora Aegyptiaco-Arabica*. Since then countless collectors and travellers have visited Arabia yet, whilst there has been considerable interest in the plants of the region, no detailed Flora of the Peninsula has been attempted. The only work covering the entire region is Blatter's *Flora Arabica* (1919–1936). However, this is of limited practical use, being little more than an annotated checklist, and is now largely out of date. Schwartz's *Flora des Tropischen Arabien* (1939) covers tropical Arabia roughly south of a line from Jeddah to Muscat. This is more up to date but is, again, primarily a checklist containing only a few descriptions of new taxa. The island of Socotra has a more complete coverage than any equivalent area on the mainland thanks to Balfour's monumental *Botany of Socotra* (1888) which contains both detailed descriptions and excellent illustrations.

Since the publication of these works, and particularly in the last fifteen years, there has been an enormous increase of interest in the Arabian flora and a number of works covering individual countries and regions have appeared. The most important of these are: Al-Rawi and Daoud's *Flora of Kuwait* (1985, 1987), Batanouny's *Ecology and Flora of Qatar* (1981), Collenette's *An Illustrated Guide to the flowers of Saudi Arabia* (1985), Cornes and Cornes' *The wild flowering plants of Bahrain* (1989), Mandaville's *Flora of Eastern Saudi Arabia* (1990), Migahid's *Flora of Saudi Arabia* (1978), Miller and Morris' *Plants of Dhofar* (1988), Phillips' *Wild flowers of Bahrain* (1988) and Western's *The Flora of The United Arab Emirates* (1989). These vary in style and completeness but none cover in any detail the floristically richest areas in the south and southwest of the Peninsula.

Some ten years ago the Royal Botanic Garden Edinburgh and the Royal Botanic Gardens Kew, recognising that the absence of a modern Flora covering the entire region was a major constraint on research concerning the natural environment of the Peninsula, began a project to produce a Flora of the Arabian Peninsula and Socotra which has resulted in the publication of the present volume.

The *Flora of the Arabian Peninsula and Socotra* aims to provide a regional framework for the flora of the countries of the Arabian Peninsula. It includes all native flowering plants and ferns as well as cultivated and amenity plants of economic importance found in the following countries: Saudi Arabia, the Republic of Yemen (including the Socotran archipelago), Oman, the United Arab Emirates, Qatar, Bahrain and Kuwait.

Preface

The Flora aims to be of use to specialist and amateur alike. Descriptions are concise and keys are constructed to be simple and easy to use. It is comprehensively illustrated and, as it is intended as a reference source, full citations have been given for species and synonyms, with references to useful works also supplied. Dot maps are provided for all species. Detailed notes on ecology, the uses of plants and local names have been excluded for the most part, as it is felt that these would be better dealt with in local Floras. No attempt has been made to solve all taxonomic problems encountered; many plants are inadequately known or only known from single gatherings. Notes have been given where groups and plants require further research.

Five further volumes are planned. The sequence of families will follow the system of Engler and Prantl: volume 2, from Leguminosae to Vitaceae (including Euphorbiaceae); volume 3, Tiliaceae to Verbenacaeae (including Umbelliferae, Asclepiadaceae, Rubiaceae, Convolvulaceae and Boraginaceae); volume 4, from Labiatae to Compositae (including Scrophulariaceae and Acanthaceae); volume 5, the monocots; and finally volume 6 will contain a key to families and monographic chapters on vegetation, phytogeography, etc. The editors would be interested to hear from anyone interested in contributing to future volumes.

ACKNOWLEDGEMENTS

The research for volume one has been undertaken largely in the Herbarium of the Royal Botanic Garden Edinburgh with regular visits to the Royal Botanic Gardens, Kew and the Natural History Museum. We wish to thank the Directors and Curators of these herbaria for providing facilities and loans of specimens, and the numerous members of staff who have provided help and advice over the years. We would also like to acknowledge the support and guidance provided by past and present members of the Flora management group, including J.F.M. Cannon, J. Cullen, I.C. Hedge, D.S. Ingram, G.Ll. Lucas, D.G. Mann, N.K.B. Robson and G. Wickens.

A number of collaborators have revised and commented upon families and genera for volume one. We are indebted to the following: L. Boulos, K. Browicz, S. Holmes, D.F. Chamberlain, R.A. Clement, A. Farjon, H.E. Freitag, I. Friis, M.G. Gilbert, B.E. Jonsell, L.E. Kers, J. Lamond, D.G. Long, D.R. McKean, L.J. Musselman, C.N. Page, E. Raadts, K.H. Rechinger, N.K.B. Robson, N.P. Taylor, M. Thomas, M. Thulin, C.C. Townsend, M. Maier-Stolte and S. J. Zmarzty.

We are also particularly grateful to L. Boulos for his critical revision of the Chenopodiaceae and for his support and help over the years; S. Chaudhary for critically commenting on all the accounts in volume one and for making many additions to the distribution maps from specimens held in the Riyadh Herbarium; S. Collenette for advice on the distributions of Saudi Arabian plants and for her useful comments on the draft accounts; R.A. Clement for invaluable assistance during the early stages of the project, and in particular for her work on the Caryophyllaceae; D.F. Chamberlain for much useful advice and support over the years; R. Cameron and D. Mitchell for looking after the living collections of Arabian plants at Edinburgh; J.R.I. Wood for many useful discussions and for generously allowing us to use the data from his card index of the North Yemen flora; F.N. Hepper for checking the Forsskål names in volume one; S. Ghazanfar, I. McLeish, S.A. Gabali, J.P. Mandaville, R.R. Mill and C. Will for reading and commenting on early drafts of the introduction and other parts of the Flora and finally to the late Dorothy Hillcoat who, for so many years and almost single-handedly, kept alive research into the Arabian flora.

The maps and illustrations were prepared by Mary Bates, A. Farrar, P. Halliday, A. Lezemore, M. Main, G. Malcolm, R. Park, O.M. Parry, G.A. Rodriguez, G. Shephard and M. Tebbs. We would like to thank A. Farjon for allowing us to use his drawings of *Juniperus* and F. White and J. Léonard for allowing us to use their Phytogeographical map of Africa and Asia. We would also like to acknowledge the

Acknowledgements

support given by the *Friends of the Royal Botanic Garden Edinburgh* who provided funds for the illustrations of Crassulaceae and Rosaceae.

The present work could not have been carried out without extensive support in the field. Many people have given us help with transport, accommodation and in many other ways. We thank all of these and would like to give our particular thanks to A. Al-Rawi, V. Armer, M. Bazara, A.S. Bilaidi, R. Daly, U. Deil, J.R. Edmondson, D. Forester, M.D. Gallagher, N. Gifri, J. Grainger, L. Guarino, J.J. Lavranos, I. McLeish, M. Morris, A. Mukred, K. Müller-Hohenstein, N. Obadi, A. Podzorski, M. and K. Stanley-Price, T. Tear, R.A. Western, R.P. Whitcombe, D. Wood and J.R.I. Wood.

Finally, the editors would like to thank Ian Davidson of Edinburgh University Press and Norma Gregory of the Royal Botanic Garden Edinburgh for editorial assistance during the production of this first volume.

The editors, whilst acknowledging the help of all those mentioned above, accept full responsibility for any mistakes.

A.G. Miller and T.A. Cope

ABBREVIATIONS

(a) Miscellaneous abbreviations

Many botanical abbreviations are derived from latin words; these are printed in italics after the abbreviated form.

aff.	*affinis*: akin to, bordering.
auct.	*auctorum*: of authors: *auct. arab.*, *auctorum arabiorum*, of authors of works on the Arabian flora; *auct. mult.*, *auctorum multorum*, of many authors.
in sched.	*in schedula*: on a herbarium sheet or label.
illustr.	illustration.
J.	Jebal (mountain).
loc. cit.	*loco citato*: on the page previously cited.
nom.	*nomen*: name; *nom. ambig*, *nomen ambiguum*, ambiguous name; *nom. illeg.*, *nomen illegitimum*, illegitimate name; *nom. cons.*, *nomen conservandum*, conserved name; *nom. confus.*, *nomen confusum*, confused name; *nom. nud.*, *nomen nudum*, name unaccompanied by a description.
n.v.	*non vidi*: not seen.
op. cit.	*opere citato*: in the work previously cited.
pers. comm.	personal communication.
p.p.	*pro parte*: in part.
s.n.	*sine numero*: without number.
s.l.	*sensu lato*: in a broad sense.
sens.	*sensu*: in the sense of the author indicated and not as originally intended.
syn.	synonym.
t., tab.	*tabula*: plate.

(b) Abbreviations of authors, books and journals

Abbreviations (except for certain exceptions noted below) of authors, journals and books follow those adopted in the following standard works.

Author names: Brummitt, R.K. & Powell, C.E. (1992). *Authors of plant names*. Royal Botanic Gardens, Kew.

Journals: Lawrence, G.H.M. et al., eds. (1968, 1991). *Botanico-Periodicum-Hunterianum* and supplement.

Books: Stafleu, F.A. & Cowan, R.S. (1976–1992). *Taxonomic Literature*. 2nd edition. 8 vols and suppl., Utrecht.

Abbreviations

The frequently used references, listed below, are shortened to author, date and page number e.g. *Gypsophila montana* Balf.f. (1888 p.19). For full references see the bibliography.

T. Anderson (1860)
Balfour (1882); (1884); (1888)
Blatter (1914-1916); (1919-1936)
Boulos (1988)
Collenette (1985)
Cornes & Cornes (1989)
Deflers (1889)
Delile (1813); (1814)
Forsskal (1775)
Mandaville (1990)
Miller & Morris (1988)
Phillips (1988)
Richard (1847); (1850)
Schwartz (1939)
Schweinfurth (1894); (1896); (1899)
Vahl (1790); (1791); (1794)
Western (1990)

Selected abbreviated references to local Floras in Arabia and Floras of adjacent countries are abbreviated as follows:

Adumbr. Fl. Aeth.	Adumbratio Florae Aethiopicae (1953–). Parts 1–32. in *Webbia* 9–33.
En. Pl. Aeth.	Cufodontis, G. (1953–1972). Enumeratio Plantarum Aethiopiae Spermatophyta. *Bull. Jard. Bot. Etat Brux.* 23–42.
Fl. Egypt	Täckholm, V. (1974). *Student's Flora of Egypt*. Ed. 2.
Fl. Ethiopia	Hedberg, I. & Edwards, S. (eds) (1989–). *Flora of Ethiopia*.
Fl. Iranica	Rechinger, K.H. (ed.) (1963–). *Flora Iranica*.
Fl. Iraq	Townsend, C.C., Guest, E. et al. (eds) (1966–). *Flora of Iraq*.
Fl. Kuwait	Daoud, H.S. (1985). *Flora of Kuwait*. Vol. 1, revised by Ali Al-Rawi. Al-Rawi, A. (1987). *Flora of Kuwait*. Vol. 2.
Fl. Libya	Ali, S., El-Gadi, A & Jafri, S. (eds) (1976–) *Flora of Libya*.
Fl. Pakistan	Nasir, E. & Ali, S.I. (eds) (1970). *Flora of Pakistan*.
Fl. Palaest.	Zohary, M. & Feinbrun-Dothan, N. (1966–1986). *Flora Palaestina*.
Fl. Qatar	Batanouny, K.H. (1981). *Ecology and Flora of Qatar*.
Fl. Trop. E. Afr.	Turrill, W.B. et al. (eds) (1952–). *Flora Tropical East Africa*.
Fl. Zamb.	Exell, A.W. et al. (eds) (1960–). *Flora Zambesiaca*.
Fl. Lowland Iraq	Rechinger, K.H. (1964). *Flora of Lowland Iraq*.

BIBLIOGRAPHY

See Miller et al. (1982) for a complete bibliography.

Al-Hubaishi, A. & Müller-Hohenstein, K. (1984). *An introduction to the vegetation of Yemen.* Eschborn. GTZ. 209 pp.
Al-Rawi, A. (1987). *Flora of Kuwait.* vol. 2. Kuwait University. (for vol. 1 see Daoud 1985).
Anderson, T. (1860). Florula Adenensis. *J. Proc. Linn. Soc. Bot.* 5, suppl. 1: 1–47.
Anon. (1992). *Natural History of Saudi Arabia.* Ministry of Agriculture and Water, Riyadh, Saudi Arabia. 212 pp.
Baierle, H.U. & Frey, W. (1986). A vegetational transect through central Saudi Arabia (at-Taif-ar-Riyad). In Kürschner, H. (ed.) 1986a: 111–136.
Balfour, I.B. (1882, 1884). Diagnoses plantarum novarum Phanerogamarum Socotrensium, etc., parts 1–4. *Proc. Roy. Soc. Edinburgh* 11: 498–514, 834–842 (1882); 12: 76–98, 402–411 (1884).
— (1888). Botany of Socotra. *Trans. Roy. Soc. Edinburgh* 31: 1–446.
Batanouny, K.H. (1981). *Ecology and Flora of Qatar.* University of Qatar.
Batanouny, K.H. & Baeshin, N.A. (1978). Studies in the Flora of Arabia 1. *Taeckholmia* 9: 67–81.
Blatter, E. (1914–1916). Flora of Aden. *Rec. Bot. Surv. India* 7: 1–418.
— (1919–1936). Flora Arabica. *Rec. Bot. Surv. India* 8: 1–519.
Boulos, L. (1988). *The weed flora of Kuwait.* Kuwait University, Kuwait. 175 pp.
Chaudhary, S.A. (1983). Vegetation of the Great Nafud. *J. Saudi Arabian Nat. Hist. Soc.* 2(3): 32–37.
— & Cope, T.A. (1983). Studies in the Flora of Arabia: VI, a checklist of grasses of Saudi Arabia. *Arab Gulf J. Sci. Res.* 1: 313–354.
Clayton, W.D. & Cope, T.A. (1980). The chorology of Old World species of Gramineae. *Kew. Bull.* 35: 135–171.
Collenette, S. (1985). *An Illustrated Guide to the flowers of Saudi Arabia.* Scorpion Publishing, London, 514 pp.
Cope, T.A. (1985). Studies in the Flora of Arabia: XX, A key to the grasses of the Arabian Peninsula. *Arab Gulf J. Sci. Res., Special Publ.* 1: 1–82.
Cornes, M.D & Cornes, C.D. (1989). *The wild flowering plants of Bahrain.* IMMEL Publishing, London. 272 pp.
Daoud, H.S. (1985). *Flora of Kuwait.* vol. 1, revised by Ali Al-Rawi. Kuwait University. (for vol. 2 see Al-Rawi 1987).
Davis, S.D. et al. (1986). *Plants in danger, what do we know?* I.U.C.N., Cambridge, 461 pp.
Deflers, A. (1889). *Voyage au Yemen. Journal d'une excursion botanique faite en 1887 dans les montagnes de l'Arabie Heureuse.* Paris. 246 pp.

— (1895a). Esquisse de géographie botanique. La végétation de l'Arabie tropicale au-dela du Yemen. *Revue d'Égypte* 1: 349–370, 400–430.
Deil, U. (1986). Die Wadivegetation der nördlichen Tihama und Gebirgstihama der arabischen Republik Jemen. In Kürschner, H. (ed.) 1986a: 167–199.
— (1991). Rock communities in tropical Arabia. *Fl. Veg. Mundi* 9: 175–187.
Delile, A. (1813). Florae aegyptiacae illustratio. *Descr. Egypte, Nat. Hist.*: 1–182.
De Marco, G. & Dinelli, A. (1974). First contribution to the floristic knowledge of Saudi Arabia. *Ann. Bot. (Roma)* 33: 209–236.
Forbes, H.O. (ed.) (1903). *The natural history of Socotra and Abd-el-Kuri.* Liverpool Museums, Liverpool. 598 pp.
Forsskal, P. (1775). *Flora Aegyptiaco-Arabica.* Copenhagen. (Posthumous work, ed. C. Niehbuhr).
Frey, W. & Kürschner, H. (1989). Die Vegetation im vorderen Orient. Erläuerungen zur Karte A VI 1. *Beih. Tübinger Atlas Vorderen Orients, A* Nr 30.
— & — (1986). Masqat area (Oman). Remnants of vegetation in an urban habitat. In Kürschner, H. (ed.) 1986a: 201–221.
Gabali, S.A. & Al-Gifri, A.N. (1990). Flora of South Yemen - Angiospermae, a provisional checklist. *Feddes Repert.* 101 (7–8): 373–383.
Ghazanfar, S.A. (1991). Floristic composition and the analysis of vegetation of the Sultanate of Oman. *Fl. Veg. Mundi* 9: 215–227.
Griffiths, J.F. (1964). A rainfall map of eastern Africa and southern Arabia. *Memoirs E. African Meteorological Dept.* 3 (10): 1–42.
Halwagy, R. (1986). On the ecology and vegetation of Kuwait. In Kürschner, H. (ed.) 1986a: 81–109.
Hepper, F.N. (1977). Outline of the vegetation of the Yemen Arab Republic. *Publ. Cairo Univ. Herb.* 7 & 8: 307–322.
— & Wood, J.R.I. (1979). Were there forests in Yemen? *Proc. Seminar Arab. Stud.* 9: 65–71.
Hooker's Icones Plantarum (1971). New or noteworthy species from Socotra and Abd al Kuri. *Hooker's Icon. Pl.* 37,4: t.3673–3700.
König, P. (1986a). Zonation in the mountainous region of south-western Saudi Arabia. In Kürschner, H. (ed.) 1986: 137–166.
— (1986b). Vegetation und Flora im südwestlichen Saudi-Arabien (Asir, Tihama). *Diss. Bot.* 101. 257pp.
— (1988). Phytogeography of south-western Saudi Arabia. *Die Erde* 119: 75–89.
Kürschner, H. (ed.) (1986a). Contributions to the Vegetation of Southwest Asia. *Beih. Tübinger Atlas Vorderen Orients,* ANr 24.
— (1986b). Omanisch-makranische Disjunktionen. Ein Beitrag zur pflanzengeographischen Stellung und zu den florengenetischen Beziehungen Omans. *Bot. Jahrb. Syst.* 106: 541–562.
Lean, G., Hinrichsen, D. & Markham, A. (1990). *Atlas of the environment.* London. 192 pp.
Léonard, J. (1988-1989). *Contribution à l'étude de la flore et de la végétation des deserts d'Iran,* fasc. 8 (Étude des aires de distribution. Les phytochories. Les chorotypes) & fasc. 9 (Considérations phytogéographiques sur les phytochories irano-touranienne, saharo-sindienne et de la Somalie-pays Masai).
Lockwood, J.G. (1985). *World Climatic Systems.* Edward Arnold, London. 292 pp.
Mandaville, J.P. (1977). Plants. In Scientific Results of the Oman Flora and Fauna Survey 1975. *J. Oman Studies Special Report* No. 1: 229–269.
— (1984). Studies in the Flora of Arabia: XI, some historical and geographical aspects of a principal floristic frontier. *Notes Roy. Bot. Gard. Edinburgh* 42: 1–15.

— (1985). A botanical reconnaissance in the Musandam Region of Oman. *J. Oman Studies* 7: 9–28.

— (1986). Plant life in the Rub' al Khali (The Empty Quarter), South-central Arabia. *Proc. Roy. Soc. Edinburgh* 89B: 146–157.

— (1990). *Flora of Eastern Saudi Arabia.* Kegan Paul International Ltd, London, New York & Riyadh. 482 pp.

Migahid, M. A. (1978). *Flora of Saudi Arabia.* ed. 2, 2 vols. Riyadh. 940 pp.

Miller, A.G., Hedge, I.C. & King, R.A. (1982). Studies in the Flora of Arabia: I, A botanical bibliography of the Arabian Peninsula. *Notes Roy. Bot. Gard. Edinburgh* 40: 43–61.

— & Morris, M. (1988). *Plants of Dhofar - the Southern Region of Oman, traditional, economic and medicinal uses.* Office for Conservation of the environment, Oman. 360 pp.

— & Nyberg, J. A. (1991). Patterns of endemism in Arabia. *Fl. Veg. Mundi* 9: 263–279.

Munton, P. (1988). Vegetation and forage availability in the [Wahiba] sands. In Scientific Results of the Oman Flora and Fauna Survey 1985–87. *J. Oman Studies Special Report* No. 3: 241–250.

Phillips, D.C. (1988). *Wild flowers of Bahrain.* Manama, Bahrain. 207 pp.

Novikova, N.M. (1970). Drawing up a preliminary vegetation map of Arabia [in Russian]. *Geobot. Kartogr.* 1970: 61–71. (English translation: British Library Russian Transl. Service 12072).

Popov, G. B. (1957). The vegetation of Socotra. *J. Linn. Soc., Bot.* 55: 706–720.

Radcliffe-Smith, A. (1980). The Vegetation of Dhofar. In Scientific results of the Oman Flora and Fauna Survey 1977 (Dhofar). *J. Oman Studies Special Report* No. 2: 59–86.

Rappenhoner, D. (1989). *Resource conservation and desertification control in the near east.* Bayreuth. 294 pp.

Richard, A. (1847, 1850). *Tentamen Florae Abyssinicae*: I (1847); II (1850).

Rudloff, W. (1981). *World-Climates.* Wissenschaftliche Verlagsgesellschaft, Stuttgart. 632 pp.

Scholte, P., Al Khuleidi, A. & Kessler, J.J. (1991). *The vegetation of the Republic of Yemen (Western Part).* Environmental Protection Council, Sana'a. 56 pp.

Schwartz, O. (1939). Flora des tropischen Arabien. *Mitt. Inst. Allg. Bot. Hamburg* 10: 1–393.

Schweinfurth, G. (1894-1899). Sammlung arabisch-aëthiopischer Pflanzen (aus der Jahren 1881–1892). *Bull. Herb. Boissier* 2, app. II: 1–113 (1894); 4, app. II: 115–266 (1896); 7, app. II: 267–340 (1899).

Stanley Price, M., Hamoud Al-harthy, A. & Whitcombe, R.P. (1988). Fog moisture and its ecological effects in Oman. In *Arid Lands Today and Tomorrow, Proceedings of an International Research and Development Conference*: 69–88. Belhaven Press, London.

UNESCO-FAO. (1968). *Vegetaton map of the Mediterranean Zone. Explanatory notes.* Paris, UNESCO; Rome, FAO. 90 pp. (Arid zone research, 30).

Vahl, M, (1790-1794). *Symbolae botanicae* : I (1790); II (1791); III (1794).

Vesey-Fitzgerald, D.F. (1955a). Vegetation of the Red Sea coast south of Jedda, Saudi Arabia. *J. Ecol.* 43: 477–489.

— (1955b). The Vegetation of the Red Sea coast north of Jedda, Saudi Arabia. *J. Ecol.* 45: 547–562.

— (1955c). The Vegetation of central and eastern Arabia. *J. Ecol.* 45: 779–798.

Vierhapper, F. (1907). Beiträge zur Kenntnis der Flora Südarabiens und der Inseln Sokotra, Sémha und 'Abd el Kuri. *Denkschr. Kaiserl. Akad. Wiss. Math.-naturwiss. Kl.* 71: 321–490.

Walter, H. (1971). *Ecology of tropical and subtropical Vegetation.* Oliver & Boyd, Edinburgh. 539 pp.

Western, A.R. (1989). *The Flora of The United Arab Emirates, an Introduction.* United Arab Emirates University. 188 pp.

Bibliography

White, F. (1983 & 1986) *The Vegetation of Africa*: a descriptive memoir to accompany the UNESCO, AETFAT, UNSO vegetation map of Africa. - Paris: UNESCO; French translation by P. Bamps (1986).

— & Léonard, J. (1991). Phytogeographical links between Africa and Southwest Asia. *Fl. Veg. Mundi* 9: 229–246.

Wickens, G.E. (1982). Studies in the flora of Arabia: III, A biographical index of plant collectors in the Arabian peninsula (including Socotra). *Notes Roy. Bot. Gard. Edinburgh* 40: 301–330.

Zohary, M. (1973). *Geobotanical foundations of the Middle East*. Gustav Fischer, Stuttgart. Vols. 1–2, 739 pp.

LIST OF FAMILIES IN VOLUME ONE

PTERIDOPHYTA

The Fern Allies

1. Psilotaceae
2. Selaginellaceae
3. Equisetaceae

The Ferns

4. Ophioglosssaceae
5. Marsileaceae
6. Parkeriaceae
7. Actiniopteridaceae
8. Adiantaceae
9. Acrostichaceae
10. Pteridaceae
11. Polypodiaceae
12. Dennstaedtiaceae
13. Oleandraceae
14. Aspleniaceae
15. Thelypteridaceae
16. Woodsiaceae
17. Dryopteridaceae

SPERMATOPHYTA

Gymnospermae

18. Cupressaceae
19. Ephedraceae

Angiospermae

20. Casuarinaceae
21. Myricaceae
22. Juglandaceae
23. Salicaceae
24. Ulmaceae
25. Barbeyaceae
26. Moraceae
27. Cannabaceae
28. Urticaceae
29. Santalaceae
30. Loranthaceae
31. Viscaceae
32. Polygonaceae
33. Nyctaginaceae
34. Aizoaceae
35. Portulacaceae
36. Basellaceae
37. Caryophyllaceae
38. Chenopodiaceae
39. Amaranthaceae
40. Cactaceae
41. Magnoliaceae
42. Annonaceae
43. Lauraceae
44. Ranunculaceae
45. Berberidaceae
46. Menispermaceae
47. Nymphaeaceae
48. Ceratophyllaceae
49. Piperaceae
50. Aristolochiaceae
51. Hydnoraceae
52. Ochnaceae
53. Guttiferae
54. Papaveraceae
55. Fumariaceae

xvii

Families in Volume One

56. Capparaceae
57. Cruciferae
58. Resedaceae
59. Moringaceae

60. Crassulaceae
61. Pittosporaceae
62. Rosaceae
63. Neuradaceae

LIST OF FIGURES

Fig. 1. Topographical transect across the Arabian Peninsula
Fig. 2. **Ferns**: *Marsilea, Ophioglossum, Psilotum, Equisetum, Selaginella.*
Fig. 3. **Ferns**: *Actiniopteris, Anogramma, Pleopeltis, Tectaria, Cystopteris, Onychium, Ceratopteris.*
Fig. 4. **Ferns**: *Pellaea, Doryopteris, Negripteris, Cheilanthes.*
Fig. 5. **Ferns**: *Adiantum.*
Fig. 6. **Ferns**: *Pteris, Acrostichum, Pteridium.*
Fig. 7. **Ferns**: *Asplenium, Ceterach.*
Fig. 8. **Ferns**: *Arthropteris, Nephrolepis, Hypodematium, Polystichum, Dryopteris.*
Fig. 9. **Cupressaceae**. *Juniperus.*
Fig. 10. **Ephedraceae**: *Ephedra.*
Fig. 11. **Myricaceae**: *Myrica.* **Salicaceae**: Salix.
Fig. 12. **Ulmaceae**: *Celtis, Trema.* **Ochnaceae**: *Ochna.*
Fig. 13. **Moraceae**: *Dorstenia, Antiaris.*
Fig. 14. **Moraceae**: *Ficus.*
Fig. 15. **Moraceae**: *Ficus.*
Fig. 16. **Moraceae**: *Ficus.*
Fig. 17. **Urticaceae**: *Laportea, Girardinia, Urtica.*
Fig. 18. **Urticaceae**: *Pilea, Debregeasia, Pouzolzia.*
Fig. 19. **Urticaceae**: *Forsskaolea, Parietaria.*
Fig. 20. **Barbeyaceae**: *Barbeya.* **Urticaceae**: *Droguetia.*
Fig. 21. **Viscaceae**: *Viscum.* **Santalaceae**: *Thesium, Osyris.*
Fig. 22. **Loranthaceae**: *Helixanthera, Plicosepalus, Oncocalyx, Tapinanthus, Phragmanthera.*
Fig. 23. **Polygonaceae**: *Persicaria, Polygonum.*
Fig. 24. **Polygonaceae**: *Rumex.*
Fig. 25. **Polygonaceae**: *Emex, Oxygonum, Atraphaxis, Pteropyrum, Calligonum.*
Fig. 26. **Nyctaginaceae**: *Boerhavia, Mirabilis, Pisonia.*
Fig. 27. **Nyctaginaceae**: *Commicarpus.*

List of Figures

Fig. 28. **Nyctaginaceae**: *Commicarpus*.

Fig. 29. **Aizoaceae**: *Limeum, Corbichonia, Mollugo, Glinus, Gisekia*.

Fig. 30. **Aizoaceae**: *Sesuvium, Aizoon, Trianthema*.

Fig. 31. **Aizoaceae**: *Mesembryanthemum, Delosperma, Tetragonia*.

Fig. 32. **Portulacaceae**: *Talinum, Portulaca*.

Fig. 33. **Caryophyllaceae**: *Cometes, Pollichia, Sphaerocoma, Gymnocarpos*.

Fig. 34. **Caryophyllaceae**: *Sclerocephalus, Pteranthus, Paronychia, Herniaria, Scleranthus*.

Fig. 35. **Caryophyllaceae**: *Haya, Polycarpaea*.

Fig. 36. **Caryophyllaceae**: *Xerotia, Polycarpaea*.

Fig. 37. **Caryophyllaceae**: *Polycarpon, Loeflingia, Spergularia, Spergula, Minuartia*.

Fig. 38. **Caryophyllaceae**: *Arenaria, Holosteum, Stellaria, Cerastium*.

Fig. 39. **Caryophyllaceae**: *Silene*.

Fig. 40. **Caryophyllaceae**: *Silene*.

Fig. 41. **Caryophyllaceae**: *Vaccaria, Gypsophila*.

Fig. 42. **Caryophyllaceae**: *Dianthus, Petrorhagia*.

Fig. 43. **Chenopodiaceae**: *Chenopodium*.

Fig. 44. **Chenopodiaceae**: *Atriplex*.

Fig. 45. **Chenopodiaceae**: *Atriplex*.

Fig. 46. **Chenopodiaceae**: *Halocnemum, Haloxylon, Anabasis, Noaea, Agathophora, Traganum, Salicornia, Agriophyllum, Arthrocnemum*.

Fig. 47. **Chenopodiaceae**: *Seidlitzia, Lagenantha, Sevada, Halopeplis, Bassia, Bienertia, Suaeda, Halocharis*.

Fig. 48. **Chenopodiaceae**: *Cornulaca*.

Fig. 49. **Chenopodiaceae**: *Cornulaca*.

Fig. 50. **Chenopodiaceae**: *Salsola*.

Fig. 51. **Chenopodiaceae**: *Salsola*.

Fig. 52. **Chenopodiaceae**: *Salsola*.

Fig. 53. **Chenopodiaceae**: *Halothamnus*.

Fig. 54. **Amaranthaceae**: *Celosia, Pupalia, Psilotrichum*.

Fig. 55. **Amaranthaceae**: *Amaranthus*.

Fig. 56. **Amaranthaceae**: *Aerva*.

Fig. 57. **Amaranthaceae**: *Achyranthes, Alternanthera, Gomphrena, Saltia*.

Fig. 58. **Ranunculaceae**: *Clematis, Ranunculus*.

Fig. 59. **Ranunculaceae**: *Thalictrum, Adonis, Delphinium*. **Berberidaceae**: *Berberis*.

Fig. 60. **Menispermaceae**: *Tinospora, Stephania, Cocculus*.

Fig. 61. **Piperaceae**: *Peperomia*. **Ceratophyllaceae**: *Ceratophyllum*.

List Of Figures

Fig. 62. **Aristolochiaceae**: *Aristolochia*. **Hydnoraceae**: *Hydnora*.

Fig. 63. **Guttiferae**: *Hypericum*.

Fig. 64. **Guttiferae**: *Hypericum*.

Fig. 65. **Papaveraceae**: *Papaver, Glaucium*.

Fig. 66. **Papaveraceae**: *Roemeria, Hypecoum*. **Fumariaceae**: *Fumaria*.

Fig. 67. **Capparaceae**: *Cleome*.

Fig. 68. **Capparaceae**: *Cleome*.

Fig. 69. **Capparaceae**: *Cleome*.

Fig. 70. **Capparaceae**: *Maerua*.

Fig. 71. **Capparaceae**: *Dhofaria, Dipterygium, Capparis*.

Fig. 72. **Capparaceae**: *Cadaba*.

Fig. 73. **Cruciferae**.

Fig. 74. **Cruciferae**: *Brassica, Erucastrum*.

Fig. 75. **Cruciferae**: *Diplotaxis, Hirschfeldia, Sinapis*.

Fig. 76. **Cruciferae**: *Eruca, Hemicrambe, Raphanus, Enarthrocarpus*.

Fig. 77. **Cruciferae**: *Dolichorhynchus, Zilla, Cakile, Physorrhynchus*.

Fig. 78. **Cruciferae**: *Cardaria, Savignya, Schouwia, Coronopus, Carrichtera, Moricandia, Lepidium*.

Fig. 79. **Cruciferae**: *Isatis, Capsella, Neslia, Biscutella, Horwoodia, Thlaspi, Lachnocapsa, Anastatica, Schimpera*.

Fig. 80. **Cruciferae**: *Farsetia*.

Fig. 81. **Cruciferae**: *Alyssum, Lobularia, Clypeola, Arabis, Cardamine*.

Fig. 82. **Cruciferae**: *Nasturtium, Rorippa, Matthiola, Morettia, Notoceras, Diceratella*.

Fig. 83. **Cruciferae**: *Malcolmia, Leptaleum, Goldbachia, Maresia, Eremobium, Eigia, Sterigmostemum*.

Fig. 84. **Cruciferae**: *Sisymbrium, Neotorularia, Arabidopsis*.

Fig. 85. **Resedaceae**: *Caylusea, Oligomeris*.

Fig. 86. **Resedaceae**: *Ochradenus*.

Fig. 87. **Resedaceae**: *Reseda*.

Fig. 88. **Moringaceae**: *Moringa*. **Pittosporaceae**: *Pittosporum*.

Fig. 89. **Crassulaceae**: *Crassula, Aeonium*.

Fig. 90. **Crassulaceae**: *Kalanchoe*.

Fig. 91. **Crassulaceae**: *Kalanchoe*.

Fig. 92. **Crassulaceae**: *Sedum, Cotyledon, Umbilicus*.

Fig. 93. **Rosaceae**: *Prunus, Cotoneaster, Rubus*.

Fig. 94. **Rosaceae**: *Potentilla, Alchemilla, Rosa*. **Neuradaceae**: *Neurada*.

INTRODUCTION

The Arabian Peninsula covers some 2.7 million km^2, an area equal to half that of Europe, 8% of Asia or approximately 2% of the world's land surface. It is more or less rectangular in shape and tilts gently from its highest part along the Red Sea in the west, downwards to the Arabian Gulf in the east. It is a region of diverse topography ranging from rugged mountains (reaching over 3000m) through vast sand and rock deserts (including some of the hottest and driest on earth) to luxuriantly vegetated, mist-covered mountains and highly productive areas of ancient terraced agriculture. There are no permanent rivers and only a few permanent streams in the highest and wettest areas of the mountains.

GEOLOGY AND TOPOGRAPHY

In geological terms the Arabian Peninsula is part of the Arabian tectonic plate which also includes Jordan, Syria, S Iraq and SW Iran. The ancient crystalline rocks of the plate are exposed mainly in the western part of the Peninsula (the Arabian Shield) and elsewhere are covered by a thick sequence of sedimentary rocks (the Arabian Platform) deposited from Cambrian times to the present. These sediments were formed when the Peninsula was inundated by a succession of shallow seas. Extensive areas of the Peninsula are also covered by aeolian sands (derived largely from the weathering of the crystalline basement rocks) and by sands and gravels of Quaternary age.

Until the late Mesozoic era the Arabian Plate formed part of the ancient southern super-continent of Gondwana. During the Cretaceous period Gondwana broke up, and the combined African and Arabian plates drifted northwards until, during the Miocene period (20–14 million years ago), they collided with the Eurasian plate. This collision resulted in the mountain building episodes which gave rise to the major mountain chains of SW Asia and S Europe (the Alpine Orogeny) and, in Arabia, to the Hajar mountains of N Oman. Some 25 million years ago, during the early Miocene, Arabia started to separate from Africa forming the Red Sea and Gulf of Aden. The spectacular escarpment mountains which run down the western margin of the Peninsula arose from the faulting and uplifting of the rift margins associated with the opening of the Red Sea. Volcanic activity (from the Tertiary to the recent past) which accompanied the rifting has resulted in extensive areas of lava flows in the west of the Peninsula overlying the basement crystalline rocks. The Arabian Plate is still slowly

INTRODUCTION

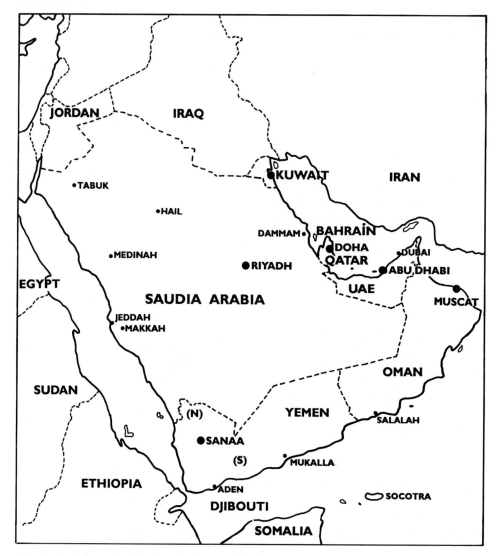

Map. 1. Countries and principal towns of the Arabian Peninsula.

moving towards Asia (causing the gradual drying up of the Arabian Gulf) and away from Africa.

The collision of the African and Arabian Plates with Eurasia during the Cenozoic was of major importance, not only in shaping the present topography of SW Asia, but also in allowing the mixing of elements of the Palaeotropical floras from the south and Holarctic floras from the north.

GEOGRAPHICAL DIVISIONS

The Arabian Peninsula can be divided into six topographic regions (map 2).

1. Western coastal plain (the Tihama)

The Tihama plain runs the length of the Red Sea coast to the west of the escarpment mountains. It is narrow in the north, broadening south of Jeddah, where it averages about 30km in width. It is covered with aeolian sands and extensive areas of fluvial deposits washed from the mountains. There are no permanent water courses crossing the Tihama but, after heavy rain in the mountains, flood waters may reach the sea. There are coralline deposits along the coast and, particularly in Yemen, large areas of coastal salt flats or sabkhah. It is a hot and humid region with low rainfall.

2. Mountainous regions of the south and west

The main mountain chains of the Arabian Peninsula (excluding the N Oman mountains) form an 'L-shaped' region which is broadest in the Yemen Highlands and has one arm, the SW escarpment mountains, running NNW (parallel with and along the length of the Red Sea) and the other arm running ENE (along the south coast of the Peninsula) and ending in the Dhofar mountains of S Oman.

The southwest escarpment mountains are composed mainly of Precambrian crystalline rocks which are overlain, particularly in the south, by large areas of limestone and sandstone of Jurassic age and extensive areas of volcanic rocks of Tertiary age. They can be treated in two sections:

i. The Asir and Yemen highlands. These form a long, more or less unbroken, chain (mostly above 2000m) of escarpments and jagged ridges running SSE from Makkah to the mountains NW and NE of Aden. They rise to almost 3000m in the Asir while in Yemen several peaks exceed 3000m, including the highest point in Arabia (J. Nabi Shu'ayb) at 3760m. To the west the highlands fall in a series of dramatic escarpments to the Tihama plain whilst in the east they slope more gradually towards the interior. The highlands are the wettest region in Arabia (above 500mm per annum in many areas) with rainfall declining rapidly to the east; the Asir mountains are generally drier than the Yemen highlands. In their southern part the highlands are heavily populated, with extensive terraced agriculture. Floristically they are the richest region in Arabia.

ii. The Hijaz mountains. These run SSE, from near Tabuk in NW Saudi Arabia, to near Makkah. They are generally more interrupted, less rugged, much drier (with rainfall between 50 and 100mm per annum), more sparsely vegetated and floristically poorer than the Yemen Highlands and the Asir.

INTRODUCTION

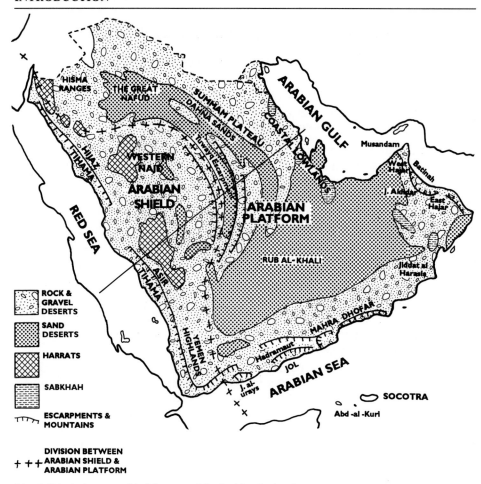

Map 2. Principal topographical features of the Arabian Peninsula.

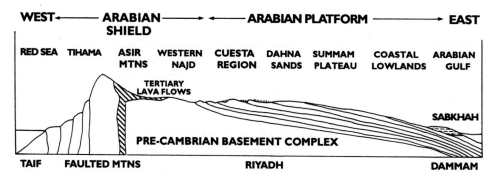

Fig. 1. Topographical transect across the Arabian Peninsula (after Anon. 1992).

Geology and Topography

The southern mountains, running roughly east and parallel to the Arabian Sea, can be divided into three sections.

i. Jebal al-'Urays. This is an isolated massif of Tertiary volcanic rocks lying approximately 100km NE of Aden. The lower and north-facing slopes are very dry but the upper, seaward-facing slopes are frequently cloud-covered and carry reasonably dense vegetation.

ii. The Jol. An inhospitable limestone plateau which averages 1000m and rises to 2200m just north of Mukalla. It is deeply dissected by the Wadi Hadramaut. Rainfall is low and vegetation is sparse except on the seaward-facing slopes.

iii. The Dhofar Mountains. A coastal range composed of limestone of Tertiary age and consisting of a relatively flat plateau, averaging between 800 and 900m along most of its length but rising to over 2100m in the east. To the south the mountains fall in steep, seaward-facing escarpments but to the north they dip gently to the gravel plains of the interior. The seaward-facing slopes capture rainfall from the SW monsoon and are densely vegetated.

INTRODUCTION

3. The mountains of N Oman

These extend some 700km from Musandam SE to Ras al Hadd. They are formed of hard sedimentary rocks (mainly very dense limestones and dolomites) with extensive areas of crystalline rocks (the ophiolitic complex) in the central part. They consist of three ranges: the Eastern Hajar reaching 2230m on J. Khadir; the Western Hajar reaching 3009m on J. al-Akhdar and the Musandam mountains reaching 1087m on J. al-Harim. They are rugged, sparsely-vegetated mountains with rainfall of 300–350mm on J. Akhdar and less on the other ranges. They are separated from the sea by the Batinah coast, a flat gravel plain.

4. The Western Najd

The Western Najd corresponds roughly to the area east of the SW escarpment mountains above the 800m contour. It extends across about half of the peninsula at its widest point and is bounded to the east by the Arabian Platform. The basement crystalline rocks are exposed over most of the region except where they are covered by extensive lava flows, locally called 'Harrats'. It is a desertic region of dissected table-lands and plains covered with thin sand and gravel deposits, from which rise isolated inselbergs. The Harrats, prior to the building of an extensive network of bulldozed tracks and the use of pick-up trucks, were largely inaccessible and formed important wildlife refugia. Rainfall is below 100mm (in many areas below 50mm) and vegetation is sparse, or in some places, more or less absent.

5. The Arabian Platform

To the east of the Arabian Shield are the sedimentary deposits of the Arabian Platform. These consist of limestones, sandstones and shales of Cambrian to Pliocene age. The Platform has a varied desertic topography including vast sand seas, bare plateaux, escarpments and plains. Vegetation is sparse on the rocky areas but can be relatively dense on the sand. Rainfall is below 100mm over most of the area. The Arabian Platform can be subdivided into a number of zones.

i. The Hisma Ranges. This area, in the northwest of Saudi Arabia, consists of buttes and canyons of Cambrian sandstone which extend to the east into a plain of low sandstone outcrops of Devonian age.

ii. The Cuesta region. This region forms a zone along the western edge of the Arabian Platform and consists of a series of low sandstone and limestone escarpments of Paleozoic to Mesozoic age. These are generally 100 to 150m in height and the longest, the Tuwayq escarpment, extends in a north to south arc of some 1200km. In depressions between these escarpments are a series of narrow belts of sand and to the east a flat gravel plateau which borders the Dahna sand belt.

iii. The Sand Seas. Huge areas of the Arabian Platform are covered by aeolian (wind blown) sands. The main regions are the Great Nafud in the north, the Rub' al-Khali or Empty Quarter in the south and, connecting them, the Dahna sands which form the eastward boundary of the Cuesta region. Other important areas of sand are the Wahiba Sands in Oman and the narrow 'stringers' of sand between the escarpments of the Cuesta region. The Rub' al-Khali has been described as the greatest area of continuous sand cover in the world. It is a region of c.500,000 km^2 of which 90% is sand-covered and about 30% is sand dunes, the highest of which attain 250m. The Great Nafud is a vast sand desert covering c.57,000 km^2 in N Saudi Arabia. The Dahna sand belt is a long, narrow, crescent-shaped body of sand which extends to some 1,200km in length and up to 50km in width. The Wahiba Sands, in the NE of Oman, cover c.9,400 km^2.

iv. The Summan plateau. This is a flat, barren, limestone plateau lying to the east of the Dahna sands and running from the Iraq border south to the Rub' al-Khali.

v. The coastal lowlands. These lie to the east of the Summan plateau and extend around the Arabian Gulf from Kuwait to the UAE. This is a region of featureless plains of low relief and covered with undulating deposits of gravel and sand.

vi. Sabkhah. Along the coastline are extensive areas of salt flats or Sabkhah. These are also found in the eastern Rub' al-Khali (where they extend over thousands of square kilometres but are largely covered with sand), in eastern Oman and along the Red Sea coast. They are generally devoid of vegetation except around their margins.

vii. Central deserts of Oman. South of the Rub' al-Khali the inland deserts of Oman and Yemen are composed mainly of limestone plains overlain with gravel.

6. The Socotran archipelago

The Socotra archipelago consists of four islands - Socotra, 'Abd al Kuri, Semhah and Darsa - situated in the northern part of the Indian Ocean due east of Somalia and lying between 12°06'–12°42'N and 52°03'–54°32'E. The islands belong to the Republic of Yemen. Socotra covers 3625 km^2, 'Abd al Kuri, Semhah and Darsa are much smaller and in total cover less than 400 km^2. Socotra is composed of a basement complex of igneous and metamorphic rocks of Precambrian age overlain by sedimentary rocks (mainly limestone and sandstone) of Eocene and Cretaceous age. Topographically the island can be divided into three zones.

i. The coastal plains. These are up to 5km in width and for the most part are covered with unconsolidated sands and gravels.

INTRODUCTION

ii. The Haggier mountains. This rugged range of granite mountains dominates the NW of the island. They are topped by a series of dramatic pinnacles, the highest of which, Mashanig, is 1519m.

iii. The limestone plateaux. Limestone plateaux cover a large part of the island; they average between 300 and 700m in height and reach 816m at their highest point in the west. They drop in steep, often almost vertical escarpments, to the coastal plain or directly to the sea.

CLIMATE

The Arabian Peninsula is part of the huge arid zone which extends across N Africa and S Asia from the Sahara to the Sind desert in India and Pakistan. This zone is characterized by practically cloudless days resulting in low relative humidity and high evaporation rates which bring maximum daytime temperatures of 50°C and over, much higher than those at the equator and among the hottest on earth. In many parts of the region the diurnal temperature range is enormous, sometimes as much as 33°C.

The latitudinal spread and varied topography of the Arabian Peninsula mean that its climate is affected by a number of different processes and in consequence climatic conditions vary considerably over the region (map 3). However, in general there are hot summers and cold or warm winters with winter and spring rainfall in the north and east, and both spring and summer rainfall in the south.

The climate of the Arabian Peninsula depends on both local topography and its position relative to the global atmospheric circulatory system. It lies beneath the Hadley Cell, a tropical atmospheric circulation system. Warm, moist air ascends in equatorial regions and moves polewards to latitudes of about 30° where it descends, is warmed adiabatically, and consequently dried. The circulation of the Cell is completed by the low level return of air in the Trade winds to equatorial regions where they meet in the Inter Tropical Convergence Zone (ITCZ). These trade winds blow continually towards the equator in both hemispheres but are deflected westwards by the rotation of the earth; thus they are from the NE in the northern hemisphere and from the SE in the southern. Beneath the descending air a stable belt of high atmospheric pressure forms. Conditions in this high pressure belt and for some distance towards the equator are unfavourable for the formation of cloud and account for the hot, dry deserts at these latitudes in both hemispheres. Rainfall only occurs when this relatively stable system is disturbed; this can happen in several ways.

i. In winter the high pressure ridge shifts south allowing mid-latitude low pressure systems to bring rain or snow to the north of the region. Low pressure systems also migrate eastwards from the Mediterranean or Red Sea bringing rainfall to the north of the region, or occasionally to the south, in winter.

ii. In winter the winds of the NE Trades (which in Arabia because of local topography blow mainly from the N or NW) cause orographic rain (the NE monsoon) in the mountains of Yemen, N Oman and Socotra.

iii. In summer the ITCZ migrates north and lies along the southern part of the

Arabian Peninsula bringing rain from the rising air masses of this belt to the southern edge of the region.

iv. In summer the NE Trade winds are displaced by those of the Indian or SW monsoon. These bring widely varying conditions to different parts of the region: in the mountains of Yemen they produce regular, orographic rainfall whereas over large parts of lowland southern Arabia and on Socotra they are dry and desiccating. However, perhaps the most dramatic effects are felt on the coastal mountains of eastern Yemen and southern Oman. Here the winds, laden with moisture gathered while crossing the Indian Ocean, are cooled to dew-point by a local upwelling of cold water and bring copious quantities of rain, drizzle and drip precipitation to the coastal mountains. In consequence, the seaward-facing slopes of the mountains are blanketed in thick cloud for several months each year.

v. Tropical storms, originating in the Arabian Sea and Bay of Bengal, bring rainfall to the coast of Arabia south of Muscat during March to April and October to November on average once every three years.

vi. Local orographic rainfall due to the high altitude of the SW escarpment mountains augments precipitation by other processes. On-shore breezes during the daytime bring moist air from the Red Sea which, as it ascends the mountains, gives rise to showers and thunderstorms in the afternoons. This rainfall tends to be less seasonal than the other forms described above.

Throughout the Peninsula rainfall is erratic and unpredictable in its timing and quantity. In some years certain areas may receive almost no rain; in other years rainfall can be relatively high. For example, in Bahrain the total rainfall in 1976 was 232mm whereas in 1970 and 1971 the totals were 9mm and 25mm respectively. No part of the Arabian Peninsula can be considered completely rainless but in some areas, for instance in the Rub' al-Khali, rain is sometimes not recorded for stretches of several years at a time. Over large parts of the Peninsula precipitation is under 100mm per annum. In the mountains rainfall is higher with over 300mm per annum common in the SW escarpment mountains. In the Ibb region of Yemen, the wettest area in Arabia, rainfall of over 1500mm per annum is typical with an impressive 2205mm recorded in 1975. Snow is not uncommon in the north of the Peninsula and has been reported in the mountains as far south as Yemen and N Oman. Dewfall is very common, particularly in coastal areas, and over large areas of desert is an important source of surface moisture.

Generally, temperatures increase from north to south in the Peninsula (with maxima in July and minima in January and February) and decrease with altitude (rising air cools at a minimum rate of 0.5°C per 100m when moist and more quickly when dry).

The general characteristics of the climates of the individual countries of the Arabian Peninsula are outlined below.

SAUDI ARABIA:

Winters (December to February) are cool to warm with occasional frosts in the north and on the mountains. Summers are very hot with temperatures commonly above

INTRODUCTION

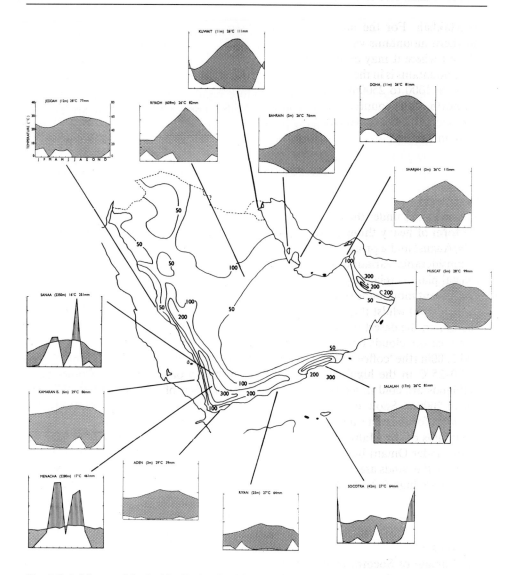

Map 3. Rainfall map of the Arabian Peninsula with selected climate diagrams.

40°C and maxima occasionally exceeding 50°C. Temperature differences between the winter and summer are greatest in the centre of the country (e.g. Riyadh). Rainfall over most of the country is highly unpredictable and variable in quantity, falling mainly in winter and spring, with summer rainfall in the SW mountains as far north as Makkah. For the most part annual rainfall is less than 100mm except in the northern mountains where it is generally above 150mm and in the mountains south of Taif where it may exceed 600mm. A large part of the summer rain falling in the Asir mountains is in the form of orographic precipitation and the western escarpments, between 1000 to 2500m, are characterised by a fog zone. Snow is not infrequent in the north of the country. Humidities are low in the interior (Riyadh 15–20%) but are high along the coast (Jeddah 55–65% and the Gulf coast 40% in summer to 70% or more in winter).

YEMEN:
Yemen comes under the influence of both SW and NE monsoons. Most rain falls, in the form of heavy thunderstorms, during the summer with peaks in April/May and July/August and a pronounced dry period during the winter. Across the country there is considerable variation in annual rainfall depending on altitude and aspect. The coastal plain is relatively dry with erratic, local rain probably less than 200mm. The escarpment mountains receive 600–800mm rising to over 1000mm in the wettest areas (around Ibb) whilst the high plateau receives 300–500mm dropping rapidly to below 100mm on the desert margins in the east. On most days, and particularly during the rainy season, clouds build up against the escarpments at an altitude of between 2000 and 2500m (the 'coffee zone'). The highest temperatures occur in summer with means of 20–25°C in the highlands rising to 30°C on the coastal plain and interior. The highlands are cold in winter with a mean temperature of 10°C and frosts occurring above 2000m. Snow occasionally falls on the higher peaks. Few climatic data are available for the eastern part of Yemen. In the extreme east of the country the coastal mountains of the Mahra region receive rain from the SW Monsoon (see the section below under Oman) but across the limestone plateau of the Jol in Hadramaut Governorate the winds are dry and desiccating. The relative humidities around the coastal regions are high (60–75%).

SOCOTRA:
The climate of Socotra is influenced by both the SW and NE monsoons. The SW monsoon blows from April until October bringing hot, dry winds which are generally desiccating but bring a little orographic rain to the mountains. Most precipitation occurs from November to March; during this period the SW winds are replaced by much lighter winds from the NE. Some temperature measurements are available for the coastal plain; these give annual temperatures of 17°–26°C (minima) and 27°–37°C (maxima). No measurements are available for the mountains although they are considerably cooler. Frost is not reported. Rainfall is very sporadic and in some years the coastal areas receive none. Average measurements for the plain are around 150mm and the mountains probably receive around 500mm. Most rain falls in winter. The mountains are frequently shrouded in clouds and heavy dews are common.

INTRODUCTION

OMAN:

There are two distinct climatic regimes in Oman. The north has cool winters (November to March) and hot summers, with rain mainly in winter and spring (Muscat at sea-level has 100mm per annum whilst the mountains at 2000m receive 300–350mm). In the monsoon-affected regions of the south precipitation (falling mainly in July and August as rain, drizzle and as drip from condensation on the vegetation) is 110mm per annum on the coast but rises to more than 500mm on the escarpment mountains. Temperatures are highest (32°C) in the south in May and June, being kept lower during the summer months by cloud cover. Humidity on the coast rises to 97% in July and August during the SW monsoon. The central deserts are very hot in summer (over 50°C in the shade) and relatively cold in winter; rainfall here is less than 50mm. Fogs are frequent in the central desert between January and May and October and November. They are caused by moist onshore winds which condense because of low temperatures in the desert. The precipitation from fog drip can be considerable: on the Jiddat al Harasis it has been calculated at 1.4 litres per square metre per day.

UAE:

Most rain falls between December and April with occasional showers in the mountains in summer. Rainfall varies from 45mm per annum at Abu Dhabi, rising to 150mm in the north and towards the mountains at Fujairah. On the coast at Abu Dhabi the mean maximum temperature is 40°C with maximum temperatures of 49°C in summer dropping to (rarely) 5°C in winter. Mean annual relative humidity is 60% on the coast (at Abu Dhabi) with the winter months generally over 70%.

QATAR:

Characterized by winter and spring rainfall, hot summers and mild winters, with a mean annual rainfall of 80mm. The hottest months are July and August with mean temperatures of 35°C. The coldest month is January with a mean of 17°C. Mean annual humidity is 60%.

BAHRAIN:

Characterized by hot summers and mild winters with rainfall mainly between November and April with mean annual rainfall of 75mm. Temperatures range from a mean daily maximum of 38°C in August to a mean daily minimum of 14°C in January. Daytime relative humidity ranges between 45 and 90% with night-time relative humidity of 90–100% not uncommon.

KUWAIT:

Characterized by hot, dry summers and cool, rainy winters. The mean annual rainfall (Kuwait City) is 111mm with great variation from year to year. The prevailing winds are from the northwest.

VEGETATION

This section gives an outline of the vegetation of the Arabian Peninsula. It includes descriptions of the main vegetation types and their principal habitats (map 4).

The greatest problem in describing the vegetation of the Arabian Peninsula is the effects man and his livestock have had for thousands of years. The potential vegetation in most areas can only be guessed at from small vestiges of apparently natural vegetation which have managed to survive in inaccessible spots.

There are several vegetation maps covering the Arabian Peninsula (none including Socotra), the most important of which are UNESCO-FAO (1968), Novikova (1970), Zohary (1973) and Frey & Kürschner (1989). All suffer from an inadequate knowledge of the flora of the region, particularly in the south, but the most recent, Frey & Kürschner, is the most accurate. There is also a large number of papers covering the vegetation of individual countries or regions. The most useful of these are listed, by country, below.

Saudi Arabia: Baierle & Frey (1986), Chaudhary (1983), König (1986a & b), Mandaville (1986, 1990), Vesey-Fitzgerald (1955, 1957a & b).
Yemen: Deil (1986), Hepper (1977), Hepper & Wood (1979), Al-Hubaishi & Müller-Hohenstein (1984), Scholte et al. (1991).
Socotra: Popov (1957), White (1983).
Oman: Ghazanfar (1991), Mandaville (1977, 1985), Munton (1988), Miller & Morris (1988), Radcliffe-Smith (1980).
UAE: Western (1989).
Qatar: Batanouny (1981).
Bahrain: Cornes & Cornes (1989).
Kuwait: Halwagy (1986).

In the following account of the vegetation of the Arabian Peninsula as far as possible the definitions used by White (1983) are followed.

1. Forest – a continuous stand of trees at least 10m tall, their crowns interlocking.
2. Woodland – an open stand of trees at least 8m tall with a canopy cover of 40% or more.
3. Bushland – an open stand of bushes usually between 3 and 7m tall with a canopy cover of 40% or more.
4. Thicket – a closed stand of bushes and climbers between 3 and 7m tall.
5. Shrubland – an open or closed stand of shrubs up to 2m tall.
6. Deserts – arid landscapes with a sparse plant cover (usually less than 15%) where the substrate contributes more to the appearance of the landscape than does the vegetation.

Using White's criteria, forest and woodland are very rare in Arabia. Woodland is found only in a few places in the SW escarpment mountains and forests are more or less absent. Bushland and thicket are generally associated with the mountains and the

INTRODUCTION

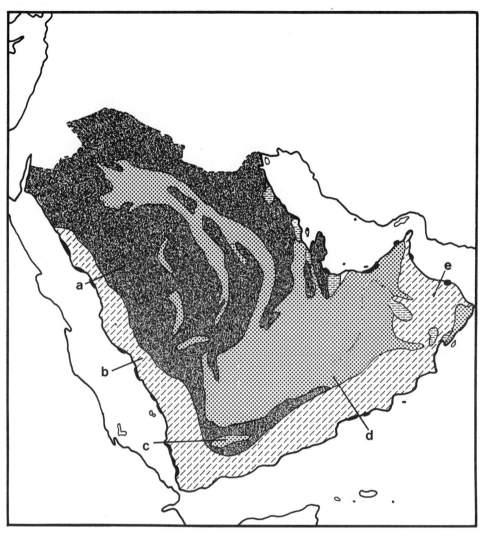

Map 4. Main vegetation types of the Arabian Peninsula and their principal habitats.

plains of the S and SW of the Peninsula. Open stands of trees and shrubs are found in wadis in the centre and east of the Peninsula, along dunes and by sabkhah (typically with *Prosopis cineraria*, *Ziziphus spina-christi*, *Acacia tortilis*, *A. gerrardii* and *Tamarix* spp.). In some parts of the Peninsula there are no native tree species at all (for example in Kuwait).

In general the vegetation of deserts and semi-deserts is difficult to treat in a simple physiognomic classification. In Arabia they tend to have a mosaic of sparse vegetation types, unlike the mountainous regions of the south and west which usually have a clear altitudinal zonation of vegetation. There is, however, a reasonable correlation between the vegetation and local edaphic features and so in these areas the vegetation has been classified in terms of major habitats. There follows a brief outline of the vegetation of the Arabian Peninsula arranged by major habitat types.

1. Vegetation of deserts and semi-deserts

In Arabia the terms desert and semi-desert can be applied to areas where there is much bare soil in evidence and where the vegetation cover is absent or very open, usually less than 15%; few areas are totally devoid of vegetation. The distinction between desert and semi-desert is difficult to define; however, distribution of vegetation can be used to differentiate between the two. In deserts the vegetation is 'contracted', that is, confined to areas such as depressions and wadis which receive water from large catchment areas or perhaps where they can tap deep underground water, whereas in semi-deserts, with increased precipitation, the vegetation, though still sparse, becomes 'diffuse', that is, more or less randomly distributed over the land surface. There is usually a very gradual transition between deserts, semi-deserts and surrounding vegetation types with no clearly defined boundaries. In N Africa, White (1983) considers the northern limit of the Sahara Desert to coincide more or less with the 100mm isohyet and the southern limit with the 150mm isohyet and further, that semi-deserts begin where rainfall drops below 250mm per annum. The situation in Arabia is apparently very similar.

A. SAND DESERTS OR ERG

The vegetation of the sand deserts is typically a very open but evenly distributed dwarf shrubland. Woodland is more or less unknown except for patches of *Prosopis* woodland along the margins of the Wahiba Sands and Rub' al-Khali. In general the sand deserts carry more plant life than the rock or gravel deserts. They occur in five main areas.

i. The Rub' al-Khali or Empty Quarter. Three main communities are identified in the Rub' al-Khali: in the central region and the northeast a community dominated by

INTRODUCTION

Cornulaca monacantha with *Cyperus conglomeratus* s.l., *Dipterygium glaucum* and *Limeum arabicum*; in the west, south and southeast a community dominated by *Calligonum crinitum* subsp. *arabicum* with *Dipterygium glaucum* and *Cyperus conglomeratus* s.l.; and lastly in the north a community dominated by *Haloxylon persicum*, *Dipterygium glaucum*, *Limeum arabicum* and *Stipagrostis drarii*.

Few areas are totally devoid of plant life; these are largely on the bare plains at the edge of the sands or on the bare floors between the sand dunes. A total of only 37 species has been recorded from the region of which ten are of importance in the vegetation. Annuals are virtually absent. This is in contrast to the sand deserts of the north where they may represent 40–75% of the flora. Mandaville (1986) reports that in 1982, after heavy rain, only rare examples of three annual species (*Eremobium aegyptiacum*, *Plantago boissieri* and *Neurada procumbens*) were found. Trees are absent except along the outer margins where open woodlands with *Acacia* spp. and *Prosopis cineraria* are found (Mandaville 1986).

ii. The Dahna sand belt. The Dahna sand belt forms a natural boundary between central and eastern Saudi Arabia. The most characteristic community of the Dahna is dominated by *Artemisia monosperma* and *Calligonum comosum* (which replaces *C. crinitum* subsp. *arabicum* in the northern part of Arabia) with *Stipagrostis drarii* and *Cyperus conglomeratus* s.l. *Limeum arabicum*, otherwise endemic to the Rub' al-Khali, penetrates the southern part of the Dahna and in the north *Scrophularia hypericifolia* penetrates from the Great Nafud. After rains, the northern part of the Dahna supports a good cover of annuals. The Central Nafuds (lying to the east and west of the Tuwayq escarpment) are a discontinuous chain of elongated sand deserts running parallel to the Dahna and have similar vegetation to the Dahna sand belt (Mandaville 1990).

iii. The Great Nafud. The Great Nafud consists of large sand dunes, hollows containing only a relatively thin cover of sand and large sheets of deep undulating sand. The two most important communities are both open, dwarf shrublands, the first dominated by *Haloxylon persicum*, *Artemisia monosperma* and *Stipagrostis drarii* and the second by *Calligonum comosum*, *Artemisia monosperma* and *Scrophularia hypericifolia* with *Moltkiopsis ciliata*, *Monsonia heliotropoides*, *Stipagrostis drarii* and *Centropodia fragilis*. The most obvious difference between the floras of the Great Nafud and the Rub' al-Khali is the relatively richer flora and dense flush of annual herbs following the winter and spring rains in the former (Chaudhary 1983).

iv. The Eastern Coastal Sands. These extend along the Arabian Gulf coastal plain east of the Summan plateau from the Jafurah sands in eastern Saudi Arabia to Kuwait. The characteristic vegetation is a dwarf shrubland with grass dominated by *Panicum turgidum* and *Calligonum comosum* with occasional bushes of *Leptadenia pyrotechnica* and *Lycium shawii* (Halwagy 1986 and Mandaville 1990).

v. The Wahiba Sands. The Wahiba Sands are an isolated sand desert in the NE of Oman. The vegetation forms a mosaic of communities with the dominant species *Calligonum crinitum* subsp. *arabicum, Dipterygium glaucum, Panicum turgidum* and *Cyperus conglomeratus* s.l. *Prosopis cineraria* forms important woodlands on the margins of the sands (Munton 1988).

B. ROCK AND GRAVEL DESERTS

The commonest physiographic features over vast areas of the Arabian Peninsula are rock and gravel deserts. These are of two main types: (i) rock and stone deserts (*hamadas*) where all fine weathering products have been removed by the wind; and (ii) gravel deserts (*regs*) where the parent material consists of heterogeneous deposits, such as alluvium, and where all the fine material has been removed by the wind.

The vegetation of these deserts forms a mosaic which is very difficult to classify. Large areas are very thinly vegetated with plants restricted to rock crevices and to drainage channels and wadis where sand and soil accumulate. An annual 'meadow' often develops in areas where soil and water collect after rain. Trees and larger shrubs (*Acacia* spp., *Lycium shawii, Ochradenus baccatus, Tamarix* spp., etc.) tend to be restricted to wadis.

On the northern and central plains two main communities can be identified (each with many local variations).

  i. Dwarf shrubland dominated by *Rhanterium epapposum* with *Astragalus* spp., *Fagonia* spp. and *Plantago* spp. This favours reasonably well-drained soil on shallow sand or amongst rocks in pockets of sand. This community covers huge areas of the Arabian Platform but does not extend far onto the Shield regions.
  ii. Dwarf shrubland dominated by *Haloxylon salicornicum* prefers deeper sand than the *Rhanterium* community but is also found on shallow sand, gravel plains and occasionally on rock surfaces. It is frequently associated with other Chenopodiaceae such as *Anabasis lachnantha* and *Agathophora alopecuroides*.

C. SABKHAH

The sabkhah are saline flats with a thin crust on the surface and saturated brine beneath. They are nearly always devoid of plant life but usually have a zone of halophytes around their margins. Vegetation can also develop where sand dunes overlay the sabkhah surface. Typical species associated with sabkhah are *Zygophyllum qatarense, Z. mandavillei, Seidlitzia rosmarinus, Halocnemum strobilaceum, Bienertia cycloptera, Aeluropus lagopoides, Salicornia europaea* and *Arthrocnemum macrostachyum*.

INTRODUCTION

2. Vegetation of the mountains and plains of the S and SW of the Peninsula

Drought-deciduous *Acacia-Commiphora* bushland (usually from 3 to 5m tall) is the dominant vegetation over the plains and lower slopes of the mountains in the S and SW of the Peninsula. To the N of Makkah and in northern Oman the *Commiphora* spp. gradually disappear and the *Commiphora-Acacia* bushland is replaced by bushland dominated by *Acacia* spp. At higher altitudes semi-evergreen bushland and thicket replace the drought-deciduous bushland. Small areas of riparian (valley) forest and secondary grassland are also found in the mountains. Over most of the region the woodlands have been severely degraded with remnants often found only in inaccessible places. However, reasonably intact woodland still survives in a few areas, for instance, the *Juniperus* woodlands in the Asir mountains of Saudi Arabia and the J. Akhdar range of northern Oman, the escarpment woodlands of Dhofar, the valley forest on the western escarpment mountains of Yemen and the bushland and thickets of Socotra.

Some of the most characteristic vegetation types of the mountains are the succulent communities dominated by species of *Aloe*, Asclepiadaceae and Euphorbiaceae. At high altitudes, where there are regular frosts, the Asclepiadaceae and Euphorbiaceae are usually replaced by succulent species of Crassulaceae. In general these succulent communities are found in three rather different habitats (Deil 1991): (i) natural habitats in rocky places in the foothills of the mountains; (ii) in a zone extending from the Yemen Highlands eastwards to S Oman on the dry, interior, north- and east-facing slopes of the mountains; (iii) in secondary habitats by abandoned settlements and other ruderal places where they are favoured by the high nitrogen levels in the soil and are afforded protection from grazing by their spines and unpalatable latex.

The vegetation of the mountains of the S and SW of the Peninsula shows a reasonable altitudinal zonation and is here described by a series of transects (see map 4). The vegetation of Socotra is treated separately at the end.

A. HIJAZ

Transect running eastwards from a point between Jeddah and Yanbu (Vesey-Fitzgerald 1957a).

1. Tihama: very open drought-deciduous bushland with *Acacia tortilis*, *A. asak* and *Maerua crassifolia* (*Acacia-Commiphora* associations are absent from the coastal plain this far north).
2. Western foothills of the Hijaz (up to 800m): open drought-deciduous bushland with *Acacia hamulosa*, *A. tortilis*, *A. ehrenbergiana* and *Commiphora* spp.
3. Western slopes of the Hijaz (800–1500m): drought-deciduous bushland with *Acacia asak* and *A. etbaica*.
4. Western slopes of the Hijaz mountains (above 1500m): open semi-evergreen bushland with *Olea europaea* ssp. *africana* and *Juniperus phoenicea* (*Juniperus* usually occurring only on isolated peaks).

5. Mountain plateaux (not exceeding 1500m): very open drought-deciduous woodland with *Acacia asak*.
6. Lower altitudes on the eastern slopes of the mountains: scattered trees of *Acacia tortilis*.
7. Dry eastern slopes of the Hijaz (down to about 1000m): scattered trees of *Acacia tortilis* and *Maerua crassifolia*

B. ASIR

Transect from Ad-Darb (Tihama) to J. Sawda, and then ENE towards the interior (König 1986a, 1986b).

1. Tihama (0–250m): *Acacia-Commiphora* drought-deciduous woodland with *A. ehrenbergiana, A. tortilis, A. mellifera, A. oerfota, A. hamulosa, Commiphora myrrha* and *C. gileadensis*.
2. (250–400m): *Acacia tortilis-Commiphora* drought-deciduous woodland with *Commiphora myrrha, C. gileadensis, C. kataf, Euphorbia cuneata, E. triaculeata, Acacia hamulosa, A. mellifera, Maerua crassifolia, Cadaba longifolia* and *Dobera glabra*.
3. (400–1100m): *Acacia asak-Commiphora* drought-deciduous woodland with *Commiphora myrrha, C. gileadensis, Euphorbia cuneata, Dobera glabra, Moringa peregrina, Grewia villosa* and *Acacia tortilis*.
4. (1100–1350m): *Acacia asak* drought-deciduous woodland.
5. (1350–1600m): *Acacia etbaica* drought-deciduous woodland.
   The following species are also commonly associated with the last two zones: *Grewia villosa, G. mollis, G. tembensis, Barleria trispinosa, B. bispinosa, Maytenus* spp., *Pyrostria* sp. and *Cadia purpurea*; *Adenium obesum, Anisotes trisulcus* and *Ecbolium viride* are found throughout the drought-deciduous woodlands:.
6. (1600–2000m): sclerophyllous scrub with *Acokanthera schimperi, Teclea nobilis, Pistacia falcata, Tarchonanthus camphoratus* and *Aloe sabaea*.
7. (2000–2400m): *Juniperus-Olea* forest with *Juniperus procera, Olea europaea* ssp. *africana, Teclea nobilis, Celtis africana, Debregeasia saenab, Rhus retinorrhoea, Maesa lanceolata, Nuxia congesta* and *Buddleja polystachya*.
8. (above 2400m): *Juniperus procera* woodland with *Acacia origena, Dodonaea viscosa* and *Euryops arabicus*.
9. (2650–2500m): *Juniperus procera* or *Acacia origena* (around terraced cultivation) woodland.
10. (2450–2300m): low shrubland with *Euphorbia schimperiana* and *Lavandula dentata*.
11. Below 2300m the vegetation thins and eventually merges into dwarf desert shrubland.

C. YEMEN ESCARPMENT MOUNTAINS

Transect from Al Luhayyah to Marib (Al-Hubaishi & Müller-Hohenstein 1984, Scholte et al. 1991).

INTRODUCTION

1. Tihama (0–300m): i. *Avicennia marina* mangrove woodland along the coast; ii. evergreen and deciduous dwarf shrubland with *Cadaba* spp., *Capparis decidua* and *Panicum turgidum* on the coastal plain; iii. drought-deciduous woodland with *Acacia ehrenbergiana, A. tortilis* and evergreens such as *Dobera glabra* and *Balanites aegyptiaca* on the plains and lower slopes of the mountains; iv. evergreen alluvial woodland with *Salvadora persica, Tamarix* spp., *Calotropis procera, Leptadenia pyrotechnica* and the grass *Desmostachya bipinnata* along the wadis.
2. Tihama foothills (300–1000m): i. *Acacia-Commiphora* drought-deciduous woodland with *Acacia tortilis, A. asak, A. abyssinica, A. mellifera, Commiphora gileadensis, C. kataf, C. myrrha, Maerua crassifolia, Delonix elata* and species of *Cadaba* and *Grewia*; ii. on rocky areas a rich succulent flora with *Adenium obesum, Adenia venenata* and species of *Euphorbia* and *Aloe*; iii. the vegetation in the larger wadis at lower altitudes resembles that of the wadis on the Tihama; at higher altitudes there is species-rich riparian woodland with *Combretum molle, Trichilia emetica, Ficus cordata, Mimusops schimperi* and *Adina microcephala*. On the lower slopes of J. Bura there is one of the few remaining areas of more or less natural woodland with *Combretum molle, Terminalia brownii, Phoenix reclinata, Acacia tortilis, A. asak, A. abyssinica, A. mellifera, Commiphora gileadensis, C. kataf* and *C. myrrha*
3. Lower Escarpment (1000–1600m): i. *Acacia-Commiphora* drought-deciduous woodland with *Acacia asak, A. mellifera, Commiphora kataf, C. myrrha, C. abyssinica, Grewia* spp. *Carissa edulis, Cadia purpurea* and large trees such as *Terminalia brownii, Cordia abyssinica* and *Adina microcephala* providing shade around the terraces and fields; a rich succulent flora develops on rocky outcrops. ii. riparian, broad-leaved evergreen woodland with *Ficus vasta, F. populifolia, Cordia abyssinica* and *Mimusops schimperi* in the wadis.
4. Higher Escarpment (1600–2200m): i. evergreen bushland and thicket characterized by *Acacia origena, A. gerrardii, A. abyssinica, Olea europaea, Buddleja polystachya, Barbeya oleoides, Ehretia abyssinica, Dodonaea viscosa, Euclea racemosa* ssp. *schimperi, Rhus abyssinica* and arborescent *Euphorbia* spp.; ii. riparian, evergreen woodland with *Ficus vasta, F. sycomorus* and *Cordia abyssinica*.
5. Highlands (2200–2700(–3700)m): drought-deciduous montane woodland with *Acacia gerrardii, A. origena, Buddleja polystachya, Cordia abyssinica, Olea europaea, Rosa abyssinica, Grewia mollis, Carissa edulis, Ehretia abyssinica, Myrsine africana, Hypericum revolutum, Nuxia congesta* and occasionally *Juniperus procera*.
6. Eastern Desert (down to 1300m): dwarf shrubland with *Euphorbia balsamifera, Kleinia odora* and a mosaic of very open *Acacia* and *Acacia-Commiphora* bushlands with *Acacia nilotica, A. gerrardii, A. oerfota, A. etbaica, A. tortilis, Maerua crassifolia, Ziziphus spina-christi* and occasionally *Dracaena serrulata*.

D. DHOFAR

Transect from Salalah to Thumrait (Miller 1988, Radcliffe-Smith 1980).

1. Coastal plain and foothills (0–50m): *Acacia-Commiphora* bushland with *Acacia*

tortilis, Commiphora habessinica, C. foliacea, C. gileadensis, Caesalpinia erianthera, Boscia arabica, Cadaba farinosa and on rocky outcrops *Adenium obesum, Sansevieria ehrenbergii, Euphorbia* sp. aff. *cactus* and *Aloe* spp.

2. Seaward-facing escarpments (50–500m): drought-deciduous thicket dominated by the endemic tree *Anogeissus dhofarica* with *Acacia senegal, Croton confertus, Maytenus dhofariensis, Euphorbia smithii, Sterculia africana* and *Commiphora* spp.
3. Seaward-facing escarpments (500–900m): semi-evergreen thicket with *Anogeissus dhofarica, Olea europaea, Dodonaea viscosa, Carissa edulis, Euclea racemosa* ssp. *schimperi* and *Rhus somalensis*.
4. Summit plateau (800–900m): grassland with scattered trees and patches of thicket with *Ficus vasta, F. sycomorus, Anogeissus dhofarica, Maytenus dhofariensis* and *Carissa edulis*.
5. North-facing slopes of the escarpment mountains (900–700m): dwarf shrubland with *Euphorbia balsamifera* ssp. *adenensis, Commiphora* spp. and *Barleria* spp.
6. Dry, north-draining wadis and slopes of the escarpment mountains (700–400m): very open bushland with *Boswellia sacra, Acacia etbaica, Maerua crassifolia, Commiphora habessinica* and *Dracaena serrulata*.
7. Desertic, gravel plains to the north of the mountains: dwarf shrubland with *Zygophyllum qatarense, Convolvulus hystrix, Heliotropium* spp. and the palm *Nannorrhops ritchieana* in the wadis.
8. Sand desert of the Rub' al-Khali: along the margins of the sands open *Prosopis cineraria* woodland and in the sands open dwarf-shrubland with *Calligonum crinitum* ssp. *arabicum, Tribulus arabicus* and *Cyperus conglomeratus* s.l.

E. JEBAL AKHDAR

Transect south from near Rustaq to J. Shams (Mandaville 1975).

1. Plains and broad wadis (0–600m): open *Acacia* bushland with *Acacia tortilis, A. ehrenbergiana, Ziziphus spina-christi, Prosopis cineraria* and *Pteropyrum scoparium*.
2. Mountain wadis (350–1050m): drought-deciduous bushland with *Acacia tortilis, Ziziphus spina-christi, Ficus cordata* ssp. *salicifolia, Prosopis cineraria, Nerium oleander* (Syn. *N. mascatense*) and *Acridocarpus orientalis*.
3. Rocky slopes (450–1350): dwarf shrubland with *Euphorbia larica, Pulicaria glutinosa* and occasional trees of *Moringa peregrina* and *Maerua crassifolia*.
4. Open or rocky slopes (1350–2300m): evergreen bushland and thicket with *Sideroxylon buxifolium* (*Monotheca buxifolia*), *Olea europaea, Sageretia spiciflora, Juniperus excelsa* ssp. *polycarpos, Ebenus stellatus* and *Dodonaea viscosa*.
5. Summit areas (2300–3050m): open evergreen woodland with *Juniperus excelsa* ssp. *polycarpos, Euryops arabicus, Lonicera aucheri* and *Cotoneaster nummularia*.

F. SOCOTRA (Popov 1957)

1. Coastal and inland plains: open deciduous shrubland dominated by *Croton socotranus* (generally below 1.5m in height) with trees of *Euphorbia arbuscula, Dendrosicyos socotranus* and *Ziziphus spina-christi* emergent.

INTRODUCTION

2. Lower slopes of the mountains, limestone plateau and escarpments: open deciduous bushland with *Croton socotranus* and *Jatropha unicostata*, and many emergent trees including *Dendrosicyos socotranus*, *Sterculia africana* var. *socotrana*, *Boswellia* and *Commiphora* spp. This is the most widespread vegetation type on the island and in some areas, particularly on seaward-facing escarpments, is dominated by succulent-stemmed trees.
3. Sheltered areas on the limestone plateau and the mid-altitude slopes of the granite mountains: sub-montane semi-evergreen thicket (generally below 6m in height) with *Rhus thyrsiflora*, *Buxus hildebrandtii*, *Carphalea obovata* and *Croton* spp.
4. Higher slopes of the granite mountains: a mosaic of thicket, low shrubland, grassland and open-rock vegetation develops; the thicket is dominated by *Rhus thyrsiflora*, *Cephalocroton socotranus* and *Allophylus rhoidiphyllus* with emergent *Dracaena cinnabari*; this merges above into a low shrubland dominated by *Hypericum* species; on the gentler slopes the vegetation has been cleared to form grassland pastures with *Themeda quadrivalvis*, *Apluda mutica*, *Heteropogon contortus*, *Dichanthium foveolatum* and *Arthraxon* spp.; the exposed rocks of the granite pinnacles are covered with lichens and cushion vegetation with *Helichrysum* spp.

3. Coastal Vegetation

a. Marine communities. 'Seagrass beds' dominated by angiosperms occur in shallow water in sheltered bays along the shores of the Arabian Gulf and Red Sea. Species include *Ruppia maritima*, *Diplanthera uninervis*, *Halophila stipulacea*, *H. ovalis*, *Cymodocea ciliata* and *C. rotundata*.

b. Mangrove woodland. Mangrove woodland with *Avicennia marina* and occasionally *Rhizophora mucronata* occurs, in scattered localities, around the coast of Arabia as far as c. 25°N. Usually the trees are stunted and shrub-like but in favourable sites they can reach 6m in height.

c. Coastal communities. Coastal vegetation varies depending on local edaphic conditions. The commonest vegetation type is dwarf-shrubland with some of the following species: *Limonium axillare*, *Cressa cretica*, *Zygophyllum* spp.; succulent Chenopodiaceae including *Suaeda* spp., *Salsola* spp., *Atriplex* spp., *Bienertia cycloptera* and *Halocnemum strobilaceum*; grasses including *Aeluropus lagopoides*, *Halopyrum mucronatum*, *Odyssea mucronata*, *Panicum turgidum*, *Urochondra setulosa* and the shrubs *Lycium shawii* and *Nitraria retusa*.

FLORISTICS

The flora of the Arabian Peninsula is still incompletely known. However, it is possible to make a rough estimate of the number of species recorded from the area. Table 1

Table 1. Floristic analysis of the Arabian Peninsula

	Families	Genera	Species
Pteridophyta	18	29	56
Gymnospermae	2	4	12
Angiospermae	c.144	c.1100	c.3400

gives a breakdown of the approximate number of families, genera and species for the whole region and Table 2 gives a breakdown by country.

The ten largest families are Gramineae (c.450 spp.), Leguminosae (c.320 spp.), Compositae (c.300 spp.), Labiatae (c.120 spp.), Euphorbiaceae (c.120 spp.), Boraginaceae (c.120 spp.), Cruciferae (c.115 spp.), Asclepiadaceae (c.110 spp.), Acanthaceae (c.110 spp.) and Scrophulariaceae (c.100 spp.). The largest genera are: *Euphorbia* (67 spp.), *Heliotropium* (46 spp.), *Cyperus* (41 spp.), *Indigofera* (37 spp.), *Caralluma* (30 spp.) and *Convolvulus* (30 spp.).

Approximately 600 (17%) of the total species are endemics, but if Socotra is excluded from the calculation then the total is reduced to c.360 (11.5%). Socotra, like many oceanic islands, has a high level of endemism (30%). Table 2 shows the percentage of endemic species in the individual countries of the Peninsula.

Table 2. Floristic richness and endemism in the countries of the Arabian Peninsula

Country or Region	Area in km sq.	Total Spp.	Native Spp.	Number of Endemic Spp.
Saudi Arabia	2,400,000	2,100	1,800	35 (2%)
Yemen (N)	190,000	1,650	1,370	60 (4%)
(S)	287,000	1,180	960	80 (8%)
Socotra	3,600	820	790	240 (30%)
Oman	270,000	1,200	1,020	75 (7%)
UAE	75,000	490	340	0 (0%)
Qatar	11,500	300	220	0 (0%)
Bahrain	660	250	195	0 (0%)
Kuwait	24,000	280	235	0 (0%)

INTRODUCTION

Table 3. Arabian families richest in endemic species

	Total species	Endemic species
Asclepiadaceae	103	55 (53%)
Acanthaceae	104	46 (44%)
Scrophulariaceae	96	35 (36%)
Labiatae	122	43 (35%)
Euphorbiaceae	121	37 (31%)
Compositae	297	57 (19%)

The endemics are not equally spread through the Peninsula but are concentrated in certain regions (Miller & Nyberg 1991). Map 5 shows the number of Arabian endemic species occurring in 100km grid squares (not endemic to the individual squares). The highest concentrations in the Peninsula are found in the mountains of the western escarpments of Saudi Arabia and Yemen (N) and in certain areas of Yemen (S), the monsoon woodlands of Dhofar and to a lesser extent the mountains of N Oman. After Socotra (for which the figure shown represents the total endemic taxa in the archipelago) the richest square, with 99 endemics, is in the Hujariyah region of Yemen.

Endemic taxa are also unequally distributed throughout the families. Table 3 shows the six families with the largest numbers of endemic species.

The Asclepiadaceae, with 53%, has the highest percentage of endemic species, largely because of the succulent stapeliad genera: *Caralluma* 30 spp. – 21 endemics; *Duvalia* 4 spp. – 2 endemics; *Echidnopsis* 4 spp. – 2 endemics; *Huernia* 6 spp – all endemic; *Rhytidocaulon* 4 spp. – all endemic. Likewise the high percentage of endemics in Euphorbiaceae (31%) and Liliaceae (37%) is in large part because of the numbers of succulent *Euphorbia* and *Aloe* species respectively. Endemism is generally very high amongst the succulents in Arabia.

There are 18 genera endemic to the area of the Flora: seven on the mainland and eleven on the Socotra archipelago. Apart from *Centaurothamnus* and *Saltia* they all have relatively restricted distributions: *Centaurothamnus* is found, mainly on cliffs, above 2000m in the western escarpment mountains; *Dolichorhynchus* occurs in small areas of sandstone gorges and buttresses at about 300m in the NW of Saudi Arabia; *Saltia* is found in *Acacia-Commiphora* bushland on foothills in the extreme SW of the Peninsula; *Cibirhiza* and *Dhofaria* occur in the monsoon woodlands of Dhofar; *Xerotia* was described from the coastal region of Hadramaut in Yemen (S) and has recently been found in central Oman; *Isoleucas*, which is only known from the coast of Hadramaut in Yemen (S), is of uncertain taxonomic status.

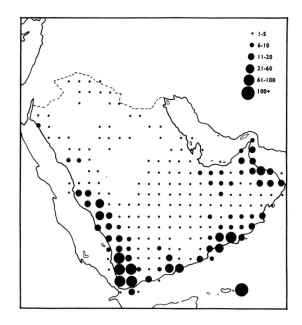

Map 5. Showing the numbers of Arabian endemic species occurring in 100km grid squares.

The majority of endemic taxa in Arabia are associated with mountainous areas. Mountains provide a rich variety of ecological niches and afford a degree of environmental stability during periods of climatic change. Many of the present distributions in Arabia date back to the Pleistocene when the area was affected by a series of major fluctuations in rainfall and temperature (Mandaville 1984, 1986). The mountains along the coasts of Arabia were buffered from the catastrophic effects of climatic change by generally more stable and humid conditions. In these more mesic conditions relict palaeo-African and palaeo-Indo-Malesian elements survived. The mountains, isolated from one another by tracts of inhospitable desert, provided ideal sites for vicariant evolution. This process can be seen in many genera whose species show similar patterns of distribution and endemism e.g. *Campylanthus*, *Farsetia*, *Kickxia*, *Lavandula*, *Maytenus*, *Ochradenus*, *Pulicaria*, *Pycnocycla*, *Schweinfurthia*, *Taverniera* and *Teucrium* (Miller & Nyberg 1991).

PHYTOGEOGRAPHY

Phytogeography attempts to divide the globe into natural floristic units. This has traditionally been accomplished using a hierarchy based on Floral Kingdoms, Regions,

INTRODUCTION

Domains, Provinces and Districts. Thus, according to Zohary (1973), Arabia falls into three phytogeographical regions: the East Saharo-Arabian Subregion of the Saharo-Arabian Region (covering the arid interior), the Nubo-Sindian Province (covering a strip around the coasts except in the extreme SW) and the Eritreo-Arabian Province (in the SW of the Peninsula and Socotra) - the last two provinces both coming within the Sudanian Region.

White (1983) proposed a new system for Africa based on regional centres of endemism separated by regional transition zones and regional mosaics. He defined a regional centre of endemism as 'a phytochorion which has at the same time more than 50% of its species confined to it and a total of more than 1000 endemic species'.

This system has been extended to cover SW Asia (based on the study of the areas of 509 species from the deserts and subdeserts of Iran) by Léonard (1989) and White and Léonard (1991). The general limits of their phytogeographical areas are outlined in map 6. Using this system Arabia is subdivided into three main phytogeographical areas: the Saharo-Sindian regional zone (SS); the Somalia-Masai regional centre of endemism (SM) and the Afromontane archipelago-like regional centre of endemism (shaded black on the map). The Saharo-Sindian regional zone is further divided into the Arabian regional subzone (SS1) and the Nubo-Sindian local centre of endemism (SS2).

Characteristic vegetation and species of the three main phytogeographical areas in Arabia are given below.

1. The Saharo-Sindian regional zone. Characteristic vegetation: dwarf shrubland. Characteristic species: *Anastatica hierochuntica*, *Asteriscus graveolens*, *Calligonum* spp., *Cornulaca* spp., *Haloxylon salicornicum*, *Moltkiopsis ciliata*, *Neurada procumbens*, *Oligomeris linifolia*, *Rhazya stricta*, *Scrophularia deserti* and *Suaeda aegyptiaca*.
2. The Somalia-Masai regional centre of endemism. Characteristic vegetation: *Acacia-Commiphora* bushland and thicket and at higher altitudes evergreen bushland and thicket. Characteristic species: *Acacia* spp., *Commiphora* spp., species of Capparaceae, *Dobera glabra*, *Euclea racemosa* ssp. *schimperi*, *Grewia* spp. and *Olea europaea*.
3. The Afromontane archipelago-like regional centre of endemism. Characteristic vegetation: *Juniperus procera* semi-evergreen bushland and thicket. Characteristic species: *Acacia origena*, *Buddleja polystachya*, *Catha edulis*, *Cussonia holstii*, *Dombeya torrida*, *Erica arborea*, *Hypericum revolutum*, *Nuxia congesta*, *Rosa abyssinica* and *Teclea nobilis*.

Floristic elements from surrounding phytochoria also extend into Arabia. For example there are enclaves of typical Mediterranean species in the mountains of

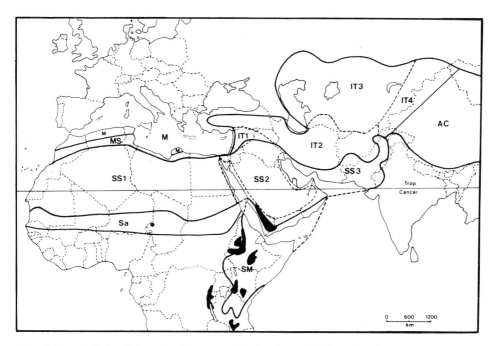

Map 6. General limits of the main phytogeographical regions of Africa and Asia: AC, Central Asiatic region; IT, Irano-Turanian regional centre of endemism further divided into four local centres of endemism (IT1–4); M, Mediterranean regional centre of endemism; MS, Mediterranean- Sahara regional transition zone; SS, Saharo-Sindian regional zone further divided into the Arabian regional sub-zone (SS1), the Nubo-Sindian local centre of endemism (SS2) and the Nubo-Sindian local centre of endemism (SS3); SM, Somalia-Masai regional centre of endemism; shaded black, Afromontane archipelago-like regional centre of endemism. After Léonard (1989) and White and Léonard (1991).

western Saudi Arabia, including species such as *Anagyris foetida*, *Hypericum hircinum*, *Juniperus phoenicea*, *Rhus tripartita* and *Velezia rigida*. Similarly, Irano-Turanian elements occur in the mountains of N Oman and in NE Arabia, for example, *Artemisia herba-alba*, *Dionysia mira*, *Prunus arabica* and *Viola akhdarensis*.

HISTORY OF BOTANICAL EXPLORATION

The history of botanical collecting in the Arabian Peninsula has been covered in detail elsewhere and so this section gives a brief guide to some of the most useful reference sources. The most up-to-date and complete biographical index of plant collectors in the Arabian peninsula (including Socotra) was compiled by Gerald Wickens (1982). It contains brief biographical details of collectors, the country and dates of their collections, the herbaria where their collections are lodged and literature references. Blatter, in his Flora Arabica (1919–1936: 451–501), includes a chapter on the botanical exploration of Arabia with notes on the lives and activities of collectors and travellers. It is now rather out of date but gives a very comprehensive review of early collectors starting with the oldest references to Arabian plants, given in the Old Testament books Genesis and Exodus, notes on the earliest Arab botanists in the 9th to 14th centuries and an extensive section covering collecting from the 16th to 20th centuries. Other useful accounts of collectors in Arabia are found in Mandaville (1990) and Batanouny & Baeshin (1978).

CONSERVATION AND THREATS

Much of the region is fertile and even land that has been classed as desert can be productive for grazing if it is not overused. However, the vegetation has been severely degraded in most areas. The principal causes of this are overgrazing on fragile rangelands, poor agricultural practices, the clearance of woodland for fuelwood, timber and charcoal, and inappropriate irrigation practices resulting in the salinization or alkalization of agricultural land. Although this destruction of the natural vegetation is not a recent process in Arabia, in most areas the rate of degradation seems to be rapidly accelerating.

Over large areas of Arabia, fuelwood and charcoal are the main energy sources and there is now a severe fuelwood shortage which in some areas is becoming acute. Improved transport linked to an increased demand by the urban population, particularly for charcoal, has led to the rapid deforestation of increasingly large areas around towns and cities in the region.

Overgrazing by goats, sheep, cattle and camels is one of the principal causes of degradation of the rangelands and is a major problem in many parts of the region. The problem is particularly acute in the interior deserts where, because of the breakdown of traditional pastoral practices, stock is staying in the desert all year round. Water and grain are supplied to the herds by tanker while trucks are used to rapidly transport the livestock to areas where vegetation has appeared after rain.

Changing agricultural practices throughout the region are causing problems of many sorts: the breakdown and abandoning of traditional rangeland management in Arabia is leading to serious overgrazing; in Yemen the former migration of rural workers to the rich Gulf States led to the virtual abandonment of the sophisticated systems of terraces in the mountains resulting in serious soil erosion on the steep slopes; in parts of Arabia large areas of rangeland have been put under the plough to grow cereals and other crops.

Very little of the area is under effective protection with many important ecosystems wholly unprotected and threatened. According to Lean et al. (1990), in most countries of SW Asia less than 1% of the land is classified as protected areas. As in most parts of the world when there is a conflict between protected areas and the industrial or strategic interest of the nation, the protected sites will lose out.

PLAN OF THE FLORA

1. COVERAGE AND SEQUENCE

All native flowering plants and ferns are dealt with as well as cultivated and amenity plants of economic importance. The family sequence follows that of Engler and Prantl (*Syllabus der Pflanzenfamilien*, ed. H. Melchior, 12th edition, 1964).

2. NOMENCLATURE

All published names based on material from the Flora area are included; other synonyms, including basionyms, are omitted. Insufficiently known species are included as Sp. A, Sp. B, etc. References to important works are included at the end of the family or generic descriptions. Subspecies and varieties are usually keyed out separately under species. Full authority and place of publication are cited for each species and all synonyms. Citations to reliable illustrations are cited from literature references where they have been found useful. Type specimens are only indicated when collected from the Flora area. The herbarium abbreviation(s) of the institute(s) holding the type is also given; these follow *Index Herbariorum*, ed. 8, Regnum Vegetabile 120: 557–650 (1990).

3. DESCRIPTIONS AND OTHER NOTES

Descriptions are short and diagnostic; where possible they have been based on Arabian material. Family and generic descriptions apply to taxa found within Arabia and it should not be assumed that they could be used outside the region. Brief notes on habitats and altitudinal ranges are given after the descriptions.

The internal distribution of all taxa is indicated in the following sequence: **Saudi Arabia**; Republic of Yemen subdivided into **Yemen (N)** (previously the Yemen Arab Republic), **Yemen (S)** (previously the People's Democratic Republic of Yemen) and **Socotra** (which includes the islands of Abd' al-Kuri, Semha and Darsa); **Oman**; **UAE** (United Arab Emirates); **Qatar**; **Bahrain**; **Kuwait**.

Specimens are not cited. However, all material examined has been separately listed and the information will be made available at a later date. External distributions are indicated concisely.

4. MAPS

Dot-maps are provided at the back of each volume. On these maps each dot represents a locality from which a specimen has been examined or cited. Solid dots represent herbarium specimens examined and open circles are based on literature records and in some cases on reliable field observations. The maps are intended to give an indication of the range of taxa. Some categories of plants, particularly weeds, cultivated plants and certain very common species are generally under-recorded and so, in these cases, the maps are sometimes rather incomplete.

PTERIDOPHYTA

A.G. MILLER

The pteridophytes (in Arabia) are herbs with free-living sporophytes containing vascular tissue and bearing sporangia which produce spores and give rise to inconspicuous gametophytes. The gametophytes are thalloid, lack vascular tissue and bear archegonia and anthemidia.

There is no general agreement among specialists regarding family limits within the Pteridophyta. Family descriptions and a key to families have therefore been omitted. Short descriptions of the Ferns and Fern Allies are given followed by a synoptical arrangement of the genera according to R.J. Johns (1991, Pteridophytes of Tropical East Africa. RBG, Kew). An artificial key to genera is provided.

THE FERN ALLIES

Leaves small, one-nerved. Sporangia borne in the axils of specialized leaves (the sporophylls) which are solitary or arranged in cone-like strobili or (in *Equisetum*) borne on the underside of stalked scales (sporangiophores) and arranged in terminal cone-like strobili. Sporangia and spores either all similar (homosporous) or dissimilar (heterosporous) with the megasporangia containing megaspores and the microsporangia containing microspores.

THE FERNS

Stems (the rhizomes) usually clothed with scales. Leaves (the fronds) simple or compound with several to many nerves, usually circinate in bud (except *Ophioglossum*), comprising a stalk (the stipe) and a lamina with a central nerve (the rhachis); lamina either simple or divided into pinnae and pinnules. Sporangia homosporous or heterosporous (*Marsilea*), usually borne on the lower surface of the fronds, usually in clusters (the sori); sori naked or covered with a specialized scale-like organ (the indusium) or by the reflexed leaf margin or on specialized organs (*Marsilea* and *Ophioglossum*).

PTERIDOPHYTA

SYNOPTICAL ARRANGEMENT OF THE FAMILIES OF FERNS AND FERN ALLIES

THE FERN ALLIES

Psilotaceae
  1. Psilotum
Selaginellaceae
  2. Selaginella

Equisetaceae
  3. Equisetum

THE FERNS

Ophioglossaceae
  4. Ophioglossum
Marsileaceae
  5. Marsilea
Parkeriaceae
  6. Ceratopteris
Actiniopteridaceae
  7. Actiniopteris
Adiantaceae
  8. Cheilanthes
  9. Negripteris
  10. Pellaea
  11. Doryopteris
  12. Anogramma
  13. Adiantum
  14. Onychium
Acrostichaceae
  15. Acrostichum
Pteridaceae
  16. Pteris

Polypodiaceae
  17. Pleopeltis
  18. Loxogramme
Dennstaedtiaceae
  19. Pteridium
Oleandraceae
  20. Nephrolepis
  21. Arthropteris
Aspleniaceae
  22. Asplenium
  23. Ceterach
Thelypteridaceae
  24. Christella
Woodsiaceae
  25. Cystopteris
Dryopteridaceae
  26. Hypodematium
  27. Tectaria
  28. Polystichum
  29. Dryopteris

KEY TO GENERA OF ARABIAN FERNS AND FERN ALLIES

1. Stems jointed, hollow, with longitudinal ribs; leaves reduced to a short toothed sheath about the nodes; sporangia arranged in terminal cone-like structures   **3. Equisetum**
+ Stems not jointed, hollow, or ribbed; leaves never arranged in a sheath about the nodes; sporangia never arranged in terminal cone-like structures   2

2. Plants leafless except for scattered, minute scale-like sporophylls; sporangia 3-lobed   **1. Psilotum**
+ Plants with well-developed leaves, sometimes minute but not scale-like; sporangia not 3-lobed   3

3. Leaves simple, numerous, small and moss-like, 1-nerved; plants heterosporous **2. Selaginella**
+ Leaves simple or dissected, usually large, never moss-like; plants heterosporous (*Marsilea*) or homosporous 4

4. Fronds 4-foliolate; leaflets ± obovate; aquatic or semi-aquatic plants **5. Marsilea**
+ Fronds various, never 4-foliolate; aquatic, semi-aquatic or terrestrial plants 5

5. Fronds dimorphic; sterile segments simple, narrowly ovate to elliptic, inserted at the base of the fertile fronds or on the petioles; fertile segments simple with two rows of sporangia arranged in a spike near the tip **4. Ophioglossum**
+ Fronds dimorphic or not; if dimorphic then the sterile segments pinnately divided; sporangia arranged on the underside of the lamina 6

6. Fronds fan- or wedge-shaped with linear segments **7. Actiniopteris**
+ Fronds various but never fan-shaped with linear segments 7

7. Ephemeral fern; fronds very small, up to 10cm long; indusia absent **12. Anogramma**
+ Perennial or rarely annual ferns; fronds rarely less than 10cm; indusia present or absent 8

8. Fronds succulent, dimorphic; ultimate segments of the fertile fronds linear and of the sterile triangular or ovate; plants of swamps or pools **6. Ceratopteris**
+ Fronds not succulent, dimorphic or not; fronds not as above; plants terrestrial 9

9. Fronds simple 10
+ Fronds pinnately divided 11

10. Sori round; lamina with scattered dark-centred peltate scales beneath **17. Pleopeltis**
+ Sori elongate; lamina without peltate scales beneath **18. Loxogramme**

11. Lamina 1-pinnate, or 1-pinnate above and the basal pinnae lobed 12
+ Lamina 2–5-pinnate 17

12. Sori marginal 13
+ Sori superficial on the veins or covering the entire undersurface of the fertile pinnae 14

13. Sori borne on the inner surface of reflexed leaf margins; pinnae with flabellate venation **13. Adiantum**
+ Sori borne on the surface of lamina, hidden by the reflexed leaf margin; pinnae with pinnate venation **16. Pteris**

PTERIDOPHYTA

14. Large fern; entire undersurface of the fertile pinnae covered with sori
    15. Acrostichum
+ Small or medium-sized ferns; sori discrete, superficial on the veins, not covering the entire undersurface of the pinnae   15

15. Sori circular, never obscured; pinnae cordate at the base   **20. Nephrolepis**
+ Sori elongate, sometimes obscured by scales; pinnae not cordate at the base   16

16. Pinnae narrowed at the base, stalked to the rhachis   **22. Asplenium**
+ Pinnae broad at the base, auriculate or decurrent along the rhachis
    23. Ceterach

17. Sori marginal or submarginal   18
+ Sori superficial on the veins   27

18. Sori oval, solitary on ultimate segments, opening towards the margin
    22. Asplenium
+ Sori numerous, discrete, partly or wholly continuous along pinnules, opening inwards from the margin   19

19. Pinnules fan- or wedge-shaped, with flabellate venation; lamina glabrous; sori borne on inner surface of reflexed leaf margins   **13. Adiantum**
+ Pinnules various (but not as above), with pinnate venation; lamina glabrous or not; sori borne on surface of lamina, hidden by the reflexed leaf margin   20

20. Parsley-like fern; fertile and sterile fronds dissimilar, irregularly pinnate
    14. Onychium
+ Not parsley-like fern; fronds all similar, regularly pinnate   21

21 Lamina hairy or covered with white powder   22
+ Lamina glabrous   24

22. Large fern, 1–2m; fronds spaced; rhizome densely hairy   **19. Pteridium**
+ Small or medium-sized ferns, up to 50(–100)cm; fronds tufted; rhizome clothed with scales   23

23. Lamina covered with white powder beneath; sporangia black or dark brown, glossy; stipe clothed with scales throughout   **9. Negripteris**
+ Lamina without white powder beneath or if so (*Cheilanthes farinosa*) sporangia pale brown and stipe clothed with scales in the lower part only
    8. Cheilanthes

24. Stipe pale brown or straw-coloured   **16. Pteris**
+ Stipe reddish brown or reddish black   25

25. Lamina pentagonal; pinnae decurrent along the rhachis   **11. Doryopteris**
+ Lamina triangular or ovate; pinnae not decurrent along the rhachis   26

26.	Small ferns, fronds to 20cm; sori discrete	**8. Cheilanthes**
+	Medium-sized ferns, fronds usually more than 30cm; sori continuous	**10. Pellaea**
27.	Sori elongate; indusia narrow, opening inwards or obsolete or absent	**22. Asplenium**
+	Sori round; indusia hood-like, peltate or reniform	28
28.	Indusia hood-like; plant ± glabrous except for a few scales at the base of the stipe	**25. Cystopteris**
+	Indusia reniform or peltate; plant hairy or clothed with scales	29
29.	Stipe clothed with scales, the scales extending to the tip of the lamina	30
+	Stipe hairy, if clothed with scales then the scales restricted to the base of the stipe	31
30.	Sori peltate; pinnules aristate	**28. Polystichum**
+	Sori reniform; pinnules dentate	**29. Dryopteris**
31.	Lamina elliptic to narrowly ovate, pinnate with the pinnae pinnatifid with entire lobes	32
+	Lamina ovate-triangular, 2–4-pinnatifid	33
32.	Pinnae articulated to the stipe; lamina often with scattered white glands above; rhizome-scales peltate; pinnae not tapering at the base	**21. Arthropteris**
+	Pinnae not articulated to the stipe; lamina without scattered white glands above; rhizome-scales not peltate; lamina tapering at the base	**24. Christella**
33.	Lamina with gemmae (or their scars) visible on the upper surface; stipe thinly clothed with scales but without a conspicuous tuft at the base; veins anastomosing	**27. Tectaria**
+	Lamina without gemmae; stipe with a conspicuous tuft of reddish-brown scales at the base; veins free	**26. Hypodematium**

Family 1. PSILOTACEAE

1. PSILOTUM Swartz

Erect herbs; rhizomes horizontal, rootless. Stems dichotomously branched, rigid. Leaves few, scale-like. Sporangia 3-locular, borne in the axils of minute bilobed bracts on the upper part of the stems.

1. P. nudum (L.) P. Beauv., Prodr. aethéogam.: 112 (1805).

Small bushy herb. Stems up to 35cm, erect, branched throughout, triangular in section, c.2mm across, glabrous. Leaves minute, widely spaced, narrowly ovate, c.1.5mm. Sporangia c.2.5mm across, 3-lobed. **Map 7, Fig. 2.**

In shade on large boulders and on cliffs in a shady gorge; 1000–1700m.

Saudi Arabia, Yemen (N). Widespread in tropical and subtropical regions.

Very rare in Arabia and known only from near Turbah in Yemen and from a deep gorge in the southern Asir region of Saudi Arabia. Outside Arabia its nearest station is in southern Kenya.

Family 2. SELAGINELLACEAE

2. SELAGINELLA P. Beauv.

Erect or prostrate herbs; stems bearing aerial roots (rhizophores). Leaves small, moss-like, each with a minute ligule at the base, either spirally arranged and similar or 2-ranked and dimorphic with a median above and a lateral on either side. Spores of two kinds: microspores (male) borne in microsporangia and megaspores (female) borne in megasporangia. Sporangia borne in the axils of specialized leaves known as sporophylls; sporophylls similar or dimorphic, arranged in terminal strobili.

1. Erect, tufted plant; leaves strongly discolorous, dark green above, pale green beneath **1. S. imbricata**
+ Prostrate, creeping plants; leaves concolorous, pale or bright green 2

2. Leaf margins white, long-ciliate; sporophylls similar **2. S. yemensis**
+ Leaf margins green, not or only shortly ciliate; sporophylls dimorphic 3

3. Stems 1–3cm long; median leaves acute **3. S. perpusilla**
+ Stems up to 40cm long; median leaves with a long setaceous tip c.$\frac{1}{2}$ as long as the leaf **4. S. goudotana** var. **abyssinica**

1. S. imbricata (Forsskal) Spring in Decne in Arch. Mus. Hist. Nat. 2: 193 (1841). Syn.: *Lycopodium imbricatum* Forsskal (1775 p.187). Illustr.: Adumbr. Fl. Aeth. 27: 560, t.3 (1975); Collenette (1985 p.458). Type: Yemen (N), *Forsskal* 620 (C).

Tufted fern-like plant, bright dark green above, pale green or brown beneath, curling into a loose ball when dry. Stems up to 30cm, leafless and unbranched below. Leaves dimorphic: laterals imbricate, oblong-elliptic, (1.5–)2–3.5 × (0.5–)1–1.3mm, obtuse, with entire margins, hyaline on the front edge towards the base; median oblong-ovate, 1.25–1.5 × 0.7–1mm, obtuse or acute and falcate, with hyaline margins. Strobili 2.5–5(–8)mm, male above, female below; sporophylls all similar, ovate-triangular, c.1.0 × 0.8mm, acute, the margin hyaline and denticulate. **Map 8, Fig. 2.**

In crevices on shady cliffs, rocks and cliff ledges; 100–1500m.

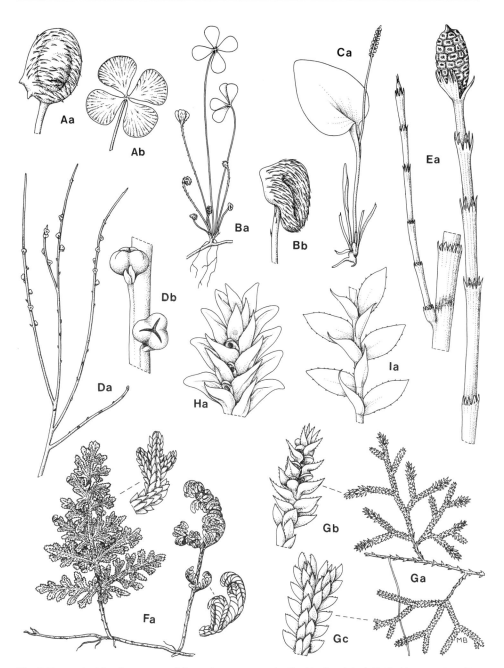

Fig. 2. Ferns. A, *Marsilea coromandeliana*: Aa, sporocarp (×8); Ab, frond (×1.5). B, *M. aegyptiaca*: Ba, habit (×1); Bb, sporocarp (×8). C, *Ophioglossum polyphyllum*: Ca, habit (×0.6). D, *Psilotum nudum*: Da, habit (×0.6); Db, part of fertile branch (×6). E, *Equisetum ramosissimum*: Ea, fertile branch (×2). F, *Selaginella imbricata*: Fa, habit (×1) and leaves (×3). G, *S. yemensis*: Ga, habit (×1.5); Gb, strobilus, (×6); Gc, stem leaves (×6). H, *S. perpusilla*: Ha, strobilus (×12). I, *S. goudotana*: Ia, stem leaves (×7).

Saudi Arabia, Yemen (N & S), Oman. Southern and eastern Africa north to Ethiopia and Madagascar.

2. S. yemensis (Swartz) Spring in Decne in Arch. Mus. Hist. Nat. 2: 191 (1841). Syn.: *Lycopodium sanguinolentum* Forsskal (1775 p.CXXV) non L. (1753); *L. yemense* Swartz, Syn. fil. 182 (1806); *S. arabica* Baker, Handb. fern-allies 38 (1887). Illustr.: Adumbr. Fl. Aeth. 27: 567 (1975); Collenette (1985 p.458). Type: Yemen (N), *Forsskal* s.n. (S).

Creeping, often mat-forming, bright green plant. Stems prostrate up to 30cm. Leaves dimorphic: laterals oblong-ovate, $1.8-2.5 \times 0.75-1.25$mm, acute or rounded, with ciliate margins especially near the base, the base unequal $\pm$ amplexicaul; median ovate, $1-2 \times 0.5-1$mm, acuminate at the tip, with ciliate margins, the base rounded or asymmetric with a triangular lobe. Strobili $(2-)3-4$mm, usually with a solitary megasporangium at the base and microsporangia above; sporophylls all similar, resembling the median leaves. **Map 9, Fig. 2.**

In shady places on rocks, cliffs and under boulders; 950–2800m.

Saudi Arabia, Yemen (N & S). Tropical NE Africa.

3. S. perpusilla Bak. in J. Bot. 23: 292 (1885).

Small, moss-like, pale green plant. Stems $0.5-1.5(-2.5)$cm long, creeping. Leaves dimorphic: laterals ovate, $1.5-1.75 \times 0.7-0.8(-1)$mm, acute, the margins minutely denticulate, the base $\pm$ rounded; median ovate-elliptic, $0.5-0.9(-1) \times 0.3-0.6$mm, acuminate at the tip, the margins minutely denticulate, the base $\pm$ rounded. Strobili $(1.5-)2-3$mm, mainly female with only occasional microsporangia at the top; sporophylls dimorphic, slightly larger than the leaves, shortly ciliate on the margins towards the base. **Map 10, Fig. 2.**

Shady, humid gully in *Anogeissus* woodland; 600m.

Oman. Zaire, E. Africa and Madagascar.

Only collected once in our area but easily overlooked. Resembles and grows with mosses on wet clay in dense shade under trees.

4. S. goudotana Spring var. **abyssinica** (Spring) Bizzarri in Adumbr. Fl. Aeth.: 585 (1975); Illustr.: op. cit., p.588.

Creeping, moss-like plant. Stems prostrate or $\pm$ erect, up to 40cm. Leaves dimorphic; laterals oblong-ovate, $2-3.5 \times 1-2$mm, apiculate, the margins denticulate, the base oblique and $\pm$ amplexicaul; median $1.5-2 \times 1$mm, ovate, drawn into a long setaceous tip which is c.$\frac{1}{2}$ as long as the leaf. Strobili $0.5-1.2$mm; sporophylls dimorphic, similar to the leaves; lateral sporophylls male only; median sporophylls male above and female below. **Map 11, Fig. 2.**

Wet rock face by a waterfall; 2900m.

Yemen (N). Tropical Africa.

In Arabia known only from a single locality in Yemen.

Family 3. EQUISETACEAE

3. EQUISETUM L.

Erect rhizomatous herbs. Stems round, hollow, ridged, jointed and often with whorls of branches at the nodes. Leaves reduced to a many-toothed sheath at each node. Sporangia borne on the undersurface of hexagonal stalked scales (sporangiophores) which are arranged in compact terminal cone-like heads.

1. E. ramosissimum Desf., Fl. atlant. 2: 398 (1799). Illustr.: Fl. Zamb., Pteridophyta: 33 (1970); Collenette (1985 p.233).

Rhizomes long-creeping, blackish brown. Stems up to 1m, branched mainly below, grey-green, the internodes with 10–25 ribs, the sheaths about twice as long as broad, the teeth blackish with hair-tips; branchlets slender, hollow, 8-ribbed. Cones terminal on the main stems, ellipsoid, 5–15(–20)mm. **Map 12, Fig. 2.**

By streams, pools and in damp places; 750–2800m.

Saudi Arabia, Yemen (N & S), Oman. Europe, Asia and Africa.

Family 4. OPHIOGLOSSACEAE

4. OPHIOGLOSSUM L.

Perennial herbs with globose or elongated corm-like rhizomes. Fronds 1-several, petiolate, with a simple sterile segment and a fertile segment inserted at the base of the sterile segment or on the petiole. Fertile segments simple, without a lamina; sporangia large, sunken, arranged in two rows on the upper part of the fertile segment, dehiscing by transverse slits.

1. O. polyphyllum A. Braun in Seubert., Fl. azor.: 17 (1844). Syn.: *O. aitchisonii* (C.B. Clarke) D'Almeida in J. Indian Bot. Soc. 3: 63 (1922); *O. capense* sensu Schwartz (1939). Illustr.: Collenette (1985 p.380).

Rhizome erect; leaves (1–)2, surrounded at the base by persistent leaf bases. Petioles 1.5(–10)cm. Sterile lamina narrowly ovate to elliptic, 2–4 × 1–1.5(–2)cm, acute or obtuse, cuneate at the base. Fertile segment 3–8(–10)cm, attached at the base of the sterile lamina, acute at the tip, with 13–30 pairs of sporangia. **Map 13, Fig. 2.**

Appearing after rain in sandy and silty depressions, 20–2900m.

Saudi Arabia, Yemen (N), Socotra, Oman, Bahrain, Kuwait. From tropical and southern Africa to Iran, Afghanistan and Northern India.

2. O. reticulatum L., Sp. pl.: 1063 (1753).

Similar to *O. polyphyllum* but the sterile lamina broadly ovate, 3–6 × 1.5–3cm, with a cordate or truncate base.

In a damp shady place in an orchard (Yemen) and on granite cliffs (Socotra); 1000–2400m.

Yemen (N), Socotra. Pantropical.

Much rarer in Arabia than the preceding species. The only record from N Yemen (*Hepper* 6313) may possibly represent an extreme, broad-leaved, form of *O. polyphyllum*.

Family 5. MARSILEACEAE

5. MARSILEA L.

Aquatic or semi-aquatic ferns; rhizomes creeping, slender. Fronds with 4 leaflets arranged in a terminal cluster. Leaflets fan-shaped with a cuneate base and the outer margins entire, sinuate or variously toothed. Heterosporous, the sporangia contained in closed sporocarps (specialized pinnae); sporocarps hard, borne on short pedicels.

A difficult genus. The species cannot be definitely named without sporocarps. Aquatic forms are normally sterile with large, entire leaflets whereas land forms tend to be fertile and have smaller, often toothed, leaflets.

1. Sporocarps broadly-elliptic to oblong-elliptic with two distinct teeth, the lateral walls ribbed; leaflets with pellucid streaks between the veins
   . . . . **1. M. coromandeliana**
+ Sporocarps ± quadrangular with a single tooth, the lateral walls with a single vertical furrow; leaflets without pellucid streaks between the veins
   . . . . **2. M. aegyptiaca**

1. M. coromandeliana Willd., Sp. pl., 5(1): 539 (1810).

Leaflets 2–10 × 2–12mm, glabrous or occasionally hairy, with pellucid streaks between the veins. Sporocarps solitary, (2–)2.5–3(–4) × 1.5–2(–2.7)mm, broadly-elliptic to oblong-elliptic with two distinct teeth, the upper acute, the lower obtuse, the lateral walls ribbed. **Map 14, Fig. 2.**

Rooted in mud at the margins of ponds and streams; 30–500m.

Socotra. Tropical and southern Africa, Madagascar and India.

2. M. aegyptiaca Willd., Sp. pl., 5(1): 540 (1810).

Leaflets 2–12(–30) × 2–12(–25)mm, glabrous or hairy, without pellucid streaks between the veins. Sporocarps solitary or grouped, (1–)1.5–2 × (1–)1.5–2mm, ± quadrangular with a single conical to acute tooth, the lateral walls with a single vertical furrow. **Map 15, Fig. 2.**

Rooted in mud at the margins of ponds and streams; 850–3100m.

Saudi Arabia, Yemen (N). North and tropical Africa, Madagascar.

Family 6. PARKERIACEAE

6. CERATOPTERIS Brongn.

Aquatic annual ferns. Rhizomes short, erect. Fronds tufted, strongly dimorphic. Sporangia sessile; sori arranged along the veins on the lower surface of the fronds, protected by the reflexed margins of the pinnules.

Lloyd, R.M. (1974). Systematics of the genus *Ceratopteris* Brongn. (Parkeriaceae) II Taxonomy. *Brittonia* 26: 139–160.

1. C. cornuta (P. Beauv.) Le Prieur in Ann. Sci. Nat. (Paris) 19: 103. pl. 4A (1830).
Syn.: *C. thalictroides* (L.) Brongn. in Bull. Sci. Soc. Philom. Paris 1821: 186 (1821).
Illustr.: Fl. Iraq 2: 71 (1966).

Aquatic fern, usually rooting in mud in shallow water. Stipe about half as long as the frond. Fronds tufted, up to 60cm, light green, succulent, rather brittle, glabrous. Lamina of sterile fronds ovate or ovate-triangular, up to 20 × 15cm, pinnate to bipinnate-pinnatifid; lobes triangular or ovate, acute or obtuse; lamina of fertile fronds ovate, up to 60 × 15cm, (2–)3–4-pinnate, with linear lobes. **Map 16, Fig. 3.**

In low altitude swamps, pools or sluggish streams; 50–1000m.

Saudi Arabia, Yemen (N), Socotra, Oman. Tropical Africa, Iraq, NE India, Myanmar, Indonesia and N Australia.

Family 7. ACTINIOPTERIDACEAE

7. ACTINIOPTERIS Link

Small ferns. Fronds tufted, green when fresh, silvery grey when dry. Rhizomes short, creeping, covered with dense linear-lanceolate scales. Fertile fronds similar or dissimilar to the sterile, often with longer lamina; stipe furrowed, glabrous or with scattered scales; lamina shorter than the stipe, wedge- or fan-shaped, dichotomously branched; segments ± linear, the sterile toothed at the apex, the fertile entire. Sori submarginal, linear-elongated, covered by the reflexed margin of the frond.

Pichi-Sermolli, R.E.G. (1962). On the fern genus *Actiniopteris* Link. *Webbia* 17: 1–32.

1. Lamina fan-shaped, with 26–48 segments; segments ± truncate and toothed at the apex **1. A. radiata**
+ Lamina wedge-shaped, with 4–18 segments; segments ± acute and laterally toothed at the apex **2. A. semiflabellata**

Fig. 3. Ferns. A, *Actiniopteris semiflabellata*: Aa, fresh and dried fronds (× 0.6); Ab, tip of frond (× 20). B, *A. radiata*: Ba, fresh and dried fronds (× 0.6); Bb, tip of frond (× 20). C, *Anogramma leptophylla*: Ca, habit (× 0.3); Cb, fertile pinna (× 4). D, *Pleopeltis macrocarpa*: Da, fertile frond (× 0.6) and peltate scales (× 40). E, *Tectaria gemmifera*: Ea, frond (× 0.25). F, *Cystopteris fragilis*: Fa, frond (× 0.6); Fb, fertile pinna (× 3); Fc, sorus (× 25). G, *Onychium divaricatum*: Ga, sterile frond (× 0.6); Gb, fertile frond (× 0.6) and enlargement. H, *Ceratopteris cornuta*: Ha, fertile and sterile fronds (× 0.25).

1. A. radiata (Swartz) Link, Fil. spec.: 80 (1841). Syn.: *A. australis* sensu Schwartz (1939) pro parte non (L.f.) Link. Illustr.: Webbia 17: 13 (1962).

Fronds 5–25(–40)cm long. Lamina sharply declinate when desiccated, fan-shaped, 1–3cm long, dichotomously divided 4–6 times into 26–48 ± linear segments; segments ± truncate and toothed at the apex. Stipe 2–5 × as long as the lamina. **Map 17, Fig. 3.**

In shady rock crevices and on rocky, scrub-covered slopes; 350–1400m.

Saudi Arabia, Yemen (N). Throughout tropical Africa to India and Sri Lanka.
Local at low altitudes on the western escarpment mountains.

2. A. semiflabellata Pichi-Sermolli, op. cit. p.24. Syn.: *A. australis* sensu Schwartz (1939) pro parte non (L.f.) Link; *A. dichotoma* sensu Balfour (1888 p.328); *Acrostichum dichotomum* Forsskal (1775 p.184). Illustr.: Collenette (1985 p.406).

Similar to *A. radiata* but the fronds (3–)5–30cm long; lamina twisted sideways, not sharply declinate when desiccated, wedge-shaped, 1–7cm long, dichotomously divided 2–4(–5) times into 4–18 segments; segments ± acute and laterally toothed at the apex; stipe 1.5–2.5 × as long as the lamina. **Map 18, Fig. 3.**

Shady crevices on cliffs, terrace-walls and in bushland; 350–2600m, and down to 50m on Socotra.

Saudi Arabia, Yemen (N & S), Socotra, Oman. NE and tropical E Africa, Madagascar, Réunion and Mauritius.

Much commoner in Arabia than the previous species.

Family 8. ADIANTACEAE

8. CHEILANTHES Swartz

Terrestrial ferns. Rhizomes short-creeping or erect; rhizome-scales linear. Fronds tufted, 1–3-pinnate, hairy, clothed with scales or glabrous. Sori marginal, usually continuous; indusia narrow, formed from the reflexed leaf margins (the soral flaps), rarely the soral flaps obsolete.

1.	Frond covered with white powder beneath	**1. C. farinosa**
+	Frond glabrous, hairy or clothed with scales beneath, never covered with white powder	2
2.	Lamina glabrous beneath	**5. C. pteridioides**
+	Lamina hairy or clothed with scales beneath	3
3.	Lamina triangular-ovate; pinnae pinnatifid, the basal pinnae with a pinnatifid lower lobe	**2. C. coriacea**
+	Lamina narrowly elliptic or narrowly ovate; basal pinnae without a pinnatifid lower lobe	4

ADIANTACEAE

4.	Lamina densely clothed with scales beneath	**3. C. marantae**
+	Lamina thinly to densely hairy beneath	5
5.	Lamina densely hairy beneath	**4. C. vellea**
+	Lamina with a few scattered hairs beneath	**5. C. pteridioides**

1. C. farinosa (Forsskal) Kaulf., Enum. filic.: 212 (1824) Syn.: ? *Pteris farinosa* Forsskal (1775 p.187); *P. decursiva* Forsskal (1775 p.186). Type: Yemen (N), *Forsskal* (?lost).

Rhizome erect; rhizome-scales dark brown. Fronds tufted. Stipe 10–23cm, reddish brown, glabrous or with a few scales below. Lamina 2–3-pinnate-pinnatifid, 12–25(–40) × 6–10(–18)cm, oblong, acuminate, dark green above, covered with white powder beneath; lower pinnae unequally triangular, the upper oblong; pinnules pinnatifid-crenate or with oblong rounded segments. Soral flaps membranous, continuous along the margins. **Map 19, Fig. 4.**

Shrub-covered cliffs, shady gullies and grassland; 2000–2900m.

Yemen (N). Tropical Africa.

See comments under *Negripteris sciona*.

2. C. coriacea Decne in Arch. Mus. Hist. Nat. 2: 190 (1841). Illustr.: Collenette (1985 p.408). Type: Yemen (N), *Botta* s.n. (P).

Rhizome short-creeping; rhizome scales reddish brown with a black midrib. Fronds tufted, leathery. Stipe 5–15cm, reddish brown, covered with pale brown scales. Lamina 2–3-pinnate-pinnatifid, (2–)3.5–6 × 2–4cm, triangular-ovate, dark green and glabrous above, brown and hairy beneath; pinnae oblong, pinnatifid; pinnules oblong, rounded, the margins obscurely crenate; lowest pinnae triangular with a pinnatifid basal segment. Soral flaps continuous along the margin, almost completely obscuring the surface of the pinnae beneath. **Map 20, Fig. 4.**

Rock crevices and terrace-walls; 700–2900m.

Saudi Arabia, Yemen (N & S), Oman. NE tropical Africa and SW Iran.

3. C. marantae (L.) Domin in Biblioth. Bot. 85: 133 (1915). Illustr.: Collenette (1985 p.408).

Rhizome creeping, densely clothed in pale reddish brown scales. Fronds closely spaced. Stipe 3–10cm, reddish brown, covered with pale brown scales. Lamina 2-pinnate-pinnatifid, 10–20cm long, narrowly oblong-elliptic, green and glabrescent above, pale brown and densely clothed with scales beneath; pinnae narrowly oblong to narrowly triangular, pinnatifid; pinnules oblong, rounded, the margins entire. Soral flaps narrow, hidden by scales. **Map 21, Fig. 4.**

Juniperus woodland, grassy slopes and exposed rocky hillsides; 1650–3200m.

Saudi Arabia, Yemen (N). Mediterranean region, Macaronesia, Crimea and the Caucasus.

8. Cheilanthes

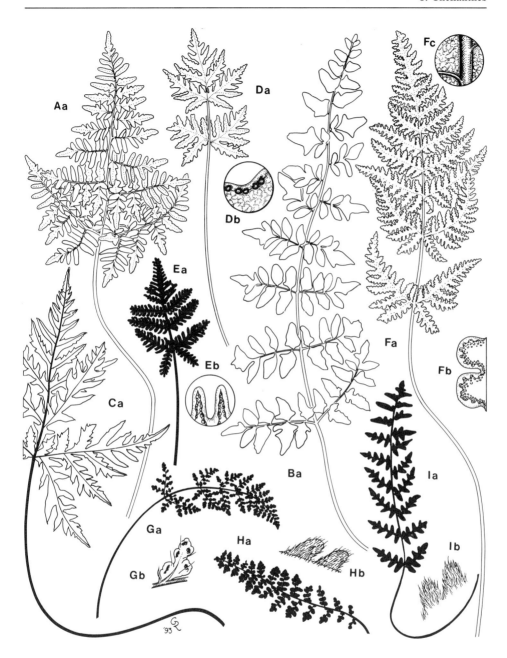

Fig. 4. Ferns. A, *Pellaea quadripinnata*: Aa, frond (×0.6). B, *P. viridis*: Ba, frond (×0.6). C, *Doryopteris concolor*: Ca, frond (×0.6). D, *Negripteris sciona*: Da, frond (×0.6); Db, part of fertile pinnule (×15). E, *Cheilanthes coriacea*: Ea, frond (×0.6); Eb, fertile pinnules (×5). F, *C. farinosa*: Fa, frond (×0.6); Fb, fertile pinnules (×10); Fc, undersurface of frond (×50). G, *C. pteridioides*: Ga, frond (×0.6); Gb, fertile pinnule (×10). H, *C. vellea*: Ha, frond (×0.6); Hb, fertile pinnule (×10). I, *C. marantae*: Ia, frond (×0.6); Ib, fertile pinnule (×10).

4. C. vellea (Aiton) F. Mueller, Fragm. 5: 123 (1866). Syn.: *C. catanensis* (Cosent) H.P. Fuchs in Brit. Fern Gaz., 9: 45 (1961); *Cosentinia vellea* (Ait.) Tod., Syn. pl. acot. vasc.: 15 (1866); *Notholaena vellea* (Aiton) R. Br., Prodr.: 146 (1810). Illustr.: Collenette (1985 p.409).

Rhizome erect, clothed in pale brown scales. Fronds tufted. Stipe short, reddish brown, densely hairy. Lamina 2–3-pinnate, (5–)8–30 × 1–4cm, narrowly elliptic, densely or thinly lanate with white or brown hairs beneath, dark green and thinly hairy above; pinnae oblong, pinnate or pinnatifid; pinnules round to oblong, with undulate margins. Soral flaps obsolete. **Map 22, Fig. 4.**

Rock crevices and amongst boulders; 500–2900m.

Saudi Arabia, Yemen (N), Oman. Mediterranean region and Macaronesia, south to Ethiopia and east to Afghanistan.

Sometimes placed in a genus of its own (*Cosentinia* Tod.) on account of its spore morphology (see Pichi-Sermolli in *Webbia* 39: 178 (1985)).

5. C. pteridioides (Reichard) C. Chr., Index filic.: 178 (1905). Syn.: *C. fragrans* Swartz, Syn. fil.: 127 (1806) nom. illegit. Illustr.: Collenette (1985 p.409).

Rhizome erect, with reddish brown scales. Fronds tufted. Stipe 2–9cm, reddish brown, shiny, with reddish brown scales. Lamina bright green, 2–3-pinnate, 1.5–10 × 2–5cm, narrowly ovate, glabrous or with scattered glandular hairs; pinnae triangular below becoming oblong above, pinnate to pinnatifid; pinnules oblong to orbicular, with undulate margins. Soral flaps whitish, fimbriate-margined. **Map 23, Fig. 4.**

Rock crevices, amongst boulders and on terrace-walls; 500–3200m.

Saudi Arabia, Yemen (N), Oman. Macaronesia, Mediterranean region and throughout SW Asia.

9. NEGRIPTERIS Pichi-Sermolli

Similar to *Cheilanthes* but the sporangia dark brown or black and glossy, not pale brown.

1. N. sciona Pichi-Sermolli in Nuovo. Giorn. Bot. Ital. 53: 131 (1946). Syn.: *Cheilanthes farinosa* sensu Balfour (1888) non (Forsskal) Kaulf.; *C. farinosa* sensu Schwartz (1939) pro parte non (Forsskal.) Kaulf. Illustr.: Adumbr. Fl. Aeth. 4: 133 (1955).

Rhizome erect with reddish brown scales. Fronds tufted. Stipe 3–11cm, dark brown, shiny, with reddish brown scales. Lamina 2–3-pinnate–pinnatifid, 3–14 × 3–7cm, narrowly ovate or ovate to pentagonal, green above, powdery-white (at least in the young fronds) beneath; pinnae 1–2-pinnatifid, triangular at the base of the frond, oblong above, the ultimate segments oblong to triangular. Sori marginal, interrupted; sporangia dark brown or black, glossy; soral flaps membranous. **Map 24, Fig. 4.**

Shady rock crevices and on soil in dense thickets, often on limestone; 40–1800m.

Yemen (N & S), Socotra, Oman. Ethiopia, Somalia and Kenya.

Negripteris sciona is often confused with *Cheilanthes farinosa*. However, it is readily distinguished by its smaller stature and the conspicuously contrasting dark brown or black sporangia on the powdery-white undersurface of the frond.

10. PELLAEA Link

Terrestrial ferns. Rhizomes short-creeping or erect; rhizome-scales linear. Fronds tufted, 1–4-pinnate, hairy, clothed with scales or glabrous. Sori marginal, continuous; indusia narrow, membranous, formed from the reflexed leaf margins (the soral flaps).

1.	Stipe and rhachis clothed with scales at maturity	**3. P. involuta**
+	Stipe and rhachis glabrous or with a few scales at the base of the stipe at maturity	2
2.	Lamina triangular, 3–4-pinnate, coriaceous	**1. P. quadripinnata**
+	Lamina ovate, 2-pinnate, sometimes pinnules on lowest pinnae lobed, herbaceous	**2. P. viridis**

1. P. quadripinnata (Forsskal) Prantl in Bot. Jahrb. Syst. 3: 420 (1882). Syn.: *Cheilanthes quadripinnata* (Forsskal) Kuhn, Filic. afr.: 74 (1868); *Pteris quadripinnata* Forsskal (1775 p.186). Type: Yemen (N), *Forsskal* (lost).

Rhizome short, creeping; rhizome scales reddish brown with a dark central stripe. Fronds tufted. Stipe 20–35(–40)cm, reddish brown, shiny, glabrous above, with linear scales at the base. Lamina coriaceous, 3–4-pinnate, 10–20(–60) × 10–18 (–40)cm, pentagonal to triangular, glabrous; pinnae ovate-triangular; ultimate segments oblong, rounded, the margins revolute and crenate. **Map 25, Fig. 4.**

Rare in grassland; 1800–2200m.

Yemen (N). Southern Africa, E tropical Africa, Cameroun, Madagascar and Comoro Is.

2. P. viridis (Forsskal) Prantl in Bot. Jahrb. Syst. 3: 420 (1882). Syn.: *Cheilanthes viridis* (Forsskal) Swartz, Syn. fil. 127 (1806); *Pteris viridis* Forsskal (1775 p. 186). Type: Yemen (N), *Forsskal* (lost).

Rhizome short, creeping; rhizome-scales minutely serrulate, pale brown and concolorous or with a dark central stripe. Fronds tufted. Stipe (2.5–)4–20(–40)cm, reddish black, glabrous or clothed with scales below. Lamina herbaceous, 2-pinnate, sometimes pinnules on lowest pinnae lobed, 20(–50) × 8(–24)cm, lanceolate to narrowly ovate, glabrous; ultimate segments ovate to triangular, simple or with 1–3 triangular to ovate basal lobes, the margins minutely crenate. **Map 26, Fig. 4.**

In deep shade on terrace-walls and in semi-deciduous thicket; 600–1800m.

Yemen (N), Socotra. Southern Africa, E tropical Africa, Madagascar and the Mascarenes.

Originally collected by Forsskal at Hadia on J. Raymah. Not now known from this locality but found in a single locality on J. Bura. On Socotra found in semi-evergreen thicket.

3. P. involuta (Swartz) Bak., Syn. fil., ed. 2: 148 (1874). Syn.: *Cheilanthes involuta* (Swartz) Schelpe & Anthony in Contr. Bolus Herb. 10: 155 (1982); *Pellaea viridis* sensu Balfour (1888) non (Forsskal) Prantl pro parte.

Similar to *P. viridis* but the rhachis clothed with scales and the rhizome-scales entire not minutely serrulate. **Map 27.**

No habitat details available. 600m.

Socotra. Southern Africa, E tropical Africa and Madagascar.

In Arabia known only from a single gathering from Socotra. *P. involuta* is closely related to, and sometimes considered to be a variety of, *P. viridis*.

11. DORYOPTERIS J.E. Smith

Terrestrial ferns. Rhizomes short-creeping; rhizome-scales linear. Fronds tufted, 2–4-pinnatifid, glabrous. Sori marginal, discrete or continuous; indusia narrow, membranous, formed from the reflexed leaf-margins (the soral flaps).

1. D. concolor (Langsd. & Fisch.) Kuhn in Decken, Reis. Ost-Afr. 3, 3: 19 (1879). Syn.: *Cheilanthes concolor* (Langsd. & Fisch.) R. & A. Tryon in Rhodora 83: 133 (1981); *Doryopteris kirkii* (Hook.) Alston in Bol. Soc. Brot., sér. 2, 30: 14 (1956); *Pellaea concolor* (Langsd. & Fisch.) Bak. in Mart., Fl. bras. 1: 596 (1870).

Rhizome short, creeping; rhizome-scales pale brown with a dark central stripe. Fronds tufted, herbaceous. Stipe longer than the lamina, reddish black, clothed with scales. Lamina pentagonal, 2–4-pinnatifid, up to 20 × 18cm, glabrous; pinnae pinnatifid, the lowest triangular, the upper oblong-ovate; ultimate segments triangular to oblong. Sori continuous or interrupted along the margins; soral flaps membranous. **Map 28, Fig. 4.**

In shade in evergreen thicket and dwarf shrubland; 900–1100m.

Socotra. Throughout the tropics.

12. ANOGRAMMA Link

Small, ephemeral, terrestrial ferns. Rhizomes minute, with hair-like scales. Fronds ± dimorphic, 2–3-pinnate, membranous. Sori borne along the veins; indusia absent.

1. A. leptophylla (L.) Link, Fil. spec. 137 (1841). Illustr.: Fl. Iraq 2: 66 (1966); Collenette (1985 p.407).

Small, ephemeral fern. Rhizome minute with pale brown scales. Fronds erect. Stipe reddish-brown, glabrous or with a few scales below. Lamina membranous, (1–)2–3-pinnate, ovate to narrowly elliptic, 1–10 × 1–3cm; pinnae ovate to triangular, glabrous;

pinnules fan-shaped or wedge-shaped, toothed or lobed, decurrent downwards to the winged rhachis. Sori in lines along the ultimate veins of the pinnules. **Map 29, Fig. 3.**

In damp, shady crevices and under boulders; 400–2050m.

Saudi Arabia. Widespread in both the New and Old Worlds.

A small and easily overlooked fern. Strictly speaking it has an annual sporophyte and in suitable habitats a perennial gametophyte.

13. ADIANTUM L.

Terrestrial ferns. Rhizomes short or long-creeping, covered with brown scales. Fronds tufted or spaced, 1–4-pinnate; stipe dark brown, reddish brown or black, polished; segments oblong, trapeziform, fan-shaped or subcircular, glabrous or hairy. Sori borne on the inner surface of the reflexed leaf margins (the soral flaps).

1.	Fronds 1-pinnate	2
+	Fronds 2–4-pinnate	4
2.	Pinnae hairy	**1. A. incisum**
+	Pinnae glabrous	3
3.	Segments fan-shaped or subcircular with a slightly asymmetric base, sub-opposite; petioles 0.5–1.5mm	**2. A. balfourii**
+	Segments oblong with a very asymmetric base, alternate; petioles 3–18mm	**3. A. philippense**
4.	Frond 1–2(–3)-pinnate; segments persistent, shallowly to deeply lobed, the margin crenate-dentate; veins of the sterile segments ending in marginal teeth	**4. A. capillus-veneris**
+	Frond 3(–4)-pinnate; segments deciduous, shallowly lobed, the margin crenate; veins of the sterile segments ending in sinuses	**5. A. poiretii**

1. A. incisum Forsskal (1775 p.187). Syn.: *A. caudatum* sensu Schwartz (1939). Illustr.: Adumbr. Fl. Aeth. 5: 671 (1957); Collenette (1985 p.407). Type: Yemen (N), *Forsskal* (C).

Rhizome short, erect; rhizome-scales narrowly lanceolate, brown. Stipe reddish-brown, hairy. Fronds tufted, dark green, pinnate. Lamina narrowly lanceolate, 15–40 × 1–3(–4)cm, long-attenuate and often rooting at the tips. Pinnae alternate, oblong to trapeziform or wedge-shaped, 5–20 × 2–10mm, with deeply or shallowly toothed margins, hairy; petioles up to 1mm. Soral flaps lunate to oblong, 1–3mm. **Map 30, Fig. 5.**

In shade on terrace-walls, rock crevices and amongst boulders; 600–2200m.

Saudi Arabia, Yemen (N & S), Oman. Tropical and southern Africa and India.

ADIANTACEAE

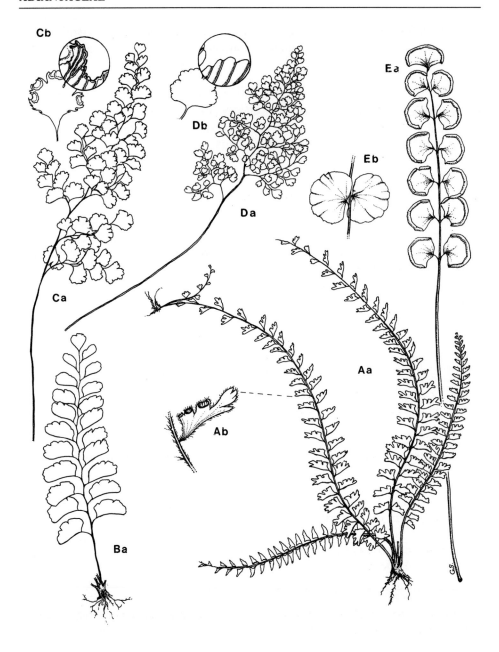

Fig. 5. Ferns. A, *Adiantum incisum*: Aa, habit (×0.3); Ab, fertile pinna (×1.5). B, *A. philippense*: Ba, frond (×0.3). C, *A. capillus-veneris*: Ca, frond (×0.3); Cb, fertile pinnule (×2) and enlargement of margin. D, *A. poiretii*: Da, frond (×0.3); Db, fertile pinnule (×3) and enlargement of margin. E, *A. balfourii*: Ea, frond (×0.5); Eb, pinnae (×0.5).

2. A. balfourii Baker, Diagn. Fil. Nov. Socotr.: 1 (1882); Baker ex Kuhn, Ber. Deutsch. Bot. Ges. 1: 258 (1883); Baker in Hooker's Icon. Pl. 17: t.1630 (1886). Illustr.: Adumbr. Fl. Aeth. 5: 659 (1957); Balfour (1888 t.99). Type: Socotra, *Balfour, Cockburn & Scott 198* (K).

Rhizome short, creeping or erect; rhizome-scales ± linear, reddish-brown; fronds densely tufted. Stipe black or dark reddish-brown, glabrous. Lamina pinnate, narrowly oblong, 8–35 × 2–3.5cm. Pinnae subopposite, fan-shaped to subcircular, 1–2 × 1–2.75 cm, the margin crenate or ± entire and sometimes with a hardened black or brown edge, rounded to cuneate at the base, glabrous; petioles up to 1.5mm. Soral flaps ± continuous around the entire margin. **Map 31, Fig. 5.**

In shade amongst limestone boulders and in rock crevices; 150–850m.

Socotra. Somalia, Ethiopia and Djibouti.

3. A. philippense L., Sp. pl.: 1094 (1753). Illustr.: Adumbr. Fl. Aeth. 5: 667 (1957).

Rhizome short, erect or creeping; rhizome-scales linear, dark brown; fronds loosely tufted. Stipe reddish-brown, glabrous. Lamina pinnate, narrowly ovate, 20–40 × 4–8cm, sometimes rooting at the tip. Pinnae alternate, oblong, the tips rounded, 1–3 (–4.5) × 0.8–2cm, the margin shallowly incised into truncate lobes, unequally and broadly cuneate at the base, glabrous; petioles 3–18mm. Soral flaps linear-oblong, 3–8mm. **Map 32, Fig. 5.**

On heavy clay soils on limestone in deciduous *Anogeissus* thicket; 200–600m.

Oman, ?Yemen (S). Pantropical.

In Arabia, found only in the monsoon-affected coastal mountains of Oman and probably extending into Yemen.

4. A. capillus-veneris L., Sp. pl.: 1096 (1753). Illustr.: Fl. Iraq 2: 64, (1966); Collenette (1985 p.406).

Rhizome short, creeping; rhizome-scales linear, pale brown; fronds shortly spaced, arching. Stipe blackish-brown or reddish-brown, glabrous. Lamina 1–2(–3)-pinnate, narrowly ovate to ovate-triangular, 10–40 × 3–12cm, drooping. Pinnules alternate, fan- or wedge-shaped, 7–20(–25) × 4–18mm, shallowly to deeply lobed, glabrous; sterile pinnules crenate-dentate, the veins ending in teeth; petiolules 1–1.5mm. Soral flaps oblong to lunate, 2–4 × 0.75–1.75mm. **Map 33, Fig. 5.**

On rocks and cliffs in shade near permanent water and seepages; 10–3050m.

Saudi Arabia, Yemen (N & S), Socotra, Oman, UAE, Bahrain. Widely distributed in warm temperate and tropical areas of the world.

5. A. poiretii Wikstr. in Köngl. Vetensk. Akad. Handl. 1825: 443 (1826). Syn.: *A. thalictroides* Willd. ex Schlechtend., Adumbr. Pl. 5: 53 (1832); *A. aethiopicum* sensu Balfour (1888) non L. Illustr.: Adumbr. Fl. Aeth. 5: 689 (1957), as *A. thalictroides*.

Rhizome far-creeping; rhizome-scales brown, lanceolate, minutely ciliate; fronds closely spaced. Stipe reddish-brown, glabrous or with a few scales below. Lamina

3(–4)-pinnate, broadly-ovate to ovate-triangular, 20–30 × 8–16cm, arching. Pinnules alternate, fan-shaped or semi-circular, 5–10 × 5–10mm, shallowly lobed, glabrous, articulated on the petiolules, deciduous; sterile pinnules crenate, the veins ending in sinuses; petiolules 2–6mm. Soral flaps lunate, 1.25–2.25 × 1cm. **Map 34, Fig. 5.**

In shade on terrace-walls and cliffs; 2000–3000m.

Yemen (N), Socotra. Tropical and southern Africa, India, Central and South America. A local fern of high-rainfall and misty areas.

14. ONYCHIUM Kaulfuss

Terrestrial ferns; rhizomes creeping, densely clothed with scales. Fronds irregularly 3–4-pinnate. Sori marginal, covered by the reflexed leaf margins.

1. O. divaricatum (Poir.) Alston in Bol. Soc. Brot. 30: 21 (1956). Syn.: *Allosorus melanolepis* Decne in Arch. Mus. Hist. Nat. 2: 189 (1841); *Onychium melanolepis* (Decne) Kunze, Farrnkräuter 2: 9, t.104, f.2 (1848). Illustr.: Adumbr. Fl. Aeth. 9: 310 (1963); Collenette (1985 p. 409).

Small parsley-like fern; rhizome-scales linear, entirely black or with a pale margin. Stipe straw-coloured, 4–18cm, glabrous or rarely with scattered scales below. Fronds tufted, dimorphic. Lamina ovate, 7–20 × 5–13cm, glabrous; ultimate segments of the sterile fronds ovate or wedge-shaped, 4–7 × 3–5mm, divided into 2–5 narrowly elliptic lobes; ultimate segments of the fertile fronds narrowly oblong, 1.5–15 × 1–1.25mm, entire, acute. Sori linear. **Map 35, Fig. 3.**

Rock crevices, terrace-walls and cliffs; 50–3000m.

Saudi Arabia, Yemen (N & S), Socotra, Oman, UAE. Trop. N & NE Africa; Iran to SE & E Asia.

Family 9. ACROSTICHACEAE

15. ACROSTICHUM L.

Large mangrove or marsh ferns. Rhizomes erect or creeping; rhizome-scales large; roots thickened. Fronds tufted, pinnate, glabrous. Sori covering the lower surface of the upper pinnae.

1. A. aureum L., Sp. pl.: 1069 (1753).

Large, tufted fern; rhizome massive; rhizome-scales c.1cm. Fronds coriaceous, up to 1.5m. Stipe 10–75cm, pale brown, glabrous. Lamina simply pinnate; pinnae narrowly oblong, (8–)15–25(–50) × 1–4(–5)cm, rounded, with entire or undulate margins, petiolate, the upper pinnae fertile, the lower sterile; sori completely covering the undersurface of the fertile pinnae. **Map 36, Fig. 6.**

By brackish seepages on cliffs and hillsides; 350–850m.

Saudi Arabia. Widespread around tropical coasts.

Acrostichum aureum is usually a plant of mangrove swamps. In Arabia it is only known from two populations in Saudi Arabia where it grows in very atypical habitats by brackish seepages on a hillside and on a cliff-ledge.

Family 10. PTERIDACEAE

16. PTERIS L.

Rhizomes erect, short- or long-creeping; rhizome-scales linear to ovate. Fronds tufted to widely spaced, 1-pinnate to 4-pinnatifid, glabrous. Sori marginal, usually continuous except for the sterile apex, covered by the reflexed membranous leaf margins (soral flaps).

1.	Lamina 1-pinnate or 1-pinnate above and with the basal pinnae 2-lobed		2
+	Lamina 2–4-pinnatifid		3
2.	Basal pinnae simple; lamina with 10–40 pairs of pinnae		**1. P. vittata**
+	Basal pinnae 2-lobed; lamina with 2–8 pairs of pinnae		**2. P. cretica**
3.	Tips of sterile pinnules minutely crenate-serrate; upper surface of lamina lacking minute spines		**3. P. dentata**
+	Tips of sterile pinnules entire; upper surface of lamina with minute spines on the midribs (costae) and veins (costules)		**4. P. quadriaurita**

1. P. vittata L., Sp. pl.: 1074 (1753). Syn.: *P. obliqua* Forsskal (1775 p.185); *P. longifolia* sensu Balfour (1888) non L.; *P. longifolia* sensu Schwartz (1939) non L.; ? *P. subciliata* Forsskal (1775 p.185). Illustr.: Collenette (1985 p.410).

Rhizome short-creeping; rhizome-scales pale-brown. Fronds tufted, leathery. Stipe usually short, 2–30cm, straw-coloured, with pale brown scales below. Lamina narrowly elliptic to elliptic-oblong, 1-pinnate, 12–80(–100) × 4–25(–40)cm, glabrous; pinnae simple, 10–40 pairs, oblong to linear-oblong, 1–15 × 0.5–1.8cm, rounded to attenuate at the tip, the sterile pinnae serrulate. **Map 37, Fig. 6.**

Terrace-walls, wadi-sides and cliffs etc., usually in shade and near permanent water, sometimes as a weed of man-made habitats; 5–2700m.

Saudi Arabia, Yemen (N & S), Socotra, Oman. Southern and tropical Africa.

2. P. cretica L., Mant. pl. 1: 130 (1767). Syn.: *P. semiserrata* Forsskal (1775 p.186).

Rhizome short-creeping; rhizome-scales brown. Fronds tufted. Stipe long, 20–40

Fig. 6. Ferns. A, *Pteris quadriaurita*: Aa, frond (×0.25); Ab, pinnules showing costal spines (×1). **B**, *P. dentata*: Ba, frond (×0.25). **C**, *P. vittata*: Ca, frond (×0.25); Cb, fertile pinna (×1). **D**, *P. cretica*: Da, frond (×0.25). **E**, *Acrostichum aureum*: Ea, fertile frond (×0.25). **F**, *Pteridium aquilinum*: Fa, part of frond (×2).

(–60)cm, straw-coloured, glabrous except for a few scales at the base. Lamina light green, ovate-triangular, 1-pinnate above and with the lowest pinnae 2-lobed, 12–20 × 12–25cm, glabrous; pinnae and lobes of the lowest pinna linear-elliptic, 6–13(–20) × 0.5–1.5(–2.5)cm, attenuate, the sterile pinnae minutely serrulate. **Map 38, Fig. 6.**

Sandstone cliffs and grassland; 2100–2800m.

Yemen (N). Widespread in the Old World tropics and subtropics.

3. P. dentata Forsskal (1775 p.186). Syn.: ? *P. regularis* Forsskal (1775 p.186); *P. serrulata* Forsskal (1775 p.187). Illustr.: Collenette (1985 p.409). Type: Yemen (N), *Forsskal* (lost), *Schweinfurth* 1402 (neotype C).

Rhizome erect; rhizome-scales reddish brown. Fronds tufted, thin-textured. Stipe 15–50cm, straw-coloured becoming reddish brown at base, glabrous or with a few scales below. Lamina ovate, 2–4-pinnatifid, 15–60(–100) × 15–25(–40)cm, glabrous; middle and upper pinnae narrowly oblong, acute, pinnatifid; lower pinnae triangular, 1–3-pinnatifid; ultimate segments angled forwards, narrowly oblong, 1–5 × 0.2–0.5cm, crenate-serrate. **Map 39, Fig. 6.**

Terrace-walls and cliffs, usually by permanent water and in shade; 900–3050m.

Saudi Arabia, Yemen (N & S). S Africa, tropical Africa and the Mascarenes.

A very common fern of the SW escarpment mountains.

4. P. quadriaurita Retz., Observ. bot. 6: 38 (1791). Syn.: *P. catoptera* Kunze in Linnaea 18: 119 (1844).

Rhizome erect; rhizome-scales reddish brown. Fronds tufted, membranous-textured. Stipe 25–50(–90)cm, straw-coloured, glabrous except for a few scales at the base. Lamina oblong-ovate, 3-pinnate-pinnatifid, c.50(–90) × 25(–60)cm, glabrous but with minute spines on the surface above; upper pinnae narrowly oblong, acute, pinnatifid; lowest pinnae with the basal pinnules pinnatifid and resembling the upper pinnae; ultimate segments patent, narrowly oblong, 1–2.5 × 0.3–0.4cm, entire. **Map 40, Fig. 6.**

Rock crevices, terrace-walls and dense montane woodland, in shade; 1100–1800m.

Yemen (N), Socotra. Tropical Africa.

A difficult species complex in need of revision. All Arabian and Socotran material is referable to subsp. *catoptera* (Kunze) Schelpe.

Family 11. POLYPODIACEAE

17. PLEOPELTIS Humb. & Bonpl. ex Willd.

Terrestrial ferns; rhizomes creeping, clothed in clathrate scales. Fronds simple, entire. Sori round, with dark-centred peltate scales (paraphyses). Indusia absent.

1. P. macrocarpa (Bory ex Willd.) Kaulf. in Berlin Jahrb. Pharm. Verbundenen Wiss. 21: 41 (1820).

Rhizome far-creeping, with brown scales. Fronds coriaceous, thickened, shortly spaced. Stipe 2–5(–8)cm, glabrous or with scattered scales. Lamina simple, narrowly elliptic, 10–12(–20) × 0.5–1.4(–1.7)cm, entire, gradually attenuate into the stipe below, glabrous above, with dark-centred peltate scales beneath. **Map 41, Fig. 3.**

Damp rocks on shaded cliffs and epiphytic on trees; 2300–2800m.

Saudi Arabia, Yemen (N), Socotra. Southern Africa, tropical Africa, Madagascar, the Mascarenes, India and America.

Easily confused with *Loxogramme lanceolata*. See comment under that species.

18. LOXOGRAMME (Blume) C. Presl.

Terrestrial ferns; rhizomes creeping, clothed in clathrate scales. Fronds simple, entire. Sori elongate. Indusia absent, without paraphyses.

1. L. lanceolata (Swartz) C. Presl, Tent. pterid.: 215, t.9 f.8 (1836).

Rhizome long-creeping, with dark greyish brown scales. Fronds coriaceous, thickened, shortly spaced. Stipe 1–3cm, glabrous. Lamina simple, narrowly elliptic, 10–12(–35) × 0.5–1.5(–2.8)cm, entire, narrowing gradually into the stipe below, glabrous. Sori linear-oblong. **Map 42.**

Damp rocks in shaded gullies; 1700–1900m.

Yemen (N). Southern Africa, tropical Africa, Madagascar and the Mascarenes.

Easily confused with *Pleopeltis macrocarpa* which, however, can be distinguished by its rounded not elongate sori and the presence of dark-centred peltate scales on the underside of the fronds.

Family 12. DENNSTAEDTIACEAE

19. PTERIDIUM Scop.

Terrestrial ferns; rhizomes long-creeping, densely clothed with hairs. Fronds spaced, 3–4-pinnate, hairy. Sori linear, continuous along the leaf margins, covered by the reflexed leaf margin and an inner indusium.

1. P. aquilinum (L.) Kuhn in Decken, Reis. Ost-Afr. 3, 3: 11 (1879).

Rhizome clothed with brown hairs. Fronds up to 2m, closely spaced. Stipe 20(–40)cm, straw-coloured, hairy and swollen at the base, glabrous above. Lamina broadly triangular to broadly ovate, glabrescent above, tomentose below. Pinnae ovate-triangular to oblong, acute; pinnules oblong, acute, pinnate to 2-pinnatifid. Indusia and reflexed leaf margins membranous and ciliate. **Map 43, Fig. 6.**

Grassland; 2000–2300m.

Yemen (N). Cosmopolitan.

Known only from an area of grassland on the outer escarpment mountains of Yemen (N). The Yemen populations belong to subsp. *aquilinum*.

Family 13. OLEANDRACEAE

20. NEPHROLEPIS Schott

Rhizomes short, erect, often forming tubers and producing thin stolons; rhizome-scales peltate. Fronds pinnate. Sori circular, intramarginal; indusia reniform.

1. N. undulata (Afzel. ex Swartz) J. Sm. in Bot. Mag. 72, Comp.: 35 (1846). Syn.: *N. cordifolia* sensu Balfour (1888) non C. Presl.

Tubers up to 2.5cm across. Fronds pinnate, up to 30–70(–100)cm, herbaceous. Stipe brown, with sparse brown scales. Lamina narrowly elliptic, 20–85 × 4–11cm, glabrous; pinnae narrowly oblong, acute, weakly crenate, cordate at the base, sessile. Sori in a row either side of the midrib; indusia membranous, entire. **Map 44, Fig. 8.**

Semi-deciduous thicket; 850m.

Socotra. Tropical Africa, Madagascar and the Mascarenes.

21. ARTHROPTERIS J. Smith

Terrestrial ferns; rhizomes creeping; rhizome-scales peltate. Fronds 2-pinnatifid. Sori circular, intramarginal; indusia reniform.

1. A. orientalis (J.F. Gmelin) Posthumus in Rec. Trav. Bot. Néerl. 21: 218 (1924). Syn.: *Dryopteris orientalis* (J.F. Gmelin) C. Chr., Index fil.: 281 (1905); *Nephrodium pectinatum* (Forsskal) Hier in Bot. Jahrb. Syst. 28: 341 (1900); *Polypodium pectinatum* Forsskal (1775 p.185).

Rhizome thin, far-creeping; rhizome-scales few, brown. Fronds widely spaced, thinly coriaceous, up to 40cm. Stipe brown, thinly hairy and with scattered scales at the base. Lamina 2-pinnatifid, narrowly ovate, 15–25(–40) × 6–10(–17)cm, acuminate, thinly hairy, often with scattered white glands above; pinnae narrowly oblong, pinnatifid into narrowly oblong and rounded lobes, the lobes 2–5(–9) × 1–2.5(–3.5)mm. Sori almost covering the under-surface of the lobes; indusia glabrous, entire. **Map 45, Fig. 8.**

Rocky banks and gullies; 1000–2000m.

Yemen (N). Tropical Africa, Madagascar and the Mascarenes.

Local in the high rainfall areas of the SW escarpment mountains.

Family 14. ASPLENIACEAE

22. ASPLENIUM L.

Terrestrial ferns; rhizomes short-creeping, clothed in clathrate scales. Fronds 2-pinnatifid (in Arabia). Sori elongate, borne along the veins; indusia narrow, opening inwards towards the midrib or rarely towards the margin, membranous or absent or obsolete.

1.	Frond 1-pinnate, the pinnae shallowly toothed or entire	2
+	Frond 2–3-pinnate or 1-pinnate with pinnatifid pinnae	3
2.	Pinnae up to 1cm long; rhachis reddish-brown to black	**1. A. trichomanes**
+	Pinnae 1–5cm long; rhachis green	**7. A. schweinfurthii**
3.	Sori appearing marginal, 1 per lobe; indusia opening towards the margin	**6. A. rutifolium**
+	Sori superficial on the lamina, not appearing marginal, 2 or more per lobe; indusia opening towards the midrib	4
4.	Frond proliferous (bearing young plants) at the tip; lamina pinnate-pinnatifid with oblong segments, but if the segments deeply lobed or serrate towards the tips then see *A. aethiopicum*	**5. A. protensum**
+	Frond not proliferous at the tip; lamina 2–3-pinnate-pinnatifid with obovate or wedge-shaped segments	5
5.	Stipe and rhachis densely covered (at least below) with a mixture of hairs and hair-like scales	**2. A. aethiopicum**
+	Stipe and rhachis glabrous or with scattered scales	6
6.	Stipe reddish-brown, green above, swollen at the base	**3. A. adiantum-nigrum**
+	Stipe green, sometimes dark brown below, not swollen at the base	**4. A. varians**

1. A. trichomanes L., Sp. pl.: 1080 (1753). Illustr.: Collenette (1985 p.407).

Rhizome erect; rhizome-scales dark brown, linear-lanceolate. Fronds dark green, wiry, tufted, persistent. Stipe short, reddish-brown to almost black, glabrous, glossy. Lamina 1-pinnate, linear-lanceolate, 5–13(–30) × 0.8–2cm; pinnae oblong, 3.5–9 × 2–5mm, rounded at the tip, crenate-dentate, unequal and broadly cuneate, subsessile at the base, glabrous. Sori linear, borne along the veins between the midrib and margin, almost completely covering the surface of the pinnae at maturity. **Map 46, Fig. 7.**

Rock crevices and cliffs in the mist-affected regions of the SW escarpment mountains; 1950–3050m.

Saudi Arabia, Yemen (N & S), Socotra. Widely distributed in temperate regions and on tropical mountains throughout the world.

Fig. 7. Ferns. A, *Asplenium rutifolium*: Aa, frond (×0.6); Ab, pinnule (×6). B, *A. aethiopicum*: Ba, frond (×0.6); Bb, pinna (×2). C, *A. varians*: Ca, frond (×0.6); Cb, pinnule (×6). D, *A. trichomanes*: Da, frond (×1). E, *A. protensum*: Ea, frond (×0.6); Eb, pinna (×6). F, *A. adiantum-nigrum*: Fa, frond (×0.6). G, *A. schweinfurthii*: Ga, frond (×0.6); Gb, pinna (×2). H, *Ceterach dalhousiae*: Ha, frond (×0.6); Hb, pinna (×2). I, *C. phillipsianum*: Ia, pinna (×2). J, *C. officinarum*: Ja, pinna (×2).

2. A. aethiopicum (Burm. f.) Becherer in Candollea 6: 23 (1935). Syn.: *Acrostichum filare* Forsskal (1775 p.184); *Asplenium filare* (Forsskal) Alston in J. Bot. 72 suppl.: 4 (1934); *A. lanceolatum* Forsskal (1775 p.185); *A. praemorsum* Swartz, Prodr.: 130 (1788). Illustr.: Collenette (1985 p.407).

Rhizome erect; rhizome-scales dark brown. Fronds dark green, thinly coriaceous, densely tufted. Stipe short, dark brown becoming green above, with mixed scales and hairs. Lamina pinnate with 1–2-pinnatifid pinnae, narrowly ovate, (5–)10–30 × 5–10cm, thinly covered with hair-like scales densely so on the rhachis; pinnae narrowly ovate to trapeziform-acuminate, 1.5–6 × 0.5–2cm, acute or acuminate; segments obovate, wedge-shaped to narrowly oblong, unevenly deeply lobed, serrate towards the tips. Sori 3–8mm, linear. **Map 47, Fig. 7.**

In shade amongst rocks and on terrace-walls; 1050–3200m.

Saudi Arabia, Yemen (N & S), Socotra. Tropical and southern Africa.

3. A. adiantum-nigrum L., Sp. pl.: 1081 (1753).

Rhizome short or creeping; rhizome-scales hair-like, blackish brown. Fronds shiny green, somewhat coriaceous, up to 25cm, tufted or shortly spaced. Stipe blackish or reddish brown, green above, often polished, equal in length to the lamina, the base swollen, with scattered hairs. Lamina 2–3-pinnate, ovate-triangular, 5–15 × 3–8cm, glabrous or with scattered hairs; pinnae ovate-triangular, (5–)10–55 × (3–)5–25mm, acute or rounded at the tip; pinnules and pinnule-lobes wedge-shaped to obovate, the outer margins dentate. Sori linear, 1–2mm, often almost covering the surface of the pinnule at maturity. **Map 48, Fig. 7.**

Rock crevices, terrace-walls, cliffs and under trees; often by water; 2000–3500m.

Saudi Arabia, Yemen (N). Europe, Africa, SW Asia, Taiwan, Hawaii and N America.

4. A. varians Wall. ex Hook. & Grev. subsp. **fimbriatum** (Kunze) Schelpe in Bol. Soc. Brot. 41: 11 (1967). Illustr.: Fl. Zamb., Pteridophyta: 171 (1970).

Rhizome erect; rhizome-scales lanceolate with long hair-points. Fronds tufted, pale green, ± herbaceous, up to 15cm. Stipe green or the base brown, shorter than the blade, with scattered hairs. Lamina 2–3-pinnate, narrowly elliptic, 5–8(–13) × 1–3.5(–5)cm, glabrous; pinnae ovate to oblong, 5–18 × 3–8mm; pinnules and pinnule-lobes obcuneate to obovate, the outer margins dentate. Sori linear, almost covering the surface of the pinnule at maturity. **Map 49, Fig. 7.**

Rocks in a deeply shaded gully; 1880m.

Yemen (N). Tropical East Africa and eastern southern Africa.

Very rare in Arabia; known only from a single, frequently mist-filled gully in the outer SW escarpment mountains.

5. A. protensum Schrad. in Gött. Gel. Anz. 1818: 916 (1818).

Rhizomes creeping; rhizome-scales lanceolate, brown. Fronds herbaceous, up to 60cm, tufted or shortly spaced, often proliferous at the tips. Stipe short, dark brown or blackish brown, densely covered with brown scales and hairs. Lamina pinnate with pinnatifid pinnae, narrowly oblong-elliptic, 20–65 × 4–10cm; pinnae narrowly oblong or trapeziform, acute or acuminate, unequal at the base, pinnately divided into oblong lobes, the lobes acute or shallowly incised at the tips, thinly pubescent and with scattered scales beneath. Sori linear. **Map 50, Fig. 7.**

Rocks in a shaded gully; 1700m.

Yemen (N). Tropical and southern Africa.

Very rare in Arabia; known only from a single, frequently mist-filled gully in the outer SW escarpment mountains.

6. A. rutifolium (Berg.) Kunze in Linnaea 10: 521 (1836). Syn.: *Asplenium achilleifolium* (Lam.) C. Chr. var. *bipinnatum* (Forsskal) C. Chr. (1922 p.30); *A. rutifolium* (Berg) Kunze var. *bipinnatum* (Forsskal) Schelpe in J. S. Afr. bot. 30: 194 (1964); *Lonchitis bipinnata* Forsskal (1775 p.184). Illustr.: Fl. Zamb., Pteridophyta: 186 (1970).

Rhizome erect; rhizome-scales lanceolate, dark brown. Fronds green, somewhat fleshy, up to 20cm, tufted. Stipe green (drying brownish green), with pale brown scales below. Lamina pinnate with deeply 1–2-pinnatifid pinnae, narrowly oblong, up to 15 × 7.5cm; pinnae narrowly oblong, the ultimate segments oblong-obtuse, glabrous or with scattered scales. Sori linear, one per lobe, appearing marginal, opening towards the margin. **Map 51, Fig. 7.**

Rocks in a shaded gully; 1800m.

Yemen (N). Southern Africa, E tropical Africa, Madagascar and the Mascarenes.

Very rare in Arabia; known only from a single, frequently mist-filled gully in the outer SW escarpment mountains.

7. A. schweinfurthii Baker, Diagn. Fil. Nov. Socotr.: 1 (1882); Baker ex Kuhn, Ber. Deutsch. Bot. Ges. 1: 258 (1883); Balfour (1888 p.328); Illustr.: Balfour (1888 t. 100). Type: Socotra, *Schweinfurth* 490 (K).

Rhizome erect or creeping; rhizome-scales narrowly triangular, blackish brown. Fronds dark green, coriaceous, 10–40cm, tufted. Stipe green, covered with brown scales. Lamina pinnate, oblong, 8–20 × 2–8cm; pinnae oblong, 10–50 × 3–12mm, acute or obtuse, with serrate margins, truncate at the base, with scattered hair-like scales below. Sori linear. **Map 52, Fig. 7.**

Rock crevices and beneath shrubs; 1050–1100m.

Socotra. Endemic.

23. CETERACH DC.

Terrestrial ferns; rhizomes short-creeping, clothed in clathrate scales. Fronds pinnatifid, glabrous or densely clothed with scales beneath. Sori elongate, borne along the veins; indusia obsolete.

Included by some authors in *Asplenium* (see Bir, Fraser-Jenkins & Lovis (1985) in *Brit. Fern Gaz.* 13: 53–65).

1. Lamina glabrous or with scattered scales beneath   **2. C. dalhousiae**
+ Lamina densely clothed with scales beneath, ± glabrous above   2

2. Pinnae triangular-ovate to oblong, decurrent at the base, the lower surface completely hidden by scales   **1. C. officinarum**
+ Pinnae oblong, auriculate or weakly decurrent at the base, the lower surface partly visible, never completely hidden by scales   **3. C. phillipsianum**

1. C. officinarum DC. in Lam. & DC., Fl. franç., ed. 3, 2: 566 (1805). Syn.: *Asplenium ceterach* L., Sp. pl.: 1080 (1753). Illustr.: Fl. Iraq 2: 75 (1966); Collenette (1985 p.408).

Rhizome-scales narrowly lanceolate, acuminate, blackish brown. Fronds coriaceous, tufted. Stipe 1–3cm, densely clothed with scales. Lamina pinnatifid, narrowly elliptic, 3–15 × 1–2cm, the lower surface hidden by overlapping pale brown or silvery scales, ± glabrous above; pinnae alternate, triangular-ovate to oblong, 5–10 × 3–6mm, rounded at the tip, with entire or weakly crenate margins, decurrent at the base. Sori linear, hidden by scales. **Map 53, Fig. 7.**

Terrace-walls and rock crevices; 1850–3300m.

Saudi Arabia, Yemen (N). Europe, N Africa, Somalia, SW and C Asia.

2. C. dalhousiae (Hook.) C. Chr., Index fil. 1: 170 (1905); Syn.: *Asplenium dalhousiae* Hook. in Hooker's Icon. Pl. 2: t. 105 (1837).

Similar to *C. officinarum* but the lower surface of the lamina glabrous or with a few scales on the midrib; stipe up to 1cm; lamina up to 20 × 3.5cm; pinnae up to 15 × 7mm. **Map 54, Fig. 7.**

In shade on cliffs and terrace-walls; 1100–2000m.

Yemen (N). Ethiopia and India.

3. C. phillipsianum Kümmerle, Bot. Közlem 6: 287 (1909). Syn.: *Asplenium phillipsianum* (Kümmerle) Bir, Fraser-Jenkins & Lovis, op. cit.: 62 (1985); *Gymnogramma cordata* sensu Balfour (1888) non (Thunb.) Schlecht.

Rhizome-scales narrowly lanceolate, acuminate, blackish brown. Fronds coriaceous, tufted. Stipe up to 5mm, densely clothed with scales. Lamina deeply pinnatifid, narrowly elliptic, 10 × 2cm, densely clothed with scales beneath but the surface visible at least at the margins, ± glabrous above; pinnae alternate, narrowly oblong, 5–12 × 5–9mm, rounded at the tip, with weakly crenate margins, auriculate or weakly decurrent at the base. Sori linear, hidden by scales. **Map 55, Fig. 7.**

In shrubland; 800m.

Socotra. Tropical and southern Africa.

In our area known only from a single, rather atypical gathering which has weakly crenate pinnae and thus resembles *C. officinarum*. Typical plants from Africa have distinctly crenate to pinnatifid pinnae.

Family 15. THELYPTERIDACEAE

24. CHRISTELLA Léveillé

Terrestrial ferns; rhizomes erect or creeping, clothed with scales; scales not peltate. Fronds 2-pinnatifid. Sori circular, intramarginal; indusia reniform.

1. C. dentata (Forsskal) Brownsey & Jermy in Brit. Fern Gaz. 10: 338 (1973). Syn.: *Cyclosorus dentatus* (Forsskal) Ching in Bull. Fan Mem. Inst. Biol. 8: 206 (1938); *Dryopteris dentata* (Forsskal) C. Chr. in Kongel. Danske Vidensk. Selsk. Naturvidensk. Math. Afh. 8, 6: 24 (1920); *D. mauritiana* sensu Schwartz (1939) non (Fée) C. Chr.; *Polypodium dentatum* Forsskal (1775 p.185); *Nephrodium molle* sensu Balfour (1888) non Desv.; *N. parasitica* sensu Vierhapper (1907) non (L.) C.B. Clarke. Type: Yemen (N), *Forsskal* (C).

Rhizome short-creeping; rhizome-scales dark brown. Fronds tufted, herbaceous, up to 80cm. Stipe pale brown, shortly hairy. Lamina pinnate with pinnatifid pinnae, elliptic to narrowly elliptic, (10–)15–60(–100) × (3–)6–18(–40)cm, acuminate, thinly hairy especially on the costae beneath; pinnae narrowly oblong, gradually decreasing in size and more widely spaced below, acuminate, deeply pinnatifid into oblong rounded lobes. Sori borne midway between the costule and margin; indusia hairy. **Map 56, Fig. 8.**

In shade on terrace-walls and cliffs, usually near water; 450–1900m.

Yemen (N), Socotra. Macaronesia, S Europe and widespread in Africa.

Family 16. WOODSIACEAE

25. CYSTOPTERIS Bernh.

Terrestrial ferns; rhizomes short-creeping, clothed with scales. Fronds tufted, 2–3-pinnate, glabrous. Sori round; indusia attached at the base of the sori, covering the sori until maturity, later shrivelling.

1. C. fragilis (L.) Bernh. in Neues J. Bot. 1, 2: 26, t.2, f.9 (1806). Illustr.: Fl. Iraq 2: 78 (1966).

Rhizome-scales narrowly ovate, pale brown. Fronds softly herbaceous, light green, up to 30cm. Stipe straw-coloured or pale green, $\frac{1}{4}-\frac{1}{3}$ as long as the lamina, clothed with scales below. Lamina narrowly ovate, thin-textured. Pinnae narrowly ovate, the longest at the middle of the lamina, the lowest pair usually distant. Pinnules ovate to

WOODSIACEAE

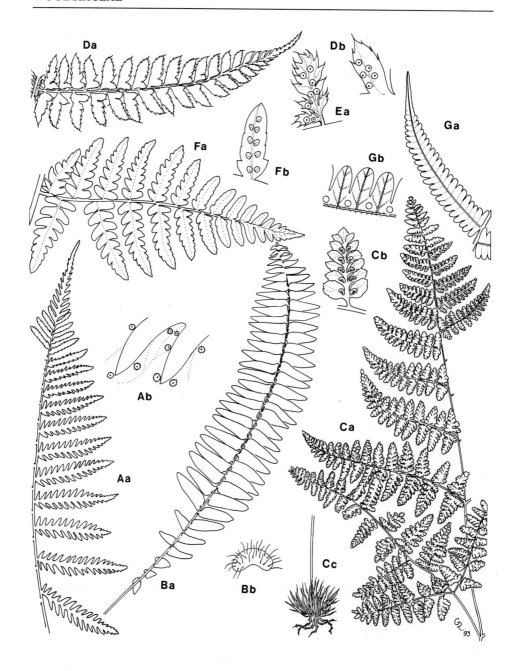

Fig. 8. Ferns. A, *Arthropteris orientalis*: Aa, frond (× 0.5); Ab, close-up of pinnules showing glands (× 10). B, *Nephrolepis undulata*: Ba, frond (× 0.5); Bb, indusium (× 10). C, *Hypodematium crenatum*: Ca, frond (× 0.5); Cb, pinnule (× 3); Cc, base of stipe showing scales. D, *Polystichum* sp. *A*: Da, pinna (× 1); Db, pinnule (× 3). E, *P. fuscopaleaceum*: Ea, pinnule (× 3). F, *Dryopteris schimperiana*: Fa, pinna (× 1); Fb, pinnule (× 3). G, *Christella dentata*: Ga, pinna (× 1), Gb, pinnules (× 4).

oblong, dentate or pinnatifid with toothed lobes. Sori in two rows; indusia membranous, inflated, hood-like, often disappearing with age. **Map 57, Fig. 3.**

Damp rocks in shaded gullies; 2200–2900m.

Saudi Arabia, Yemen (N). Cosmopolitan.

Family 17. DRYOPTERIDACEAE

26. HYPODEMATIUM Kunze

Terrestrial ferns; rhizomes creeping, densely clothed with scales. Fronds tufted, 2–4-pinnatifid, pilose. Stipe densely clothed with scales at the base. Sori subcircular; indusia reniform.

1. H. crenatum (Forsskal) Kuhn in Decken, Reis. Ost-Afr. 3, 3: 37 (1879). Syn.: *Nephrodium crenatum* (Forsskal) Bak., Fl. Mauritius: 497 (1877); *Polypodium crenatum* Forsskal (1775 p.185); *Dryopteris crenata* (Forsskal) O. Kuntze, Revis. gen. pl. 2: 811 (1891). Type: Yemen (N), *Forsskal* (?lost).

Rhizome short; rhizome-scales very dense, golden brown, c.1cm long. Fronds tufted, softly herbaceous. Stipe 20–23cm, straw-coloured, with a dense tuft of scales at the base, hairy above. Lamina ovate-triangular, very finely 2–4 times pinnately divided, 12–30(–33) × 15–25(–30)cm, hairy. Lowest pinnae triangular-ovate, 1–3-pinnate, sometimes enlarged; upper pinnae pinnate, oblong-acute; ultimate segments oblong, shortly incised with crenate-serrate margins. Sori in two rows on the pinnules; indusia hairy. **Map 58, Fig. 8.**

Shady cliffs and terrace-walls; 100–2600m.

Saudi Arabia, Yemen (N & S), Socotra, Oman. Tropical and southern Africa eastwards to Japan.

Arabian plants are all referable to the diploid subsp. *crenatum*.

27. TECTARIA Cav.

Terrestrial ferns. Rhizome creeping to erect, densely clothed with large scales. Fronds tufted, 3-pinnatifid. Sori circular, borne at vein intersections or terminal on the veins; indusia reniform.

1. T. gemmifera (Fée) Alston in J. Bot. 77: 288 (1939). Illustr.: Fl. Zamb., Pteridophyta: 227 (1970).

Medium-sized fern, often with gemmae (or their scars) on the upper surface of the lamina. Rhizome erect; rhizome-scales dense, dark brown. Fronds tufted, arching, herbaceous. Stipe light-brown, thinly hairy with minute hairs and with a few scales at the base. Lamina triangular-ovate, up to 50(–90) × 30(–60)cm, 2-pinnate with pinnatifid pinnules below and bipinnatifid above, the lobes crenate, glabrescent, the costa hairy; upper pinnae oblong with acuminate tips; lowest pair of pinnae unequally triangular, up to 15(–48) × 10(–38)cm. Sori up to 2mm across, circular; indusia c.1mm across, reniform. **Map 59, Fig. 3.**

On rocks in a shaded gully; 1700m.

Yemen (N). E tropical Africa, southern Africa and Madagascar.

In Arabia known only from a single, frequently mist-filled gully on the outer SW escarpment mountains.

28. POLYSTICHUM Roth

Terrestrial ferns; rhizomes erect or creeping, densely clothed with scales. Fronds pinnate to 4-pinnatifid; ultimate segments asymmetric at the base, often dentate-aristate. Sori circular; indusia peltate.

1.	Pinnules pinnatisect, at least at the base; the margins serrate with bristles more than 1mm	**1. P. fuscopaleacum**
+	Pinnules not pinnatisect, the margins serrate with bristles less than 1mm	**2. P. sp. A**

1. P. fuscopaleaceum Alston in Bol. Soc. Brot. ser. 2, 30: 22 (1956).

Rhizome creeping; rhizome-scales brown. Fronds tufted, herbaceous, up to 100cm. Stipe pale-brown or greenish brown, with a mixture of broad and hair-like scales. Lamina 2-pinnate, narrowly ovate, 20–55 × 7–25cm, glabrous except for hair-like scales on the veins; pinnae narrowly oblong, attenuate to the tips; pinnules (at least the basal ones) pinnatisect, narrowly oblong, acute, serrate with long aristate bristles, the bristles more than 1mm, the base asymmetric with an enlarged lobe. Sori in two rows, one each side of the midrib; indusia pale brown, disappearing at maturity. **Map 60, Fig. 8.**

Densely shaded gullies; 2900m.

Yemen (N). Tropical Africa.

According to Fraser-Jenkins (pers. comm.) the Arabian plants, here placed in *P. fuscopaleaceum*, may in fact be referable to *P. yunnanense* Christ. or *P. woronovii* Christ.

2. P. sp. A

Similar to *P. fuscopaleaceum* but generally with fewer scales, the pinnules never pinnatisect and the marginal teeth terminating in short (less than 1mm) bristles. **Map 61, Fig. 8.**

Densely shaded gully; 1700m.

Yemen (N). Apparently only known from a single gully in Yemen.

29. DRYOPTERIS Adans.

Terrestrial ferns; rhizomes erect or creeping, often massive, densely clothed with broad soft scales. Fronds 2–4-pinnate. Sori circular; indusia reniform.

1. D. schimperiana (A. Br.) C. Chr., Index fil.: 291 (1905). Syn.: *Dryopteris rigida* sensu Schwartz (1939) non (Hoffm.) Und.

Rhizome creeping; rhizome-scales dense, reddish-brown. Fronds tufted, herbaceous, up to 75cm. Stipe straw-coloured, thinly clothed with pale reddish brown scales. Lamina 2-pinnate with pinnatifid pinnules, narrowly to broadly oblong-ovate, 20–50(–90) × 12–28(–33)cm, glabrous except for hair-like scales on the veins; pinnae narrowly oblong acuminate; pinnules narrowly oblong, pinnatifid; lobes rounded, weakly crenate-serrate. Sori in two rows, one either side of the mid-vein of the pinnule; indusia brown, membranous. **Map 62, Fig. 8.**

Terrace-walls, cliffs and grassy slopes; 2000–3000m.

Yemen (N). NE tropical Africa.

SPERMATOPHYTA

GYMNOSPERMAE
Family 18. CUPRESSACEAE

A.G. MILLER

Evergreen, monoecious or dioecious, resinous trees and shrubs. Leaves scale-like or needle-like. Inflorescences of axillary and terminal cones; male cones with scales (microsporophylls) bearing 2–6 pollen sacs on their lower surface; female cones with seed scales bearing 1–many ovules at the base of the upper surface. Fruit woody or fleshy and berry-like.

1. Ripe female cones fleshy, the cone scales united at maturity; at least the juvenile leaves needle-like **1. Juniperus**
+ Ripe female cones woody, the cone scales separating at maturity; all leaves scale-like (cultivated) 2

2. Ripe female cones sub-globose; cone-scales peltate, valvate, each with a central umbo; seeds winged **2. Cupressus**
+ Ripe female cones ovoid; cone-scales basally attached, overlapping, each with a hook-like boss near the apex; seeds unwinged **3. Thuja**

1. JUNIPERUS L.

Aromatic trees and shrubs, usually dioecious. Leaves opposite or in whorls of 3, needle-like and spiny in juvenile and mature growth or becoming scale-like when mature. Male strobili small, terminal; scales bearing 2–6 pollen sacs. Female cones subtended by small scale-like bracts; seed scales 3–8, becoming united into a berry-like fruit; seeds 1–12 in each cone, without wings.

Farjon, A. (1992). Taxonomy of multiseed junipers (Juniperus sect. Sabina) in Southwest Asia and East Africa (Taxonomic notes on Cupressaceae I). *Edinb. J. Bot.* 49(3): 251–283; Kerfoot, O. & Lavranos, J. (1984). Studies in the Flora of Arabia 10: *Juniperus phoenicea* L. & *J. excelsa* M. Bieb. *Notes Roy. Bot. Gard. Edinb*urgh 41: 483–489.

1. Juvenile leaves abundant, needle-like; adult leaves scale-like with a distinct, denticulate scarious border and an oblong, depressed gland on the dorsal surface; ripe cone bright red to dark red, not glaucous; seeds 3–9
1. J. phoenicea

+ Juvenile leaves few in number; adult leaves scale-like, border neither denticulate nor scarious, with a conspicuous oval to circular or linear-elliptic gland on the dorsal surface; ripe cones brown to blackish purple, often glaucous; seeds rarely more than 6 2

2. Ultimate branchlets with adult leaves 0.6–1mm diam.; leaves mostly with free apices; ripe cones 3–7mm diam., less than 4-seeded **2. J. procera**

+ Ultimate branchlets with adult leaves 1–1.3mm diam.; leaves mostly appressed; ripe cones 7–12mm diam., (2–)5–6(–7)-seeded
3. J. excelsa subsp. polycarpos

1. J. phoenicea L., Sp. pl.: 1040 (1753). Illustr.: Collenette (1985 p.215).

Tree or shrub up to 8m; bark (on old wood) dark grey-brown, with longitudinal fissures and peeling in narrow papery strips. Juvenile leaves needle-like, abundant. Mature leaves rhomboid-ovate, c.1mm long, closely imbricate with an oblong depressed gland on the dorsal surface; margins denticulate, scarious. Male strobili pale brown, ovoid, c.4 × 3mm; scales suborbicular, 1.5–2mm diam. Female cones at pollination subglobose, c.3mm diam, blackish violet; seed scales 6–8. Ripe cones globose, c.1cm diam., dark red to orange-red, 3–9-seeded. **Map 63.**

Dry montane woodland; (1400–)1900–2350m;

Saudi Arabia. Mediterranean region, Sinai and Jordan.

More drought-tolerant than the following species. Fairly widespread in the northern and central Hijaz mountains of Saudi Arabia as far south as the Taif escarpment where it grows together with *J. procera*. South of Taif it is replaced by *J. procera*.

2. J. procera Hochst. ex Endl., Syn. Conif.: 26 (1847). Illustr.: Fl. Trop. E. Afr., Gymnospermae: 15 (1958); Collenette (1985 p.214) as *J. excelsa*.

Tree up to 8m; bark (on old wood) pale brown or grey-brown, with deep longitudinal fissures, peeling in narrow strips. Juvenile leaves few in number, needle-like, 8–10 × c.1mm; mature leaves scale-like, up to 0.5–1 × 0.5–0.7mm, closely imbricate with a conspicuous linear-elliptic gland on the dorsal surface, mostly with free apices; margins entire, not scarious. Male strobili greenish, turning orange-brown, ellipsoid to subglobose, 3–5 × 2–3mm; scales peltate, c.2.5mm across. Female cones at pollination subglobose, c.2mm diam., bluish green; seed scales 4(–6); mature cones globose, 3–7mm diam., brown to purplish black, bluish or pruinose, (1–)2–3(–4)-seeded. **Map 64, Fig. 9.**

Juniperus woodland and *Juniperus procera-Olea europaea* woodland; 2100–3300m.

Saudi Arabia, Yemen (N). Highlands of E Africa from Sudan and Ethiopia south to eastern Zaire and northeastern Zimbabwe.

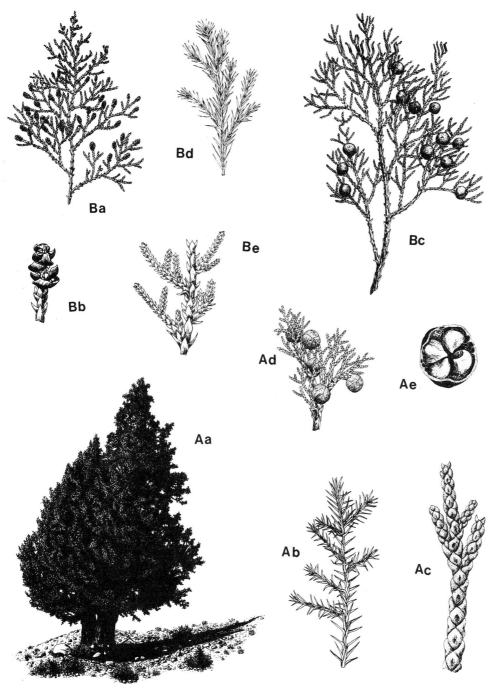

Fig. 9. Cupressaceae. A, *Juniperus procera*: Aa, tree; Ab, branchlet with juvenile leaves (×0.5); Ac, ultimate branchlet with scale leaves (×2.5); Ad, branchlet with cones (×0.5); Ae, cone with seeds (×1.5). B, *J. excelsa* subsp. *polycarpos*: Ba, branchlet with male strobili (×0.5); Bb, male strobili (×2.5); Bc, branchlet with cones (×0.5); Bd, branchlet with juvenile leaves (×2.5); Be, ultimate branchlet with scale leaves (×2.5). (after Farjon, 1992)

Juniperus procera woodland is well-developed in the high-rainfall areas of the Asir mountains of Saudi Arabia where it forms an almost unbroken woodland along the edge of the escarpment from Taif south to the Yemen border. In Yemen it is characteristic of lower rainfall areas (c.250–600mm) at higher altitudes but never forms well-developed woodland. Friis (1992 p.93) considers it an Afromontane near-endemic.

For notes on the taxonomy of this and the following species see notes under *J. excelsa* subsp. *polycarpos* below.

3. J. excelsa M. Bieb. subsp. **polycarpos** (K. Koch) Takhtajan, Fl. Yerevana: 53 (1972). Syn.: *J. macropoda* Boiss., Fl. orient. 5: 709 (1884).

Tree up to 10m; bark (on old wood) purplish to reddish brown, with longitudinal fissures, peeling in narrow papery strips. Juvenile leaves few in number, needle-like, 8–10 × c.1mm; mature leaves scale-like, 1.2–1.6 × 0.8–0.9mm, closely imbricate with a conspicuous oval or ± circular gland on the dorsal surface, mostly appressed; margins entire, not scarious. Male strobili greenish, turning yellowish, ellipsoid to subglobose, 3–4 × 2–3mm; scales peltate, c.2.5mm across. Female cones at pollination subglobose, 2–3mm diam., purplish green to blue; seed scales 4(–6); mature cones globose, 7–12mm diam., purplish brown to blackish purple, often pruinose, (2–)5–6(–7)-seeded. **Map 65, Fig. 9.**

A frequent associate in *Monotheca buxifolia-Olea europaea* open woodland and, above 2300m, the dominant tree in open woodland with *Ephedra pachyclada*, *Euryops arabicus* and *Daphne mucronata*; 1450–3000m.

Oman. Eastern Turkey eastwards to C Asia and Pakistan (subsp. *polycarpos*); *J. excelsa* subsp. *excelsa* extends from the Balkans to the Caucasus mountains.

In Arabia known only from the western Hajar mountains in northern Oman. Individual specimens may grow up to 10m tall.

J. excelsa belongs to sect. *Sabina*, the members of which are characteristic of the dry mountainous areas of SW Asia. Various species of dubious taxonomic value have been described from the region. The group as a whole has recently been revised by Farjon (1992). Kerfoot (op. cit.) includes the Afromontane species *J. procera* Endl. within *J. excelsa*. However, Friis (1992 p.92), whilst noting that the two species are closely related, retains *J. procera* as a distinct taxon because of its smaller (4–7mm versus 7–12mm diam.) fruiting cones which contain fewer (1–4 versus 5–7) seeds. Recent chemical analysis of plants from E Africa and Europe supports the view that the two species should be kept separate (Adams, R. P. in *Biochem. Syst. Ecol.* 18 (4): 207–210 (1990) and Farjon op. cit). Examination of herbarium material from Arabia has shown that plants from the SW escarpment mountains of Saudi Arabia and Yemen match *J. procera* in cone characters whereas those from the Hajar mountains of Oman match *J. excelsa* subsp. *polycarpos*. The two species are therefore kept separate in this account.

2. CUPRESSUS L.

Aromatic monoecious trees or rarely shrubs. Leaves scale-like. Male strobili small, terminal; scales numerous, bearing 2–6 pollen sacs. Female cones at pollination subglobose; seed scales bearing numerous ovules in several rows at the base. Mature female cones subglobose; scales 8–14, woody, peltate, valvate, separating when ripe; seeds narrowly winged.

1. C. sempervirens L., Sp. pl.: 1002 (1753).

Tree up to 30m; bark grey-brown, lightly fissured. Leaves c.1mm, obtuse, with an obscure resin gland. Male strobili yellow-brown, up to 7mm long. Mature female cones 2–3cm diam., ripening in the second year; scales brownish grey, irregular, each with a central umbo. **Map 66.**

Cultivated as an ornamental garden and street tree.

Saudi Arabia, Yemen (N). Mediterranean region to Iran, planted elsewhere.

3. THUJA L.

Similar to *Cupressus* but mature female cones ovoid, with 6–12 basally attached, overlapping scales each with a hook-like boss near the apex; seeds wingless.

1. T. orientalis L., Sp. pl.: 1002 (1753).

Shrub or small tree to 10m; bark reddish brown. Leaves broadly ovate, distinctly grooved on the back, those of the branchlets c.1mm, those of the main shoots c.2mm. Mature female cones c.1.5 × 1cm, glaucous; scales thick, with a strongly hooked boss near the apex.

Cultivated as an ornamental garden and street tree; c.2000m.

Saudi Arabia. A native of China, now widely planted.

Family 19. EPHEDRACEAE

H. FREITAG & M. MAIER-STOLTE

Dioecious, or rarely monoecious, erect or climbing spartoid shrubs. Leaves opposite or whorled, fused towards the base, usually reduced to membranous sheaths, rarely linear. Flowers in small cones. Male flowers each subtended by a bract and consisting of two united scales surrounding a staminal column bearing (2–)3–8(–9) sessile or stipitate anthers. Female flowers solitary or in groups of 2–3 subtended by 2–4 pairs of free or fused bracts; ovules with the integument prolonged into a slender tube; bracts scarious or swollen and fleshy in fruit.

Meyer, C.A. (1846). *Vers. Monogr. Ephedra* 225–226; Stapf, O. (1889). *Akad. Wiss. Wien., Math.-Naturwiss. Kl., Denkschr.* 56 (2): 1–112; Freitag, H. & Maier-Stolte, M. (1989). The Ephedra species of Forsskal: identity and typification. *Taxon* 38: 545–556; Freitag, H. & Maier-Stolte, M. (1992). A new species and combination in the genus Ephedra from Arabia. *Edinb. J. Bot.* 49: 89–93 (1993).

EPHEDRACEAE

EPHEDRA L.

Description as for the family.

1. Pith dark brown; plant erect with conspicuously parallel branches   2
+ Pith white; plant erect with divaricate branches or procumbent or climbing   3

2. Margins of the bracts and leaf sheaths glabrous; female cones usually 1-seeded with the inner bracts fused for $\frac{1}{3}-\frac{1}{2}$ their length; male flowers with 6–8 anthers   **6. E. pachyclada**
+ Margins of the bracts and leaf sheaths minutely ciliate; female cones usually 2-seeded with the inner bracts fused for $\frac{1}{2}-\frac{3}{4}$ their length; male flowers with 6–8 anthers   **7. E. milleri**

3. Stems and main branches erect, divaricately branched; anthers distinctly (0.5–1.0mm) stipitate (if anthers sessile see *E. aphylla*)   4
+ Stems and main branches climbing, spreading or procumbent; anthers sessile   5

4. Shrub up to 1m tall; twigs coarse, always more than 1mm thick; bracts of the female cones free to the base, with broad scarious and erose margins at maturity   **1. E. alata**
+ Small shrub up to 0.5m tall; twigs usually thinner, at least some less than 1mm thick; bracts of the female cones fused for more than $\frac{1}{2}$ their length, margins ciliate   **2. E. transitoria**

5. Margins of the leaf sheaths and bracts glabrous; peduncles of the female cones usually conspicuously curved; anthers 4–6 per flower   **4. E. foeminea**
+ Margins of the leaf sheaths and bracts minutely ciliate; peduncles of the female cones usually straight; anthers 3–4 per flower   6

6. Leaves often up to 10–15(–40)mm long; immature female cones ovoid, (1–)2(–3)-seeded   **5. E. foliata**
+ Leaves up to 3mm long; immature female cones narrowly cylindrical, 1(–2)-seeded   **3. E. aphylla**

1. E. alata Decne in Ann. Sci. Nat. Bot., sér. 2, 2: 239 (1834). Illustr.: O. Stapf, op. cit.: t.1/1 as *E. alata* var. *decaisnei*; Fl. Iraq 2: 85 (1966); Fl. Palaest. 1: t.20 (1966); Collenette (1985 p.230).

Erect shrub up to 1m, forming dense clumps by creeping rhizomes; stems divaricately branched; twigs coarse, more than 1mm thick, smooth or scabridulous. Leaves up to 3(–6)mm long. Margins of leaf sheaths and bracts minutely ciliate. Male cones usually sessile, in dense axillary clusters; flowers with 4–6 anthers, most or all anthers distinctly (0.5–1.0mm) stipitate. Female cones usually sessile, 2-seeded; immature cones ovoid; bracts spreading, free to the base, with broad scarious and erose margins at maturity. **Map 67, Fig. 10.**

Desert plains and sandy runnels; (350–)500–1000m.

Saudi Arabia. From Algeria to Iraq, including Sinai, Negev and S Jordan.

Ephedra

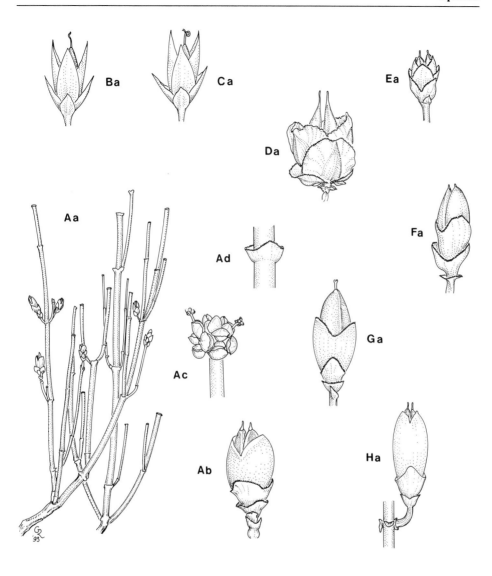

Fig. 10. Ephedraceae. A, *Ephedra milleri*: Aa, habit, female plant (× 0.6); Ab, fruit (× 4); Ac, male cone (× 4); Ad, leaf (× 4). Ba–Ha, *Ephedra* fruits (all × 4); B, *E. pachyclada* subsp. *pachyclada*; C, *E. pachyclada* subsp. *sinaica*; D, *E. alata*; E, *E. transitoria* (immature); F, *E. foliata*; G, *E. aphylla*; H, *E. foemina*.

2. E. transitoria H. Riedl in Anz. Oest. Akad. Wiss, Math.-nat. Kl. 98: 27 (1961). Illustr.: Fl. Iraq 2: 87 (1966); Collenette (1985 p.232).

Erect shrub up to 0.5m, forming loose clumps by creeping rhizomes; stems divaricately branched; twigs (at least some) less than 1mm thick, scabridulous. Leaves up to 3(–7)mm long. Margins of leaf sheaths and bracts minutely ciliate. Male cones at the tips of side branches or in axillary clusters; flowers with 4(–5) distinctly (0.5–1.0mm) stipitate anthers. Female cones usually at the tips of side branches, 2-seeded; immature cones ovoid; innermost bracts fused for over $\frac{1}{2}$ their length; ripe cones fleshy, red. **Map 68, Fig. 10.**

A rare plant of sandy runnels on limestone in the extreme north of Arabia; 800–900m.

Saudi Arabia. Syria, Jordan and Iraq.

3. E. aphylla Forsskal (1775 p.170). Syn.: *E. alte* C.A. Mey. (in part - male specimens) op. cit.: 265 (1846). Illustr.: C.A. Mey., op. cit.: t.3/4 (1846) male specimen as *E. alte*; Fl. Palaest. 1: t.21 (1966) as *E. alte*.

Erect or hanging, rarely climbing shrub up to 1.5m; stems usually scabridulous. Leaves up to 3mm long. Margins of leaf sheaths and bracts minutely ciliate. Male cones in dense axillary clusters or on side branches; flowers with 3–4 sessile anthers. Female cones axillary or on side branches, 1(–2)-seeded; immature cones narrowly cylindrical; innermost bracts fused for at least $\frac{3}{4}$ their length; ripe cones fleshy, red. **Map 69, Fig. 10.**

Shallow runnels in soft limestone; 700m.

Saudi Arabia. From Libya to Lebanon, Syria and E Jordan.

In Arabia a rare plant occurring only in the extreme north of Saudi Arabia.

4. E. foeminea Forsskal (1775 p.219). Syn.: *E. campylopoda* C.A. Mey., op. cit.: 263 (1846). Illustr.: C.A. Mey., op. cit.: t.2 (1846) as *E. campylopoda*; O. Stapf, op. cit.: t.2/12 as *E. fragilis* var. *campylopoda*; Fl. Palaest. 1: t.22 (1966) as *E. campylopoda*; Collenette (1985 p.231) as *E.* sp. aff. *foliata* 1410.

Climbing or hanging shrub up to 5m; stems scabridulous. Leaves up to 2.5mm long. Margins of leaf sheaths and bracts glabrous. Male cones usually in axillary clusters; flowers with 4–6 sessile anthers; often with (1–)2 female flowers at the tips of the male cones. Female cones in axillary clusters, usually on conspicuously curved stalks, (1–)2-seeded; immature cones narrowly cylindrical; innermost bracts fused for at least $\frac{3}{4}$ their length; ripe cones fleshy, bright red. **Map 70, Fig. 10.**

Cliffs, ravines and in *Juniperus* woodland; 1700–2800m.

Saudi Arabia, Yemen (N). E Mediterranean region, including Sinai and Jordan; Ethiopia and Somalia.

Rare in the Hijaz mountains; common in the Asir mountains. It requires more humid conditions than the other Arabian species.

5. E. foliata Boiss. ex C.A. Mey., op. cit.: 297 (1846). Syn.: *E. ciliata* C.A. Mey., op. cit.: 290 (1846); *E. peduncularis* Boiss., Fl. orient. 5: 717 (1884). Illustr.: C.A. Mey., op. cit.: t.3/4 (1846) female specimen as *E. alte*; O. Stapf, op. cit.: t.2/10; Fl. Palaest. 1: t.23 (1966) as *E. peduncularis*; Fl. Pakistan 186: 29 (1987) as *E. ciliata*; Collenette (1985 p.231 as E. aff. *foliata* 2443 & p.232 as *Ephedra* sp. 1481).

Climbing or prostrate shrub up to 4m; stems smooth or minutely hispidulous. Leaves up to 10–15(–40)mm long. Margins of leaf sheaths and bracts ciliate. Male cones at the tips of slender branches; flowers with 3–4 sessile anthers. Female cones usually on long loosely branched twigs, (1–)2(–3)-seeded; immature cones ovoid; innermost bracts fused for over half their length; ripe cones fleshy, translucent white. **Map 71, Fig. 10.**

Rocky slopes and wadi-sides; often scrambling over trees and shrubs; 0–1500 (–2100)m.

Saudi Arabia, Yemen (N & S), Oman, UAE, Qatar, Bahrain, Kuwait. N Africa east to India and C Asia south to Somalia

A common and widespread species in the deserts of Arabia. An abnormal form with many-flowered male cones occurs sporadically. It has been described from S Iran as *E. polylepis* Boiss. & Hausskn.

6. E. pachyclada Boiss., Fl. orient. 5: 713 (1884). Syn.: *E. alte* sensu Schwartz (1939 p.22); *E. fragilis* sensu Schwartz (1939 p.23).

Erect, densely branched shrub up to 0.75(–1)m; stems scabridulous. Margins of leaf sheaths and bracts glabrous. Male cones in dense axillary clusters; flowers with (5–)6–8(–9) sessile or shortly stipitate anthers. Female cones in axillary clusters or shortly stalked, 1-seeded; immature cones ovoid; innermost bracts fused up to a third their length; ripe cones fleshy, red. **Map 72, Fig. 10.**

subsp. **pachyclada**. Illustr.: O. Stapf, op. cit.: t.2, 14; Fl. Pakistan 186: 29 (1987) as *E. intermedia*.

Micropyle 1–2mm, straight or with the upper part densely twisted.

Open rocky slopes and cliffs in *Juniperus* woodland; (1000–)1500–3000m.

Oman. S Iran to N Pakistan.

subsp. **sinaica** (H. Riedl) Freitag & Maier-Stolte in Edinb. J. Bot. 49: 92 (1992). Syn.: *E. sinaica* H. Riedl in Notes Roy. Bot. Gard. Edinburgh 38: 291 (1980). Illustr.: H. Riedl op. cit.: 292; Collenette (1985 p.232) as *E. intermedia*.

Micropyle 2–2.5(–3)mm, loosely and irregularly coiled.

Open rocky slopes and cliffs; 1750–3300m.

Saudi Arabia, Yemen (N). Sinai.

Rare in the higher Hijaz mountains; more common in the Asir mountains and N Yemen where it is frequent on the high plateau and inner rain-shadow regions.

7. E. milleri Freitag & Maier-Stolte in Edinb. J. Bot. 49: 89 (1992). Illustr.: Freitag & Maier-Stolte op. cit.: 90, 91. Type: Oman, *A.G.Miller* 7667B (fem.) (E, K).

Erect, densely branched shrub up to 80cm; stems thick, tuberculate. Leaves up to 1.5–2(–2.5)mm long. Margins of leaf sheaths and bracts minutely ciliate. Male cones pedunculate, solitary or 2–3 in axillary clusters; flowers usually with 5 sessile anthers. Female cones solitary or 2–3 in shortly pedunculate axillary clusters, 2-seeded or 1-seeded by abortion; immature cones ovoid; innermost bracts fused up to $\frac{1}{2}$ to $\frac{2}{3}$ their length; ripe cones fleshy, red. **Map 73, Fig. 10.**

Rocky slopes in open *Acacia-Commiphora* shrubland; 900–1200m.

Oman, ?Yemen (S). Endemic.

A sterile, stunted specimen (22 iii 1854, *Perrotet* in *Deflers*) from near Aden probably belongs here. *E. milleri* can be expected to be found in dry, rocky areas in the Mahra and Hadramaut Governorates of Yemen.

ANGIOSPERMAE
DICOTYLEDONES
Family 20. CASUARINACEAE

A.G. MILLER

Monoecious or dioecious trees or shrubs. Branches "*Equisetum*-like", slender, striate, jointed. Leaves reduced to a whorl of 4–16 fused scales at each node. Male flowers in terminal narrowly cylindrical spikes, solitary in the axils of whorled bracts, each subtended by two bracteoles; perianth rudimentary; stamen solitary. Female flowers in lateral or terminal capitate heads, subtended by a bract and two bracteoles; perianth absent; ovary superior, unilocular, ovules 2; style short, with 2 linear stigmas. Fruit a 1-seeded samara with a small terminal wing, enclosed by persistent, woody valve-like bracteoles, crowded into a cone-like structure.

CASUARINA L.

Description as for the family

1. C. equisetifolia L., Amoen. acad. 4: 143 (1759). Illustr.: Fl. Trop. E. Afr.: 6 (1985); Fl. Iraq 4, 1: 64 (1980).

Monoecious tree, 7–25m tall, with slender, pendent branches. Branchlets with whorls of (6–)7–8 uniformly pale green scale-leaves. Male spikes 10–30(–40) × 1–2mm. Female heads ovoid or subglobose, 3–5mm long. Infructescence subglobose or cylindrical, (10–)30 × (10–)12–18mm; bracteoles c.5–8 × 1.5–4mm, longitudinally ridged on the back; samara 5–7 × 2–3mm, pale brown, glossy, the wing translucent. **Map 74.**

Cultivated for shelter and as an ornamental.

Saudi Arabia, Yemen (N & S), Oman, UAE. SE Asia and Australasia; possibly also native in Madagascar and tropical E Africa.

Commonly cultivated throughout Arabia although under-collected hence the scarce records.

Several species are cultivated in the Middle East and tropical NE Africa and various other species (including *C. cunninghamiana* Miq., *C. cristata* Miq., *C. glauca* Sieb. ex Spreng., *C. lehmanniana* Miq., *C. obesa* Miq. and *C. verticillata* Lam.) have been recorded from Arabia. The hybrid *C. cunninghamiana* × *C. equisetifolia* has also been recorded from Kuwait. Keys to the most commonly found species are found in Fl. Iraq 4, 1: 62 (1980) and Fl. Trop. E. Afr.: 3–5 (1985).

They are quick-growing trees particularly suitable for cultivation in dry areas. In Arabia they are used for wind-breaks, timber, soil stabilisation, firewood and as ornamentals.

Family 21. MYRICACEAE

A.G. MILLER

Trees or shrubs, monoecious or dioecious, evergreen, aromatic. Leaves alternate, simple. Stipules absent. Flowers in dense catkin-like axillary spikes, each usually subtended by scale-like bracts and bracteoles. Male flowers: perianth absent, with 2–12 stamens. Female flowers: ovary superior, 1-locular with a single basal ovule; style short with 2 slender branches. Fruit a drupe.

MYRICA L.

Description as for the family.

1. M. humilis Chamisso & Schlechtendal in Linnaea 6: 535 (1831). Syn.: *M. salicifolia* A. Rich. (1850 p.277). Illustr.: Fl. Ethiopia 3: 260 (1989); Collenette (1985 p.372).

Tree up to 7m. Leaves coriaceous, oblong-elliptic to oblong-ovate, 5–12 × 1.5–4.5cm, acute or obtuse, the margin entire or undulate to obscurely serrate, the base rounded to cuneate, covered with small yellow glands especially beneath; petiole 0.5–2cm. Male spikes 0.5–2.5 × c.0.5cm, tomentose; bracts broadly-ovate, c.1.5 × 2mm. Female spikes 3–15mm, elongating to 2–3cm in fruit; bracts triangular-ovate, c.1 × 1–2mm. Fruit subglobose to ellipsoid, 3–4 × 2–3mm, covered with flaking white warts and small yellow glands. **Map 75, Fig. 11.**

Wadi-sides, steep slopes and ravines (degraded evergreen woodland); 1525–2900m.

Saudi Arabia, Yemen (N). Tropical Africa.
An Afromontane species.

MYRICACEAE

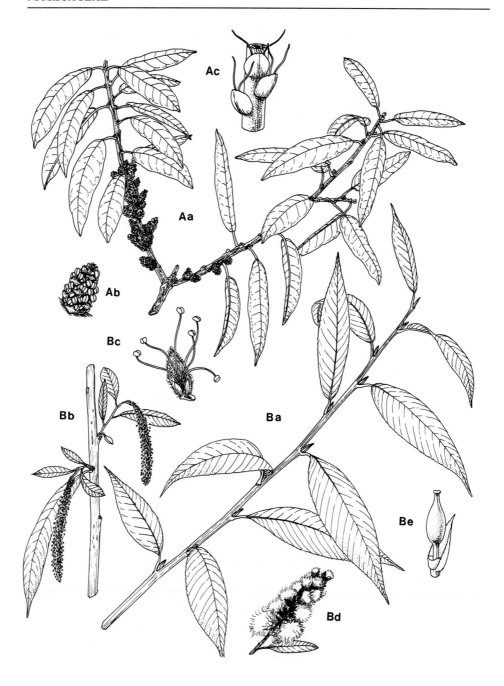

Fig. 11. Myricaceae. A, *Myrica humilis*: Aa, male flowering branch (×0.6); Ab, male flowers (×2); Ac, female flowers (×2). **Salicaceae.** B, *Salix mucronata*: Ba, sterile branch (×0.6); Bb, male catkins (×0.6); Bc, male flower (×7); Bd, female catkin (×0.6); Be, female flower (×7).

Family 22. JUGLANDACEAE

D.G. LONG

Monoecious trees. Leaves deciduous, alternate, pinnate. Stipules absent. Male flowers in pendulous catkins; perianth 3–6-lobed, adnate to the bract; stamens 6–40. Female flowers 1–3 in short terminal racemes, bracteate, each with 4 lanceolate perianth segments; ovary inferior, 1-locular, with a solitary basal ovule; style with 2 plumose branches. Fruit an ovoid drupe; seeds solitary, 2–4-lobed, the cotyledons often contorted.

JUGLANS L.

Description as for the family.

1. J. regia L., Sp. pl.: 997 (1753). Illustr.: Fl. Iraq 4 (1): 59 (1980).
Large tree. Leaflets in 2–5 pairs, oblong-ovate to oblong-elliptic, 2.5–15 × 0.8–7cm, acute or acuminate at the tip, the margins entire, the base rounded, glabrous except for tufts of hairs in the vein axils beneath. Male catkins 7–15cm. Fruit 4–5 × 3–4cm; seed with brain-like lobes. **Map 76.**

Cultivated in the mountains; 2400m.

Yemen (N), Oman. A native of SE Europe and temperate Asia, now cultivated throughout the temperate areas of the world.
The Walnut tree.

Family 23. SALICACEAE

D.R. McKEAN

Dioecious, deciduous trees or shrubs. Leaves usually alternate, simple, entire or toothed. Stipules present. Flowers bracteate, in erect or pendulous spikes or racemes (catkins). Bracts entire or toothed. Perianth much reduced, represented by a small cup-like nectary or by 1–2 nectary scales. Male flowers with 2–many stamens; filaments free or united. Female flowers with a superior, unilocular ovary; ovules numerous; placentation basal or parietal; style short with a branched stigma. Fruit a many-seeded, 2-valved capsule. Seeds with a tuft of long silky hairs from the funicle.

SALIX L.

Dioecious trees. Leaves usually narrowly lanceolate or narrowly elliptic; petioles short. Catkins cylindric, appearing with the leaves (in Arabia), on short leafy shoots.

Bracts persistent, entire or notched. Male flowers with 2–10 stamens. Female flowers with a flask-shaped, stipitate ovary.

1. Leaves with silky hairs, at least beneath; male flowers with two stamens; female flowers with 1 or 2 oblong or linear nectary scales **3. S. excelsa**
+ Leaves sometimes thinly villous at first but soon becoming glabrous; male flowers with 5–10 stamens; female flowers with a cup-shaped nectary 2

2. Capsules 4.5–6mm long; fruiting stipes 1.5–3.5mm long; male flowers with (5–)6–8(–10) stamens **1. S. mucronata**
+ Capsules 2–4(–5)mm long; fruiting stipes 0.5–1(–1.8)mm long; male flowers with 4–5(–7) stamens **2. S. acmophylla**

1. S. mucronata Thunb., Prodr. pl. cap.: 6 (1794). Syn.: *S. subserrata* Willd., Sp. pl. 4: 671 (1806). Illustr.: Collenette (1985 p.436); Fl. Ethiopia 3: 259 (1989); Fl. Trop. E. Afr.: 2 (1985) all as *S. subserrata*.

Small tree or shrub up to 12m; bark dark grey or brown, deeply fissured; twigs reddish brown, thinly villous at first becoming glabrous. Leaves oblong-elliptic to linear-lanceolate, up to 14 × 3cm, acute or acuminate, olive-green, ± glossy above, often glaucous beneath, glabrous or sometimes thinly sericeous at first; mature leaves usually glandular-serrulate; petiole up to 1.5cm; stipules minute, ovate-acuminate, soon falling. Catkins 2–9 × 0.4–1cm; bracts ovate, 1.5–3.5 × 1–1.7mm, yellowish green, densely villous especially on the margins. Male flowers with (5–)6–8(–10) stamens; filaments 3.5–6mm, villous at the base; nectaries 2, unequal, entire or lobed. Female flowers: ovary ovoid, 2–2.5mm, glabrous; styles 0.5–1mm; stigma 2-lobed; nectaries cup-shaped. Capsules ovoid, 4.5–6mm long; fruiting stipe 1.5–3.5mm. **Map 77, Fig. 11.**

By pools, in wadis, and as a shade tree near cultivation; 2000–2150m.

Saudi Arabia, Yemen (N). Southern and tropical Africa northwards to Egypt.

Usually found near cultivation and possibly introduced. See comments under *S. acmophylla*.

2. S. acmophylla Boiss., Diagn. pl. orient. sér. 1, 1(7): 98 (1846).

Similar to *S. mucronata* but capsules 3–4(–5)mm long; fruiting stipe 0.5–1(–1.8)mm long; male flowers with 4–5(–7) stamens. **Map 78.**

Damp sand in deep ravines, rocky wadis, by irrigation channels and ?planted in oases; 50–1800m.

Saudi Arabia, Oman. SW and C Asia.

The distributions of *Salix acmophylla* (from SW and C Asia) and *S. mucronata* (from Africa) meet in Arabia and, because of an overlap of characters, it has sometimes been difficult to assign Arabian material to one species or the other. Sterile material is apparently indistinguishable – the degree of toothing on the leaf margin (apparently more pronounced in *S. mucronata*) does not work satisfactorily when a range of material is examined. Female plants of *S. acmophylla* can be separated by their larger fruits and longer stipes but distinguishing male plants is more problematic. There is, however, an apparent difference in stamen number: *S. acmophylla* has 4–5(–7) stamens whereas *S. mucronata* has (5–)6–8. Plants from Oman and NW Saudi Arabia are fruiting and therefore it has been possible to assign them to *S. acmophylla*. However, the identity of specimens from the mountains of SW Arabia is not certain; only male trees have been seen and these have flowers with (5–)6–7 stamens. Therefore, taking into account distribution and stamen number, they have been provisionally assigned to *S. mucronata*. The relationships and status of these species in Arabia need further study and efforts should be made to collect both male and female specimens.

Mandaville (1990 p.125) notes that in the Eastern Province of Saudi Arabia *S. acmophylla* grows by irrigation ditches and in oases and comments that its status, whether native or introduced, is uncertain. Likewise the plants from Oman are of uncertain status. However, the plants from NW Saudi Arabia are found in "deep ravines and are difficult to reach" (*Collenette* 7578) and are undoubtedly native.

Salix wilhelmsiana Bieb. has been recorded from Oman (Fl. Turkey 7: 716 (1982)). However, all the Omani material examined has proved to be narrow-leaved forms of *S. acmophylla*. *S. wilhelmsiana* can be distinguished from *S. acmophylla* by its densely sericeous ovaries and the male flowers containing only two stamens.

3. S. excelsa S.G.Gmel., Reise Russland 3: 308, t.34, f.2 (1774).

Tree up to 7m; bark greyish, deeply fissured; twigs olive- or reddish brown, sericeous at first becoming glabrous. Leaves narrowly elliptic, up to 12 × 3cm, acuminate, margins finely serrate, dull green above, sericeous beneath at first becoming glabrous; petiole up to 1.5cm; stipules minute soon falling. Catkins up to 40 × 8mm; bracts oblong-ovate, hairy. Male flowers with 2 stamens and 2 nectaries. Female flowers: ovary ovoid, glabrous; style short; stigma short and bifid; nectaries 1 or 2. Capsules broadly flask-shaped, 5–7mm long when ripe; fruiting stipe up to 1mm. **Map. 79.**

By permanent streams and planted as a shade tree by cultivation; 2200–3000m.

Yemen (N). SW and C Asia.

Records of *S. excelsa* in Arabia have previously been confused with *S. alba*. However, *S. excelsa* can be distinguished by its heavier branches, broader leaves, larger capsules and broader catkins. Only male trees are known from Arabia and they have almost certainly been introduced.

Family 24. ULMACEAE

A.G. MILLER & J.A. NYBERG

Deciduous trees or shrubs. Leaves alternate, entire or toothed, often with an oblique base, pinnately nerved with 3(–5) main nerves from the base (in Arabia). Flowers small, male, female or bisexual, actinomorphic, solitary or clustered in the leaf axils. Sepals (4–)5(–8), free or shortly fused. Petals absent. Stamens as many as and opposite the sepals. Ovary superior, 1–2-locular; ovules solitary, pendulous; styles 1–2, linear or 2–3-branched. Fruit a drupe (in Arabia) or samara.

1.	Leaves entire or serrate in the upper two thirds	**1. Celtis**
+	Leaves finely serrate almost to the base	**2. Trema**

1. CELTIS L.

Trees or shrubs. Leaves entire or serrate in the upper two thirds. Stipules linear. Flowers brownish or greenish, in male or bisexual axillary clusters. Sepals (4–)5, free or fused at the base. Ovary 1-locular. Styles 2–5mm, simple or 2–3-branched. Fruit a thinly-fleshed drupe.

1.	Mature leaves serrate in the upper part; style simple	**1. C. africana**
+	Mature leaves entire; style 2–3-branched	**2. C. toka**

1. C. africana Burm. f., Prodr. fl. cap.: 31 (1768). Syn.: *C. kraussiana* Bernh. in Flora 28: 87 (1845).

Tree up to 7(–20)m; bark smooth, grey. Leaves ovate to narrowly ovate, 3–8 (–12) × 2–4(–6)cm, acuminate, serrate in the upper two-thirds, thinly scabrid to subglabrous, with tufts of hairs in the angles of the nerves beneath; petioles 1.5–4mm. Flowers in axillary clusters, either all male or the upper with a few bisexual flowers; pedicels elongating in fruit. Sepals 4–5, 2.5–3.5mm. Styles simple, 2.5–4mm. Fruit yellow or orange, sub-globose, 6–10 × 5–8mm, pubescent; pedicels 2–2.5cm. **Map 80, Fig. 12.**

Valley forest and field edges; (800–)1000–2700m.

Saudi Arabia, Yemen (N). Tropical and southern Africa.

2. C. toka (Forsskal) Hepper & J.R.I. Wood in Kew Bull. 38 (1): 86 (1983). Syn.: *Ficus toka* Forsskal (1775 p.219); *Celtis integrifolia* Lam., Encycl. 4: 140 (1797). Type: Yemen (N), *Forsskal* (C).

Tree up to 20m with a large spreading crown; bark smooth or scaly. Leaves ovate to oblong-elliptic, 3–10 × 2–6cm, abruptly acuminate, entire (toothed on seedlings and suckers), rounded to cordate below; thinly scabridulous and with tufts of hairs in the axils of the nerves beneath; petioles 3–8mm. Flowers in axillary clusters, either all male or with a few longer-pedicelled female or bisexual flowers. Sepals 5, 1.5–2(–3)mm. Styles 2–3-branched, 3–5mm. Fruits yellow, subglobose, 8–12 × 6–10mm, puberulous. **Map 81, Fig. 12.**

Valley forest; 400–750m.

Yemen (N). W, C and NE tropical Africa.

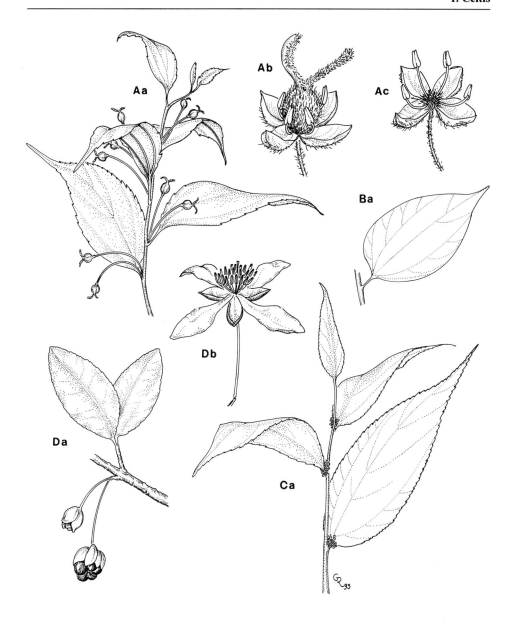

Fig. 12. **Ulmaceae**. A, *Celtis africana*: Aa, branch with young fruits (×0.6); Ab, hermaphrodite flower (×5); Ac, male flower (×5). B, *C. toka*: Ba, leaf (×0.6). C, *Trema orientalis*: Ca, flowering shoot (×0.6). **Ochnaceae**. D, *Ochna inermis*: Da, fruits and leaves (×0.6); Db, flower (×5).

2. TREMA Lour.

Trees or shrubs. Leaves finely serrate. Stipules linear-lanceolate. Flowers small, greenish, male, female or bisexual, in axillary cymes. Sepals (4–)5, fused at the base. Ovary 1-locular. Styles 2, 0.5–2mm, simple. Fruit a drupe.

1. T. orientalis (L.) Blume, Mus. bot. 2 (4): 62 (1856). Syn.: *T. guineensis* (Schum.) Engl. var. *hochstetteri* (Buchinger) Engl., Pflanzenwelt Afrikas 3 (1): 14 (1915); *T. hochstetteri* (Buchinger) Engl., Hochgebirgsfl. Afrikas: 190 (1892). Illustr.: Fl. Ethiopia 3: 269 (1989).

Tree or shrub up to 10m; bark smooth, whitish or grey, the young shoots tomentose. Leaves oblong-lanceolate, 8–15(–22) × 3–7cm, acuminate, finely serrate except for the entire base, rounded to cordate below, pubescent to scabrid above, sericeous becoming tomentose or pubescent beneath; petioles 5–10mm. Calyx 1–2mm; styles reddish brown-tomentose. Fruits black, ovoid to globose, (3–)4–5mm, glabrescent. **Map 82, Fig. 12.**

Fields margins, terrace-walls, wadi-sides and rocky gorges in drought-deciduous woodland; (100–)800–1100m.

Saudi Arabia, Yemen (N), Oman. Tropical Africa, Madagascar and tropical Asia.

Family 25. BARBEYACEAE

A.G. MILLER

Dioecious shrubs or trees. Leaves opposite and decussate, simple, entire, densely white tomentose beneath, glabrous above; stipules absent. Flowers small, regular, in axillary cymes. Male flowers: calyx with 3–4 segments, fused at the base, valvate; petals absent; stamens 6–12. Female flowers: calyx of 3–4 segments, enlarging and becoming wing-like and membranous in fruit; petals absent; staminodes absent; ovary superior, 1-locular with a solitary pendent ovule; style linear. Fruit dry and dehiscent, ellipsoid.

A monotypic family, sometimes included in the Ulmaceae.

BARBEYA Schweinf.

Description as for the family

1. B. oleoides Schweinf. in Malpighia 5: 332 (1891). Illustr.: Fl. Ethiopia 3: 270 (1989); Collenette (1985 p.75).

Shrub or small tree, up to 5m, resembling an olive. Leaves oblong-elliptic to narrowly ovate, 2–5 × 0.5–2cm, acute or mucronate, the base rounded or attenuate into a short petiole. Male flowers in dense clusters; calyx segments cream, c.3 × 2mm; stamens c.3mm, the filaments very short. Female flowers in loose clusters; calyx segments yellow-green, narrowly elliptic to elliptic-oblong, 7–8mm, enlarging up to 20 × 8mm in fruit. **Map 83, Fig. 20.**

Rocky slopes in *Juniperus-Olea* woodland and *Acacia-Commiphora* bushland; 800–2150m.

Saudi Arabia, Yemen (N). Ethiopia and Somalia.

Family 26. MORACEAE

A.G. MILLER

Monoecious or dioecious trees, shrubs or herbs; sap milky or watery. Leaves alternate or subopposite, simple or lobed; stipules present. Inflorescences bisexual or unisexual, of spikes or heads, on disc-shaped or clavate receptacles or inside hollow receptacles (figs). Flowers unisexual, actinomorphic; perianth segments (2–)4(–6) or absent; male flowers with 1–4 stamens; female flowers with an inferior or superior 1-celled ovary; ovule solitary, pendulous; stigmas 1–2. Fruit an achene or drupe, sometimes enclosed or immersed in the fleshy receptacle.

1. Herbs or shrubs with swollen, succulent stems or lacking an aerial stem but with an underground tuber; flowers on a flattened receptacle surrounded by ray-like appendages   **3. Dorstenia**
+ Woody trees or shrubs; flowers in various sorts of inflorescence but not as above   2
2. Flowers borne inside a globose or obovate hollow receptacle (the fig)   **4. Ficus**
+ Flowers borne externally on catkin-like spikes or on discoid or clavate receptacles   3
3. Female flowers borne in capitate heads; male flowers in spikes; leaves 3-nerved from the base   **1. Morus**
+ Female flowers solitary on clavate receptacles; male flowers numerous on discoid receptacles; leaves pinnately nerved   **2. Antiaris**

1. MORUS L.

Dioecious trees. Leaves alternate, entire, 3-nerved from the base; stipules small, lateral, free. Flowers in unisexual, axillary, pendunculate spikes. Male flowers in catkin-like spikes; perianth segments 4, fused at the base; stamens 4. Female flowers in short capitate spikes; perianth segments 4; stigmas 2. Fruit a drupe, enclosed by the enlarged fleshy perianth and forming a cylindrical or ovoid compound fruit.

Morus macroura Miq. is recorded from Saudi Arabia where it is rarely cultivated for its edible fruits. It is distinguished from *M. alba* and *M. nigra* by its larger (more than 5cm long) yellowish white fruits and its longer (more than 5cm) male spikes.

1. Ripe fruits dark purple to black; leaves usually unlobed; perianth of female flowers with long, soft hairs **1. M. nigra**
+ Ripe fruits white or rarely purplish black; leaves usually lobed; perianth of female flowers glabrous or shortly pubescent **2. M. alba**

1. M. nigra L., Sp. pl.: 986 (1753). Illustr.: Fl. Pakistan 171: 49 (1985).

Tree to up 10m. Leaves broadly ovate, usually simple, sometimes 2–3-lobed, 6–12 × 6–12cm, shortly acuminate, serrate, cordate at the base, scabrous above, pubescent beneath. Male spike 1.5–3cm long. Styles and perianths of the female flowers covered with long soft hairs. Mature fruit 2–2.5cm diam., dark purple to black. **Map. 84.**

Planted as a street tree.

Saudi Arabia, Yemen (N), Qatar, Bahrain. Widely cultivated in C & S Europe, N Africa, SW Asia, NW India and C Asia. Wild origin obscure.

The "Black Mulberry" – cultivated for shade and its edible fruit.

2. M. alba L., Sp. pl.: 986 (1753). Syn.: *M. alba* var. *arabica* Bureau in DC., Prodr. 17: 144 (1873); *M. arabica* (Bureau) Koidzumi in Bot. Mag. Tokyo 31: 35 (1917). Illustr.: Fl. Pakistan 171: 51 (1985).

Tree up to 5m. Leaves ovate, very polymorphic, usually at least some 3-lobed, rarely all simple, up to 10 × 8cm, acute or shortly acuminate, crenate-dentate, cordate at the base, glabrous or pubescent on the veins beneath. Male spike c.1cm. Styles and perianth of the female flowers glabrous or shortly pubescent. Mature fruits 1–2.5cm diam., white or rarely purplish black. **Map. 85.**

Apparently naturalized in irrigated date gardens; 50–500m.

Oman. A native of China and E Asia, now widely cultivated in warm areas of the world.

The "White Mulberry" – food plant of the silkworm (*Bombyx mori*).

2. ANTIARIS Lescheu.

Monoecious or dioecious trees with milky latex. Leaves alternate. Stipules free, semi-amplexicaul. Flowers on receptacles on short axillary branches. Male flowers many, on a disc-shaped, pedunculate receptacle surrounded by 1–3 rows of bracts; perianth segments 2–7, free, stamens 2–4. Female flower solitary on a clavate, bracteate receptacle; perianth 4-lobed; stigmas 2. Fruit fused with the enlarged fleshy, orange receptacle.

1. A. toxicaria Lescheu. in Ann. Mus. Natl. Hist. Nat. Paris 16: 478, t. 22 (1810). Syn.: *A. challa* (Schweinf.) Engl., Veg. Erde 3(1): 33, t. 20 (1915); *Ficus challa* Schweinf. (1896 p.44). Illustr.: Fl. Trop. E. Afr.: 14 (1989); Fl. Ethiopia 3: 301 (1989).

Shrub or tree up to 20(–30)m. Leaves elliptic to oblong-elliptic or obovate, 10–16 × 5–10cm, shortly acuminate or obtuse, subentire, the base obtuse to subcordate, scabrous above, pubescent or scabrous beneath; petioles 3–9mm. Male inflorescence 0.5–0.8cm diam.; peduncle 0.5–1cm. Female inflorescence ellipsoid, 0.3–0.5cm × c.0.3cm; peduncle 0.5–1cm. Infructescence ellipsoid, ovoid or globose, 1–1.5 × 0.8–1cm. **Map 86, Fig. 13.**

Valley forest; 600–900m.

Yemen (N). Tropical Africa and tropical Asia.

Arabian material is all referable to subsp. *toxicaria* var. *africana* A. Chev. which is distinguished by its scabrous leaves with prominent venation beneath.

3. DORSTENIA L.

Monoecious herbs or shrubs; stems either succulent or plants stemless with an underground tuber; latex milky. Leaves alternate, crowded at the ends of branches or rising directly from the tuber, entire or toothed, rarely palmately lobed. Flowers minute, sunk into a flattened pedunculate receptacle which is fringed by linear or rounded, ray-like appendages. Male flowers with a 2–3 lobed perianth and 2–3 stamens. Female flowers less numerous; perianth with rudimentary lobes; style simple; stigma unbranched (in Arabia). Fruit a minutely tuberculate achene, explosively expelled.

Friis, I. (1983). The acaulescent and succulent species of *Dorstenia* sect. *Kosaria* from NE tropical Africa and Arabia. *Nord. J. Bot.* 3: 533–538.

1.	Plants with underground tubers, lacking aerial stems; leaves cordate to reniform sometimes lobed	2
+	Plants with succulent aerial stems; leaves narrowly elliptic to broadly obovate or circular	3
2.	Receptacle narrowly elliptic	**1. D. barnimiana**
+	Receptacles circular	**2. D. socotrana**
3.	Herbs up to 50cm; stems cylindrical or arising from a tuberous basal part	**3. D. foetida**
+	Shrubs up to 2.5m; stems bottle-shaped	**4. D. gigas**

1. D. barnimiana Schweinf., Pl. quaed. nilot.: 36, t. 12 (1862). Syn.: *D. barnimiana* Schweinf. var. *ophioglossoides* (Bureau) Engl., Monogr. afrik. Pflanzen-Fam. 1: 25 (1898). Illustr.: Fl. Ethiopia 3: p.279 (1989).

Perennial herb, stemless with an underground tuber; tuber 2–6cm across. Leaves cordate to reniform, entire or rarely palmately divided, 2–10 × 3.5–11cm, rounded or bluntly acute, entire or irregularly denticulate, cordate or rounded at the base, glabrescent to puberulous; petioles 4–15cm. Receptacles 1–2 in the leaf axils, narrowly elliptic, 1–3.5 × 0.3–1cm; appendages linear, 0.4–6cm, evenly distributed or restricted to a few at the base and one at the apex; peduncles 4–12cm. **Map 87, Fig. 13.**

Amongst rocks on wadi-sides, terrace-walls, rocky slopes; 800–2500m.

Yemen (N & S). Tropical NE Africa.

2. D. socotrana A.G. Miller in Edinb. J. Bot. 53 (1996). Type: Socotra, *Miller et al.* 12647 (E).

Perennial herb, stemless with an underground tuber; tuber c.1cm across. Leaves cordate to reniform, entire, 0.8–2.5 × 0.8–2.8cm, rounded, entire or irregularly denticulate, cordate at the base, glabrous or minutely puberulous; petioles 1–8cm. Receptacles solitary in the leaf axils, circular, 3–5mm diam. (excl. appendages); appendages linear, 1–3mm, evenly distributed; peduncles 1–3cm. **Map 87.**

Cracks in limestone boulders; c.160m.

Socotra. Endemic.

3. D. foetida (Forsskal) Schweinf. (1896 p.120). Syn.: *D. radiata* Lam., Encycl. 2: 315 (1786); *Kosaria foetida* Forsskal (1775 pp. CXXI, 164). Illustr.: Fl. Ethiopia 3: p.279 (1989). Type: Yemen (N), *Forsskal* (C).

Succulent herb; stems swollen, succulent, cylindrical from a swollen tuberous base or branched and tuberous, marked with pronounced leaf scars. Leaves very variable, narrowly obovate to ovate or suborbicular, 2–18 × 1–2.5cm, acute to rounded, entire to crenate or dentate, often ± crisped, long-attenuate to rounded at the base, glossy green, glabrous to puberulous; petioles 1–3.5cm. Receptacle circular, 0.5–1.5cm diam., (excl. appendages); appendages linear, up to 3cm: peduncles 1–6cm. **Map 88, Fig. 13.**

D. foetida is an exceptionally variable species within which some rather ill-defined infraspecific taxa have been recognized. Plants from Arabia belong to subsp. *foetida* of which two varieties have been recognized.

1. Leaves narrowly ovate, obovate or elliptic, the margin crenate to toothed; petiole equalling or slightly shorter than the lamina   var. **foetida**
+ Leaves broadly ovate to suborbicular, the margin entire; petiole equalling or longer than the lamina   var. **obovata**

var. **foetida**. Syn.: *D. arabica* Hemsley in Hooker's Icon. pl. 26: t.2503 (1897).

Cliffs and rocky slopes in dry deciduous woodland and *Euphorbia balsamifera* shrubland; 100—1600m.

Saudi Arabia, Yemen (N & S), Oman. Sudan, Somalia and Ethiopia.

var. **obovata** (A. Rich.) Schweinf. & Engl. in Engl., Monogr. afrik. Pflanzen-Fam. 1: 27 (1898).

Cliffs and rocky slopes; 900m.

Yemen (N). Ethiopia and Sudan.

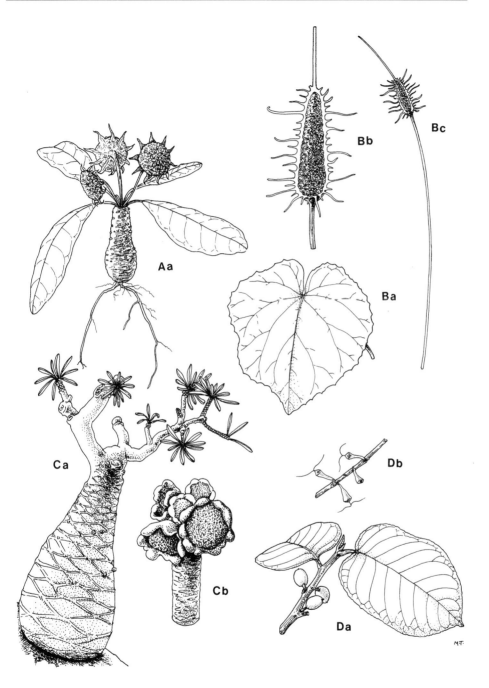

Fig. 13. Moraceae. A, *Dorstenia foetida*: Aa, habit (× 0.6). B, *D. barnimiana*: Ba, leaf (× 1); Bb, inflorescence (× 1); Bc, inflorescence (× 0.5). C, *D. gigas*: Ca, habit (× 0.15); Cb, inflorescences (× 1.5). D, *Antiaris toxicaria*: Da, fruiting branch (× 0.5); Db, female inflorescence (× 0.5).

4. D. gigas Schweinf. ex Balf.f. (1883 p.95). Syntypes: Socotra, *Balfour* 638 (E); *Schweinfurth* 737 (K).

Succulent shrub up to 2.5m; stems bottle-shaped, swollen, up to 1.5m across. Leaves dark green, bullate, narrowly obovate, 3–15 × 1–3cm, acute or obtuse, obscurely crenate, attenuate into a short petiole, pubescent on the veins. Receptacle circular, yellowish green, 5–10(–2.5)mm diam.; appendages rounded, 1.5–3 × 1–3mm; peduncle 5–20mm. **Map 89, Fig. 13.**

On steep, often inaccessible, limestone and granite cliffs; 600–1100m.

Socotra. Endemic.

4. FICUS L.

Monoecious or dioecious trees or shrubs, sometimes epiphytic. Leaves alternate, rarely opposite, entire or lobed. Stipules enclosing the terminal bud, free or partly fused, often caducous and leaving a prominent scar. Flowers minute, numerous, borne inside a specialised hollow receptacle - the fig; figs with a small apical opening (the ostiole) fringed by small bracts; bracts also found in a whorl on the peduncle or scattered over the surface of the fig. Flowers of three kinds: male, female and gall. Perianths 2–6-lobed; male flowers with 1–3 stamens; female flowers with a filiform style; gall flowers with a reduced style and stigma. Achenes enclosed within the fig.

A number of Asiatic species are cultivated as ornamentals and street trees in Arabia. A key to those so far recorded in Arabia is provided below. Useful keys to the exotic species of *Ficus* are found in Corner, E.J.H. *Gard. Bull. Singapore* 21 (1): 1–186 (1965) and Condit, I.J. *Ficus; the exotic species* (1969).

Key to cultivated species

1. Leaves 3–5-lobed; fruits obovoid (recorded from most countries in Arabia) **F. carica** L.
+ Leaves entire; figs globose or ellipsoid ... 2

2. Leaves pandurate, the lamina 15–30 × 12–20cm (Saudi Arabia) **F. lyrata** Warb.
+ Leaves elliptic, ovate or obovate, never lyrate ... 3

3. Leaves finely pubescent beneath, at least when young; leaf lamina 10–20 × 8–15cm (Saudi Arabia, Yemen (N), Oman, UAE) **F. benghalensis** L.
+ Leaves glabrous ... 4

4. Leaf tips rounded ... 5
+ Leaf tips acute, acuminate or caudate ... 6

5. Black glands present in the axils of the veins on the undersurface of the leaves; leaf lamina polymorphic, broadly obovate to deltate, 2.5–7.5 × 1.5–5cm (Saudi Arabia) **F. deltoidea** Jack.
+ Black glands absent; leaf lamina obovate or elliptic to obovate, (3–)5–10 × (1.5–)2–5 (–6)cm (Saudi Arabia) **F. microcarpa** L.f.

6. Leaves broadly ovate with a caudate tip; leaf lamina 6–25 × 4–10cm (Saudi Arabia, Oman) **F. religiosa** L.
+   Leaves ovate, elliptic or obovate, the tip acute or shortly acuminate   7

7. Leaves with more than 20 pairs of lateral veins; leaf lamina 15–30 × 5–15cm (Saudi Arabia, Oman, Qatar) **F. elastica** Roxb. ex Hornem.
+   Leaves with up to 15 pairs of lateral veins   8

8. Leaf venation obscure, the lateral veins parallel; leaf tips acuminate; leaf lamina 5–12 × 2–6cm (Saudi Arabia) **F. benjamina** L.
+   Leaf venation distinct, the lateral veins not parallel, ± ascending; leaf tips acute or shortly acuminate; leaf lamina 5–10 × 3–7cm (Saudi Arabia)
    F. amplissima J.E. Smith.

Key to native and naturalized species

1. Leaves lobed, if unlobed leaves also present then the margins serrate   2
+   Leaves simple (see also *F. sycomorus* in which the leaves on sucker-shoots are rarely lobed), the margins entire or repand, never serrate   5

2. Sap watery; bracts scattered over the surface of the fig and peduncle; figs globose   **4. F. exasperata**
+   Sap milky; bracts restricted to the ostiole and a single whorl at the base of the fig; figs pear-shaped   3

3. Figs 3–5cm diam; leaves up to 30 × 30cm   **1. F. carica**
+   Figs less than 2cm diam.; leaves up to 15 × 15cm   4

4. Tip of leaves and leaf lobes obtuse to rounded or if acute then the lobes themselves lobed   **3. F. johannis**
+   Tip of leaves and leaf lobes acute or acuminate, the lobes serrate but never lobed   **2. F. palmata**

5. Figs 2–4cm diam., borne on leafless branchlets on the older wood   6
+   Figs 0.5–1.5(–2)cm diam., borne in the leaf axils   7

6. Twigs and petioles minutely puberulous and with long simple hairs; leaves usually with rounded tips   **5. F. sycomorus**
+   Twigs and petioles with only one type of hair, either puberulous, tomentose or hirsute; leaves with acute or acuminate tips   **6. F. sur**

7. Leaves scabrous, simple or sometimes the juvenile leaves lobed; sap watery; bracts scattered over the surface of the figs and peduncles   **4. F. exasperata**
+   Leaves smooth, simple; sap milky; bracts restricted to the ostiole and the base of the fig   8

8. Leaves more than 3 × as long as broad; ostiole round, bracts visible   9
+   Leaves less than 3 × as long as broad; ostiole a slit; bracts not visible   10

9.	Leaves usually lanceolate, glabrous, usually less than 3cm broad; basal lateral veins unbranched, ± curved, running almost parallel to the leaf-margin	
		7. F. cordata subsp. **salicifolia**
+	Leaves ovate to elliptic, scabrous, pilose or glabrous, 4–7cm broad; basal lateral veins usually branched, almost straight, not running parallel to the leaf margin	**8. F. ingens**
10.	Figs pedunculate; leaf tip acuminate	**10. F. populifolia**
+	Figs sessile; leaf tip rounded or acute	11
11.	Leaves oblong; ostiole 2-lipped; leaves without a glandular patch	**9. F. glumosa**
+	Leaves cordate; ostiole 3-lipped; leaves with a glandular patch on the lower surface near the base of the midrib	**11. F. vasta**

1. F. carica L., Sp. pl.: 1059 (1753). Illustr.: Fl. Ethiopia 3: 281 (1989).

Shrub or small tree up to 10m. Sap milky. Leaves 3–5-lobed, broadly ovate in outline, c.10–20(–30) × 10–20(–30)cm, the leaves and lobes with acute to rounded tips, serrate, cordate at the base, scabrous above, scabrous, glabrescent or puberulous beneath. Stipules semi-amplexicaul. Figs 1–2 in the leaf axils, obovoid, 3–5cm diam., reddish, sessile or pedunculate; peduncles up to 3cm, bearing a whorl of 3–4 bracts. Ostiolar bracts visible. **Map 90, Fig. 14.**

Cultivated, occasionally naturalized.

Saudi Arabia, Yemen (N & S), Oman, Qatar, Bahrain.

Cultivated throughout the world for its edible fruit. Not known to occur naturally in the wild. See comments under *F. palmata*.

2. F. palmata Forsskal (1775 p.179).

Shrub or small tree up to 5m. Sap milky, Leaves simple or 2–3-lobed, ovate or suborbicular in outline, 2–15 × 1.5–8(–15)cm, the apex rounded or obtuse to acute or acuminate, serrate, truncate or subcordate at the base, scabrous above, scabrous to densely pubescent or glabrescent beneath. Stipules semi-amplexicaul, caducous. Figs 1–2 in the leaf axils, obovoid to globose, 1–2cm diam., pubescent, pinkish purple; peduncles 0.5–1.5cm, bearing a whorl of 3–4 bracts. Ostiole round, the ostiolar bracts visible. **Fig. 14.**

F. palmata, F. johannis and *F. carica* are a difficult complex of polymorphic species where species limits are often far from clear. *F. carica*, the cultivated fig, is thought to be of possible hybrid origin with the putative parents *F. johannis* and *F. carica* subsp. *rupestris* (see Browicz in Fl. Iranica 153: 9 (1982)).

In Arabia *F. palmata* is represented by two taxa: subsp. *palmata*, distinguished by its acute or acuminate leaves, occurs in the south and south west of the peninsula and outside Arabia extends into tropical NE Africa; and subsp. *virgata*, which has broader, rounded leaves, and is found in the extreme NW of Saudi Arabia, and outside Arabia occurs in Iran, Afghanistan, Pakistan and NW India. Subsp. *virgata* is apparently indistinguishable from *F. pseudosycomorus*, a species from Sinai, Palestine and Jordan.

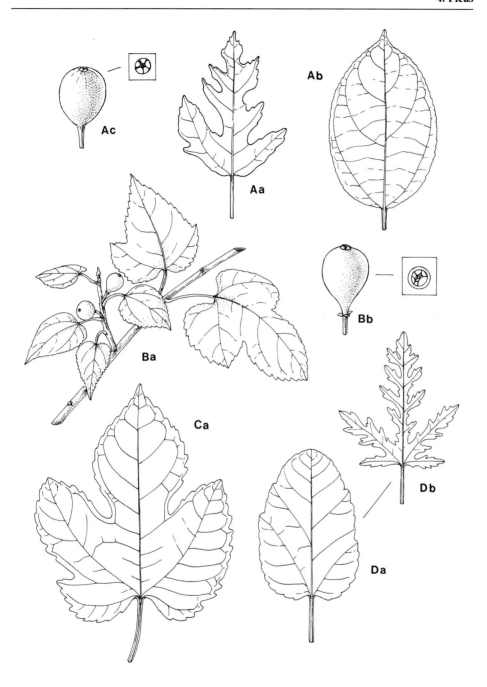

Fig. 14. Moraceae. A, *Ficus exasperata*: Aa, juvenile leaf (× 0.5); Ab, adult leaf (× 0.5); Ac, fig and ostiole (× 1). B, *Ficus palmata*: Ba, fruiting branch (× 0.5); Bb, fig and ostiole (× 1). C, *F. carica*: Ca, leaf (× 0.5). D. *F. johannis*: Da–Db, entire and dissected leaves (× 1).

F. pseudosycomorus has, in this account, been provisionally placed in the synonymy of *F. palmata* subsp. *virgata*.

There are a number of problematic specimens. Certain forms of *F. palmata*, with large, deeply-lobed leaves, can sometimes be difficult to distinguish from *F. carica* and several gatherings (e.g. *Collenette* 4566; *Fayed* 1209; *Grainger* 531) appear to be somewhat intermediate between *F. palmata* and *F. carica* in leaf size and shape but are apparently wild and have been provisionally treated here as *F. palmata*. A cultivated specimen from Oman (*Western* 1191), which has large, unlobed leaved with deeply cordate bases, may represent an introduction of *F. carica* subsp. *rupestris* (Hausskn. ex Boiss.) Browicz.

1. Leaves simple or lobed, ovate in outline, the tip acute or acuminate
   subsp. **palmata**
+ Leaves simple, rarely shallowly lobed, broadly ovate or suborbicular in outline, the tip rounded or obtuse to shortly acute subsp. **virgata**

subsp. **palmata**. Syn.: *F. morifolia* Forsskal (1775 p.179); *F. forskalaei* Vahl, Enum. pl. 2: 196 (1805). Illustr.: Fl. Ethiopia 3: 282 (1989); Collenette (1985 p.368) as *F. carica*. Type: Yemen (N), *Forsskal* (C). **Map 91, Fig. 14.**

Terrace-walls, cliffs and rocky slopes; 150–2700m.

Saudi Arabia, Yemen (N & S), Oman. NE tropical Africa.

subsp. **virgata** (Roxb.) Browicz in Fl. Iranica 153: 12 (1982). Syn.: ?*F. pseudosycomorus* Decne in Ann. Sci. Nat. Bot. Sér. 2 (2): 242 (1834). Illustr.: Fl. Iranica 153: t.6 (1982); Collenette (1985 p.368) as *F. carica* forma; Fl. Palaest. 1: t.34 (1966). **Map 92, Fig. 14.**

In drifted sand against cliffs and in narrow ravines; 950–2350m.

Saudi Arabia. Sinai, Palestine and Jordan; Iran, Afghanistan, Pakistan and NW India.

3. F. johannis Boiss., Diagn. pl. orient. sér 1, 1 (7): 96 (1846). Syn.: *F. carica* sensu Western (1989) non L.; *F. geraniifolia* Miq. in Hook., J. Bot. 7: 255 (1848). Illustr.: Western (1989 p.31) as *F. carica*; Fl. Iranica 153: t.5 (1982).

Shrub or small tree up to 8m. Sap milky. Leaves 3–5-lobed, rarely simple, sometimes highly dissected with ± linear lobes, ovate in outline, 1–7 × 1–7cm, truncate or cordate at the base, scabrous or smooth; lobes acute, obtuse or rounded, serrate or sinuately lobed; petioles 3–30mm. Figs 1–2 in the leaf axils, obovoid, 1–2cm diam., green ripening purple; peduncles 5–20mm, bearing a whorl of 3–4 bracts. Ostiole round, the ostiolar bracts visible. **Map 93, Fig. 14.**

Rocky slopes, cliffs, ravines, walls, rocky wadi-beds and field margins; 100–2500m.

Oman, UAE. Iran, Afghanistan, Pakistan and adjacent regions of C Asia.

F. johannis shows considerable variation in leaf shape, often with deeply dissected and more or less entire leaves on the same plant. It is frequently heavily grazed. A note on a herbarium specimen from Oman (*Whitcombe* 422) records that the sap is supposed to cause an allergic skin reaction. Arabian material is all referable to subsp. *johannis*.

4. F. exasperata Vahl, Enum. pl. 2: 197 (1805). Syn.: *F. serrata* Forsskal (1775 p.179). Illustr.: Fl. Ethiopia 3: 284 (1989).

Shrub or tree up to 25m. Sap watery. Leaves entire or the juvenile foliage 3–5-lobed, ovate to elliptic or obovate in outline, 5–27 × 4–18cm, acute or obtuse to shortly acuminate, entire or repand to subserrate, rounded to cuneate or subcordate at the base, scabrous; petiole 0.5–4cm. Stipules amplexicaul, caducous. Figs 1–2 in the leaf axils, globose to obovoid, 1–1.5cm diam., scabrous, yellow, orange or red; peduncle 0.5–1(–1.5)cm; bracts scattered on the peduncles and the surface of the fig. Ostiole round, the ostiolar bracts visible. **Map 94, Fig. 14.**

Riverine woodland and deciduous shrubland; 650–1700m.

Yemen (N). Tropical Africa, S India and Sri Lanka.

5. F. sycomorus L., Sp. pl.: 1059 (1753). Syn.: *F. chanas* Forsskal (1775 p.219). Illustr.: Collenette (1985 p.370).

Tree up to 20m. Sap milky. Twigs and petioles minutely puberulous and also with long straight hairs. Leaves simple, ovate to subcircular or rarely lobed (see note below), 3–14 × 2.5–11cm, rounded to obtuse, entire or repand, cordate to rounded at the base, scabrous or smooth above, pubescent to glabrescent beneath; petioles 0.5–7cm. Stipules amplexicaul, free, caducous. Figs on leafless branchlets borne on the trunk and larger branches, obovoid to subglobose, (1.5–)2–3cm diam., puberulous to tomentose, yellowish to reddish when ripe; peduncles 0.5–2.5cm, bearing 3 basal bracts. Ostiole rounded, with several ostiolar bracts visible. **Map 95, Fig. 15.**

Cliffs and rocky slopes, often near water; 100–1800m.

Saudi Arabia, Yemen (N & S), Oman. Tropical and southern Africa to Egypt and Syria.

A specimen from Oman (*Radcliffe-Smith* 5330) is unusual in having lobed leaves (resembling those of *F. palmata*) on the sucker-shoots.

6. F. sur Forsskal (1775 p.180). Syn.: *Sycomorus sur* (Forsskal) Miq. in Verh. Kon. Akad. Wet. Amster. Nat., ser. 3 (1): 121 (1849); *Ficus capensis* Thunb., Ficus: 13 (1786). Illustr.: Fl. Ethiopia 3: 288 (1989); Fl. Trop. E. Afr.: 57 (1989). Type: Yemen (N), *Forsskal* (C).

Tree up to 20m. Sap milky. Twigs and petioles puberulous, tomentose or hirsute. Leaves ovate to elliptic or lanceolate, 4–22 × 3–9cm, acute to acuminate, entire or repand, rounded to cordate at the base, glabrous above, sparsely puberulous or glabrous beneath; petioles 1–8cm. Stipules amplexicaul, free, caducous. Figs on leafless branchlets borne on the trunk and larger branches, obovoid to subglobose, 2–4cm diam., glabrous to densely tomentose, orange-grey when ripe; peduncles 0.5–

2cm, bearing a whorl of 3 basal bracts. Ostiole rounded, with several ostiolar bracts clearly visible. **Map 96, Fig. 15.**

Riverine woodland and fields margins; 300–2700(–3000)m.

Yemen (N). Tropical and southern Africa.

7. F. cordata Thunb. subsp. **salicifolia** (Vahl) C.C. Berg in Kew Bull. 43: 82 (1988). Syn.: *F. ambiguum* Forsskal (1775 p.219); *F. indica* sensu Forsskal non L.; *F. salicifolia* Vahl (1790 p.82 t.23). Illustr.: Fl. Ethiopia 3: 290 (1989); Collenette (1985 p.369); Western (1989 p.31) all as *F. salicifolia*. Type: Yemen (N), *Forsskal* 780 (C).

Shrub or small tree up to 8m. Sap milky. Leaves lanceolate, 2–16(–27) × 1–3(–6)cm, acute to acuminate, entire, rounded or slightly cordate at the base, glabrous; petioles 1–5cm. Stipules fully amplexicaul, free. Figs 1–3 in the leaf axils, globose, 7–10mm diam., glabrous to puberulous, white to purplish when ripe, sessile or pedunculate; peduncles up to 4mm, bearing 3 basal bracts. Ostiole circular, with 3 bracts visible. **Map 97, Fig. 15.**

Rocks and cliffs, often near water; 0–2500m.

Saudi Arabia, Yemen (N & S), Socotra, Oman, UAE. Distributed throughout Africa.

8. F. ingens (Miq.) Miq. in Ann. Mus. Bot. Lugduno-Batavum 3: 288 (1867). Syn.: *F. ingentoides* Hutch. in Bull. Misc. Inform 1915: 319 (1915); *F. lutea* sensu Miller & Morris (1988) and Schwartz (1939) non Vahl. Illustr.: Fl. Trop. E. Afr.: 62 (1989); Fl. Ethiopia 3: 291 (1989); Collenette (1985 p.369).

Shrub or small tree up to 10m. Sap milky. Leaves narrowly ovate to oblong-lanceolate, 7–22 × 4–10cm, acute to acuminate, entire, cordate or rounded at the base, glabrous; petioles 1–5cm; stipules fully amplexicaul, free, caducous. Figs 1–2 in the leaf axils, globose, 10–15mm diam., glabrous to tomentose, white to pink when ripe, sessile or pedunculate; peduncles up to 4mm, bearing 3 basal bracts. Ostiole circular, with 3 bracts visible. **Map 98, Fig. 15.**

Boulders, cliffs and rocky slopes, in drought-deciduous woodland; (20–)150–2300m.

Saudi Arabia, Yemen (N & S), Oman. Tropical and southern Africa.

9. F. glumosa Del., Voy. Méroé: 63 (1826). Syn.: *F. glumosa* Del. var. *glaberrima* Martelli, Fl. bogos.: 76 (1886). Illustr.: Fl. Ethiopia 3: 293 (1989); Collenette (1985 p.368).

Shrub or tree up to 30m, starting as an epiphyte or on rocks. Sap milky. Leaves oblong-elliptic or oblong-obovate, 5–15 × 3–8cm, rounded or shortly acuminate, entire, cordate or rounded at the base, glabrous or thinly pubescent above, glabrescent to densely pubescent beneath; petioles 2–4cm. Stipules fully amplexicaul. Figs 1–2 in the leaf axils, globose, 6–7(–10)mm diam., glabrous to densely tomentose, pink or red when ripe; peduncles up to 3mm, bearing 2(–3) basal bracts. Ostiole slit-shaped, 2-lipped, with no bracts visible. **Map 99, Fig. 16.**

4. Ficus

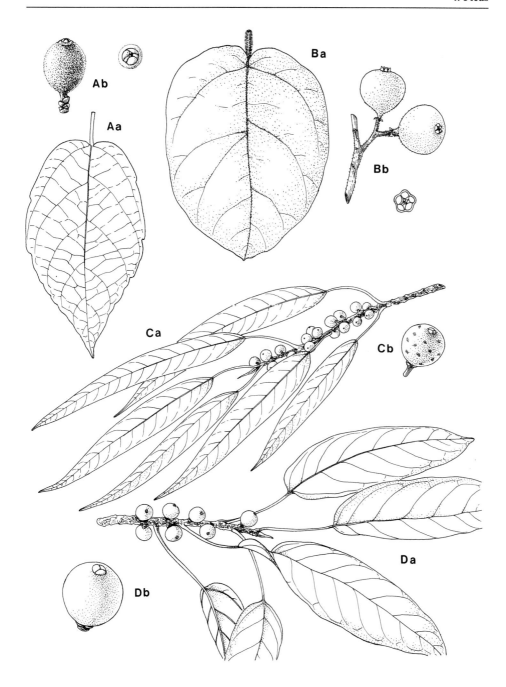

Fig. 15: Moraceae. A, *Ficus sur*: Aa, leaf (×0.6); Ab, fig and ostiole (×1.5). B, *F. sycomorus*: Ba, leaf (×0.6); Bb, figs and ostiole (×0.6). C, *F. cordata* subsp. *salicifolia*: Ca, fruiting branch (×0.6); Cb, fig (×1.5). D, *F. ingens*: Da, fruiting branch (×0.6); Db, fig (×1.5).

On boulders, cliffs and rocky slopes in drought-deciduous woodland; 5–1300m.

Saudi Arabia, Yemen (N). Tropical and southern Africa.

10. F. populifolia Vahl (1790 p.82 & t.22). Syn.: *F. religiosa* sensu Forsskal non L. Illustr.: Fl. Ethiopia 3: 295 (1989); Collenette (1985 p.369). Type: Yemen (N), *Forsskal* (C).

Tree up to 12m. Sap milky. Leaves broadly ovate to cordate, 5–20 × 5–16cm, long acuminate, entire, cordate at the base, glabrous; petioles 5–14cm. Stipules fully amplexicaul, caducous. Figs 1–2 in the leaf axils, subglobose, 0.5–1(–1.5)cm diam., minutely puberulous to glabrous, green with red spots when ripe; peduncles 0.5–2cm, bearing 2 bracts. Ostiole 3-lipped, with no bracts visible. **Map 100, Fig. 16.**

Rocky slopes and ravines in drought-deciduous woodland; 5–2400m.

Saudi Arabia, Yemen (N & S). W, C, E, and NE tropical Africa.

11. F. vasta Forsskal (1775 p. 179). Syn.: *F. socotrana* Balf. f. (1883 p.96). Illustr.: Fl. Ethiopia 3: 297 (1989); Collenette (1985 p.370). Type: Yemen (N) *Forsskal* (C).

Tree up to 20m, starting as an epiphyte or on rocks, often with aerial roots. Sap milky. Leaves broadly ovate to suborbicular, 7–30 × 6–20cm, rounded or shortly acuminate, entire, cordate at the base, glabrous to glabrescent above, glabrescent to velutinous beneath; petioles 4–10cm, with a large area of glandular tissue at the base of the midrib beneath. Stipules fully amplexicaul, caducous. Figs 1–2 in the leaf axils, subglobose, 1–1.5(–2)cm diam., tomentose to glabrescent, green with paler spots when ripe; sessile or pedunculate; peduncles up to 6mm, bearing a whorl of basal bracts which fall leaving a hairy rim; ostiole prominent, 2-lipped. **Map 101, Fig. 16.**

Cliffs and terrace-walls; drought-deciduous bushland and wooded grassland; 350–2300m.

Saudi Arabia, Yemen (N & S), Socotra, Oman. E & NE tropical Africa.

Species incompletely known

F. taab Forsskal (1775 p.219).

Family 27. CANNABACEAE

A.G. MILLER

Dioecious annual or perennial herbs. Stems erect or climbing. Leaves alternate or opposite below, simple or palmately divided; stipules free or fused. Male flowers in pendulous axillary panicles; tepals 5, free; stamens opposite the tepals. Female flowers in short erect leafy spikes; perianth unlobed, thin, membranous, enveloping the ovary;

4. Ficus

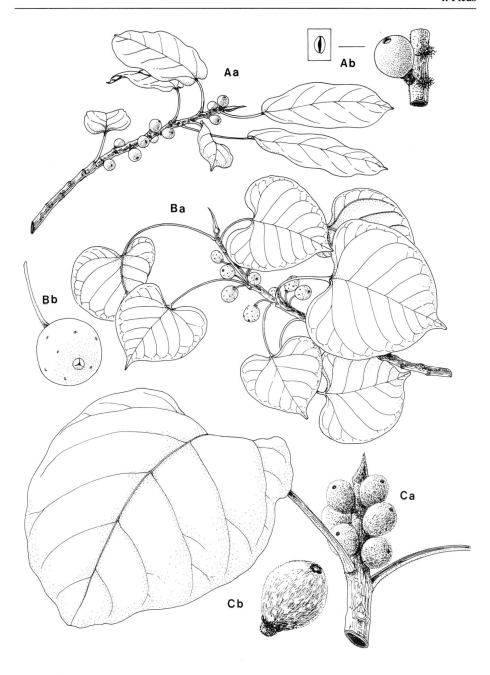

Fig. 16. Moraceae. A, *Ficus glumosa*: Aa, fruiting branch (×0.6); Ab, fig and ostiole (×2). B, *F. populifolia*: Ba, fruiting branch (×0.6); Bb, fig (×2). C, *F. vasta*: Ca, leaf and figs (×0.6); Cb, fig (×1.25).

ovary superior, 1-locular with a solitary pendulous ovule; style short; stigmas 2, filiform. Fruit an achene, enclosed in the persistent perianth.

CANNABIS L.

Description as for the family.

1. C. sativa L., Sp. pl.: 1027 (1753). Illustr.: Fl. Ethiopia 3: 328 (1989).

Herb up to 1(–5)m. Leaves 3–7(–11)-foliolate; leaflets narrowly lanceolate, 2.5–15 × 0.3–1.5cm, acuminate, serrate. Male inflorescence up to 18cm; tepals c.3mm. Female inflorescence shorter, resinous-glandular. Achene ovoid, 3–3.5mm. **Map 102.**

Cultivated.

Yemen (N). Cultivated throughout the world.
  Grown as a source of fibre and oil.

Family 28. URTICACEAE

A.G. MILLER

Monoecious or dioecious herbs or shrubs, often bearing stinging hairs. Leaves alternate or opposite, entire or variously toothed, simple or lobed, pinnately or palmately 3-nerved from the base; stipules usually present, free or fused. Flowers minute, usually unisexual, actinomorphic or zygomorphic, 2–5-merous, arranged in axillary clusters, panicles or cymes, globose heads or on receptacles. Male flowers 2–5-merous; perianth deeply lobed or with free segments; stamens opposite to and equal in number to the perianth lobes, sometimes a rudimentary ovary present. Female flowers 3–5-merous; lobes equal or unequal, free or fused, sometimes absent; ovary superior, 1-locular with a solitary basal ovule; style absent or very short; stigma filiform or brush-like. Fruit an achene, free or enclosed in the persistent, dry or fleshy perianth.

1.	Leaves opposite	2
+	Leaves alternate or rarely opposite towards the base of the stem	5
2.	Leaves entire	**5. Pouzolzia**
+	Leaves crenate-serrate to coarsely dentate	3
3.	Plants clothed in stinging hairs; leaves coarsely dentate	**1. Urtica**
+	Plants lacking stinging hairs; leaves crenate-serrate	4
4.	Trailing herb with stems up to 1.5m; upper leaves not in a whorl; stipules lateral, clearly free to the base	**9. Droguetia**
+	Delicate herb with stems up to 10cm, forming dense carpets; upper pairs of leaves so crowded as to appear almost whorled; stipules intrapetiolar, connate to the tip	**4. Pilea**

5.	Shrubs	6
+	Herbs	7
6.	Flowers in dense globose heads; leaves crenate, never entire, densely woolly beneath	**6. Debregeasia**
+	Flowers in axillary clusters; leaves entire and woolly beneath, if crenate then thinly hairy beneath	**5. Pouzolzia**
7.	Plants clothed with stinging hairs	8
+	Plants lacking stinging hairs	9
8.	Plant, especially in the region of the inflorescence, clothed with stout, spinescent hairs up to 5mm long; female perianth sac-like; leaves often lobed	**3. Girardinia**
+	Plant with weak hairs up to 3mm long; female perianth of 2 pairs of ± free, unequal segments; leaves simple	**2. Laportea**
9.	Flowers arranged on axillary receptacles	10
+	Flowers arranged in axillary clusters	11
10.	Receptacles densely woolly within; flowers surrounded by an involucre of 2–8 free bracts	**8. Forsskaolea**
+	Receptacles glabrous within; flowers surrounded by an involucre of bracts fused almost to the apex, giving a toothed margin	**9. Droguetia**
11.	Diffusely branched prostrate or ascending herb; stipules absent	**7. Parietaria**
+	Shrub or erect herb; stipules present	**5. Pouzolzia**

1. URTICA L.

Annual or perennial, monoecious or rarely dioecious herbs, clothed with stinging hairs. Leaves opposite, dentate to serrate; stipules free. Inflorescence of axillary clusters, spike-like panicles or globose heads. Flowers 4-merous with free tepals. Male flowers with the tepals subequal; stamens 4; rudimentary ovary present. Female flowers with 2 unequal pairs of tepals, the inner pair membranous and enclosing the achene in fruit. Achenes smooth.

1.	Female and male flowers mixed in spike-like inflorescences	**1. U. urens**
+	Female flowers in globose, pedunculate heads, the male in lax, axillary panicles	**2. U. pilulifera**

1. U. urens L., Sp. pl.: 984 (1753). Illustr.: Collenette (1985 p.492).

Erect, monoecious annual, 15–50cm. Leaves ovate, 1–5 × 0.5–3.5cm, acute or obtuse, coarsely dentate, rounded or obtuse at the base; petioles 1–2cm. Panicles bisexual, spike-like, c.1cm long, clustered in the leaf axils. **Map 103.**

Weed of cultivation and waste places; 20–3000m.

Saudi Arabia, Yemen (N & S), Socotra, UAE, Qatar, Kuwait. Widespread in temperate regions and on tropical mountains.

2. U. pilulifera L., Sp. pl.: 983 (1753). Illustr.: Fl. Iraq 4 (1): 95 (1980); Collenette (1985 p.492).

Erect, monoecious annual; stems simple or branched from the base, 30–60cm. Leaves ovate, 2–8 × 2–6cm, acute, coarsely dentate, truncate at the base; petioles 1–5cm. Male panicles lax, 2–5cm. Female flowers in globose heads; heads pedunculate, c.0.5–1.5cm diam. **Map 104, Fig. 17.**

Weed of cultivation and waste places; 1700–2200m.

Saudi Arabia, Yemen (N). Europe, N and tropical NE Africa, W Asia.

2. LAPORTEA Gaudich.

Monoecious, annual herbs clothed with stinging hairs. Leaves alternate or rarely opposite, dentate or serrate; stipules connate. Inflorescences bisexual, axillary panicles. Flowers green, 4–5-merous, sessile or shortly pedicellate. Male flowers: perianth lobes equal, often hooded at the apex; 3–5 stamens; rudimentary ovary present. Female flowers: perianth lobes ± free, unequal, the lateral pair enlarged; stigma sessile, filiform, simple or 3-fid. Achenes ovoid, compressed, with a raised marginal ridge enclosing a warty depression, dispersed with the persistent perianth.

Chew, W.-L. (1969). A monograph of Laportea. *Gard. Bull. Straits Settlem.* 25: 111–178.

1. Inflorescence of remote clusters of flowers along an unbranched peduncle or of axillary cymes; stems and inflorescences thinly hairy with stinging hairs; stigmas 3-fid   **1. L. interrupta**
+ Inflorescence a branched panicle; stems and inflorescences densely covered with long glandular hairs and shorter stinging hairs; stigma simple   **2. L. aestuans**

1. L. interrupta (L.) Chew op. cit. 21: 200 (1965). Illustr.: Chew, op. cit. 25: 147 (1969).

Erect or ascending annual herb, sparsely clothed with stinging hairs; stems usually simple, up to 0.1(–1.5)m. Leaves alternate or rarely opposite, triangular-ovate, 2–12 × 2–7cm, acuminate, crenate-serrate, truncate or rounded at the base; petioles 1.5–5(–12)cm. Inflorescence 5–30cm, the flowers in remote clusters along an unbranched, elongated peduncle or (in young plants) in ± axillary cymes. Stigma 3-fid. **Map 105, Fig. 17.**

Damp, shady areas in drought-deciduous *Anogeissus* woodland; 150–1000m.

Oman. Old World tropics.

2. L. aestuans (L.) Chew op. cit. 21: 200 (1965). Syn.: *Fleurya aestuans* (L.) Gaudich. in Freyc., Voy. Uranie: 497 (1830); *Urtica divaricata* Forsskal (1775 p.160). Illustr.: Chew, op. cit. 25: 165 (1969).

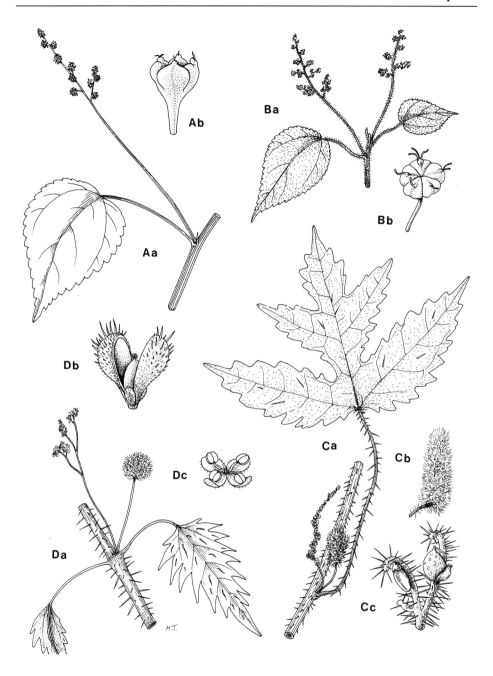

Fig. 17. Urticaceae. A, *Laportea interrupta*: Aa, leaf and inflorescence (×0.6); Ab, male flower (×10). B, *L. aestuans*: Ba, leaf and inflorescence (×0.6); Bb, male flower (×10). C, *Girardinia diversifolia*: Ca, leaf and inflorescence (×0.5); Cb, infructescence (×0.5); Cc, detail of infructescence (×5). D, *Urtica pilulifera*: Da, shoot with male and female inflorescences (×1); Db, female flower (×10); Dc, male flower (×10).

Erect annual herb; stems simple or branched, up to 50(–100)cm, clothed with long glandular and shorter stinging hairs. Leaves ovate, 2–6(–15) × 1–4(–12)cm, acuminate, serrate-dentate, rounded at the base; petiole 2–6(–15)cm, thinly hairy with stinging hairs. Inflorescence a branched panicle, 5–10(–20)cm, glandular-hairy. Stigma simple. **Map 106, Fig. 17.**

Weed of coffee plantations and in damp, shady places; 600–1200m.

Yemen (N). Pantropical.

3. GIRARDINIA Gaudich.

Monoecious or dioecious herbs, bearing long stinging hairs throughout. Leaves alternate, coarsely serrate, simple or 3–5(–7)-lobed, 3-nerved from the base; stipules connate, 2-fid. Male flowers in dense clusters along spike-like panicles; perianth of 4–5 free segments; stamens 4–5. Female flowers in clusters along condensed spike-like or branched axillary panicles, the clusters fiercely armed with stinging hairs; perianth fused, sac-like, enclosing the ovary, with or without a membranous tepal; stigma sessile, filiform. Achene flattened, suborbicular.

Friis, I. (1981). A synopsis of *Girardinia* (Urticaceae). *Kew Bull.* 36: 143–157.

1. G. diversifolia (Link) I. Friis op. cit.: 145 (1981). Syn.: *G. condensata* (Steud.) Wedd. in Ann. Sci. Nat. Bot., Sér. 4, 1: 181 (1854); *Urtica palmata* Forsskal (1775 p.159). Illustr.: I. Friis, op. cit.: 150; Fl. Trop. E. Afr.: 14 (1989); Fl. Ethiopia 3: 307 (1989). Type: [*Urtica palmata*] Yemen (N), *Forsskal* (C).

Erect herb, up to 2m tall. Leaves elliptic to broadly ovate, simple or shallowly to deeply lobed, 5–15 × 2.5–15cm, rounded to cordate at the base; petioles 2–12cm; stipules ovate. Flowers c.1.5mm diam. Achenes blackish, c.2.5mm diam. **Map 107, Fig. 17.**

900–1500m.

Yemen (N). Tropical Africa and Asia W to Indonesia and Korea.

Known only from a single area, on J. Raymah in the SW escarpment mountains, where it is locally frequent. *G. bullosa* (Steud.) Wedd. from tropical NE Africa may possibly occur in Arabia; it differs from *G. diversifolia* in its larger and broader leaves and in the presence of a horn-like appendage on the tepals of the male flowers.

4. PILEA Lindley

Monoecious or dioecious, annual herbs. Leaves opposite, crenate-serrate, with linear cystoliths. Stipules intrapetiolar, connate to the tip. Flowers minute, in axillary cymes or panicles. Male flowers: perianth segments 2–4, free; stamens 2–4. Female flowers 3–4-lobed, the median lobe larger than the laterals; stigma sessile, brush-like. Achenes ovoid, compressed.

Friis, I. (1989). A revision of *Pilea* in Africa. *Kew Bull.* 44(4): 557–600.

1. P. tetraphylla (Steud.) Blume, Mus. bot. 2: 50 (1856). Illustr.: Fl. Trop. E. Afr.: 28 (1989); Fl. Ethiopia 3: 311 (1989).

Dioecious herb, often forming dense carpets; stems erect, up to 10cm, glabrous. Leaves ovate, 7–30 × 5–20mm, acute or obtuse, broadly cuneate at the base, the upper 2 pairs crowded and appearing ± whorled, glabrous or sparsely hairy. Flowers in dense axillary panicles forming dense ± cruciform terminal panicles of mainly female flowers; male inflorescences in the axils of the lower leaves. Achenes c.0.75mm. **Map 108, Fig. 18.**

Wet and shady boulders and tree trunks in drought-deciduous *Anogeissus* woodland; 200–850m.

Oman. Tropical Africa and Madagascar.

5. POUZOLZIA Gaudich.

Monoecious herbs or shrubs. Leaves alternate, rarely opposite, entire or crenate-serrate; stipules free. Flowers in axillary, bisexual clusters. Male perianth 4–5-lobed; stamens 4–5. Female perianth tubular, enclosing the ovary, toothed at the apex; stigma filiform, deciduous. Achene enclosed in the persistent perianth.

Friis, I. & Jellis, S. (1984) A synopsis of *Pouzolzia* in tropical Africa. *Kew Bull.* 39: 587–601.

1.	Leaves crenate-serrate	**3. P. parasitica**
+	Leaves entire	2
2.	Leaves white-tomentose beneath	**1. P. mixta**
+	Leaves pubescent beneath	**2. P. auriculata**

1. P. mixta Solms-Laub. in Schweinf., Beitr. Fl. Aethiop.: 188 (1867). Syn.: *P. arabica* Deflers (1889 p.206); *P. hypoleuca* Wedd. in DC., Prodr. 16(1): 227 (1869). Illustr.: Fl. Ethiopia 3: 318 (1989).

Shrub up to 2m. Leaves alternate, ovate, 2–7 × 1.5–4.5cm, acuminate, entire, cuneate to rounded at the base, densely and shortly hairy above, densely and shortly white-woolly beneath especially when young; petioles 1–3cm. Flower clusters c.2–3mm across. Male flowers numerous; perianth campanulate, (4–)5-lobed; female flowers few. **Map 109, Fig. 18.**

Cliffs, terrace-walls and rocks; 600–1500m.

Yemen (N). Widespread on the drier mountains of tropical Africa.

2. P. auriculata Wight, Ic. pl. Ind. orient. 6: 10, t.1980 (1853).

Herb up to 75cm. Leaves alternate or rarely opposite, narrowly to broadly ovate, 2.5–10(–12) × 0.75–4cm, acute or acuminate, entire, cuneate to rounded at the base, pubescent; petioles 2.5–4(–7)cm. Flower clusters c.5mm across. Male perianth 5-lobed. **Map 110.**

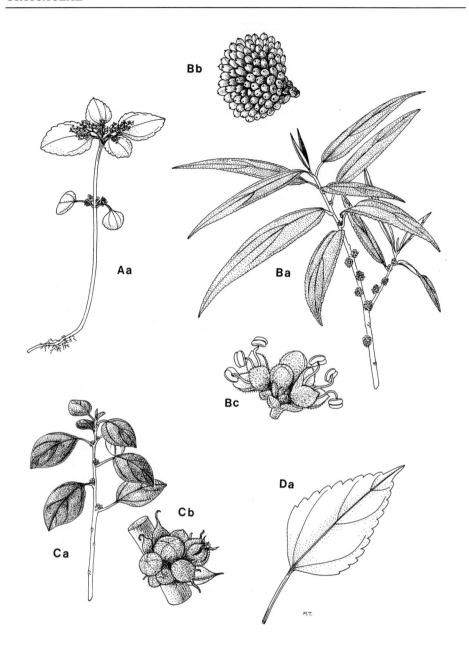

Fig. 18. Urticaceae. A, *Pilea tetraphylla*: Aa, habit (×1). B, *Debregeasia saeneb*: Ba, flowering shoot (×0.6); Bb, female inflorescence (×5); Bc, male inflorescence (×4). C, *Pouzolzia mixta*: Ca, leafy shoot (×0.3); Cb, inflorescence (×15). D, *P. parasitica*: Da, leaf (×0.5).

Amongst rocks in evergreen bushland; 650–850m.

Socotra. India and Sri Lanka.

3. P. parasitica (Forsskal) Schweinf. (1896 p.145). Syn.: *Urtica parasitica* Forsskal (1775 p.160); *Urtica muralis* Vahl (1790 p.77) nom. illegit. Illustr.: Fl. Ethiopia 3: 318 (1989). Type: Yemen (N), *Forsskal* 770 (C).

Shrub up to 1(–2)m. Leaves alternate, ovate to broadly ovate, 4–11 × 2–7cm, acuminate, crenate-serrate becoming entire towards the base, cuneate to rounded at the base, 3-nerved, sparsely hispidulous above, thinly pubescent beneath especially on the nerves; petioles 2–5cm. Flower clusters 5–10mm across. Male flowers numerous; perianth campanulate, (3–)4-lobed; female flowers few. **Map 111, Fig. 18.**

On rocks in damp gullies; 1000–2000m.

Yemen (N). Widespread in the mountains of tropical Africa.

6. DEBREGEASIA Gaudich.

Monoecious shrubs. Leaves alternate, finely serrate, 3-veined at the base; stipules connate, 2-fid. Flowers in dense globose heads; heads usually unisexual, axillary, sessile or in shortly stalked clusters. Male flowers: perianth 3–4(–5)-lobed; stamens equal in number to the perianth lobes. Female flowers much smaller than the male; perianth ovoid becoming fleshy in fruit, minutely toothed at the apex; stigma sessile, brush-like. Achenes enclosed in the succulent perianths.

1. D. saeneb (Forsskal) Hepper & J.R.I. Wood in Kew Bull. 38: 86 (1983). Syn.: *D. bicolor* (Roxb.) Wedd. in A.DC., Prodr. 16: 235[25] (1869); *D. salicifolia* (D. Don) Rendle in Prain, Fl. Tr. Afr. 6 (2): 295 (1916–17); *Rhus saeneb* Forsskal (1775 p.206). Illustr.: Fl. Ethiopia 3: 319 (1989). Neotype: Yemen (N), *Schweinfurth* 436 (K).

Shrub or small tree up to 5m. Leaves lanceolate, 3–10 × 0.5–1.5cm, acuminate, acute or rounded at the base, sparsely hairy becoming glabrous above, densely white woolly beneath; petioles up to 1cm. Flowering heads 3–5mm diam. Male flowers pink; perianth lobes broadly ovate-triangular, c.1 × 1.25mm. Female flowers c.1mm long. Fruiting heads globose, 4–8mm diam. **Map 112, Fig. 18.**

Rocky slopes and stream sides; 600–2450m.

Saudi Arabia, Yemen (N). Ethiopia and the western Himalayas.

7. PARIETARIA L.

Monoecious annual or perennial herbs. Leaves alternate, simple, entire; stipules absent. Flowers green, (3–)4-merous, bracteate, in 1–several-flowered axillary clusters, unisexual or bisexual. Stigma sessile, brush-like, protruding from the perianth. Achene ovoid, enclosed in the persistent perianth.

1. Perianths of the female flowers covered in clavate hairs; bracts of the female flowers strongly accrescent and hiding the flowers in fruit   **1. P. alsinifolia**
+ Perianth of female flowers with straight or hooked (never clavate) hairs; bracts of female flowers not strongly accrescent, not hiding the flowers in fruit   2

2. Fruiting perianths of the bisexual and female flowers similar, cup-shaped with erect lobes; bracts often glandular; usually a weak-stemmed plant with rather inconspicuous flowers   **2. P. debilis**
+ Fruiting perianths of bisexual flowers strongly accrescent, tubular, with inflexed lobes; fruiting perianths of female flowers ovoid, with connivent lobes; bracts not usually glandular; plants usually not particularly weak-stemmed, the flowers conspicuous   3

3. Perianths of fruiting bisexual flowers 1.8–2mm long; perianths of female flowers covered with long spreading hairs at the lobes and with round papillae towards the base; fruits narrowly ovoid, c.0.8 × 0.4mm   **3. P. umbricola**
+ Perianths of fruiting bisexual flowers 2.8–3.5mm long; perianths of female flowers thinly white-hairy; fruits ovoid, c.1 × 0.7mm   **4. P. judaica**

1. P. alsinifolia Delile, Descr. Egypte, Hist. nat. 137 (1813). Illustr.: Fl. Iraq 4 (1): 96 (1980).

Annual herb; stems weak, prostrate, ascending or erect, 10–45cm, villous. Leaves ovate to circular, 0.5–35 × 0.5–33mm, acute or obtuse and often constricted towards the tip, rounded at the base, thinly hairy; petioles up to 40mm. Flowers in axillary clusters of female and bisexual flowers. Bracts linear-lanceolate to broadly ovate, often cordate at the base, equalling or longer than the flowers, strongly accrescent and enclosing the female flowers in fruit. Male flowers apparently absent. Female flowers ovoid with ovate lobes, 1–1.2 × c.0.8mm, covered with white clavate hairs, slightly accrescent in fruit; fruiting perianths ovoid with the lobes connivent at the tip. Bisexual flowers cup-shaped with narrowly oblong-ovate lobes, c.1.25mm, externally thinly hairy on the lobes and glabrous or papillose below, slightly accrescent in fruit; fruiting perianth cup-shaped with erect lobes, 1.5–2mm long. Fruit reddish brown, ovoid, 0.9–1 × 0.4–0.6mm, enclosed in membranous bracts. **Map 113, Fig. 19.**

In shade under trees, rock crevices and amongst boulders; 0–2150m.

Saudi Arabia, Oman, UAE, Qatar. N Africa, SW and C Asia.

The strongly accrescent fruiting bracts of *P. alsinifolia* usually readily distinguish it from the other Arabian species. The bracts are very variable in size, shape and texture: varying from almost linear-lanceolate to broadly ovate, often with a distinct cordate base and either thin- or relatively thick-textured.

2. P. debilis G. Forster, Fl. ins. austr.: 73 (1786). Illustr.: Fl. Trop. E. Afr.: 53 (1989); Fl. Ethiopia 3: 321 (1989).

Annual or perennial herb; stems weak, prostrate, erect or ascending, up to 50cm, pubescent. Leaves broadly ovate to circular, 0.5–20 × 0.3–15mm, acute or rounded and often constricted towards the tip, rounded or subcordate at the base, thinly hairy;

petioles 5–15(–25)mm. Flowers in 1–several-flowered axillary clusters of mainly female and bisexual flowers. Bracts lanceolate, equalling or shorter than the flowers, often glandular-hairy. Male flowers rare. Female flowers ovoid. Bisexual flowers cup-shaped. Perianths of male and bisexual flowers c.1mm long, with ovate lobes, covered in straight and hooked hairs, slightly accrescent in fruit; fruiting perianths cup-shaped with erect lobes. Fruit brownish, ovoid, c.1 × 0.6mm. **Map 114, Fig. 19.**

Shady rock crevices; (550–)2000–3000m [the lower altitude is for Socotra].

Yemen (N & S), Socotra. Tropical Africa, Madagascar, tropical Asia, Australia, New Zealand and South America.

Australina capensis Wedd., a southern African species, was recorded from Socotra in the last century. The specimen upon which the record was based cannot be traced but the occurrence of this species on Socotra seems highly unlikely. It was possibly mistaken for *Parietaria debilis* which it superficially resembles.

3. P. umbricola A.G. Miller in Edinb. J. Bot. 51 (1): 46 (1994). Type: Yemen (N), *Miller* 190 (E).

Annual or perennial herb; stems prostrate or ascending, up to 25cm, villous. Leaves ovate to broadly elliptic or subcircular, 5–20 × 5–13mm, acute or obtuse and often constricted towards the tip, cuneate to rounded at the base, pubescent to villous; petioles 3–18mm. Flowers ± sessile in axillary clusters, the clusters typically consisting of a solitary bisexual flower surrounded by several female flowers. Bracts free or shortly fused, ovate to narrowly oblong-ovate, ± equalling or shorter than the flowers. Male flowers apparently absent. Female flowers ovoid with narrowly ovate lobes, c.0.8 × 0.5mm, externally with long spreading hairs above and round papillae below, accrescent; fruiting perianth ovoid, 1.3–1.5mm long, the lobes connivent at the tip. Bisexual flowers cup-shaped with ovate-triangular lobes, c. 1 × 1–1.2mm, externally with long spreading hairs above and glabrous below, accrescent; fruiting perianth tubular with inflexed lobes, 1.8–2mm long. Fruits reddish brown, narrowly ovoid, c.0.8 × 0.4mm. **Map 115, Fig. 19.**

In shade on cliffs and amongst boulders; 1200–2200m.

Saudi Arabia, Yemen (N). Endemic.

4. P. judaica L., Fl. Palaest.: 32 (1756). Illustr.: Fl. Iraq 4 (1): 96 (1980).

Perennial herb; stems ascending or procumbent, up to 30cm. Leaves broadly ovate to elliptic, 10–35(–70) × 5–20(–30)mm, acute to bluntly acuminate, cuneate to truncate at the base, villous; petioles 2–10(–25)mm. Flowers sessile in axillary clusters, the clusters typically consisting of a solitary female flower and several bisexual flowers. Bracts free or shortly fused, ovate, ± equalling or shorter than the flowers. Male flowers rare. Female flowers ovoid with narrowly ovate lobes, c.1.5 × 0.5mm, externally white-hairy, scarcely accrescent; fruiting perianths ovoid, the lobes connivent and forming a narrow neck. Bisexual flowers cup-shaped with ovate-triangular lobes, c. 1 × 1–1.3mm, externally white-hairy, accrescent; fruiting perianths tubular with inflexed lobes, 2.8–3.5mm long. Fruits black or dark brown, ovoid, c.1.0 × 0.7mm. **Map 116, Fig. 19.**

URTICACEAE

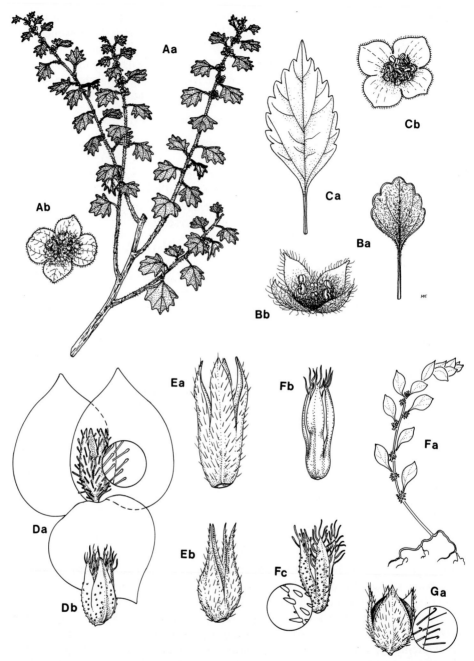

Fig. 19. Urticaceae. A, *Forsskaolea griersonii*: Aa, flowering branch (× 0.6); Ab, inflorescence (× 4). B, *F. tenacissima*: Ba, leaf (× 1); Bb, inflorescence (× 4). C. *F. viridis*: Ca, leaf (× 1); Cb, inflorescence (× 4). D. *Parietaria alsinifolia*: Da, fruiting female flower and enlarged bracts (× 25) with detail of hairs; Db, fruiting hermaphrodite flower (× 25). E. *P. judaica*: Ea, fruiting hermaphrodite flower (× 25); Eb, fruiting female flower (× 25). F. *P. umbricola*: Fa, habit (× 0.6); Fb, fruiting hermaphrodite flower (× 25); Fc, fruiting female flower (× 25) with detail of hairs. G. *P. debilis*: Ga, fruiting hermaphrodite flower (× 25) with detail of hairs.

Rocky slopes and walls; 2000–2400m.

Yemen (N). Europe, N Africa, SW Asia and N India.
In Arabia known only from in and around the town of Manakhah.

8. FORSSKAOLEA L.

Monoecious annual or perennial herbs or shrublets, often clinging because of hooked hairs. Leaves alternate, dentate, pinnately-veined; stipules free. Flowers arranged on axillary receptacles; receptacles subsessile, densely woolly within, surrounded by an involucre of 2–8 bracts. Male flowers forming an outer ring; perianth 3-lobed; stamen 1. Female flowers in the centre of the receptacle; perianth absent; stigma sessile, filiform. Fruits enclosed in an involucre of bracts.

1. Bracts narrowly spathulate, 1–2mm broad **1. F. tenacissima**
+ Bracts broadly ovate to broadly obovate or orbicular, 3–5mm broad 2

2. Leaves glabrous or thinly scabrous, sometimes thinly woolly beneath, the margins crenate-serrate **2. F. viridis**
+ Leaves white-woolly beneath, scabrous above, with 2–3 acute teeth on either side **3. F. griersonii**

1. F. tenacissima L., Opobalsamum: 18 (1764). Syn.: *Caidbeja adhaerens* Forsskal (1775 p.82). Illustr.: Collenette (1985 p.490); Fl. Ethiopia 3: 323 (1989). Syntypes: Egypt and Yemen (N), *Forsskal* (C).

Annual or perennial herb, often woody below; stem erect or ascending, 10–100cm, thinly woolly and with long rigid and hooked hairs. Leaves ovate to broadly obovate or orbicular, 0.5–4.5 × 0.3–3.5cm, obtuse, serrate, cuneate at the base, shortly petiolate, discolorous, densely white-woolly beneath, green and with short hooked hairs above. Bracts narrowly spathulate, acute, 5–8 × 1–2mm, densely hairy with stiff hairs. **Map 117, Fig. 19.**

Dry rocky slopes, roadsides and disturbed ground; 10–2600m.

Saudi Arabia, Yemen (N & S), Oman, UAE, Qatar. N Africa, Sudan, Ethiopia, Palestine, Iran, Pakistan, Afghanistan and India.

2. F. viridis Ehrenb. in Hook., Niger Fl.: 179 (1849). Illustr.: Fl. Ethiopia 3: 323 (1989); Fl. Trop. E. Afr.: 55 (1989); Collenette (1985 p.491).

Annual herb; stems erect, up to 1m, scabridulous to tomentose with hooked hairs. Leaves ovate to rhombic, 1–9 × 1–4cm, acute, crenate-serrate, truncate or cuneate at the base, long-petiolate, concolorous, green, glabrous to thinly scabrid or sometimes thinly woolly beneath. Bracts broadly ovate to broadly obovate, 3–8 × 4–5mm, acute, stiffly hairy at the base, becoming membranous in fruit. **Map 118, Fig. 19.**

URTICACEAE

Wadi-bottoms, in the shade of boulders, rocky slopes, fossil coral etc.; 10–1370m.

Saudi Arabia, Yemen (S), Socotra, Oman. Tropical E & NE Africa; Namibia and Angola.

3. F. griersonii A.G. Miller in Edinb. J. Bot. 51 (1): 44 (1994). Type: Yemen (S), *Grierson* 152 (BM, E).

Subshrub; stems erect, up to 1m, scabrid with white, bulbous-based hairs. Leaves subsucculent, ovate to broadly ovate, 5–15 × 3–15mm, acute, the margins revolute with 2–3 acute teeth on either side, narrowing gradually below into a short petiole, discolorous, white-woolly beneath, green and scabrid above. Bracts 4, broadly obovate, acute, the margin entire or sinuate, 4 × 3–4mm, scabrid and with stiff hairs at the base. **Map 119, Fig. 19.**

Rocky volcanic slopes and *Euphorbia balsamifera*, shrubland; 100–1400m.

Yemen (S). Endemic.

Only known from J. al-'Urays, a frequently mist-covered volcanic massif east of Aden.

9. DROGUETIA Gaudich.

Monoecious, or sometimes dioecious, annual or perennial herbs or subshrubs. Leaves opposite or alternate, crenate-serrate, 3-veined at the base; stipules free. Flowers arranged in sessile axillary involucres (receptacles); involucres bowl-shaped or campanulate, with toothed margins, containing only female or both male and female flowers. Male flowers with a tubular, 3-lobed perianth and a solitary stamen. Female flowers: perianth absent; stigma sessile, filiform. Achene ovoid, compressed, included in the persistent involucre.

1. D. iners (Forsskal) Schweinf. (1896 p.146). Syn.: *Urtica iners* Forsskal (1775 p.160). Illustr.: Fl. Ethiopia 3: 324 (1989). Type: Yemen (N), *Forsskal* (C).

Perennial herb; stems prostrate or ascending, up to 1.5m. Leaves mainly opposite, ovate, c.1(–8) × 0.5–1(–5)cm, acuminate, rounded or cuneate at the base, thinly appressed-hairy above, hairy on the nerves beneath; petiole slender, 0.5–2cm. Inflorescences clustered or solitary, either bisexual or female: bisexual globular, up to 4mm diam., with a bowl-shaped involucre; female smaller, with an ovoid involucre, 1–2-flowered. Achene shining, brownish, c.1.5mm long. **Map 120, Fig. 20.**

Stream-banks and terrace-walls in the SW escarpment mountains; 1400–2500m.

Yemen (N). Tropical and southern Africa; India and Java.

Species doubtfully recorded

Migahid (1978) records *D. debilis* Rendle, a tropical African species, from Saudi Arabia. The record seems very doubtful. It is a weak-stemmed ascending herb with alternate, scabrous leaves.

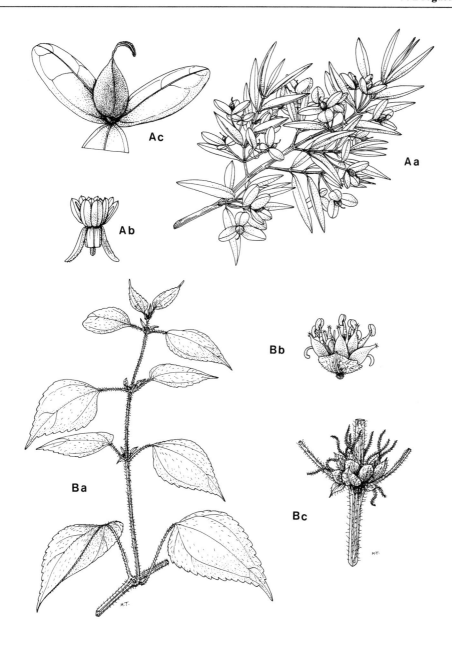

Fig. 20. Barbeyaceae. A, *Barbeya oleoides*: Aa, flowering branch (× 0.6); Ab, male flower (× 3); Ac, female flower (× 3). **Urticaceae**. B, *Droguetia iners*: Ba, flowering shoot (× 1); Bb, bisexual inflorescence (× 5); Bc, female inflorescence (× 2).

Family 29. SANTALACEAE

A.G. MILLER

Trees, shrubs or herbs, sometimes root parasites, often dioecious. Leaves alternate, simple, entire. Stipules absent. Flowers usually small, greenish or white, actinomorphic, hermaphrodite or unisexual, solitary or in spikes, racemes or cymes. Perianth 1-whorled, 3–5-lobed. Stamens attached to the perianth tube opposite the lobes. Disc epigynous. Ovary inferior, 1-locular; ovules 1–3; style simple; stigma capitate or lobed. Fruit a 1-seeded nut or drupe.

Balfour (1888 p.268) records *Thesidium* sp. (*Balfour, Cockburn & Scott* 354) from Socotra; the specimen cannot be traced and the record remains a mystery.

1. Herbs; flowers bisexual, (4–)5-merous; leaves scale-like or linear **1. Thesium**
+ Dioecious shrubs; flowers 3(–4)-merous; leaves elliptic to elliptic-oblong **2. Osyris**

1. THESIUM L.

Annual or perennial herbs, often woody-based. Leaves linear or scale-like. Flowers in spikes or racemes, bisexual, ± sessile, subtended by a bract and 2 bracteoles. Perianth 4–5-lobed. Stigma capitate. Fruit a nut, sometimes with a fleshy exocarp, surmounted by the persistent perianth.

1. Leaves scale-like, up to 2mm **1. T. stuhlmannii**
+ Leaves linear, more than 4mm 2
2. Annual; stems erect or ascending; fruit a dry, greenish nut **2. T. humile**
+ Perennial; stems creeping; fruit a nut with a fleshy red exocarp **3. T. radicans**

1. T. stuhlmannii Engl., Pflanzenw. Ost-Afrikas C: 168 (1895).

Erect woody-based perennial; stems 20(–30)cm, ridged, branched mainly from the base. Leaves scale-like, 1–2mm, acute. Flowers crowded on short axillary branches. Bracts and bracteoles narrowly ovate, acute, shorter than the nut. Perianth white, campanulate, c.2mm, 5-lobed; lobes narrowly ovate, c.1mm. Nut dull green, ovoid, 1.5 × 1.25mm, obscurely longitudinally veined; persistent perianth c.1.5 × as long as the nut. **Map 121, Fig. 21.**

Montane grassland; 1900m.

Yemen (N). Ethiopia, Kenya, Tanzania and Malawi.

In Arabia known only from a single area of grassland on J. Raymah in the SW escarpment mountains

2. T. humile Vahl (1794 p.43). Illustr.: Fl. Palaest. 1: t.45 (1966).

Erect or ascending annual; stems 7–20cm, branched. Leaves linear, 10–35 (–50) × 0.5–1.5mm. Flowers subsessile, in racemes. Bracts (2–)3–6 × as long as the nuts, 2–3 × as long as the bracteoles. Perianth greenish white, campanulate, 1.25–1.5(–1.75)mm, 5-lobed; lobes triangular, c.0.5mm. Nut dull green, oblong-ellipsoid, 2.5–3(–3.5) × 1.2–2mm, reticulately nerved; persistent perianth 0.25–0.5 × as long as the nut. **Map 122, Fig. 21.**

Rocky ravines, old grassland and a weed of cultivation; 500–1600m.

Saudi Arabia. Mediterranean region eastwards to Iraq.

3. T. radicans Hochst. in A. Rich. (1851 p.235). Illustr.: Fl. Ethiopia 3: 381 (1989).

Low-creeping perennial; stems up to 30cm long, branched. Leaves linear to lanceolate, fleshy, 4–20 × 0.5–1mm, acute, flattened. Flowers solitary or on short 1–3-flowered branches in the leaf axils. Bracts and bracteoles leaflike, shorter than or rarely longer than the fruit. Perianth white, campanulate, c.1.5mm, 5-lobed; lobes triangular, obtuse, hooded. c.0.7mm. Nut green becoming bright red and fleshy, ovoid, 3–3.5 × 2.75–3mm, indistinctly 5-ribbed and reticulate, persistent perianth c.0.5 × as long as the nut. **Map 123, Fig. 21.**

Grassland and rocky slopes; 1800–3100m.

Saudi Arabia, Yemen (N). Somalia, Ethiopia and Kenya.

Species doubtfully recorded

T. viride A.W.Hill and *T. schweinfurthii* Engl. were recorded from Saudi Arabia by Migahid (1978). They are natives of tropical Africa and their occurrence in Arabia seems doubtful.

2. OSYRIS L.

Dioecious, evergreen trees or shrubs. Leaves elliptic to elliptic-oblong. Flowers subtended by minute caducous bracts, in axillary, pedunculate, umbellate cymes or in terminal panicles. Perianth 3(–4)-lobed with a lobed disc. Female flowers with a 3(–4)-lobed stigma, the ovary adnate to the perianth tube and sterile stamens. Male flowers with 3(–4) stamens, without an ovary. Fruit a drupe.

1. Tree or large shrub, glabrous throughout; leaves 2–7cm long
   1. O. quadripartita
+ Small shrub; puberulous throughout; leaves 3–12mm long **2. O. sp. A.**

1. O. quadripartita Decne in Ann. Sci. Nat. Bot. sér. 2, 6: 65 (1836). Syn.: *O. abyssinica* A. Rich. (1851 p.236); *O. arborea* Wall., Numer. List n.4035 (1831) nom. inval.; *O. lanceolata* Hochst. & Steud. ex DC., Prodr. 14: 633 (1857); *O. pendula* Balf.f. (1883 p.93). Illustr.: Fl. Ethiopia 3: 383 (1989); Collenette (1985 p.438) as *O. abyssinica*.

Glabrous shrub or small tree up to 5m, much branched; branches sometimes pendent, ridged. Leaves 2–7 × 0.5–4cm, acute or mucronate, the margin slightly thickened, attenuate below, subsessile, coriaceous. Female flowers solitary or in 2–3-flowered cymes; male flowers in 5–13-flowered cymes. Perianth yellowish-green; lobes triangular-ovate, 1–1.5 × 1–2.25mm. Fruits globose to obpyriform, 5–10 × 5–8mm, becoming bright red. **Map 124, Fig. 21.**

Dry, rocky slopes; 600–2350m.

Saudi Arabia, Yemen (N & S), Socotra. Africa and southern Asia.

2. O. sp. A.

Puberulous shrub, to c. 60cm, much branched; branches stout, glaucous. Leaves 3–12 × 2–4mm, acute or mucronate, shortly petiolate, coriaceous. Female flowers

SANTALACEAE

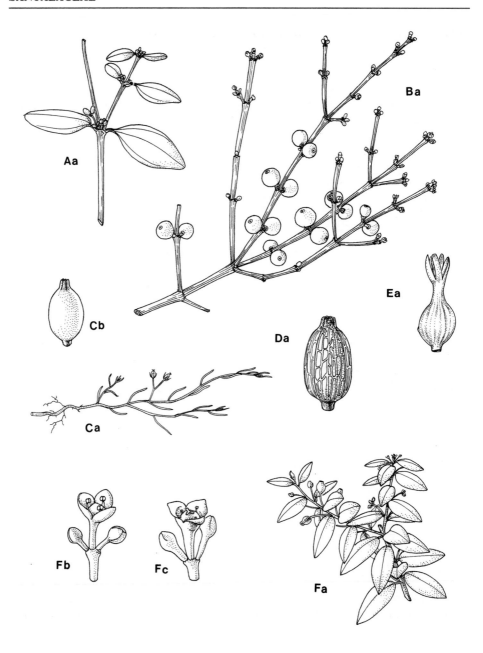

Fig. 21. Viscaceae. A, *Viscum triflorum*: Aa, leaves (×0.6). B, *V. schimperi*: Ba, flowering and fruiting branch (×0.6). **Santalaceae.** C, *Thesium radicans*: Ca, habit (×0.6); Cb, nutlet (×8). D, *T. humile*: Da, nutlet (×8). E, *T. stuhlmannii*: Ea, nutlet (×8). F, *Osyris quadripartita*: Fa, fruiting branch (×0.5); Fb, male flower (×8); Fc, female flower (×8).

solitary; male flowers not known. Perianth yellowish-green; lobes triangular-ovate, c. 1 × 1.25mm. Fruits unknown. **Map 124.**

Dry rocky hills; c. 900m.

Oman.

This very peculiar plant, known only from a single bush found growing near the Oman/Yemen border, has been tentatively placed in *Osyris*. The only specimens available (*Collenette* 8939 and *McLeish* 2773) bear only a single female flower between them and further material is needed before its generic position can be assessed with certainty. It differs from *O. quadripartita* in its low stature, smaller leaves and in being densely puberulous throughout. The area in Oman where it was found is poorly known botanically and the dry mountains across the border, in the Mahra Governorate of Yemen (S), are completely unexplored.

Family 30. LORANTHACEAE
(excluding Viscaceae)

A.G. MILLER & J.A. NYBERG

Shrubs, hemiparasitic on the branches of trees and shrubs. Stems brittle, not articulated. Leaves opposite or alternate, simple, entire, often leathery; stipules absent. Flowers usually large and colourful, actinomorphic, bisexual, in racemes or umbels. Calyx reduced, forming a narrow rim. Petals 3–6, free or united, the corolla tube sometimes split down one side. Stamens opposite to and attached to the petals. Ovary inferior, 1-celled; style simple or absent. Fruit a 1-seeded berry; seeds sticky.

1.	Flowers 4-merous, in 1-sided racemes	**2. Helixanthera**
+	Flowers 5-merous, in fascicles or umbels	2
2.	Corolla with 5 free petals or 5-lobed, the corolla tube never split down one side; stamens straight or slightly curved after anthesis; corolla lobes with paired folds towards the base within	**1. Plicosepalus**
+	Corolla 5-lobed with the corolla tube split down one side; stamens strongly inflexed or incurved after anthesis; corolla lobes without paired folds towards the base within	3
3.	Corolla lobes longer than the corolla tube	**3. Oncocalyx**
+	Corolla lobes shorter than the corolla tube	4
4.	Plant glabrous	**4. Tapinanthus**
+	Plant rusty stellate-tomentose throughout	**5. Phragmanthera**

1. PLICOSEPALUS Tieghem

Hemiparasitic shrubs. Leaves leathery, alternate or subopposite. Flowers 5-merous, regular, shortly pedicelled, arranged in pedunculate umbels. Receptacle cup-shaped. Calyx a narrow rim. Corolla free or fused into a tube at the base; lobes linear-spathulate, with c.5 paired folds at their base within. Stamens straight or slightly curved after anthesis.

1. Corolla lobes fused at the base, becoming reflexed; umbels 2-flowered; leaves oblanceolate to oblong-ovate **1. P. acaciae**
+ Corolla lobes free, curved; umbels 4–6-flowered; leaves narrowly oblong **2. P. curviflorus**

1. P. acaciae (Zucc.) Wiens & Polh. in Nord. J. Bot. 5: 221 (1985). Syn.: *Loranthus acaciae* Zucc. in Abh. Math.-Phys. Cl. Königl. Bayer. Akad. Wiss. 3: 249 (1840); *L. arabicus* Deflers (1889 p.197); *Tapinostemma arabicum* (Deflers) Tieghem in Bull. Soc. Bot. France 42: 258 (1895). Illustr.: Collenette (1985 p.354).

Shrub up to 1m, glabrous throughout. Leaves oblanceolate to oblong-ovate, 7–45 × 2–20mm. Umbels 2-flowered. Flowers green becoming bright red or scarlet. Corolla fused at the base; tube 3–10mm; lobes becoming twisted and reflexed, 20–35mm. Nectary scales oblong to triangular, 2–4 × 2mm. Fruits bright scarlet, ovoid, 5–7 × 10–12mm. **Map 125, Fig. 22.**

Succulent *Euphorbia* shrubland and *Acacia-Commiphora* bushland; recorded on *Euphorbia, Tamarix and Ziziphus* spp.; 450–2000m.

Saudi Arabia, Yemen (N & S), Oman. Ethiopia, Sudan, Egypt and Palestine.

2. P. curviflorus (Benth. ex Oliv.) Tieghem in Bull. Soc. Bot. France 41: 504 (1894). Syn.: *Loranthus curviflorus* Benth. ex Olive in Hooker's Icon. Pl. 14: pl.1304 (1880); *L. faurotii* Franch. J. Bot. (Morot) 1: 135 (1887); *Plicosepalus faurotii* (Franchet) Tieghem loc. cit. Illustr.: Collenette (1985 p.354).

Shrub up to 1m, glabrous throughout. Leaves leathery, somewhat glaucous, linear to narrowly oblong, 30–70 × 3–12mm. Umbels 4–6-flowered. Flowers orange at the base, scarlet above. Corolla lobes free to the base, (25–)30–37mm, curved. Nectary scales absent. Fruits scarlet, ellipsoid, 10–15 × 5–8mm. **Map 126, Fig. 22.**

Acacia-Commiphora bushland; only recorded on *Acacia tortilis*; 50–2100(–2500)m.

Saudi Arabia, Yemen (N & S). NE, E and C tropical Africa and the Middle East.

2. HELIXANTHERA Lour.

Hemiparasitic shrubs. Leaves alternate. Flowers 4-merous, arranged in terminal and axillary 1-sided racemes. Bracts scale-like. Receptacle campanulate. Calyx short, truncate. Corolla 4-lobed, divided to the base; lobes linear, twisting near the middle, with paired folds near the base within. Stamens straight after anthesis.

1. H. thomsonii (Sprague) Danser in Verh. Kon. Ned. Akad. Wetensch., Afd. Natuurk., Tweede Sect., 39 no. 6: 60 (1933). Syn.: *Loranthus thomsonii* Sprague, Fl. Tr. Afr., 6(1): 276 (1910).

Shrub. Leaves obovate to spathulate, (4–)8–20(–25) × (1.5–)2–4(–6)cm, rounded, attenuate below, glabrous. Racemes 3–4cm. Corolla green or greenish yellow; lobes 1.5–3cm. Fruit bright red, obovoid, c.7mm long. **Map 127, Fig. 22.**

Rocky hills in *Acacia-Commiphora* bushland; only recorded on *Commiphora gileadensis*; 1200m.

Yemen (N). Ethiopia, Somalia and Kenya.

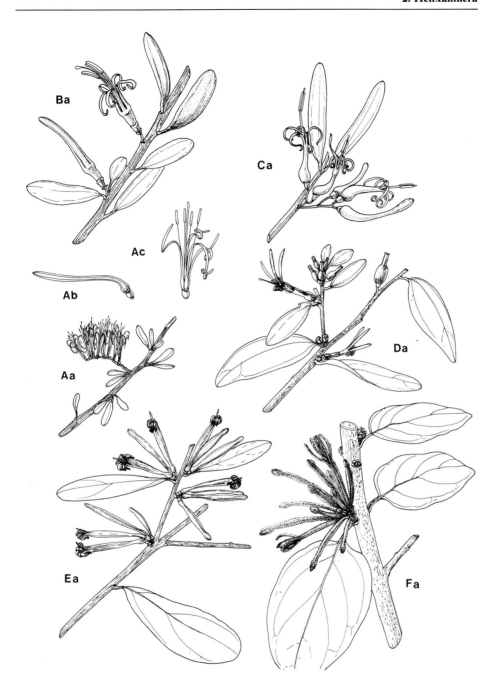

Fig. 22. **Loranthaceae**. A, *Helixanthera thomsonii*: Aa, flowering branch; Ab, flower in bud (×1); Ac, opened flower (×1). Ba–Fa, flowering branches (all ×0.6): B, *Plicosepalus acaciae*; C, *P. curviflora*; D, *Oncocalyx schimperi*; E, *Tapinanthus globiferus*; F, *Phragmanthera austroarabicus*.

3. ONCOCALYX Tieghem

Hemiparasitic shrubs. Leaves leathery, alternate. Flowers 5-merous, sessile, solitary or in fascicles. Bracts cup-like. Receptacle tubular. Calyx a ciliate rim. Corolla 5-lobed, the tube split down one side; lobes erect or reflexed, linear-oblong. Stamens strongly incurved at anthesis.

1.	Corolla lobes reflexed; plant entirely glabrous	**3. O. glabratus**
+	Corolla lobes erect; plant hairy	2
2.	Leaves narrowly oblong to ovate or broadly elliptic, puberulous or glabrescent	**1. O. schimperi**
+	Leaves broadly ovate, densely tomentose	**2. O. doberae**

1. O. schimperi (Hochst. ex A. Rich.) M.G. Gilbert in Nord. J. Bot. 5: 222 (1985). Syn.: *Loranthus schimperi* Hochst. ex A. Rich. (1847 p.341). Illustr.: Collenette (1985 p.353).

Shrub up to 1m. Leaves narrowly oblong to ovate or broadly elliptic, $1.5-5 \times 1-3$cm, rounded, 3–5-veined, puberulous or glabrescent. Umbels sessile. Bracts puberulous. Flowers yellow. Receptacle 4–7mm. Corolla lobes 13–32mm. Fruit ellipsoid, c.8×5mm. **Map 128, Fig. 22.**

Rocky slopes and sandy wadis in *Acacia-Commiphora* bushland; recorded on *Tamarix*, *Boscia*, *Maerua*, *Olea*, *Salvadora* and *Moringa* spp.; 500–2300m.

Saudi Arabia, Yemen (N & S), Oman. Ethiopia, Somalia.

2. O. doberae (Schweinf.) A.G. Miller & J. Nyberg in Edinb. J. Bot. 51 (1): 39 (1994). Syn.: *Loranthus doberae* Schweinf. (1896 p.151). Syntypes: Yemen (N), *Schweinfurth* 216 & 940 (B, n.v.).

Similar to *O. schimperi* but leaves densely tomentose, broadly ovate. **Map 129.**

On *Dobera glabra*; silty soils on the coastal plains and in the mountains; 300–2150m.

Yemen (N & S). Endemic.

This species is closely related to *O. schimperi* differing only in the characters covered in the key. It is apparently host specific to *Dobera glabra* whereas *O. schimperi* is parasitic on a range of trees including *Dobera glabra*.

3. O. glabratus (Engl.) M.G. Gilbert in Nord. J. Bot. 5: 222 (1985).

Glabrous shrub. Leaves ovate-elliptic, $3.5-5 \times 2-3$cm. Umbels pedunculate. Flowers yellow becoming orange-brown. Fruits orange-red, 10×6mm. **Map 130.**

Parasitic on *Olea*; 1980m.

Saudi Arabia. Ethiopia and Somalia

In Arabia only collected twice, near J. Ibrahim in Saudi Arabia.

4. TAPINANTHUS Blume

Hemiparasitic shrubs, glabrous or with simple or branched hairs. Leaves leathery, opposite. Flowers 5-merous, sessile, arranged in axillary umbels. Bract a scale or cup-like. Receptacle cup-shaped. Calyx an entire or toothed rim. Corolla 5-lobed, with the tube split down one side; lobes linear-spathulate, much shorter than the tube. Stamens strongly inrolled after anthesis.

1. T. globiferus (A. Rich.) Tieghem in Bull. Soc. Bot. France 42: 267 (1895). Syn.: *Loranthus globiferus* A. Rich. (1847 p.341). Illustr.: Collenette (1985 p.355).

Shrub up to 1m, glabrous throughout. Leaves narrowly obovate to ovate or oblong, $4-20 \times 1-3.5$cm, obtuse or rounded, cuneate at the base. Flowers pale pinkish green with rose-coloured tips. Calyx an entire rim. Corolla 30–45mm; lobes 7–10mm, the tube abruptly constricted towards the base; buds with a swollen ovoid tip. Fruit orange red, subglobose, c.9mm long. **Map 131, Fig. 22.**

Stony wadis and limestone hills in *Acacia-Commiphora* bushland; recorded on *Ficus*, *Acacia* and *Commiphora* spp.; 400–1700m.

Saudi Arabia, Yemen (N & S). N Africa, W, C and NE tropical Africa

5. PHRAGMANTHERA Tieghem

Hemiparasitic shrubs; young shoots reddish tomentose with stellate hairs. Leaves opposite or subopposite. Flowers 5-merous, sessile and arranged in dense axillary umbels. Receptacle and calyx cup-like. Corolla 5-lobed with the tube deeply split down one side; lobes linear-spathulate, much shorter than the tube. Stamens strongly inrolled at anthesis.

1. P. austroarabica A.G. Miller & J. Nyberg in Edinb. J. Bot. 51 (1): 37 (1994). Syn.: *Loranthus regularis* auctt. arab. non Sprague; *L. rufescens* auctt. arab. non DC. Illustr.: Collenette (1985 p. 394) as *L.* sp. aff. *rufescens*. Type: Yemen (N), *Miller & Long* 3226 (E).

Shrub up to 1.5m diam. Leaves ovate, $5-13 \times 3-7$cm, acute or obtuse, rounded or cordate at the base, densely clothed with rust-coloured stellate-tomentose hairs becoming glabrescent; petioles 1.5–3.5cm. Flowers rusty outside, creamy orange inside, in dense axillary fascicles. Calyx a toothed rim. Corolla 35–45mm; lobes c.10mm, linear with an ovoid tip. Fruit greenish brown, obovoid crowned by the persistent calyx, $6-14 \times 4-8$mm, glabrescent to densely tomentose. **Map 132, Fig. 22.**

Acacia-Commiphora bushland and succulent *Euphorbia* shrubland; recorded on *Acacia* spp., *Ficus vasta*, *F. cordata*, *Ziziphus spina-christi*; 1200–2500m.

Saudi Arabia, Yemen (N & S). Endemic.

Widespread in the mountains of SW Arabia. In the Arabian literature previously incorrectly identified as either *L. regularis* or *L. rufescens*; both species restricted to tropical Africa.

Family 31. VISCACEAE

A.G. MILLER & J.A. NYBERG

Dioecious or monoecious shrubs, hemiparasitic on branches of trees and shrubs. Stems green, brittle, articulated. Leaves opposite, simple, sometimes reduced to scales; stipules absent. Flowers minute, actinomorphic, unisexual, solitary, clustered or in 1–3-flowered dichasia at the nodes. Perianth of 3–4 valvate lobes. Male flowers with stamens opposite to and attached to the perianth lobes. Female flowers without staminodes; ovary inferior, 1-celled; style simple or absent. Fruit a 1-seeded berry; seeds sticky.

Gilbert (Fl. Ethiopia 3: 377 (1989)) records the genus *Korthalsella* Tieghem from Socotra. However, no specimen could be traced and the record must be treated with caution.

VISCUM L.

Glabrous, hemiparasitic shrubs. Stems jointed, the internodes rounded or flattened. Leaves opposite, with a broad lamina or reduced to scales. Flowers minute, in unisexual or hermaphrodite dichasia each subtended by a pair of scale-like bracts. Male flowers with anthers dehiscing by many pores. Female flowers with a globose stigma. Fruits translucent, ovoid or globose.

1. Leaves reduced to scales — **3. V. schimperi**
+ Leaves well-developed with a broad lamina — 2
2. Leaves with well-defined veins; fruits greenish white; male and female flowers 2–2.5mm long; monoecious — **1. V. triflorum**
+ Leaves with obscure veins; fruits red; male flowers 6–9mm long, female flowers 1–2mm long; dioecious — **2. V. cruciatum**

1. V. triflorum DC., Prodr. 4: 279 (1843).

Monoecious shrub up to 50cm, glabrous throughout; internodes terete or the upper flattened. Leaves narrowly ovate to broadly ovate, 3–5 × 0.5–2cm, acute to rounded, the margin undulate, with well-defined venation. Flowers yellowish, c.2.5mm. Fruit greenish white, ovoid, c.5mm diam. **Map 133, Fig. 21.**

Recorded on *Teclea nobilis* and *Celtis africana*; 1150–1550m.

Saudi Arabia. Tropical Africa.

Plants from Arabia are referable to subsp. *nervosum* (A. Rich.) M.G. Gilbert.

2. V. cruciatum Sieb. ex Boiss., Voy. bot. Espagne 1 (9): 274 (1840). Illustr.: Fl. Palaest. 1: pl.47 (1966).

Dioecious shrub up to 75cm, glabrous throughout; internodes terete. Leaves oblong-elliptic, 15–35 × 5–12mm, rounded, the margin entire, tapering at the base, obscurely 3-veined. Flowers green, the male 6–9mm with the anthers many-celled and concrescent with the perianth lobes, the female 1–2mm. Fruit red, globular, 4–6mm diam. **Map 133.**

Olea-Juniperus woodland; only recorded on *Olea*; 1980m.

Saudi Arabia. Portugal, Spain, Algeria, Morocco and Palestine.

In Arabia very rare and known only from a single locality where it parasitizes several olive trees.

3. V. schimperi Engl. in Bot. Jahrb. Syst. 20: 132 (1894). Illustr.: Collenette (1985 p.355).

Monoecious shrub up to 75cm, glabrous throughout; stems jointed, the lower internodes terete, the upper flattened. Leaves reduced to minute paired scales. Flowers greenish yellow, c.2.3mm. Fruits reddish orange, globose, c.8mm diam. **Map 134, Fig. 22.**

Recorded on *Acacia* spp.; 1600–2000m.

Saudi Arabia, Yemen (N). Ethiopia, Somalia, Kenya and Uganda.

Family 32. POLYGONACEAE

J.A. NYBERG & A.G. MILLER

Herbs or shrubs, sometimes aquatic, rarely climbing. Leaves usually alternate, simple or lobed, sometimes absent. Stipules connate forming an often membranous sheath (the ocrea). Flowers bisexual or unisexual, actinomorphic, in racemes, panicles or clusters. Perianth with 3–6 segments, free or fused, often accrescent and enclosing the fruit, sometimes winged or toothed or bearing spines. Stamens 6–15. Ovary superior, 1-celled with a solitary, basal ovule; styles 2–4. Fruit a trigonous or biconvex nutlet, sometimes winged or beset with bristles, sometimes hidden by the accrescent perianth lobes.

Antigonum leptopus Hook. & Arn. is cultivated as an ornamental in Arabia. It is a shrubby climber with cordate leaves, tendrils and white or pink, papery perianth segments surrounding the nutlets.

1.	Shrubs; leaves less than 10mm long	2
+	Annual or perennial herbs; or if shrubs then leaves more than 20mm long	4
2.	Leaves ovate; fruits covered by the accrescent, inner pair of perianth lobes	**8. Atraphaxis**
+	Leaves linear or minute and soon deciduous; Fruit winged or covered with stiff setae, not covered by the accrescent perianth lobes	3
3.	Fruit 3-winged; leaves linear, persistent	**9. Pteropyrum**
+	Fruit 4-winged or covered with stiff setae; leaves minute, soon deciduous	**7. Calligonum**

4.	Perianth segments indurated in fruit and bearing 3 rigid spines	5
+	Perianth segments sometimes accrescent but not indurated in fruit, sometimes toothed but not bearing rigid spines	6
5.	Leaves simple; flowers in axillary clusters	**6. Emex**
+	Leaves sinuously lobed; flowers in elongated spikes	**3. Oxygonum**
6.	Leaves large, all basal, palmately-nerved	**4. Rheum**
+	Leaves small or medium-sized; stem leaves usually present, not palmately-nerved	7
7.	Inner perianth segments accrescent and winged or toothed in fruit	**5. Rumex**
+	Inner perianth segments never accrescent, nor winged or toothed in fruit	8
8.	Flowers in axillary clusters sometimes forming lax spikes; ocreae 2-lobed, lacerate	**1. Polygonum**
+	Flowers in globose heads or spikes; ocreae cylindric, truncate or fringed with bristles	**2. Persicaria**

1. POLYGONUM L.

Annual or perennial herbs. Ocreae 2-lobed, membranous, lacerate. Leaves narrow, entire. Flowers 1–6 in axillary clusters. Perianth 4–5-lobed. Stamens 4–5, not dilated at the base. Styles 2–3. Nuts trigonous.

1.	Bushy, clump-forming perennial; stems up to 1.5m	**2. P. palaestinum**
+	Erect, ascending or prostrate annual herbs; stems up to 60cm	2
2.	Leaves linear-spathulate, up to 10(–15)mm; nuts c.1mm; stems prostrate	**3. P. corrigioloides**
+	Leaves elliptic-lanceolate or narrowly elliptic-oblong, 10–50mm; nuts 2–3.5mm; stems erect, ascending or prostrate	3
3.	Inflorescence becoming leafless and spike-like above; all leaves deciduous at length; pedicels longer than the flowers	**1. P. argyrocoleum**
+	Inflorescence leafy; leaves persistent; pedicels shorter than the flowers	**4. P. aviculare**

1. P. argyrocoleum Steud. ex Kunze in Linnaea 20: 17 (1847). Syn.: *P. patulum* sensu Collenette (1985) non M. Bieb. Illustr.: Collenette (1985 p.403) as *P. patulum*.

Glabrous annual; stems erect or ascending, up to 60cm. Ocreae brownish at the base. Leaves narrowly elliptic-lanceolate, 10–50 × 2–5mm, acute, attenuate below, deciduous at length. Inflorescence of 2–6-flowered clusters, leafy below, becoming leafless and spike-like above; pedicels longer than the flowers. Perianth 2.2–2.5mm; lobes green with pinkish margins. Nuts c.2mm, smooth, shining. **Map 135, Fig. 23.**

Roadsides, waste ground and sometimes a serious weed of cultivation; 10–2300m.

Saudi Arabia, UAE, Qatar, Kuwait. Iraq, Iran, Afghanistan, Armenia and C Asia.

P. argyrocoleum is related to and often confused with *P. equisetiforme* Sibth. & Sm., a perennial species and native of the Mediterranean region. Immature plants of *P. argyrocoleum*, with leafy inflorescences, can be confused with *P. aviculare* L. Records of *P. patulum* M. Bieb. from Kuwait and possibly *P. equisetiforme* from Qatar are referable to *P. argyrocoleum*.

2. P. palaestinum Zoh., Fl. Palaest. 1: 341 (1966).

Bushy, glabrous, perennial herb; stems erect, up to 1.5m. Ocreae brownish at the base. Leaves narrowly elliptic-lanceolate, 10–30(–60) × 2–5(–8)mm, acute, attenuate below, soon deciduous. Inflorescence of 2–3-flowered clusters, forming an interrupted spike, leafless above; pedicels shorter than the flowers. Perianth 3–4mm; lobes green with white margins. Nuts 2–3mm, smooth, shining. **Map 136.**

Sandy wadi-beds amongst lava flows; 700–800m.

Saudi Arabia. Palestine.

In Arabia confined to a small area in the north of Saudi Arabia where it is locally common.

3. P. corrigioloides Jaub. & Spach, Ill. pl. or. 2: 34 (1845).

Glabrous, annual herb; stems prostrate, up to 30cm. Leaves linear-spathulate, 5–10(–15) × 0.5–1.5mm, obtuse, attenuate below. Inflorescence leafy, of 2–6-flowered clusters; pedicels longer than the flowers. Perianth 1–1.5mm; lobes green with white or pinkish margins. Nuts c.1mm, smooth, shining. **Map 137.**

Muddy areas and pond margins; 1800–2800m.

Yemen (N). Iraq, Iran, Pakistan, Afghanistan and C Asia.

4. P. aviculare L., Sp. pl.: 362 (1753).

Glabrous, annual herb; stems erect or prostrate, up to 30cm. Leaves narrowly elliptic-oblong, 10–40 × 2–10mm, acute or obtuse, cuneate or attenuate below; stem leaves often larger than those on the branches. Inflorescence leafy, of 1–6-flowered clusters; pedicels shorter than the flowers. Perianth 2–3mm; lobes green with pinkish margins. Nuts 2–3.5mm, smooth, shining. **Map 138.**

Weed of cultivated ground; 1800–2900m.

Saudi Arabia, Yemen (N). Cosmopolitan.

2. PERSICARIA L.

Annual or perennial herbs, sometimes aquatic. Ocreae cylindrical, truncate or with a terminal fringe of bristles. Leaves oblong-ovate or lanceolate, entire. Flowers in globose heads or spikes. Perianth 4–5-lobed. Stamens 5–8, dilated at the base. Styles 2–3. Nuts trigonous or biconvex.

Wilson, K.L. (1990). Some widespread species of *Persicaria* and their allies. *Kew Bull.* 45(4): 621–636.

1. Flowers in globose heads subtended by an involucral leaf   **1. P. nepalensis**
+ Flowers in spikes, not subtended by an involucral leaf   2

2. Leaves oblong-ovate; base rounded or cordate   **2. P. amphibia**
+ Leaves lanceolate to elliptic-lanceolate; base attenuate, cuneate or rarely rounded   3

3. Ocreae with a terminal fringe of fine bristles; leaves not gland-dotted   4
+ Ocreae without a terminal fringe of fine bristles, rarely shortly ciliate; leaves minutely gland-dotted   6

4. Spikes slender, interrupted, 'zig-zagging' when immature; leaves narrowly lanceolate   **8. P. decipiens**
+ Spikes densely-flowered, never 'zig-zagging'; leaves lanceolate to elliptic-lanceolate   5

5. Ocreae sub-equal or shorter than the terminal bristles; bristles up to 1.5cm   **7. P. barbata**
+ Ocreae much longer than the terminal bristles; bristles up to 5mm   **6. P. maculosa**

6. Nuts not dimpled; inflorescence glabrous or at most with a few scattered glands   **3. P. glabra**
+ Nuts dimpled on both surfaces; inflorescence glandular   7

7. Erect, robust perennial; petioles 1–5cm; perianths 3–5mm; flowers often exceeding the bracts   **5. P. senegalensis**
+ Erect or ascending annual; petioles up to 1cm; perianths 1.7–3mm; flowers never exceeding the bracts   **4. P. lapathifolia**

1. P. nepalensis (Meissner) Gross in Bot. Jahrb. Syst. 49: 277 (1913). Syn.: *Polygonum nepalense* Meissner, Monogr. Polyg.: 84 (1826).

Prostrate annual herb; stems up to 30cm. Ocreae brown, 5–10mm, truncate, not fringed, sparsely brown-pubescent. Leaves ovate or elliptic, up to 5×3cm, acute, abruptly narrowed and auriculate at the base, sparsely pubescent; petiole winged. Flowers in globose heads; heads 5–10mm diam., 1–3 from the upper leaf axils, subtended by a sessile involucral leaf; peduncles with stalked glands. Perianth pink or white, c.3mm. Nut biconvex, minutely pitted, c.2mm, dull. **Map 139, Fig. 23.**

Rare weed of cultivation; 2600–3100m.

Yemen (N). Tropical Africa and tropical Asia.

2. P. amphibia (L.) S.F. Gray, Nat. Arr. Brit. Pl. 2: 268 (1821). Syn.: *Polygonum amphibium* L., Sp. pl.: 361 (1753). Illustr.: Collenette (1985 p.402).

Glabrous, aquatic* perennial; stems rooting at the nodes. Ocreae brown, up to 2cm, truncate. Leaves floating, long-petiolate, oblong-ovate, 7–10(–17) × 2–4cm,

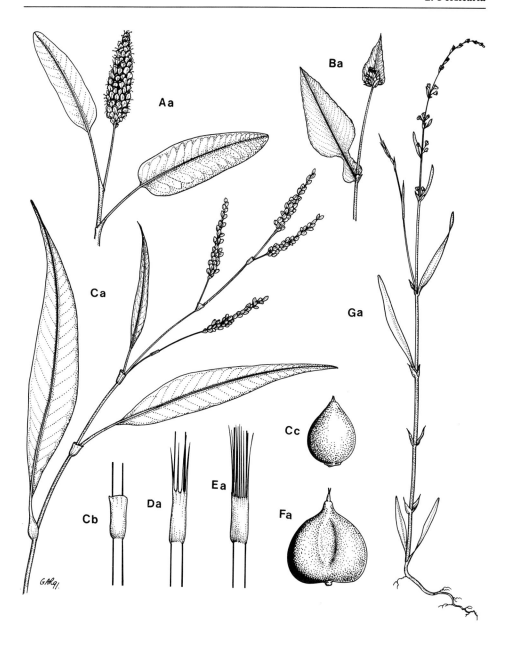

Fig. 23. Polygonaceae. A, *Persicaria amphibia*: Aa, flowering shoot (×0.6). B, *P. nepalensis*: Ba, flowering shoot (×1). C, *P. glabra*: Ca, flowering shoot (×0.3); Cb, ocrea (×2); Cc, nutlet (×10). D, *P. maculosa*: Da, ocrea (×2). E, *P. barbata*: Ea, ocrea (×2). F, *P. senegalensis*: Fa, nutlet (×10). G, *Polygonum argyrocoleum*: Ga, habit (×1).

rounded at the tip, rounded or cordate below. Racemes solitary or sometimes paired, densely flowered, 3–4cm. Perianth white or pink, c.3.5mm. Nut biconvex, c.2.5mm, shining. **Map 140, Fig. 23.**

*A terrestrial form also occurs; this has erect stems and lanceolate, subsessile, pubescent leaves.

Ponds and pond margins; 300–2650m.

Saudi Arabia, Yemen (N). Europe, temperate Asia, N Africa and N America.

3. P. glabra (Willd.) M. Gomez in An. Inst. Segunda Enseñanza Habana 2: 278 (1896). Syn.: *Polygonum glabrum* Willd., Sp. pl. 2 (1): 447 (1799).

Robust, glabrous, perennial herb; stems erect, up to 1m. Ocreae reddish brown, up to 3cm, truncate, not fringed. Leaves lanceolate, up to 25 × 5cm, acuminate, cuneate below, minutely yellow-glandular. Racemes 2–4 at the ends of branches, densely flowered, 5–10cm; peduncles and racemes glabrous or very sparsely glandular. Perianth pink, c.3.5mm. Nut biconvex or trigonous without dimpled faces, c.2.5mm, shining. **Map 141, Fig. 23.**

Margins of ponds and streams; 100–2300m.

Saudi Arabia, Yemen (N & S), Socotra, Oman. Tropics of Africa, Asia and America.

4. P. lapathifolia (L.) S.F. Gray, Nat. Arr. Brit. Pl. 2: 270 (1821). Syn.: *Polygonum lapathifolium* L., Sp. pl.: 360 (1753).

Similar to *P. glabra* but an erect or ascending, branched annual; leaves up to 15 × 3cm, often marked with a large blackish blotch, glabrous or pubescent; ocreae up to 2cm, sometimes shortly fringed; peduncles and racemes yellow-glandular; perianth white or pink, c.2mm; nut biconvex, depressed on both faces. **Map 142.**

Slow-moving streams; 2000–2400m.

Yemen (N). Europe and Asia.

5. P. senegalensis (Meissner) Soják in Preslia 46(2): 155 (1974). Syn.: *Polygonum senegalense* Meissner, Monogr. Polyg.: 54 (1826).

Robust, glabrous to white-tomentose, perennial herb; stems erect, up to 2.5m. Ocreae reddish brown, up to 3.5cm, truncate, not fringed. Leaves lanceolate, up to 25 × 6cm, acuminate, cuneate below, minutely yellow-glandular. Racemes 2–3 at the ends of branches, densely flowered, 3–10cm; peduncles and racemes densely yellow-glandular. Perianth white or pink, c.3.5mm. Nut biconvex or flattened with dimpled faces, c.3mm, shining. **Map 143, Fig. 23.**

Wadi-sides and pond margins; 950–1400m.

Yemen (N). Tropical Africa and Palestine.

6. P. maculosa S.F. Gray, Nat. Arr. Brit. pl. 2: 270 (1821). Syn.: *Persicaria dolichopoda* (Ohki) Sasaki, List pl. Formos.: 168 (1928). Syn.: *Polygonum persicaria* L., Sp. pl.: 361 (1753).

Erect or ascending annual herb; stems up to 70cm. Ocreae brown, up to 1.5cm, sparsely appressed-setulose, fringed with fine bristles up to 5mm. Leaves elliptic-lanceolate, up to 12 × 3cm, acuminate, attenuate below, the margins setulose. Racemes 1–3 at the ends of branches, densely flowered, 1–3.5cm; bracts fringed with short (up to 1mm) cilia. Perianth pink or whitish, 2.5–3mm. Nut trigonous, c.2mm, shiny. **Map 144, Fig. 23.**

Waste ground, in slow-moving water and in mud by wadi-sides and pond margins; 1800–2800m.

Saudi Arabia, Yemen (N). Temperate N hemisphere.

7. P. barbata (L.) Hara, Fl. E. Himal.: 70 (1966). Syn.: *Polygonum barbatum* L., Sp. pl.: 362 (1753); *Polygonum setulosum* sensu Collenette (1985) non A. Rich. Illustr.: Collenette (1985 p.403) as *Polygonum setulosum*.

Erect or ascending herb; stems up to 50cm. Ocreae brown, up to 2cm, appressed-setulose, fringed with fine bristles up to 1.5cm. Leaves lanceolate to narrowly elliptic, up to 15 × 2cm, acuminate, attenuate below, appressed-setulose beneath. Racemes 3–6 at the ends of branches, densely flowered, 2–5cm; peduncles appressed-setulose; bracts fringed with long (1–2mm) cilia. Perianth white or pinkish, c.2mm. Nut trigonous, 1.5–2mm. **Map 145, Fig. 23.**

Wadi-sides; 600–2600m.

Saudi Arabia, Yemen (N), Socotra. Tropics of Africa, Asia and Australia.

8. P. decipiens (R.Br.) K.L. Wilson in Telopea 3: 178 (1988). Syn.: *P. salicifolia* (Brouss. ex Willd.) Assenov in Fl. Reipubl. Popul. Bulgar. 3: 243 (1966); *Polygonum salicifolium* Brouss. ex Willd., Enum. pl.: 428 (1809); *P. serrulatum* Lagasca, Gen. sp. pl.: 14 (1816), non Webb & Moq.

Erect or decumbent perennial herb; stems up to 70cm. Ocreae brown, up to 2cm, appressed-setulose, fringed with fine (up to 2cm) bristles. Leaves narrowly lanceolate, up to 15 × 2(–3)cm, acuminate, attenuate to rounded below, the margins setulose. Racemes 2–5 at the ends of branches, slender, 2–9cm, sometimes interrupted, 'zig-zagging' when immature; bracts fringed with short (up to 1mm) cilia. Perianth pinkish or whitish, 2.5–3mm. Nut trigonous or biconvex, 2–3mm, shiny. **Map 146, Fig. 23.**

Marshes and wadi-sides; 1700–1900m.

Yemen (N). Widespread in the Old World.

3. OXYGONUM Burchell

Annual or perennial, polygamous herbs. Ocreae tubular, fringed with fine setae. Leaves alternate, sinuously lobed (in Arabia). Flowers pink or white, in axillary spike-like racemes. Male flowers: perianth segments 4–5, fused at the base. Hermaphrodite flowers: perianth segments 5, tubular; stamens 8; styles 3. Fruiting perianth enclosing the nutlet, fusiform with 3 spines arising from near the middle (in Arabia).

1. O. sinuatum (Hochst. & Steud.) Dammer, Pflanzenw. Ost-Afrikas C: 170 (1895). Syn.: *O. atriplicifolium* (Meissner) Martelli var. *sinuatum* (Hochst. & Steud.) Baker & C.H. Wright in Fl. Tr. Afr. 6(1): 101 (1909). Illustr.: Collenette (1985 p.402).

Glabrous or pubescent annual herb; stems decumbent to ascending, 10–30cm. Leaves ovate to lanceolate, 2–5 × 1–3cm, with 1–3 pairs of acute or rounded lobes; petiole 1–2cm. Inflorescence up to 25cm, the flowers distant. Flowers 2–3mm. Fruiting perianth 5–6mm long and broad (including spines); spines spreading, 1–2mm long. **Map 147, Fig. 25.**

Weed of field-borders, roadsides and stony hills; (50–)600–1700m.

Saudi Arabia, Yemen (N), Socotra. Tropical and southern Africa.

4. RHEUM L.

Robust herbs with thick rhizomes. Leaves large, all basal, palmately-nerved. Ocreae membranous. Inflorescence a thick, much-branched panicle. Flowers hermaphrodite, perianth segments 6, fused at the base. Stamens 9. Styles 3. Fruit 3-winged.

1. R. palaestinum Feinbrun in Palestine J. Bot., Jerusalem Ser. 3: 117 (1944). Illustr.: Fl. Palaest. 1: pl.65 (1966); Collenette (1985 p.403).

Leaves shortly petiolate, cordate, up to 75cm diam, the margin undulate. Flowers c.2mm. Fruit 1–1.4cm; wings c.4mm broad. **Map 148.**

Sand on a desertic undulating limestone plateau; 885m.

Saudi Arabia. Palestine.

Known only from a single area in the extreme NW of Saudi Arabia. No flowering material has been seen so the identification of the Arabian plants remains provisional. The Arabian plants were found in a similar habitat to that recorded for *R. palaestinum* in Palestine. The description has been completed from Palestinian material.

5. RUMEX L.

Annual or perennial herbs or shrubs. Ocreae cylindrical. Leaves entire or pinnately lobed. Flowers bisexual or unisexual, whorled in racemose or paniculate inflorescences. Perianth segments in 2 whorls of 3; inner whorl strongly accrescent and enveloping the nutlets in fruit, scarious and wing-like or coriaceous and entire or toothed, sometimes bearing a basal wart; outer whorl small and inconspicuous. Stamens 6. Styles 3. Nutlets trigonous.

1.	Leaves sinuously lobed	**3. R. pictus**
+	Leaves entire	2
2.	Inner perianth segments scarious and wing-like in fruit, entire or minutely denticulate	3
+	Inner perianth segments coriaceous, not wing-like, entire or toothed	6
3.	Shrubs with narrowly elliptic to oblanceolate leaves	4
+	Herbs with ovate-triangular leaves	5
4.	Inner fruiting perianth segments orbicular-reniform, 3–7 × 4–7mm	**1. R. nervosus**
+	Inner fruiting perianth segments broadly reniform, c.3.5 × 6mm	**2. R. limoniastrum**
5.	Inner fruiting perianth segments minutely denticulate with a conspicuous marginal nerve	**5. R. cyprius**
+	Inner fruiting perianth segments entire, without a conspicuous marginal nerve	**4. R. vesicarius**
6.	Inner fruiting perianth segments entire	**6. R. conglomeratus**
+	Inner fruiting perianth segments toothed	7
7.	Inner fruiting perianth segments prominently warted, the margin with long, straight teeth; teeth 2.5–4mm long	**9. R. dentatus**
+	Inner fruiting perianth segments obscurely warted or warts absent, the margin with short straight teeth; teeth c.1mm or if longer then hooked	8
8.	Inner perianth segments with hooked teeth; teeth (0.5–)1.5–3mm	**7. R. steudelii**
+	Inner perianth segments with mostly straight teeth; teeth c.1mm	**8. R. sp. A**

1. R. nervosus Vahl. (1790 p.27). Syn.: *R. persicarioides* Forsskal (1775 p.76) non L. Illustr.: Collenette (1985 p.404). Type: Yemen (N), *Forsskal* (C).

Monoecious or dioecious shrub, glabrous throughout; stems erect or ascending, up to 2m. Leaves narrowly elliptic to oblanceolate, 20–80 × 3–20mm, acute, entire, attenuate below, 3-nerved; petioles 4–25mm. Inflorescence paniculate. Flowers unisexual. Inner fruiting perianth segments orbicular-reniform, with scarious wings, 3–7 × 4–7mm, entire, cordate at the base, ripening red or straw-coloured; warts minute, reflexed. **Map 149, Fig. 24.**

Terrace-walls, cliffs and rocky mountain slopes; 900–2900m.

Saudi Arabia, Yemen (N & S). Ethiopia, Somalia, Kenya and Tanzania.

One of the commonest and most distinctive plants of the SW escarpment mountains where it is characteristic of terrace-walls.

2. R. limoniastrum Jaub. & Spach, Ill. pl. or. 2: t.106 (1844). Type: Oman, *Aucher-Eloy* 5280 (BM).

Similar to *R. nervosus* but the leaves generally smaller, 33–50 × 5–10mm, obscurely 3-nerved and apparently thinner textured; inner fruiting perianth segments broadly reniform, c.3.5 × 6mm; warts minute, 0.5 × 0.3mm. **Map 150.**

No habitat details available.

Oman. Endemic.

R. limoniastrum was described from a single gathering, made in the last century, from J. Akhdar in northern Oman. It is undoubtedly closely related to *R. nervosus* and may prove to be conspecific. However, the broadly reniform fruiting perianth segments on the type specimen are very distinct and unlike those on any of the numerous specimens of *R. nervosus* examined. Further gatherings are needed so that the importance of the fruit character can be assessed. Until then the status of this species remains doubtful. It is surprising that such a showy plant has not been re-collected.

3. R. pictus Forsskal (1775 p.77). Syn.: *R. lacerus* Balbis in Mem. Acad. Turin 7: 19 (1804–6). Illustr.: Mandaville (1990 pl.54); Fl. Kuwait 1: pl.160.; Collenette (1985 p.404); Western (1989 p.33).

Glabrous annual; stems decumbent or ascending, up to 25cm. Leaves sinuously pinnately-lobed, ovate to lanceolate, 2–4 × 0.5–2cm; petioles 2–4cm. Inflorescence racemose, of 1–3-flowered whorls. Flowers bisexual. Inner fruiting perianth segments broadly cordate, with scarious wing, up to 1cm diam., entire, pinkish ripening yellow, each segment with a conspicuous elongate wart. **Map 151, Fig. 24.**

Coastal and inland sand; 10–850m.

Saudi Arabia, UAE, Kuwait. Egypt, Syria, Palestine and Jordan.

4. R. vesicarius L., Sp. pl.: 336 (1753). Illustr.: Collenette (1985 p.405); Fl. Kuwait: 1 pls.158 & 159; Western (1989 p.34); Fl. Qatar pl.18; Mandaville (1990 pl.55).

Glabrous annual; stems erect or decumbent, up to 30cm. Leaves ovate-triangular, 1.5–6 × 1–4cm, acute or obtuse, truncate or subcordate below; petioles 3–5cm. Inflorescence racemose, of 1–4-flowered whorls. Flowers bisexual, solitary or paired, when paired the primary flower completely obscuring the secondary in fruit. Inner perianth segments cordate to orbicular, with scarious wings, up to 2cm diam., entire, yellow ripening red; warts absent or small. **Map 152, Fig. 24.**

Sandy soil, gravel, rocky slopes and wadi-beds; 10–2400m.

Saudi Arabia, Yemen (N), Oman, UAE, Qatar, Bahrain, Kuwait. N Africa and Palestine eastwards to Iraq, Iran, Afghanistan and Pakistan.

R. simpliciflorus Murb., a native of N Africa, has been recorded from Saudi Arabia. It is similar to *R. vesicarius* but differs in having solitary rather than paired flowers on each pedicel, and several pedicels at each node. However, *R. vesicarius* is very variable, also with both solitary and paired flowers on each pedicel. All Arabian material examined seems to come within the natural variation of *R. vesicarius*.

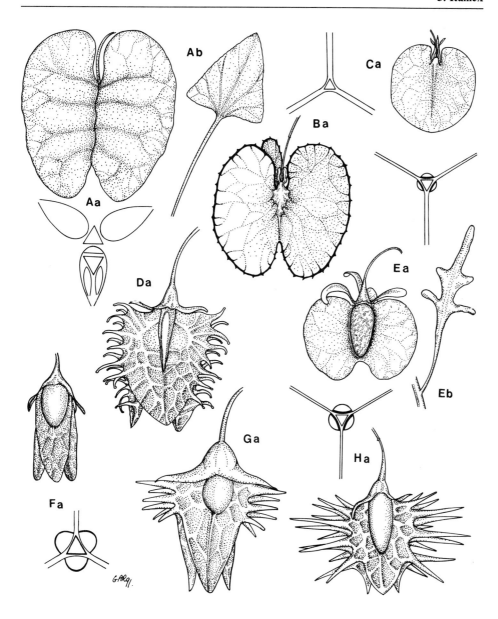

Fig. 24. Polygonaceae. A, *Rumex vesicarius*: Aa, fruit (×5); Ab, leaf (×1). B, *R. cyprius*: Ba, fruit (×5); C, *R. nervosus*: Ca, fruit (×5). D, *R. steudelii*: Da, fruit (×7). E, *R. pictus*: Ea, fruit (×5); Eb, leaf (×1.5). F, *R. conglomeratus*: Fa, fruit (×7). G, *R.* sp. *A*: Ga, fruit (×7). H, *R. dentatus*: Ha, fruit (×7).

5. R. cyprius Murb. in Acta Univ. Lund. 2, 2 (14): 20 (1907). Illustr.: Fl. Cyprus 2: 1407 (1985).

Similar to *R. vesicarius* but with the inner fruiting perianth segments minutely denticulate on the margins and with a conspicuous marginal nerve. **Map 153, Fig. 24.**

In pockets of sand in limestone rocks; 50m.

Saudi Arabia, Qatar. Cyprus and Egypt eastwards to Iran.

Recorded as rare in Qatar (Fl. Qatar, 1981) and otherwise in Arabia collected only once in the extreme north of Saudi Arabia.

6. R. conglomeratus Murr., Prodr. Fl. Goett.: 52 (1770).

Glabrous perennial herb; stems erect, up to 50cm. Basal leaves oblong-lanceolate, 10–15 × 3–5cm, acute or obtuse, the base truncate or sub-cordate; upper leaves smaller; petioles up to 15cm. Inflorescence a panicle, of numerous racemose branches each of distant many-flowered whorls. Flowers bisexual. Inner fruiting perianth segments oblong-ovate, coriaceous, 2–3 × 1–2mm, rounded, entire, ripening reddish brown, each with a conspicuous wart. **Map 154, Fig. 24.**

Disturbed ground near water; 600m.

Saudi Arabia, Oman. A widespread plant in temperate regions of the world.

In Arabia a rare plant of disturbed habitats. A single specimen (*McLeish* 2097) with immature fruits from near Sayq in the mountains of N Oman may be referable to this species. It is, according to the collector, a recent introduction which is now becoming quite common.

7. R. steudelii Hochst. ex A. Rich. (1850 p.229). Syn.: *R. nepalensis* sensu Schwartz (1939 p. 32) non Sprengel. Illustr.: Collenette (1985 p.405).

Glabrous perennial herb; stems erect, 40–200cm. Basal leaves oblong-lanceolate, 10–20 × 2.5–4.5cm, acute or obtuse, truncate, cordate or cuneate below; upper leaves smaller; petioles up to 10cm. Inflorescence a panicle, of numerous racemose branches each of distant many-flowered whorls. Flowers bisexual. Inner fruiting perianth segments ovate-triangular, coriaceous, reticulate, 5 × 3–3.5mm, rounded, truncate at the base, the margin with hooked teeth, these (0.5–)1.5–3mm long, ripening reddish brown; warts absent or obscure. **Map 155, Fig. 24.**

Cultivated ground, waste places and marshy ground by ditches and ponds; 1800–3100m.

Saudi Arabia, Yemen (N & S). Ethiopia, Somalia and southern Africa.

A common species of high rainfall areas on the SW escarpment mountains.

8. R. sp. A. Syn.: *R. obtusifolius* sensu Schwartz (1939) non L.

Similar to *R. steudelii* but a stouter plant with larger leaves; inner fruiting perianth segments with 3–5 straight teeth in the lower part, the teeth c.1mm long; each segment with an obscure, elongate wart. **Map 156, Fig. 24.**

Weed, usually by water; 2400–3000m.

Saudi Arabia, Yemen (N). Europe and western Asia.

Related to *R. steudelii* but generally a stouter plant and, according to J. Wood (pers. comm.), largely replacing it on the high plateau of Yemen and Saudi Arabia.

9. R. dentatus L., Mant. pl. 2: 226 (1771). Illustr.: Mandaville (1990 pl.53). Illustr.: Collenette (1985 p.404) as *R. conglomeratus*.

Similar to *R. steudelii* but an annual; inner fruiting perianth segments ovate, 4–5mm × 2–2.5mm, subacute, rounded at the base, the margin with straight teeth, these 2.5–4mm long, ripening yellowish green; each segment prominently warted. **Map 157, Fig. 24**.

Weed of wet and cultivated areas, waste ground and irrigation ditches; 10–2550m.

Saudi Arabia, Yemen (N), Oman, UAE, Qatar. Europe and Asia.

Chaudhary & Zawawi (1983) record *R. pulcher* as an occasional weed in Central and Eastern Saudi Arabia; these records refer to *R. dentatus*.

6. EMEX Neck.

Monoecious annual herbs, glabrous throughout. Ocreae membranous. Leaves alternate, simple. Flowers green or greenish red in axillary clusters. Male flowers: perianth segments 3–6, fused at the base; stamens 4–6. Female flowers: perianth segments 6, tubular below, accrescent and indurating in fruit, the outer 3 becoming spinescent; styles 3. Nutlets trigonous, enclosed within the hardened pitted and spinescent perianth.

1. Fruiting perianth 4–5 × 3–4mm (including spines); spines recurved, 1–3mm long; inner fruiting perianth segments not spinescent at the tip **1. E. spinosus**
+ Fruiting perianth 4–8 × 9–13mm (including spines); spines spreading, 5–6mm; inner perianth segments spinescent at the tip **2. E. australis**

1. E. spinosus (L.) Campderá, Monogr. Rumex 58 (1819). Illustr.: Collenette (1985 p.402); Fl. Qatar, pl. 18; Fl. Kuwait 1: pls.156 & 157.

Rather fleshy annual herb; stems prostrate or decumbent, 5–30cm. Leaves oblong-ovate to ovate or triangular, 2.5–8 × 1.5–4.5cm, rounded, truncate or cordate below; petioles 2–10cm. Male flowers pedicellate, c.1.5mm long. Female flowers sessile, 5–6mm long. Fruiting perianth 4–5 × 3–4mm (including spines), pitted and with 3 terminal spines; spines recurved, 1–3mm. **Map 158, Fig. 25**.

Stony, sandy and silty soils in deserts and also a weed of cultivation; 0–2900m.

Saudi Arabia, Yemen (N & S), Socotra, Oman, UAE, Qatar, Bahrain, Kuwait. Mediterranean region eastwards to Pakistan; introduced as a weed elsewhere.

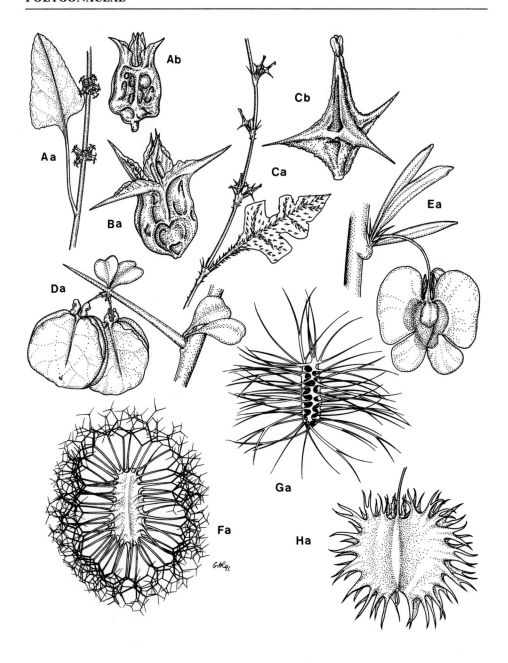

Fig. 25. Polygonaceae. A, *Emex spinosus*: Aa, flowering shoot (× 1); Ab, fruit (× 7). B, *E. australis*: Ba, fruit (× 7). C, *Oxygonum sinuatum*: Ca, fruiting shoot (× 1); Cb, fruit (× 8). D, *Atraphaxis spinosus*: Da, fruiting shoot (× 5). E, *Pteropyrum scoparium*: Ea, fruit (× 5). F, *Calligonum comosum*: Fa, part of fruit (× 4). G, *C. crinitum* subsp. *arabicum*: Ga, part of fruit (× 4). H, *C. tetrapterum*: Ha, fruit (× 4).

2. E. australis Steinh. in Ann. Sci. Nat. Bot, sér. 2, 9: 195 (1838). Illustr.: Fl. Trop. E. Afr.: 2 (1958).

Similar to *E. spinosus* but the fruiting perianth larger, 4–8 × 9–13mm (including spines); spines 5–6mm; inner perianth segments spinescent at the tip in fruit. **Map 159, Fig. 25.**

Introduced weed; 1700m.

Yemen (N). Southern Africa, introduced elsewhere.

Recorded only once from Arabia but possibly overlooked.

7. CALLIGONUM L.

Desert shrubs; main branches erect or ascending, rigid, jointed, swollen at the nodes, whitish grey; young branches green, thin, flexuous, often fascicled. Leaves minute, soon falling. Ocreae short. Flowers bisexual in axillary fascicles. Perianth segments 5, free, pale pink or whitish green with a darker mid-vein. Stamens c.15. Styles 4. Fruit indurated, winged or densely covered with stiff, simple or branched setae.

1. Fruit with 4 membranous wings, the wings toothed **3. C. tetrapterum**
+ Fruit densely covered with setae 2

2. Setae arising from 4 pairs of well-defined wings running the length of the fruit; flowering pedicels equal to or longer than the flowers **1. C. comosum**
+ Setae arising directly from the fruit surface, not from well-defined wings, sometimes flattened at the base; flowering pedicels shorter than the flowers **2. C. crinitum**

1. C. comosum L'Hér. in Trans. Linn. Soc. London 1: 180 (1791). Syn.: *C. polygonoides* subsp. *comosum* (L'Hér.) Soskov in Novosti Sist. Vyssh. Rast. 12: 153 (1975). Illustr.: Mandaville (1990 pls.56–58).

Shrub up to 120cm. Leaves subulate, 1–5mm, soon falling. Perianth lobes c.3mm; pedicels equalling or longer than the flowers. Fruits red or pale yellowish green, ellipsoid, 7–8 × 5–6mm; setae branched, 3–5mm, arising from 4 pairs of longitudinal wings. **Map 160, Fig. 25.**

Sand-dunes; 0–700m.

Saudi Arabia, ?Yemen (S), ?Socotra, Oman, UAE, Bahrain, Kuwait. Deserts from Egypt eastwards to Pakistan.

A characteristic plant of deep sand and in some areas the dominant species on sand-dunes. It is an important species for dune stabilization. Mixed populations of plants with either red or yellow fruits occur.

A single, sterile specimen from Socotra (*Virgo* 63) is probably referable to this species and another specimen (*Wissmann* 3222) from near Aden is somewhat intermediate with *C. crinitum* subsp. *arabicum*.

2. C. crinitum Boiss. subsp. **arabicum** (Soskov) Soskov, Novosti Sist. Vyssh. Rast. 12: 152. Syn.: *C. arabicum* Soskov, op. cit. 10: 134 (1973). Illustr.: Collenette (1985 p.401); Mandaville (1990 pl.59). Type: UAE, *Codrai* 34 (K).

Similar to *C. comosum* but the flowering pedicels shorter than the flowers; fruits 7–10 × 3–5mm, with less dense but generally longer (6–20mm) setae, the setae arising directly from rounded ridges on the fruit surface. **Map 161, Fig. 25.**

Sand-dunes; 10–2300m.

Saudi Arabia, Yemen (N & S), Oman, UAE. Endemic.

C. crinitum subsp. *arabicum* generally replaces *C. comosum* in sandy deserts in the south of the Peninsula and like that species also has forms with either yellow or red fruits.

3. C. tetrapterum Jaub. & Spach, Ill. pl. or. 5: t.471 (1856). Illustr.: Mandaville (1990 pl.60).

Similar to *C. comosum* but the fruits 4-winged, 10(–15) × 10(–15)mm; wings toothed, without setae. **Map 162, Fig. 25.**

Shallow sand on stony ground and in wadi-beds; 300–450m.

Saudi Arabia, Oman. Turkey, Iraq and Iran to C Asia.

Much rarer than the preceding species and, unlike them, not found on deep sand.

8. ATRAPHAXIS L.

Small shrubs. Ocreae bifid. Leaves alternate, simple, entire. Flowers bisexual, pink, in short axillary racemes. Perianth segments 4 (in Arabia), the inner pair accrescent and surrounding the fruit. Stamens 6. Stigmas 2. Nutlet lenticular.

1. A. spinosa L., Sp. pl.: 333 (1753). Illustr.: Collenette (1985 p.401).

Small shrub up to 1m; old branches whitish brown, becoming spinescent. Leaves ovate to orbicular, 3–5(–11) × 2–3(–11)mm, rounded, shortly petiolate. Inner pair of fruiting perianth segments scarious, ovate to orbicular, 4–5(–8) × 5–6(–11)mm. **Map 163, Fig. 25.**

Open hillsides and rocky slopes in deserts; 1650–2000m.

Saudi Arabia. SE Europe, SW and C Asia.

9. PTEROPYRUM Jaub. & Spach

Shrubs. Leaves alternate or fascicled, linear, entire. Ocreae small. Flowers bisexual, in axillary fascicles. Perianth segments 5, free; outer pair reflexed in fruit, the inner 3 appressed to the nutlet. Stamens 8. Styles 3. Nutlets trigonous, 3-winged; wings scarious, rounded, divided by a deep sinus above the middle.

1. P. scoparium Jaub. & Spach, Ill. pl. or. 2: t.109 (1844). Type: Oman, *Aucher-Eloy* 5720 (BM, K).

Much-branched shrub; branches whitish-grey. Leaves linear, 3–10 × 0.5–1mm. Flowers white or cream with green veins, 1.5–2mm long; pedicels 4–6mm. Fruits cream at first, ripening red, reflexed, ovate-orbicular in outline, 6–7 × 6–7mm, cordate at the base. **Map 164, Fig. 25.**

Gravelly and sandy plains, wadi-sides, open mountain slopes and open *Acacia* bushland; 0–650m.

Oman, UAE. Endemic.

Closely related to, and possibly conspecific with, *P. aucheri* Jaub. & Spach from Iran, Afghanistan and Pakistan.

Family 33. NYCTAGINACEAE

A.G. MILLER

Herbs or climbing shrubs. Leaves alternate or opposite, simple, exstipulate. Flowers bracteate, in terminal or axillary cymes, sometimes whorled, umbellate or glomerulate; actinomorphic, unisexual or bisexual. Perianths 5-lobed, tubular, petaloid, the lower part persistent around the ovary. Stamens 1–10. Ovary superior, 1-celled, 1-ovulate; style simple. Fruit (the anthocarp) a 1-seeded achene enclosed by the persistent part of the perianth.

Species of *Bougainvillea* are commonly cultivated as ornamentals in Arabia. They are woody climbers with insignificant flowers concealed by large, showy bracts.

1.	Woody climbers or small trees; stems armed with curved spines	**4. Pisonia**
+	Herbs; stems not spiny	2
2.	Flowers showy, 3.5–4cm long, surrounded by a calyx-like involucre of bracts	**1. Mirabilis**
+	Flowers small, up to 1cm; bracts inconspicuous	3
3.	Fruits 5-ribbed, glabrous or glandular-hairy	**2. Boerhavia**
+	Fruits obscurely 10-ribbed, with conspicuous sessile or stalked, wart-like glands	**3. Commicarpus**

1. MIRABILIS L.

Perennial herb. Leaves opposite. Flowers bisexual, in crowded terminal cymes, showy; bracts united into a persistent 5-lobed and calyx-like involucre. Perianths trumpet-shaped. Stamens 5–6, exserted. Stigma capitate, exserted. Fruit ribbed, eglandular.

1. M. jalapa L., Sp. pl.: 177 (1753). Illustr.: Fl. Pakistan 115: p.2 (1977).

Robust herb; stems erect or sprawling, up to 1m. Leaves triangular-ovate, 5–12 × 2–7cm, acuminate, the base truncate or cordate; petiole 1–3cm. Involucre c.1cm long, campanulate, divided to the middle into 5 acute teeth. Perianths red, rarely white or yellow, 3.5–4cm; limb 5-lobed, 2–3.5cm across. Fruit subglobose, 7–8 × 5mm. **Map 165, Fig. 26.**

Mainly near villages and abundant in high rainfall areas or near permanent water; (0–) 1000–2500m, the lower altitudes refer to cultivated material.

Saudi Arabia, Yemen (N & S), Kuwait. A native of tropical America, cultivated as an ornamental throughout the tropics and widely naturalized; now rarely cultivated in Arabia.

2. BOERHAVIA L.

Annual or perennial herbs, sometimes woody-based. Leaves opposite, entire. Flowers bisexual in terminal or axillary umbellate heads or open panicles; bracts minute. Perianths funnel-shaped or campanulate. Stamens 1–3, shortly exserted. Fruit 5-ribbed, glabrous or glandular-hairy.

1. Stems erect; flowers solitary or rarely paired on capillary peduncles; inflorescence a diffuse, much-branched, terminal panicle **1. B. elegans**
+ Stems prostrate, ascending or sprawling; flowers in (2–)3–many-flowered pedunculate clusters which are either axillary or form a loose terminal panicle by reduction of the upper leaves 2
2. Stems prostrate; flowers in simple axillary pedunculate clusters; clusters shorter than or barely exceeding the leaves; fruits frequently crowned by the persistent perianth **2. B. repens**
+ Stems sprawling or ascending; flower clusters axillary or terminal, often in panicles; peduncles exceeding the leaves; fruits not usually crowned by the persistent perianth **3. B. diffusa**

1. B. elegans Choisy in DC., Prodr. 13 (2): 453 (1849). Syn.: *B. rubicunda* Steud., Nomencl. bot., ed. 2, 1: 213 (1841) nom. nud. Type: Saudi Arabia, *Schimper* 744 (E, K).

Woody-based perennial; stems erect, up to 50cm, puberulous below. Leaves narrowly lanceolate to oblong-ovate, 20–50(–60) × 3–10(–17)mm; petioles up to 10mm. Flowers solitary or rarely paired, on capillary peduncles in erect, much-branched, diffuse panicles. Perianths pink or reddish purple, 2–3mm long. Fruit clavate, 3–4.5 × 1–1.25mm, puberulous in the furrows. **Map 166, Fig. 26.**

1. Inflorescence ± regularly dichotomously dividing; pedicels usually more than 1.5cm long; leaves usually narrowly lanceolate to narrowly oblong
subsp. **stenophylla**

\+ Inflorescence with smaller secondary branches arising from a ± continuous primary axis; pedicels usually less than 1cm; leaves ovate to ovate-lanceolate
subsp. **elegans**

subsp. **elegans**

Dry rocky slopes and wadi-beds in semi-desert; usually on basalt; 0–650m.

Saudi Arabia, Yemen (S). Endemic.

subsp. **stenophylla** (Boiss.) A.G. Miller in Edinb. J. Bot. 51 (1): 40 (1994). Syn.: *B. elegans* var. *stenophylla* Boiss., Fl. Orient. 4: 1046 (1879). Illustr.: Fl. Pakistan, 115: p.8 (1977) as *B. rubicunda*; Collenette (1985 p.375); Western (1989 p.44). Type: Oman, *Aucher-Eloy* 5250 (G).

Open rocky slopes, wadi-beds and roadsides in semi-desert; usually on limestone; 100–900m.

Saudi Arabia, Yemen (S), Oman, UAE. Ethiopia, Sudan, Kenya, Tanzania, Uganda, Iran, Pakistan and India.

The two varieties of *B. elegans* are clearly distinguished by their inflorescences: in subsp. *stenophylla* the inflorescence is an open, dichotomising panicle with the fruits borne on long, delicate pedicels giving the plants a mist-like appearance when viewed from a distance; subsp. *elegans* has a more or less monochasially branching panicle and shorter pedicels which never give the mist-like appearance so characteristic of subsp. *stenophylla*. Subsp. *stenophylla* is widespread in the SE of the Arabian peninsula but in the west is restricted to the dry, limestone plateaus of the interior; outside Arabia it is distributed from Pakistan to E Africa. Subsp. *elegans* is endemic to Arabia where it has a disjunct distribution being found in two small areas, apparently always on basalt, around Jeddah in Saudi Arabia and around Aden in Yemen.

2. B. repens L., Sp. pl.: 3 (1753). Syn.: *B. diandra* L., Sp. pl.: 1194 (1753); *B. repens* L. var. *glabra* Choisy in DC., Prodr. 13(2): 453 (1849). Illustr.: Fl. Pakistan, 115 p.6 (1977) as *B. diandra*.

Annual or perennial herb; stems prostrate, up to 75cm, glabrous or sparsely puberulous. Leaves ovate-triangular or oblong-ovate, 5–20 × 5–15mm, often whitish beneath. Flowers in short-stalked axillary clusters; peduncles 5–8mm; clusters 1–6-flowered. Perianths pale pink or white, c.1mm. Fruit 3 × 1mm, thinly pubescent and glandular, usually crowned with the persistent perianth. **Map 167, Fig. 26.**

Sandy, silty and stony ground, roadsides, wadis etc.; 50–1700m.

Saudi Arabia, Yemen (N & S). Widespread in tropical Africa and Asia.

3. B. diffusa L., Sp. pl.: 3 (1753). Syn.: *B. ascendens* Willd., Sp. pl. 1: 19 (1797); *B. coccinea* Miller, Gard. dict. ed. 8, 4 (1768); *B. diffusa* var. *viscosa* (Lagasca & Rodriguez) Heimerl in Beitr. Syst. Nyctag.: 27 (1897); *B. glutinosa* sensu Deflers (1889 p.192) non Vahl; *B. repens* sensu Balf.f. (1888) non L.; *B. repens* var. *diffusa* (L.)

NYCTAGINACEAE

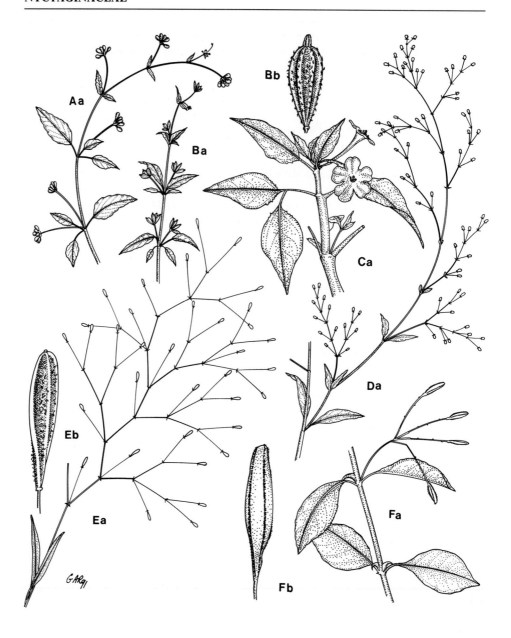

Fig. 26. Nyctaginaceae (1) A, *Boerhavia diffusa*: Aa, flowering shoot (× 1). B, *B. repens*: Ba, flowering shoot (× 1); Bb, fruit (× 7). C, *Mirabilis jalapa*: Ca, flowering shoot (× 0.5). D, *Boerhavia elegans* subsp. *elegans*: Da, flowering shoot (× 0.6). E, *B. elegans* subsp. *stenophylla*: Ea, flowering shoot (× 0.6); Eb, fruit (× 10). F, *Pisonia aculeata*: Fa, flowering shoot (× 0.5); Fb, fruit (× 3).

Boiss., Fl. orient. 4: 1046 (1879); *B. repens* var. *viscosa* Choisy in DC., Prodr. 13(2): 453 (1849); *B. repens* sensu Collenette (1985) non L. Illustr.: Collenette (1985 p.375) as *B. repens*.

Annual or perennial herb. Stems prostrate or sprawling, up to 1m long, glabrous, sparsely pubescent or sparsely to densely glandular, sometimes viscid. Leaves ovate or oblong-ovate to subcircular, 10–50 × 5–45mm, often paler beneath. Flowers in long-stalked axillary clusters or (by reduction of the upper leaves) ± leafless panicles; peduncles 1–3cm; clusters 3–12-flowered. Perianths usually pink, rarely magenta, 1.25–1.5mm long. Fruit 3–4 × 1.25–1.5mm, puberulous and usually glandular-viscid, rarely ± glabrous. **Map 168, Fig. 26**.

Found in a wide variety of habitats including sandy and gravelly plains, wadis, rocky slopes, *Acacia-Commiphora* bushland, field borders and roadsides; 0–2300m.

Saudi Arabia, Yemen (N & S), Socotra, Oman, UAE. Pantropical.

A confusing complex badly in need of a world-wide revision. I have been unable to arrive at a satisfactory treatment for the Arabian plants, and therefore this account must be considered provisional. Within *B. diffusa* s.l. different authors have recognized a number of taxa at specific, subspecific and varietal level. Extreme forms are fairly distinct, but the numerous intermediates make it impossible to define recognisable taxa. For alternative treatments see Codd in *Bothalia* 9 (1): 113–121 (1966) and Stannard in Fl. Zamb. 9 (1): 20–25 (1988).

3. COMMICARPUS Standley

Annual or perennial herbs, sometimes woody-based. Leaves opposite, entire or sinuously lobed. Flowers bisexual, in dense heads, umbels or whorls; bracts minute. Perianths funnel-shaped, tubular below. Stamens 2–6, exserted. Stigma capitate, exserted. Fruit 10-ribbed, studded with conspicuous, sessile or stalked wart-like glands.

Meikle, R.D. (1978). A key to *Commicarpus*. *Notes Roy. Bot. Gard. Edinburgh* 36: 235–249.

Measurements referring to the perianth exclude the basal anthocarp.

1.	Flowers in shortly pedunculate axillary umbels which are shorter than or equalling the subtending leaves; peduncles up to 5mm	**14. C. reniformis**
+	Flowers in inflorescences longer than the subtending leaves; peduncles more than 10mm	2
2.	Flowers numerous, sessile or shortly pedicellate, crowded into a dense head borne on a stout peduncle	**1. C. pedunculosus**
+	Flowers in umbels or whorls but not in dense heads	3
3.	Stems, at least in the region of the inflorescence, conspicuously glandular-pilose and very sticky	4
+	Stems glabrous, scabridulous, puberulous or clothed with a short indumentum of crispate hairs, not sticky	6

4.	Buds densely covered with stalked glands; flowers usually in umbels; inflorescence-region leafy	**2. C. grandiflorus**
+	Buds glabrous or thinly pilose; flowers in whorls or umbels; inflorescence-region leafless	5
5.	Flowers in whorls; stems with crisped hairs below; stamens 3	**3. C. stenocarpus**
+	Flowers in umbels or rarely in whorls; stems glabrous or scabridulous below; stamens 3–5	**8. C. heimerlii**
6.	Stems, at least in the lower part, closely crispate-pubescent or puberulous	7
+	Stems glabrous, or at most sparsely scabridulous	9
7.	Leaves sinuate or lobed; flowers sessile or very shortly pedicellate	**6. C. sinuatus**
+	Leaves entire; flowers distinctly pedicellate	8
8.	Leaves fleshy; stems crispate-pubescent; fruit pubescent between the glands	**9. C. mistus**
+	Leaves not fleshy; stems sparsely puberulous; fruit glabrous between the glands	**4. C. arabicus**
9.	Inflorescence of umbels	10
+	Inflorescence of whorls	11
10.	Inflorescence-region a leafless mass of slender, intricately branched peduncles and pedicels; fruit fusiform, inconspicuously glandular	**7. C. simonyi**
+	Inflorescence an open panicle of umbels; fruit clavate, prominently glandular at the tip	**4. C. arabicus**
11.	Fruits clavate	12
+	Fruits fusiform	14
12.	Flowers sessile or very shortly pedicellate; pedicels usually less than 4mm (those in the lowest whorl sometimes longer); perianths up to 2.5mm long	**13. C. helenae**
+	Flowers distinctly pedicellate, pedicels often more than 1cm; perianths 3–5mm long	13
13.	Inflorescence normally of 1–2(–3) whorls of flowers; perianths widely infundibuliform, without a distinct basal tube; stamens 3(–4)	**10. C. boissieri**
+	Inflorescence of 3–4(–5) whorls of flowers; perianths narrowly infundibuliform, with a distinct basal tube; stamens 2, rarely 3	**11. C. ambiguus**
14.	Perianths 2–3mm, broadly infundibuliform, without a basal tube; stamens 2	**12. C. adenensis**
+	Perianths 7–10mm, narrowly infundibuliform, with a well-developed basal tube; stamens 3	15
15.	Perianths white; inflorescences mainly of whorls	**5. C. plumbagineus**

\+   Perianths mauve, pink or purple; inflorescence-region a leafless mass of slender, intricately branched peduncles and pedicels   **7. C. simonyi**

1. C. pedunculosus (A. Rich.) Cuf., En. Pl. Aeth.: 79 (1953). Syn.: *Boerhavia pedunculosa* A. Rich. (1850 p.210).

Stems finely pubescent. Leaves ovate to ovate-triangular, up to 4cm, entire. Flowers sessile or shortly pedicellate, crowded into dense heads at the tips of stout peduncles. Perianths magenta, c.6mm long, narrowly funnel-shaped with a well-developed basal tube, externally thinly puberulous. Stamens 3–6. Fruit clavate, prominently gland-warted at the apex, smooth below and lacking glands. **Map 169, Fig. 27.**

Cliffs, stony plains and roadsides, often on volcanic rocks; 2100–2600m.

Yemen (N). Rwanda, Somalia, Sudan, Ethiopia, Uganda, Kenya and Tanzania.

All specimens examined from Arabia have 3–4 not 5–6 stamens as stated by Meikle (op. cit.).

2. C. grandiflorus (A. Rich.) Standley in Contr. U.S. Natl. Herb. 18: 101 (1916). Syn.: *Boerhavia grandiflora* A. Rich. (1850 p.209); *B. plumbaginea* Cav. var. *grandiflora* (A.Rich.) Schweinf. (1896 p.167). Illustr.: Collenette (1985 p.375).

Stems glandular-pilose, viscid. Leaves ovate to ovate-triangular, up to 6cm, entire. Flowers pedicellate, in umbels; pedicels 3–8mm. Perianths pink, mauve or purple, 6–8mm long, narrowly funnel-shaped with a well-developed basal tube, densely glandular externally. Stamens 3. Fruits clavate, puberulous, prominently gland-warted over their entire surface. **Map 170, Fig. 27.**

Rocky slopes and roadsides; 1100–2800m.

Saudi Arabia, Yemen (N). Chad, E and NE tropical Africa and India.

3. C. stenocarpus (Chiov.) Cuf., Enum. Pl. Aeth.: 81 (1953). Syn.: *Boerhavia stenocarpa* Chiov., Fl. Somala: 283 (1929).

Stems shortly glandular-pilose, viscid, with crispate hairs below. Leaves ovate or subcircular to transversely ovate, up to 4cm, entire. Flowers shortly pedicellate, in whorls; pedicels 1–4mm. Perianths mauve or purple, (4–)7–8mm long, narrowly infundibuliform with a well-developed basal tube, glabrous or thinly pilose externally. Stamens 3. Fruits fusiform, with gland-warts prominent at the tip and sessile or inconspicuous below. **Map 171, Fig. 27.**

Dry rocky slopes, rocky and sandy plains and wadi-beds; 0–1000m.

Yemen (S), Oman, UAE. Iran, Pakistan and Somalia.

4. C. arabicus Meikle in Kew Bull. 32: 474 (1978). *Boerhavia plumbaginea* Cav. var. *forskalei* Schweinf. (1896 p.167). Type: Yemen (N), *Hepper* 5659 (K).

Stems glabrous or sparsely scabridulous. Leaves not fleshy, ovate, up to 5cm, entire. Flowers distinctly pedicellate, in umbels, very rarely in whorls or irregular. Perianths mauve, pink or purple, c.5–7mm long, narrowly infundibuliform with a well-

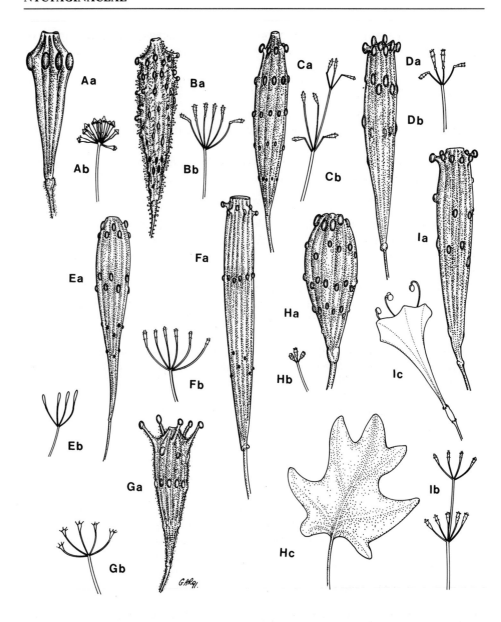

Fig. 27. Nyctaginaceae. A, *Commicarpus pedunculosus*. B, *C. grandifolius*. C, *C. stenocarpus*. D, *C. arabicus*. E, *C. heimerlii*. F, *C. simonyi*. G, *C. mistus*. H, *C. sinuatus*: Hc, leaf (×3). I, *C. plumbagineus*: Ic, flower (×10) - a, fruits (all ×7); b, inflorescences (all ×1).

developed basal tube, externally glabrous, sparsely glandular or puberulous. Stamens 3. Fruits clavate, prominently gland-warted at the tip, ± smooth and inconspicuously glandular below. **Map 172, Fig. 27.**

Rocky and disturbed ground, roadsides etc.; 1500–2900m.

Saudi Arabia, Yemen (N & S). Endemic.

The records of this species from Saudi Arabia need confirmation.

5. **C. plumbagineus** (Cav.) Standley in Contr. U.S. Natl. Herb. 18: 101 (1916). Syn.: *Boerhavia dichotoma* Vahl, Enum. pl. 1: 290 (1804); *B. plumbaginea* Cav., Icon. 2: 7, t.112 (1793); *B. plumbaginea* Cav. var. *dichotoma* (Vahl) Asch. & Schweinf. (1896 p.167); *B. plumbaginea* Cav. var. *glabrata* Boiss., Fl. orient. 4: 1044 (1879); ?*B. scandens* Forsskal (1775) non L.; *B. verticillata* Poiret in Lam., Encycl. 5: 55 (1804); *Commicarpus verticillatus* (Poiret) Standley in Contr. U.S. Natl. Herb. 18: 101 (1916); *Valeriana scandens* sensu Forsskal (1775 p.12) non Loefl. Illustr.: Collenette (1985 p.376).

Stems glabrous or at most sparsely scabridulous. Leaves ovate to subcircular, up to 5cm, entire or sinuate. Flowers distinctly pedicellate, mainly in whorls. Perianths white, 8–10mm long, narrowly infundibuliform with a well-developed basal tube, externally puberulous. Stamens 3. Fruit narrowly fusiform, prominently gland-warted at the tip, with scattered inconspicuous sessile glands below. **Map 173, Fig. 27.**

Rocky and sandy plains, wadi-beds, walls, *Euphorbia* shrubland and as a weed of irrigated ground; 400–2400m.

Saudi Arabia, Yemen (N & S), Bahrain. Spain, tropical and southern Africa and Madagascar.

6. **C. sinuatus** Meikle in Kew Bull. 29: 83 (1974). Syn.: *Boerhavia plumbaginea* Cav. var. *viscosa* Boiss., Fl. orient. 4: 1044 (1879). Illustr.: Collenette (1985 p.376).

Stems puberulous throughout. Leaves sinuate or lobed, up to 4cm. Flowers sessile or shortly pedicellate, in umbels. Perianths pink, 6–9mm long, narrowly infundibuliform with a well-developed basal tube, externally puberulous. Stamens 3–4. Fruit clavate, puberulous and covered with sessile glands. **Map 174, Fig. 27.**

Dry rocky slopes, sandy wadis and disturbed ground; 100–2600m.

Saudi Arabia, Yemen (N & S). Sinai, Somalia and Ethiopia.

7. **C. simonyi** (Heimerl & Vierh.) Meikle in Hooker's Icon. Pl. 37: t.3694 (1971). Syn.: *Boerhavia scandens* sensu Balf.f. (1888) non L.; *B. simonyi* Heimerl & Vierh. in Oesterr. Bot. Z. 53: 435 (1903); *B. plumbaginea* Cav. var. *socotrana* Heimerl. Type: Yemen (Socotra), *Simony* (WU).

Stem glabrous, or at most very sparsely scabridulous. Leaves ovate-triangular to subcircular, up to 5cm, entire. Flowers distinctly pedicellate, in umbels or rarely and abnormally in whorls or irregular, forming a leafless mass of slender, intricately branched peduncles and pedicels; pedicels up to 20mm. Perianths mauve, pink or purple, 7–7.5mm long, narrowly infundibuliform with a well-developed basal tube. Stamens 3(–5). Fruit narrowly fusiform, inconspicuously glandular. **Map 175, Fig. 27.**

Rocky slopes in semi-deciduous thicket on both limestone and granite; 200–850m.

Socotra. Endemic.

8. C. heimerlii (Vierh.) Meikle in Hooker's Icon. Pl. 37: t.3694 (1971). Syn.: *Boerhavia heimerlii* Vierh. in Oesterr. Bot. Z. 53: 435 (1903); *B. ? scandens* sensu Balfour (1888) non. L. Type: Yemen (Socotra), *Simony* (WU).

Stem glabrous towards the base, viscid above. Leaves fleshy, ovate to subcircular, up to 5cm, entire. Flowers distinctly pedicellate, in umbels, very seldom in whorls or irregular; pedicels 5–7mm. Perianths pink or purple, 6–7.5mm long, narrowly infundibuliform with a well-developed basal tube, externally glabrous. Stamens 3–5. Fruit clavate, prominently glandular at the tip, inconspicuously glandular below. **Map 176, Fig. 27.**

Rocky slopes in semi-deciduous thicket on limestone; 15–350m.

Socotra. Endemic.

9. C. mistus Thulin in Nord. J. Bot. 10(4): 405 (1990). Syn: *C. squarrosus* sensu auctt. non (Heimerl) Standley. Illustr.: Collenette (1985 p.376) as *C. squarrosus*.

Stem crispate-pubescent towards the base. Leaves fleshy, ovate, up to 3(–5)cm, entire. Flowers distinctly pedicellate, in umbels, very seldom in whorls or irregular; pedicels 5–10mm. Perianths pink or purple, 4–5(–8)mm long, narrowly infundibuliform with a well-developed basal tube, externally puberulous. Stamens 2–3. Fruits clavate, puberulous with prominent long-stalked glands at the tip and inconspicuously glandular below. **Map 177, Fig. 27.**

Dry rocky hills, particularly on limestone; 900–2700m.

Saudi Arabia, Yemen (N & S), Oman. Ethiopia, Somalia and Kenya.

A specimen from Saudi Arabia (*Collenette* 6653) differs from typical plants in its larger flowers, fruits and leaves.

10. C. boissieri (Heimerl) Cuf., Enum. Pl. Aeth.: 79 (1953). Syn.: *Boerhavia boissieri* Heimerl in Akad. Wiss. Wien Math.-Naturwiss. Kl., Denkschr. 71: 346 (1907).

Stems glabrous. Leaves ovate to subcircular, up to 7cm, entire or sinuate. Flowers distinctly pedicellate, in whorls; pedicels 6–15mm. Perianths pink, purple or magenta, 3–5mm long, widely infundibuliform, with a very short, inconspicuous basal tube, externally glabrous. Stamens usually 3(–4). Fruits clavate, prominently gland-warted at the tip, with scattered prominent glands below. **Map 178, Fig. 28.**

Dry rocky hills and *Euphorbia balsamifera* shrubland; 0–1000m.

Socotra, Oman. Pakistan and India.

11. C. ambiguus Meikle in Kew Bull. 38: 481 (1983).

Stems glabrous. Leaves broadly ovate, 3–6cm, entire or obscurely sinuate. Flowers in whorls with long, slender pedicels up to 20mm. Perianths purple, 3–5(–8)mm long, narrowly infundibuliform with a short, but distinct, basal tube, externally glabrous. Stamens 2(–3). Fruits clavate, with prominent, long-stalked glands at the tip and scattered prominent glands below. **Map 179, Fig. 28.**

Dry rocky and stony hills; 1220–2300m.

Saudi Arabia, Yemen (N). Ethiopia and Somalia.

12. C. adenensis A.G. Miller in Edinb. J. Bot. 51 (1): 39 (1994). Type: Yemen (S), *Miller et al* 8067 (E).

Stems glabrous or occasionally minutely scabridulous. Leaves ovate or rounded, up to 5cm, entire or sinuously lobed. Flowers shortly pedicellate (the pedicels up to 10mm in fruit), in irregular panicles of 2–3 whorls. Perianths pink or mauve, 2–3mm, widely infundibuliform, with a short inconspicuous basal tube, externally glabrous. Stamens 2. Fruit fusiform, prominently glandular at the tip and with inconspicuous glands below. **Map 180, Fig. 28.**

Dry volcanic slopes; 0–1100m.

Yemen (S). Endemic.

13. C. helenae (Roemer & Schultes) Meikle in Hooker's Icon. Pl. 37: t. 3694 (1971). Syn.: *Boerhavia helenae* Schultes in Roemer & Schultes, Mant. 1: 73 (1822); *B. verticillata* sensu Schwartz (1939) non Poiret; *Commicarpus stellata* Berhaut in Bull. Soc. Bot. France 100: 51 (1953). Illustr.: Collenette (1985 p.376).

Stems glabrous or sparsely scabridulous. Leaves ovate-triangular, up to 5(–9)cm, entire or sinuously lobed. Flowers sessile or shortly pedicellate, in narrow whorls; pedicels up to 4mm. Perianths pale pink or purple, 1.5–2.5mm long, widely infundibuliform, with a very short, inconspicuous basal tube, externally glabrous. Stamens usually 2. Fruits clavate, with prominent, long-stalked glands at the tip, inconspicuously glandular below. **Map 181, Fig. 28.**

Stony ground, roadsides and a weed of cultivation; 10–2300m.

Saudi Arabia, Yemen (N & S), Socotra, Oman, UAE. Tropical Africa, Egypt, Palestine, Iran and India.

An inadequate specimen (*Hillcoat* 395) from near Riyadh in Saudi Arabia may be referable to this species.

14. C. reniformis (Chiov.) Cuf. in Bull. Jard. Bot. Brux. 23, Suppl.: 80 (1953).

Stems crisped-puberulous. Leaves reniform, up to 3.5 × 4cm, glabrous or thinly puberulous. Flowers in shortly pedunculate axillary umbels; peduncles up to 5mm; pedicels up to 9mm. Perianth mauve or purple, 8–9mm, narrowly infundibuliform

NYCTAGINACEAE

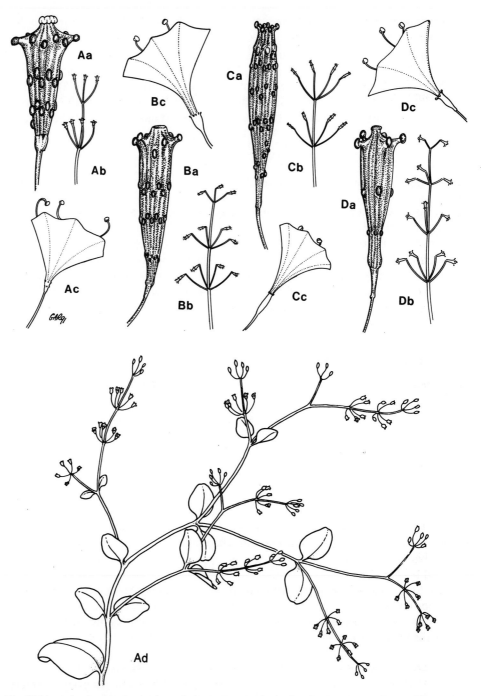

Fig. 28. Nyctaginaceae. A, *Commicarpus boissieri*. B, *C. ambiguus*. C, *C. adenensis*. D, *C. helenae* - a, fruits (all × 7); b, inflorescences (all × 1); c, flowers (all × 10); d, habit (× 1).

with a well-developed basal tube. Stamens 2–3. Fruits clavate, glabrous with prominent long-stalked glands at the tip and inconspicuously glandular below.

Gypsum hills; 1020m.

Yemen (S). Somalia.
In Arabia only collected once in an area of gypsum hills north of Mukalla.

4. PISONIA L.

Dioecious climbing shrub or small tree; stems armed with curved spines. Leaves opposite or alternate. Flowers in dense axillary pedunculate cymes which expand in fruit; bracts minute. Perianths infundibuliform or campanulate. Stamens 6–8, exserted. Stigma fimbriate, exserted. Fruit ribbed with rows of glandular hairs along the ribs.

1. P. aculeata L., Sp. pl.: 1026 (1753). Illustr.: Fl. Zamb. 9 (1): 26 (1988).

Stems up to 8m, with opposite spreading branches. Leaves elliptic to obovate or ovate, up to 10 × 7cm, acute or obtuse, the base obtuse to cuneate, tomentose beneath; petiole up to 2.5cm. Flowers cream or yellow-green, 5-lobed; male flowers infundibuliform, c.3mm long; ovary vestigial; female flowers campanulate, c.2mm long; stamens absent. Fruit narrowly ellipsoid, 6–10 × 2–2.5mm. **Map 182, Fig. 26.**

Riverine woodland; 500–1400m.

Saudi Arabia, Yemen (N). Widespread in Old and New World tropics.

Family 34. AIZOACEAE
(including Gisekiaceae, Molluginaceae and Tetragoniaceae)

A.G. MILLER

Succulent or subsucculent annual or perennial herbs or rarely subshrubs. Leaves opposite, alternate or whorled, simple, entire; stipules absent or membranous. Flowers regular, bisexual, axillary or terminal, solitary or in cymes or fascicles. Sepals 5, free or united below. Petals small or absent or replaced by petaloid staminodes. Stamens 5–many. Staminodes often present, sometimes showy and petaloid. Ovary superior or inferior, 1–5-celled; carpels 2–5, fused or rarely free; placentation axile, parietal, basal or apical; ovules 1–many per cell. Fruit a capsule opening by valves or circumscissile, sometimes indehiscent or splitting into 1-seeded mericarps.

1. Delicately branched erect or ascending annual herbs; stem-leaves linear, whorled or if absent then the leaves all in a basal rosette **5. Mollugo**
+ Prostrate or decumbent annual or perennial herbs or shrublets; stem-leaves broader, alternate, opposite or whorled, never in a basal rosette 2

2.	Petals or petaloid staminodes present	3
+	Petaloid staminodes absent (sepals sometimes petaloid in *Aizoon*)	7
3.	Plants glandular-pubescent; fruit splitting into 2 indehiscent 1-seeded mericarps	**2. Limeum**
+	Plants glabrous or papillose; fruit a many-seeded capsule	4
4.	Leaves alternate, elliptic to broadly obovate, flat in section	5
+	Leaves opposite, ± narrowly oblong, terete in section, very fleshy	6
5.	Petaloid staminodes many (petals absent), pink or magenta; stamens many; capsule not surrounded by the persistent calyx	**3. Corbichonia**
+	Petals 5, white; stamens 5; capsule surrounded by the persistent calyx	**13. Telephium**
6.	Annual herbs; petaloid staminodes white; plants of saline coastal areas and salt flats	**11. Mesembryanthemum**
+	Perennial herbs; petaloid staminodes pink; plants of rocky mountainous areas	**10. Delosperma**
7.	Leaves alternate	8
+	Leaves opposite	9
8.	Fruit dry and indehiscent; ovary inferior; prostrate, diffusely branched herb	**12. Tetragonia**
+	Fruit a many-seeded capsule dehiscing by 5 valves; ovary superior; prostrate or ascending herbs, not diffusely branched	**9. Aizoon**
9.	Carpels free; fruit a cluster of 5 achenes; plant marked throughout with numerous linear raphides	**1. Gisekia**
+	Carpels united; fruit a capsule; plant without linear raphides	10
10.	Plants softly tomentose; sepals free	**4. Glinus**
+	Plants glabrous or papillose; sepals united into a tube below	11
11.	Capsule dehiscing by 5 valves	**9. Aizoon**
+	Capsule circumscissile or dehiscing by a 2-valved lid	12
12.	Style 1	**7. Trianthemum**
+	Styles 2–5	13
13	Flowers clustered; fruit a 4-seeded capsule dehiscing by means of a 2-valved lid, the valves usually separating	**8. Zaleya**
+	Flowers solitary; fruit a several-seeded (usually more than 4) capsule dehiscing by a circumscissile lid	**6. Sesuvium**

1. GISEKIA L.

Annual herbs, marked throughout with numerous linear raphides. Leaves opposite, sub-succulent; stipules absent. Flowers in axillary, sessile or pedunculate fascicles. Sepals 5, free. Petaloid staminodes absent. Stamens 5–20. Ovary superior, of 5 free carpels, the carpels 1-ovuled. Fruit a cluster of 5 achenes.

1. G. pharnaceoides L., Mant. pl. 2: 562 (1771). Syn.: *Pharnaceum occultum* (Forsskal, 1775 p.58). Illustr.: Fl. Trop. E. Afr.: 4 (1961); Collenette (1985 p.37); Western (1989 p.35).

Glabrous herb; stems prostrate, up to 30cm. Leaves subsessile, linear-oblong or narrowly oblanceolate to elliptic, 5–30(–40) × 1–8(–12)mm, obtuse or acute. Flowers greenish or reddish, in congested fascicles; fascicles sessile or on up to 15(–50)mm peduncles; pedicels 2–10mm. Sepals ovate, c.2mm. Stamens 5. Achenes c.1.25mm long, papillose. **Map 183, Fig. 29.**

Sandy and stony areas, stabilized dunes and on granite; cultivated and waste ground, and roadsides; 15–1050m.

Saudi Arabia, Yemen (N & S), Socotra, Oman, UAE. Africa, Palestine, Iran, Pakistan, India, Sri Lanka and the Mascarenes.

2. LIMEUM L.

Herbs or shrublets, frequently glandular-hairy. Leaves opposite or alternate; stipules absent. Flowers solitary or in axillary cymes. Sepals 5, free, with broad membranous margins. Stamens c.7; staminodes c.5, membranous, petaloid, clawed, toothed at the tip. Ovary superior, 2-celled; styles 2. Fruit splitting into two 1-seeded mericarps.

Friedrich, H.C. (1956). Revision der Gattung Limeum L. *Mitt. Bot. Staatssamml. München* 2: 133–166.

1.	Mericarps reticulately rugose, greyish brown	**1. L. arabicum**
+	Mericarps smooth, brown	**2. L. obovatum**

1. L. arabicum Friedrich, op. cit.: 156 (1956). Illustr.: Collenette (1985 p.37); Western (1989 p.34). Type.: Yemen (S), *Popov et al.* 4195 (BM, K).

Intricately branched shrublet or perennial herb, glandular-hairy throughout; stems up to 80cm. Leaves sub-opposite, elliptic or ovate to sub-orbicular, 2–10 × 2–8mm, obtuse or apiculate. Flowers solitary or paired. Sepals broadly ovate, 4–5 × 1.5–3.5mm. Staminodes oblong, 3–4.5 × 1.5–2.2mm. Fruit recurved; mericarps reniform, 2.5–3.5mm long, reticulately rugose, greyish brown. **Map 184, Fig. 29.**

Sand-dunes and sandy areas in wadis etc.; 50–850m.

Saudi Arabia, Yemen (S), Oman, UAE. Endemic.

2. L. obovatum Vicary in J. Asiat. Soc. Bengal 16: 1163 (1847). Syn.: *L. humile* sensu Mandaville (1990) non Forsskal (1775) [= *Andrachne telephioides* sensu Vahl]; *L.*

AIZOACEAE

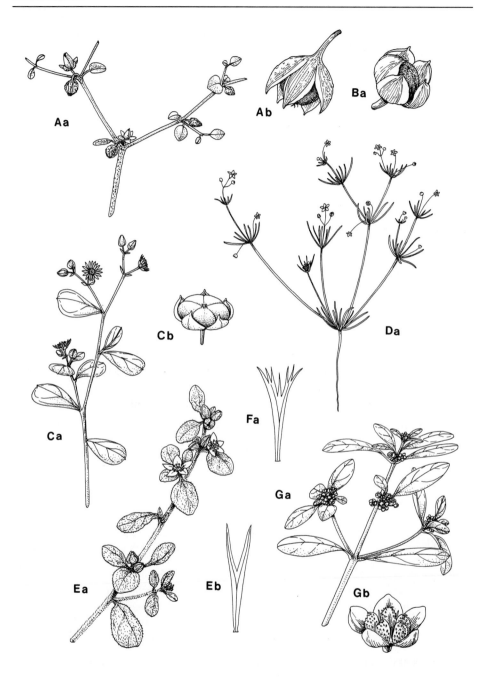

Fig. 29. Aizoaceae. A, *Limeum arabicum*: Aa, flowering shoot (×0.6); Ab, fruit (×6). B, *L. obovatum*: Ba, fruit (×6). C, *Corbichonia decumbens*: Ca, flowering shoot (×0.6); Cb, fruit (×3). D, *Mollugo cerviana*: Da, habit (×0.6). E, *Glinus lotoides*: Ea, flowering shoot (×0.6); Eb, staminode (×7). F, *G. setiflorus*: Fa, staminode (×7). G, *Gisekia pharnaceoides*: Ga, habit (×0.6); Gb, fruit (×7).

indicum Stocks ex T. Anderson (1860 p.30). Illustr.: Fl. Pakistan 41: 9 (1973) as *L. indicum*; Collenette (1985 p.38).

Similar to *L. arabicum* but usually a prostrate herb, sometimes woody below; sepals 3–4.5mm; staminodes c.2 × 1–1.4mm; mericarps smooth, brown. **Map 185, Fig. 29.**

Sandy and gravelly plains, flood plains and wadi-beds; 0–850m.

Saudi Arabia, Yemen (N & S), Oman, UAE. SW Africa, Sudan, Ethiopia and Pakistan.

3. CORBICHONIA Scop.

Annual or perennial herbs. Leaves alternate, petiolate; stipules absent. Flowers in terminal or leaf-opposed cymes. Sepals 5, free. Petals absent. Staminodes petaloid, free, many. Stamens many, in 2 rows. Ovary superior, 5-locular; ovules many; styles 5, free. Fruit a many-seeded, 5-valved capsule.

1. C. decumbens (Forsskal) Exell in J. Bot. 73: 80 (1935). Syn.: *Orygia decumbens* Forsskal (1775 p.103). Illustr.: Fl. Trop. E. Afr.: 10 (1961); Collenette (1985 p.37). Type: Yemen (N), *Forsskal* (BM, C).

Annual or perennial herb, glabrous; stems prostrate or decumbent, up to 30cm, often pink-tinged. Leaves glaucous, subsucculent, elliptic to broadly obovate, 8–30 (–50) × 4–30mm, apiculate, cuneate below; petiole 1–4mm. Cymes 5–10-flowered, the flowers opening late afternoon. Sepals green with a white membranous margin, ovate, c.4mm. Staminodes pink or magenta, many, delicate, soon perishing, becoming longer than sepals. Fruit globose, 4–5mm diam. **Map 186, Fig. 29.**

Stony and rocky areas, amongst limestone boulders, plains and wadi-beds; 0–1220m.

Saudi Arabia, Yemen (N & S), Socotra, Oman. Widespread in drier parts of tropical Africa, Iran, Pakistan and India.

4. GLINUS L.

Annual herbs. Leaves opposite or verticillate; stipules absent. Flowers in axillary fascicles. Sepals 5, free. Petals absent. Staminodes present. Stamens 11–30. Ovary superior, (3–)5-locular; ovules many; styles (3–)5, free. Fruit a (3–)5-valved many-seeded capsule.

1.	Staminodes white, linear	**1. G. lotoides**
+	Staminodes yellow, multifid	**2. G. setiflorus**

1. G. lotoides L., Sp. pl.: 463 (1753). Syn.: *Mollugo glinus* A. Rich. (1847 p.48); *Mollugo hirta* Thunb., Prodr. fl. cap.: 24 (1794). Illustr.: Fl. Trop. E. Afr.: 14 (1961); Collenette (1985 p.37).

Annual herb, softly stellate-tomentose throughout; stems prostrate or decumbent, up to 50cm. Leaves elliptic or broadly obovate to suborbicular, 5–30 × 3–20mm, obtuse or rounded, the margins entire or obscurely undulate, cuneate below; petioles 3–12mm. Fascicles 2–10-flowered, the flowers opening in the afternoon; pedicels 1–6mm. Sepals greenish white, ovate, 4–8mm. Staminodes white, c.15, linear, deeply bifid at the apex. Capsule oblong-ovoid, c.5 × 3mm. **Map 187, Fig. 29.**

Irrigated and cultivated areas, in silty depressions and on rocky slopes; 0–400m.

Saudi Arabia, Yemen (N & S), Socotra, Oman. Widespread in the tropics and subtropics.

2. G. setiflorus Forsskal (1775 p.95). Type: Yemen (N), *Forsskal* (C).
Similar to *G. lotoides* but the staminodes yellow, multifid. **Map 188, Fig. 29.**

Yemen (N). Kenya, Tanzania, Ethiopia and Somalia.
In Arabia a rare plant collected only once by Forsskal in the 18th century.

5. MOLLUGO L.

Glabrous annual herbs. Leaves opposite or verticillate, sometimes with a basal rosette; stipules small or absent. Flowers in clusters or cymes. Sepals 5, free. Petaloid staminodes absent. Stamens 3–5. Ovary superior, 3–5-locular; ovules many; stigmas 3. Fruit a capsule opening by 3 valves.

1. Both cauline and a basal rosette of leaves present; cauline leaves linear; basal leaves up to 10 × 3mm, soon withering   **1. M. cerviana**
+ Leaves all in a basal rosette; cauline leaves absent; basal leaves up to 30(–60) × 10(–15)mm, persistent   **2. M. nudicaulis**

1. M. cerviana (L.) Ser. in DC., Prodr. 1: 392 (1824). Syn.: *Pharnaceum umbellatum* Forsskal (1775 p.58). Illustr.: Fl. Pakistan 40: 4 (1973); Fl. Trop. E. Afr.: 14 (1961); Collenette (1985 p.38).

Delicately branched annual herb; stems ascending, up to 12cm. Leaves sessile, glaucous; basal leaves forming a rosette, linear to spathulate, up to 10 × 3mm, soon withering; cauline leaves linear, up to 15mm. Flowers axillary or terminal, in sessile or pedunculate 1–4-flowered fascicles; pedicels 2–10(–15)mm. Sepals greenish with white margins, 1–3mm, obtuse, persistent. Stamens 5. Capsule subglobose, 1.5–2mm. **Map 189, Fig. 29**.

Sandy and stony places by roadsides, in wadi-beds etc. and as a weed of cultivation; 30–1500m.

Saudi Arabia, Yemen (N & S). Widespread in the tropics and subtropics of the Old World.

2. M. nudicaulis Lam., Encycl. 4: 221 (1797). Illustr.: Fl. Pakistan 40: 2 (1973).

Annual herb; stems erect or ascending, up to 25cm. Leaves all in a basal rosette, obovate to oblanceolate, up to 40(–60) × 10(–15)mm, rounded, attenuate below.

Flowers in dichasial cymes forming a leafless panicle; pedicels 2–15mm. Sepals white, 2–3mm, obtuse, persistent. Stamens 3–5. Capsule subglobose to oblong-ovoid, 2.5–3mm. **Map 190.**

Stony and sandy soils; 100–1300m.

Saudi Arabia, Yemen (N), Socotra. Pantropical.

6. SESUVIUM L.

Annual or perennial herbs. Leaves opposite; stipules absent. Flowers axillary, solitary or clustered. Calyx 5-lobed; lobes triangular, shortly aristate on the back below the tip. Petaloid staminodes absent. Stamens 5–many, free or fused into a ring at the base. Ovary 2–5-celled; styles 2–5. Fruit a circumscissile capsule, the lid not splitting; seeds several.

1.	Flowers pale pink or yellow; stamens 5(–7); leaves narrowly oblong elliptic, 5–15mm long	**1. S. sesuvioides**
+	Flowers bright pinkish-mauve; stamens many; leaves narrowly oblanceolate or spathulate, 1.5–3cm long	**2. S. verrucosum**

1. S. sesuvioides (Fenzl) Verdc. in Kew Bull. 12: 349 (1957). Syn.: *Trianthema polysperma* Oliver, Fl. trop. Afr. 2: 588 (1871). Illustr.: Fl. Pakistan 41: 4 (1973); Collenette (1985 p.39).

Annual or perennial herb; stems prostrate, up to 20cm, papillose. Leaves fleshy, narrowly oblong-elliptic, 5–15 × 1–5mm, acute or obtuse, shortly petiolate. Flowers pale pink or yellow, solitary or in clusters of 2–3, sessile. Calyx lobes c.4mm. Stamens 5(–7), not fused at the base. Capsule c.5mm long. **Map 191, Fig. 30.**

Salt flats and waste ground; 0–15m.

Saudi Arabia, Yemen (N). Tropical and SW Africa, Iran, Pakistan and India.

2. S. verrucosum Raf., New Fl. 4: 16 (1838). Illustr.: Phillips (1988 p.63).

Annual or perennial herb; stems erect or prostrate, up to 1.5m, glabrous. Leaves fleshy, narrowly oblanceolate to spathulate, 1–3 × 0.1–1cm, rounded, long-attenuate below. Flowers bright pinkish-mauve, solitary, sessile. Calyx lobes 4–6mm. Stamens many, fused into a ring at the base. Capsule c.5mm long. **Map 192.**

Weed of irrigated and waste ground, apparently naturalized on salt flats; 0–20m.

Saudi Arabia, UAE, Bahrain. A native of S America.

For a discussion of the occurrence of this species in Arabia see Kew Bull. 40: 208 (1985).

7. TRIANTHEMUM L.

Annual or perennial herbs. Leaves opposite, unequal, with membranous margins at the base, often connate in pairs; stipules absent. Flowers sessile or shortly pedicellate, axillary, solitary or in clusters. Calyx 5-lobed, with a subterminal dorsal appendage. Stamens 5–many. Petaloid staminodes absent. Ovary 1–2-celled; ovules 1–many; styles 1, free. Fruit a circumscissile capsule.

1. Flowers solitary, hidden by the membranous base of the petiole; leaves usually more than 1cm long   **4. T. portulacastrum**
+ Flowers clustered, clearly visible; leaves usually less than 1cm long   2

2. Leaves linear to narrowly obovate, ± circular in section   **3. T. triquetrum**
+ Leaves elliptic, ovate or obovate, flat in section   3

3. Prostrate annual or perennial herbs; flowers green; leaves grey-green when fresh, ± black when dry   **1. T. crystallinum**
+ Bushy perennial; flowers yellow; leaves pale green when fresh, blackish green when dry   **2. T. sheilae**

1. T. crystallinum (Forsskal) Vahl (1790 p.32). Syn.: *Papularia crystallina* Forsskal (1775 p.69). Illustr.: Collenette (1985 p.39). Type: Yemen (N), *Forsskal* (C).

Prostrate annual or perennial herb, sometimes woody below; stems white, up to 15cm, papillose. Leaves ovate to obovate, 2–5(–10) × 1–3(–7)mm, obtuse or rounded, grey-green when fresh, black with a white reticulate pattern when dry. Flowers green, clustered. Calyx lobes triangular, unequal, 1–2mm; appendage 0.5–1mm. Capsule rounded, depressed in the centre around the style. **Map 193, Fig. 30.**

Sandy and alluvial soils, coarse gravel and lava slopes; 0–1550m.

Saudi Arabia, Yemen (N & S). Ethiopia and Somalia.

2. T. sheilae A.G. Miller in Edinb. J. Bot. 51 (1): 33 (1994). Illustr.: Collenette (1985 p.39). Type: Saudi Arabia, *Collenette* 4718 (E).

Bushy annual or perennial herb; stems white, ascending, up to 20(–90)cm, papillose and somewhat glandular above. Leaves elliptic to obovate, 2–10 × 1–6mm, acute to rounded, pale green when fresh, blackish green with a white reticulate pattern when dry. Flowers yellow, clustered. Calyx lobes triangular, 1–2.5mm; appendage 0.5–1mm. Capsule rounded, depressed in the centre around the style. **Map 194.**

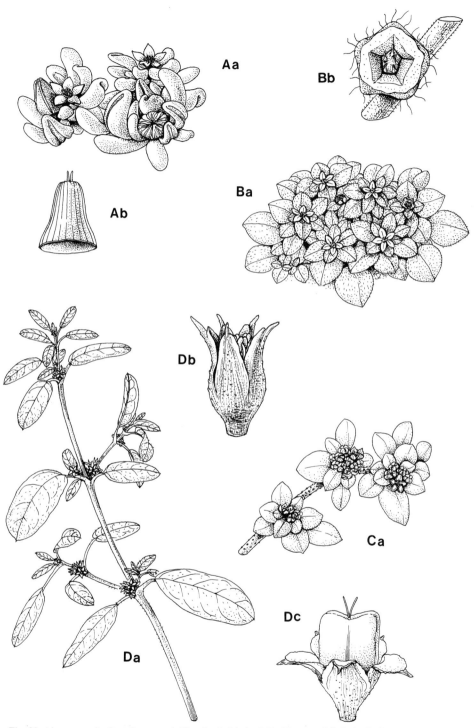

Fig. 30. Aizoaceae. A, *Sesuvium sesuvioides*: Aa, habit (×0.6); Ab, cap of fruit (×6). B, *Aizoon canariense*: Ba, habit (×0.6); Bb, fruit (×4). C, *Trianthema crystallina*: Ca, habit (×0.6). D, *Zaleya pentandra*: Da, flowering shoot (×0.6); Db, flower (×7); Dc, fruit (×7).

Open stony places, dry gravel plains, coral sand, lava blocks and near alkaline seepages; 0–1400m.

Saudi Arabia, Yemen (S). Ethiopia and Djibouti.

3. T. triquetrum Willd. in Ges. Naturf. Freunde Berlin Neue Schriften 4: 181 (1803). Syn.: *T. sedifolia* Vis., Pl. Aegypti 19, t.3/1 (1836). Illustr.: Collenette (1985 p.40).

Succulent annual herb; stems prostrate, up to 30cm, often reddish tinged, papillose. Leaves linear or narrowly elliptic to narrowly obovate, almost circular in section, up to 5(–10) × 0.5(–1)mm, acute, attenuate below into a short petiole. Flowers sessile, in clusters. Calyx lobes triangular, c.1mm, thickened at the tip. Capsule rounded, depressed in the centre around the style. **Map 195.**

Sandy soils, coastal sands and roadsides; 0–800(–1500)m.

Saudi Arabia, Yemen (N & S), Oman. Egypt, tropical and southern Africa, and Asia from Iran to Australia.

4. T. portulacastrum L., Sp. pl.: 223 (1753). Syn.: *Trianthema monogyna* L., Mant. pl.: 69 (1767). Illustr.: Fl. Pakistan 41: 4 (1973); Collenette (1985 p.39).

Annual herb; stems prostrate or ascending, up to 50cm, glabrous or sparsely pubescent. Leaves ovate or broadly obovate to ± orbicular, 5–30(–50) × 5–25 (–45)mm, rounded or apiculate, cuneate below; petiole 2–20(–25)mm. Flowers white or pink, sessile, solitary, hidden by the leaf bases. Calyx lobes narrowly ovate, c.3mm, sub-apically mucronate. Capsule 3–5-seeded, the lid flattened with a raised rim. **Map 196.**

Weed of cultivation; 0–100m.

Saudi Arabia, Yemen (N & S). Pantropical weed.

8. ZALEYA Burm. f.

Annual or perennial herbs. Leaves opposite, subsucculent; stipules absent. Flowers subsessile in congested axillary clusters. Calyx 5-lobed, the lobes with membranous margins. Stamens 5. Petaloid staminodes absent. Ovary superior, 2-celled; ovules 2 per cell; stigmas 2, free. Fruit a 4-seeded capsule, dehiscing by means of a 2-valved lid, the valves usually separating.

1. Z. pentandra (L.) C. Jeffrey in Kew Bull. 14: 238 (1960). Syn.: *Trianthema pentandra* L., Mant. pl.: 70 (1767); *Rocama prostrata* Forsskal (1775 pp. CVIII & 71). Illustr.: Fl. Trop. E. Afr.: 27 (1961); Fl. Pakistan 41: 4 (1973); Collenette (1985 p.40); Western (1989 p.36). Type: Yemen (N), from a specimen cultivated at Uppsala from seed sent by *Forsskal* (LINN).

Prostrate annual or perennial herb; stems up to 30cm, glabrous or minutely papillose. Leaves elliptic or oblong-elliptic to obovate, 10–30 × 3–20mm, obtuse; petiole 3–15mm, membranous-winged and sheathing at the base. Calyx lobes greenish or pinkish tinged with a white margin, oblong-ovate, c.2mm. Capsule black, rectangular, 1.5–4 × 1.5–2mm. **Map 197, Fig. 30.**

Stony and sandy places, in fields, wadis and irrigated areas, and as a weed of cultivation; 10–2000m.

Saudi Arabia, Yemen (N & S), Socotra, Oman, UAE. Tropical Africa, Palestine and Madagascar.

9. AIZOON L.

Annual or perennial herbs. Leaves alternate or opposite, succulent, sessile or petiolate; stipules absent. Flowers solitary or in groups, axillary or in stem forks. Calyx 5-lobed. Petals absent. Stamens many, in 5 bundles, inserted on the calyx tube. Ovary superior, (4–)5-locular; ovules many; styles (4–)5, free. Fruit a many-seeded capsule, dehiscing by 5 valves.

1. Leaves orbicular to obovate, petiolate; calyx lobes 2–3.5mm long; prostrate, pilose herb   **1. A. canariense**
+ Leaves narrowly oblong to oblong-oblanceolate, sessile; calyx lobes 5–9mm long; erect or ascending, papillose herb   **2. A. hispanicum**

1. A. canariense L., Sp. pl.: 488 (1753). Syn.: *Glinus chrystallinus* Forsskal (1775 p.95). Illustr.: Fl. Trop. E. Afr.: 33 (1961); Collenette (1985 p.36); Mandaville (1990 pl.14); Western (1989 p.35).

Annual or perennial herb, densely to thinly pilose; stems prostrate, up to 30cm. Leaves alternate, fleshy, orbicular to obovate, 10–35 (including petiole) × 4–20mm, rounded or bluntly subacuminate, attenuate below; petioles 3–15mm. Flowers greenish or yellowish with a yellow centre, solitary in leaf axils, sessile. Calyx lobes triangular, 2–3.5 × c.2mm. Stamens 12–25. Fruit red or pink, (4–)5-angled, depressed centrally, 4–6mm diam., the valves not recurving on dehiscence. **Map 198, Fig. 30.**

Very common on sandy and gravelly ground, rocky slopes, wadis, irrigated areas and waste ground; 0–2000m.

Saudi Arabia, Yemen (N & S), Socotra, Oman, UAE, Qatar, Bahrain, Kuwait. Tropical and N Africa, Palestine, Iraq, Iran and Pakistan.

2. A. hispanicum L., Sp. pl.: 488 (1753). Illustr.: Collenette (1985 p.36); Mandaville (1990 pl.15).

Papillose annual herb; stems dichotomously branched, erect or ascending, up to 20cm. Leaves opposite or sometimes alternate below, narrowly oblong to oblong-lanceolate, 10–45 × 1–10mm, obtuse, sessile. Flowers white, sessile in stem forks. Calyx lobes narrowly triangular, 5–9mm. Stamens numerous. Fruit 5-angled, 5–6mm diam., the valves strongly recurving on dehiscence. **Map 199.**

Sandy and gravelly areas; 0–650m.

Saudi Arabia, Oman, Bahrain, Kuwait. S Europe, N Africa, Palestine, Iraq and Iran.

10. DELOSPERMA N.E.Br.

Succulent perennial herbs, densely papillose throughout. Leaves opposite, succulent; stipules absent. Flowers solitary, terminal or axillary. Calyx of 5 unequal succulent lobes. Stamens many, inserted on the calyx. Staminodes many, petaloid. Ovary inferior, 5-celled; stigmas 5. Fruit a 5-valved, many-seeded capsule.

1. D. harazianum (Deflers) Poppend. & Ihlenf. in Mitt. Inst. Allg. Bot. Hamburg 16: 184 (1978). Syn.: *Mesembryanthemum harazianum* Deflers (1889 p.140). Illustr.: Collenette (1985 p.36). Type: Yemen (N), *Deflers* 337 (P).

Succulent herb forming clumps up to 20cm across. Leaves glistening, banana-shaped, up to 10–20 × c.4mm, acute. Flowers pink shading to white in the centre, c.1.5cm diam. Calyx lobes 4–7mm. Staminodes c.5mm. **Map 200, Fig. 31.**

Exposed rocky slopes and plains; 2350–2900m.

Saudi Arabia, Yemen (N & ?S). Endemic.

D. harazianum is very similar to, and perhaps conspecific with, *D. abyssinicum* (Regel) Schwantes, a species from Ethiopia and northern Kenya.

11. MESEMBRYANTHEMUM L.

Annual or perennial succulent herbs, covered with shining papillae. Leaves opposite or subopposite, succulent, terete, sessile; stipules absent. Flowers solitary, axillary. Calyx with 5 unequal lobes. Petaloid staminodes numerous. Stamens numerous. Ovary half-inferior, 5-celled; stigmas free or connate at the base. Fruit a many-seeded capsule, opening by an apical star-shaped slit.

1. Staminodes longer than the calyx; leaves 5–15mm thick   **1. M. forsskalei**
+ Staminodes shorter than or barely exceeding the calyx; leaves 2–4mm thick
    2. M. nodiflorum

1. M. forsskalei Hochst. ex Boiss., Fl. orient. 2: 765 (1872). Syn.: *M. cryptanthum* Hook. f. in Hooker's Icon. pl. t.1034 (1868); *Opophytum forskahlii* (Hochst.) N.E. Br. in Gard. Chron. ser 3, 84: 253 (1828). Illustr.: Mandaville (1990 pl.16); Collenette (1985 p.38) as *Opophytum forsskahlei*.

Annual herb; stems erect or ascending, up to 25cm. Leaves fleshy, subterete, ± linear-oblong, up to 5 × 1.5cm. Staminodes white or cream, yellowish at the base, longer than the calyx. Capsule 12–15mm long. **Map 201, Fig. 31.**

Coastal sands and salt flats; 0–850m.

Saudi Arabia, Qatar, Bahrain. Libya, Egypt and Palestine.

2. M. nodiflorum L., Sp. pl.: 481 (1753). Syn.: *Cryophytum nodiflorum* (L.) L. Bolus in S. African Gard. 17: 327 (1927). Illustr.: Fl. Kuwait 1: pl.152 & 153; Mandaville (1990 pl.17); Collenette (1985 p.38).

Annual herb; stems decumbent or ascending, up to 20cm. Leaves fleshy, terete, linear, up to 5–15 × 1–1.5mm, ciliate at the base. Staminodes white or cream, yellowish at the base, shorter than or barely exceeding the calyx. Capsule 5–8mm long. **Map 202, Fig. 31.**

Coastal sands and salt flats; 0–850m.

Saudi Arabia, Oman, UAE, Qatar, Bahrain, Kuwait. N Africa, S Europe and SW Asia.

12. TETRAGONIA L.

Annual or perennial herbs. Leaves alternate; stipules absent. Flowers minute, paired in the leaf axils. Calyx 5-lobed, tubular below. Petaloid staminodes absent. Stamens 5 (in Arabia), alternating with the calyx lobes. Ovary inferior, 1-celled; styles 3. Fruit dry, indehiscent, 1-seeded.

1. T. pentandra Balf.f. (1884 p.404). Type: Yemen (Socotra), *Balfour, Cockburn & Scott* 37 (K).

Glabrous annual herb; stems prostrate, up to 60cm. Leaves broadly elliptic or ovate, 1–2 × 0.5–1.5cm, obtuse, contracted abruptly below into a petiole; petiole 0.5–1cm. Calyx lobes triangular, c.0.5mm. Fruit obconical, c.1.75mm diam., 5-angled, smooth. **Map 203, Fig. 31.**

No habitat details available.

Socotra. Somalia.

In Socotra known only from a solitary collection made last century; recently also found in Somalia.

AIZOACEAE

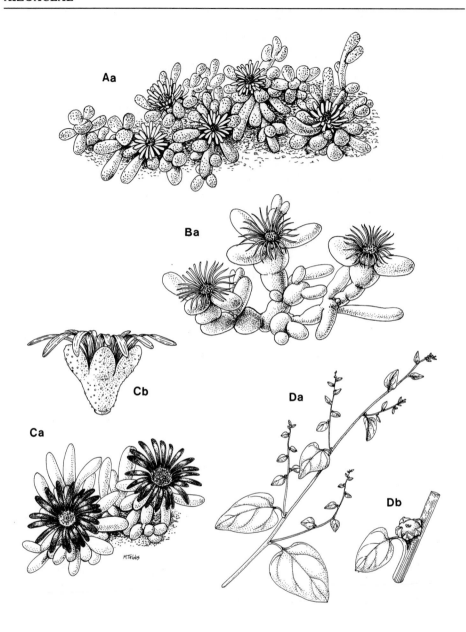

Fig. 31. Aizoaceae. A, *Mesembryanthemum nodiflorum*: Aa, habit (× 1.5). B, *M. forsskalei*: Ba, habit (× 1.5). C, *Delosperma harazianum*: Ca, habit (× 1.5); Cb, flower (× 3). D, *Tetragonia pentandra*: Da, habit (× 0.6); Db, fruit (× 4).

13. TELEPHIUM L.

Annual or perennial herbs. Leaves alternate; stipules small and membranous. Flowers in dense terminal cymes. Sepals 5. Petals 5, white, free, entire, inserted on a small disk. Stamens 5. Ovary superior, 1-locular or incompletely 3–4-loculed below; ovules many; styles 3, free. Fruit a many-seeded loculicidally dehiscing capsule.

1. T. sphaerospermum Boiss., Diagn. pl. orient. sér. 1, 1 (10): 12 (1849).

Glaucous, glabrous, annual or perennial herb; stems prostrate, 3–15cm, numerous, branched at the base but usually unbranched above. Leaves somewhat fleshy, obovate to elliptic, 5–15 × 1–5mm, acute or obtuse, attenuate below. Sepals oblong, 3–4mm, green with a white membranous margin. Petals white, as long as the sepals. Fruit trigonous-ovoid, 3–5 × 2.5–3mm, partly exserted from the persistent calyx. **Map 204.**

Dry rocky slopes, wadi-beds, sandy depressions and roadsides; 600–2600m.

Saudi Arabia, Yemen (N & S), Oman. N Africa.

Family 35. PORTULACACEAE

A.G. MILLER

Annual or perennial herbs or subshrubs, often succulent. Leaves opposite or alternate, simple, entire. Stipules absent or represented by a tuft of hairs, otherwise plants glabrous. Flowers bisexual, actinomorphic, solitary or in clusters at the ends of branches or in terminal racemes or panicles. Sepals 2, free or united below. Petals 4–6, free or united below. Stamens 4–numerous, free or attached to the base of the petals. Ovary inferior or semi-inferior, 1-celled; ovules many; placentation free-central or basal; style simple with 2–9 branches. Fruit a capsule, dehiscing by valves or circumscissile, 1–many seeded.

1.	Fruit dehiscing longitudinally by valves; flowers pedicellate; stipules absent	**1. Talinum**
+	Fruit circumscissile; flowers sessile; stipules hair-like	**2. Portulaca**

1. TALINUM Adans.

Perennial herbs. Leaves alternate, succulent. Stipules absent. Flowers pedicellate, in terminal racemose panicles. Petals 5. Stamens c.25. Ovary superior; style filiform, 3-branched. Capsule dehiscing longitudinally by 3 valves, many-seeded.

1. T. portulacifolium (Forsskal) Asch. ex Schweinf. (1896 p.172). Syn.: *Orygia portulacifolia* Forsskal (1775 p.103); *Portulaca cuneifolia* Vahl (1790 p.33); *Talinum cuneifolium* Willd., Sp. pl. 2: 864 (1799). Illustr.: Collenette (1985 p.411). Type: Yemen (N), *Forsskal* (C).

Perennial herb; stems erect, up to 2m, from a thickened rootstock. Leaves fleshy, subsessile, narrowly to broadly obovate, 2–6(–9) × 1–3cm, rounded or obtuse and mucronate at the tip, the base cuneate. Sepals ovate, 4–6mm. Petals crimson, pink or purple, obovate, 9–12mm. Capsule globose, 5–7mm diam. **Map 205, Fig. 32.**

On rocky outcrops and terrace-walls in dry, open shrubland; 10–1500m.

Saudi Arabia, Yemen (N & S) Socotra, Oman. Widely distributed throughout Africa and also in India.

Schwartz (1939) also records *T. triangulare* (Jacq.) Willd. from Yemen (S). This is a native of the New World and undoubtedly confused with the above species.

2. PORTULACA L.

Annual or perennial herbs. Leaves alternate or opposite, fleshy. Stipules represented by a tuft of hairs. Flowers sessile, solitary or in 2–10-flowered clusters at the ends of branches, surrounded by 2–several leaves. Petals 4–6, free or united at the base. Stamens 7–15, inserted at the base of the corolla. Ovary semi-inferior; style simple, 3–6-branched. Capsule circumscissile, many-seeded.

A very provisional treatment with several of the Arabian species requiring further study.

1. Leaves opposite; stipular hairs intra- and interpetiolar (surrounding the nodes) **2. P. quadrifida**
+ Leaves alternate; stipular hairs axillary 2

2. Leaves obovate to spathulate, flattened; axillary hairs inconspicuous; sepals distinctly keeled **1. P. oleracea**
+ Leaves linear, cylindrical; axillary hairs conspicuous or inconspicuous; sepals not keeled 3

3. Flowers yellow or orange; stipular hairs numerous around the flowers but few at the nodes **3. P. foliosa**
+ Flowers red; stipular hairs numerous both at the flowers and at the nodes 4

4. Stamens 10 **4. P. kermesina**
+ Stamens 15–30 **5. P. pilosa**

1. P. oleracea L., Sp. pl.: 445 (1753). Illustr.: Fl. Qatar pl.20; Collenette (1985 p.410); Cornes (1989 p.89); Western (1989 p.36).

Annual herb; stems prostrate or spreading, up to 30cm. Leaves alternate, subsessile or shortly petiolate, obovate or spathulate, $10–30 \times 2–10$mm, rounded at the tip, flattened; stipular hairs few, soon deciduous. Flowers solitary or in up to 10-flowered clusters. Sepals united below, oblong-ovate, 2–4mm. Petals yellow, obovate, 4–8mm. Seeds minutely tuberculate with star-shaped tubercles. **Map 206, Fig. 32.**

A common weed of cultivation, waysides and waste places; 0–2400m.

Saudi Arabia, Yemen (N & S), Socotra, Oman, UAE, Qatar, Bahrain, Kuwait. A widespread weed of the tropics and subtropics.

P. oleracea is a polyploid complex within which nine subspecies have been described. Two of these are recorded from Arabia: subsp. *oleracea* ($2n = 54$); and subsp. *granulato-stellulata* (Poelln.) Danin & H.G.Baker ($2n = 36$). The subspecies are apparently autogamous and recognized mainly by their seed-coat sculpturing (see Danin et al. (1978) in *Israel J. Bot.* 27: 177–211).

2. P. quadrifida L., Syst. nat. ed. 12, 2: 328 & Mant. pl. 1: 73 (1767); Syn.: *P. hareschata* Forsskal (1775 pp.CXII, 92); *P. imbricata* Forsskal (1775 p.92); *P. linifolia* Forsskal (1775 p.92). Illustr.: Collenette (1985 p.411); Cornes & Cornes (1989 p.89).

Annual herb; stems prostrate or ascending, up to 25cm, rooting at the nodes. Leaves opposite, subsessile, oblong-elliptic to ovate, $3–10 \times 1–4$mm, acute or obtuse at the tip, flattened; stipular hairs numerous, 3–5mm long, persistent. Flowers solitary or in up to 4-flowered clusters. Sepals united below, triangular, 2–4mm. Petals yellow or pink, elliptic to ovate. Seeds minutely tuberculate with rounded tubercles. **Map 207, Fig. 32.**

A weed of cultivated ground, gardens, and rocky slopes in *Acacia* shrubland; 0–3000m.

Saudi Arabia, Yemen (N & S), Socotra, Oman, Bahrain. A widespread weed of the tropics and subtropics.

Two forms of *P. quadrifida* occur in Arabia: a widespread form with small flowers (4–5mm across) and four petals; and a less common form (recorded only from Saudi Arabia) which has larger flowers (c.15mm diam.) and four to six petals. Both are illustrated in Collenette (loc. cit.). There are also several plants, similar in facies to *P. quadrifida*, but differing in being perennials with tuberous roots, stems which do not root at the nodes and distinct seed-surface sculpturing. These have been collected in the Dhofar region of Oman, on the island of Abd al Kuri and on Socotra and may represent two or possibly three new species.

3. P. foliosa Ker Gawl. in Edward's Bot. Reg. 10: t.793 (1824).

Robust annual or perennial herb; stems prostrate or erect, sometimes woody-based, up to 35cm. Leaves alternate, sessile, linear, up to 25×3mm, acute or obtuse at the

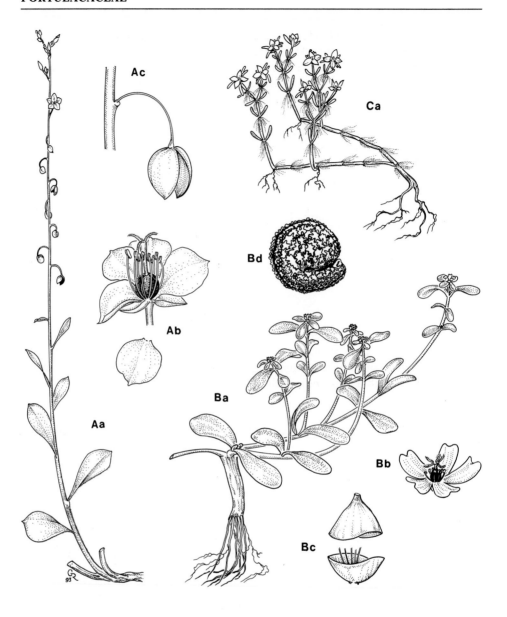

Fig. 32. Portulacaceae. A, *Talinum portulacifolium*: Aa, habit (× 0.6); Ab, flower (× 3); Ac, fruit (× 4); B, *Portulaca oleracea*: Ba, habit (× 0.6); Bb, flower (× 8), Bc, fruit (× 8); Bd, seed (× 50). C, *P. quadrifida*: Ca, habit (× 0.6).

tip, subcylindric; stipular hairs few, inconspicuous. Flowers c.1.3cm across, in (1–)2–6-flowered clusters, surrounded by tufts of hair. Sepals c.3mm. Petals yellow or orange, narrowly ovate, up to 8mm long. Capsule c.5 × 2.5mm. Seeds minutely tuberculate with stellate tubercles and a distinctive metallic sheen. **Map 208, Fig. 32.**

Over-grazed hillsides; 500–1300m.

Saudi Arabia, Yemen (N). Tropical Africa.

This species was first recorded from Arabia on J. Bura (Yemen) by Schweinfurth in the 19th century. It has recently been collected again from the same area and also from Saudi Arabia. *Portulaca foliosa* is basically a West African species which extends eastwards to western Ethiopia. Its status in Arabia needs further investigation. According to M. Gilbert (pers. comm.) the material of *P. foliosa* from Saudi Arabia is in fact referable to *P. grandiflora* Hook.

4. P. kermesina N.E.Br. in Bull. Misc. Inform. 1909: 91 (1909).

Similar to *P. foliosa* but with stipular hairs numerous both at the flowers and nodes; flowers smaller (c.4mm across), red. **Map 209, Fig. 32.**

In gravel pans amongst large granite boulders; 1300m.

Saudi Arabia. Tropical Africa.

In Arabia only from a single area (J. Shada in Saudi Arabia) where it is locally common. The Arabian plants are ephemerals with very small flowers and differ somewhat from typical African plants.

5. P. pilosa L., Sp. pl.: 445 (1775).

Similar to *P. kermesina* but an erect or ascending annual; leaves cylindrical, 5–20 × 2–4mm; petals purple-pink, 3–6 × 2.5–4.5mm. **Map 209.**

Weed; 20m.

Saudi Arabia. A native of tropical and subtropical America, naturalized in other parts of the world.

Mandaville (1990) reports that *P. pilosa* L. has spread and become a common weed around Dhahran in eastern Saudi Arabia since its introduction in the 1970s. It is an ascending herb with alternate linear, cylindrical leaves, conspicuous stipular hairs and red petals.

Family 36. BASELLACEAE

A.G. MILLER

Glabrous climbing herbs. Leaves alternate, entire, fleshy. Stipules absent. Flowers bisexual or unisexual, actinomorphic, in axillary spikes or racemes, subtended by a minute deciduous bract and 2 bracteoles which are united with the perianth. Perianth segments 5, fused at the base. Stamens 5, opposite to and borne on the perianth segments. Ovary superior, 1-locular, 1-ovulate; style simple or 3-branched. Fruit a drupe, surrounded by the persistent fleshy perianth.

Basella alba L., which is cultivated as a vegetable and used medicinally throughout SW Asia, is also likely to be found in Arabia. It is similar to *Anredera cordifolia* but the flowers are arranged in stout spikes, and the filaments are erect and straight in bud.

ANREDERA Jussieu

Climbing or trailing perennials. Flowers bisexual or unisexual, in racemes. Stamens with filaments outwardly curved in bud.

1. A. cordifolia (Tenore) van Steenis, Fl. Malesiana 5: 303 (1957). Syn.: *Boussingaultia cordifolia* Tenore in Ann. Sci. Nat. Bot. sér. 3, 19: 355 (1853).

Stems up to 6m. Leaves ovate to lanceolate, 1–11 × 1–8cm, cordate at the base; petioles up to 2.5cm. Racemes 4–30cm. Flowers white, 3.5–6mm diam. Style 3-branched. Fruits not known. **Map 210.**

Walls; 2400m.

Yemen (N). A native of South America now commonly cultivated as an ornamental (the 'Madeira Vine') in the warmer parts of the world.

In Arabia known only from Manakhah in Yemen where it is abundant on walls in the town.

Family 37. CARYOPHYLLACEAE
(Syn.: Illecebraceae)

D.F. CHAMBERLAIN

Annual or perennial herbs or subshrubs. Leaves opposite, sometimes fasciculate or appearing whorled, entire. Stipules scarious or absent. Flowers actinomorphic, bisexual, usually 5-merous, usually subtended by bracts, typically in simple or compound dichasial cymes, sometimes clustered or umbellate, occasionally solitary and terminal. Sepals free or united. Petals free or rarely absent, entire or variously dissected, often clawed. Stamens 5–10. Ovary superior, 1-locular or 2–3-locular below; styles free or connate; placentation free-central; ovules 2–many. Fruit a few- to many-seeded capsule or indehiscent and 1-seeded, rarely baccate.

This account of the Caryophyllaceae owes much to the research carried out by R.A. King, a summary of which is provided in the cited publication. First draft accounts for about half the species were derived from that research though the editor takes responsibility for some of the identifications from which the corresponding maps have been prepared.

King, R.A. & Kay, K.J. (1984). The Caryophyllaceae of the Arabian Peninsula: a check-list and key to taxa (Studies in the Flora of Arabia XII). *Arab Gulf. J. Scient. Res.* 2 (2): 391–414.

1.	Fruit one-seeded, indehiscent (rarely dehiscent in 10. *Haya*)	2
+	Fruit a few- to many-seeded capsule, dehiscing by valves or teeth	12
2.	Subshrubs or small shrubs, with woody branches	3
+	Herbs, not woody (rarely so at the base)	6
3.	Inflorescence subtended by feather-like bracts which conspicuously lengthen in fruit	**1. Cometes**
+	Inflorescence without feather-like bracts	4
4.	Stipules conspicuous, more than 2mm long, scarious; fruits surrounded by a fleshy, red receptacle formed from the bracts and peduncles; leaves flat	**2. Pollichia**
+	Stipules inconspicuous, less than 2mm long, scarious or not; fruits not as above; leaves terete	5
5.	Sepals narrowly triangular to oblong-lanceolate with a membranous margin and a shortly hooded, mucronate apex	**3. Gymnocarpos**
+	Sepals ovate to ovate-oblong, with a broadly scarious, fimbriate margin and a conspicuous deflexed apical awn, not hooded	**4. Sphaerocoma**
6.	Bracts silvery-white, completely scarious, conspicuous	**5. Paronychia**
+	Bracts green or brown to pink, sometimes with narrow scarious margins, conspicuous or not	7
7.	Leaves linear, sometimes the basal leaves linear-spathulate	8
+	Leaves elliptic to suborbicular	10
8.	Inflorescences borne on broad, flattened peduncles (c.5mm broad); flowers 4-merous	**6. Pteranthus**

| + | Inflorescences subsessile or borne on terete peduncles (c.1mm broad); flowers 5-merous | 9 |

| 9. | Inflorescence at the fruiting stage a spherical hard spinose head; stipules distinct, lanceolate | **7. Sclerocephalus** |
| + | Inflorescence at the fruiting stage not as above; stipules absent or indistinct | **8. Scleranthus** |

| 10. | Leaves less than 9mm long, narrowly elliptic to suborbicular, densely covered with stiff hairs; plant prostrate | **9. Herniaria** |
| + | Leaves (6–)10–37mm long, obovate to broadly obovate, glabrous; plant ascending to erect | 11 |

| 11. | Inflorescences stalked, with feather-like bracts | **1. Cometes** |
| + | Inflorescences sessile, without feather-like bracts | **10. Haya** |

| 12. | Leaves with stipules | 13 |
| + | Leaves without stipules | 18 |

| 13. | Sepals with bristle-like appendages on either side | **14. Loeflingia** |
| + | Sepals entire | 14 |

| 14. | Styles free to the base | 15 |
| + | Styles united, at least at the base, or solitary | 16 |

| 15. | Stipules free; calyx glabrous | **15. Spergula** |
| + | Stipules connate; calyx glandular-hairy | **16. Spergularia** |

| 16. | Dwarf shrub | **12. Xerotia** |
| + | Annuals or perennials, rarely with a woody rootstock | 17 |

| 17. | Sepals keeled, hooded; leaves not mucronate | **13. Polycarpon** |
| + | Sepals not keeled or hooded; leaves mucronate | **11. Polycarpaea** |

| 18. | Sepals free to the base | 19 |
| + | Sepals fused for at least one third of their length | 23 |

| 19. | Leaves setaceous to linear, 0.5mm broad or less; capsule dehiscing by 3 valves | **17. Minuartia** |
| + | Leaves linear-lanceolate to suborbicular, at least 1mm broad; capsule dehiscing by 4 or more teeth or valves | 20 |

| 20. | Inflorescence umbellate | **18. Holosteum** |
| + | Inflorescence not umbellate | 21 |

21.	Petals entire or emarginate	**19. Arenaria**
+	Petals bifid	22
22.	Lower leaves sessile; plant ± densely glandular-hairy; capsule dehiscing by 8 or more teeth	**20. Cerastium**
+	Lower leaves petiolate; plant lacking glandular hairs; capsule dehiscing by 6 valves	**21. Stellaria**
23.	Calyx with commissural veins (ribs) alternating with the midveins of the sepals; styles 3	**22. Silene**
+	Calyx without commissural veins or ribs; styles 2	24
24.	Small shrub with needle-like, spiny leaves	**28. Acanthophyllum**
+	Annual, biennial or perennial herbs; leaves not needle-like	25
25.	Calyx intervals membranous, hyaline, or scarious	26
+	Calyx intervals not membranous, hyaline, or scarious	27
26.	Seeds auriculate, with a lateral hilum	**24. Gypsophila**
+	Seeds peltate, with a facial hilum	**26. Petrorhagia**
27.	Calyx inflated below, with 5 conspicuously winged veins	**23. Vaccaria**
+	Calyx not inflated below, lacking winged veins	28
28.	Calyx up to 1mm wide without an epicalyx of bracteoles; plant with rigid dichotomously branched stems	**25. Velezia**
+	Calyx more than 2mm wide, with an epicalyx of bracteoles enclosing the base; plant not branching as above	**27. Dianthus**

1. COMETES L.

Annual or perennial herbs or subshrubs. Leaves opposite, linear to obovate or spathulate. Stipules inconspicuous. Inflorescences of 3(–5)-flowered pedunculate heads, terminal and axillary, subtended by feather-like bracts; bracts cream to reddish-brown, conspicuously lengthening in fruit, the ultimate segments becoming needle-like. Sepals 5, free, ± equal, hooded, mucronate, with a narrow scarious margin. Petals 5. Stamens 5. Style solitary, 3-lobed at the apex. Fruit one-seeded, indehiscent, not exceeding the sepals.

1.	Annual; leaves obovate to spathulate, usually more than 5mm broad; needle-like segments of the maturing bracts deflexed	**1. C. surattensis**
+	Subshrub; leaves linear to lanceolate, rarely ovate or obovate, usually less than 5mm broad; needle-like segments of the maturing bracts spreading	**2. C. abyssinica**

1. C. surattensis L., Mant. pl. 1: 39 (1767). Illustr.: Collenette (1985 p.101); Western (1989 p.37).

Annual. Stems ± erect, 4–25cm, greenish, densely pilose above, glabrescent below, the hairs sometimes glandular. Leaves obovate to elliptic or ± spathulate, 10–37 × 5–21mm, sparsely hairy mainly on the margins and midrib. Bracts lengthening to 10mm in fruit, their needle-like segments becoming deflexed. Sepals oblong, 3.5–4.5mm, somewhat papillose. Petals white, linear-oblong, $\frac{3}{4}$ × as long as the sepals. **Map 211, Fig. 33.**

Stony or rocky slopes and wadis, often on limestone; 10–1900m.

Saudi Arabia, ? Yemen (S), Oman, UAE. Egypt, Sinai, Iraq, S Iran and S Pakistan.
Reported from Yemen (S) but the record has not been confirmed.

2. C. abyssinica (R.Br.) Wallich, Plant. asiat. rar. 1: 18, t.18 (1830). Syn.: *C. abyssinica* subsp. *suffruticosa* Wagner & Vierhapper (1907 p.352). Illustr.: Collenette (1985 p.101).

Bushy subshrub, apparently sometimes flowering in its first year. Stems ascending to erect, 10–35cm, minutely scabrid. Leaves linear to lanceolate, rarely ovate or obovate, 7–28(–38) × 1.2–4.8(–5.5)mm, indumentum similar to that of the stem. Bracts lengthening to 15mm in fruit, their needle-like segments spreading. Sepals oblong, 4–4.5mm, papillose. Petals white, oblong, $\frac{2}{3}$ to as long as the sepals. **Map 212, Fig. 33.**

Sandy and rocky slopes and crevices, near the coast and inland; 0–2800m.

Saudi Arabia, Yemen (N & S), Socotra, Oman, UAE. Somalia, Ethiopia and Sudan.

2. POLLICHIA Ait.

Subshrub; stems woody, densely woolly. Leaves flat, opposite or in pseudowhorls. Stipules conspicuous, scarious, lanceolate with an erose margin. Flowers sessile, perigynous, in dense subsessile axillary cymes; bracts broadly ovate-lanceolate, scarious and ciliate at anthesis, the basal parts becoming enlarged and fleshy in fruit; pedicels absent. Sepals 5, free, thick with a rounded and hooded apex, red, sparsely lanate. Petals hooded. Stamens 1–2. Style solitary, obscurely 2-lobed at the apex. Fruits 1-seeded, indehiscent, surrounded by a fleshy red receptacle formed from the bracts and peduncles. Monotypic.

1. P. campestris Ait., Hort. kew. ed.1 (1): 5 (1789). Illustr.: Fl. Trop. E. Afr.: 12 (1956).

Stems 25–40 cm. Leaves linear-elliptic to elliptic, 8–15 × 1–3 mm, hairy at first especially on the margins, soon glabrescent. Sepals c.1.5mm long. Petals c.1mm. **Map 213, Fig. 33.**

Amongst rocks, also a weed of waste places, fields and vineyards; 1500–2400m.

Saudi Arabia, Yemen (N). Tropical and southern Africa.

3. GYMNOCARPOS Forssk.
Lochia Balf. f.

L. PETRUSSON & M. THULIN

Shrublets with ± tortuously spreading branches. Leaves opposite, often fascicled on younger branches, sessile or shortly petiolate, fleshy, terete, mucronate. Stipules interpetiolar, scarious, ovate-triangular with two keels, first connate, later splitting lengthwise into two. Inflorescences terminal or subterminal, shortly pedunculate; bracts scarious, ovate to suborbicular, small and stipule-like to conspicuous and equalling or exceeding the flowers, or sometimes ± leaf-like with well developed to vestigial lamina. Flowers sessile or sometimes the middle ones pedicelled, with an obconical receptacle shorter than the sepals. Sepals 5, narrowly triangular to oblong-lanceolate with a membranous margin, shortly hooded, mucronate. Petals absent. Stamens 5, alternating with 5 staminodes; filaments filiform from a broad base. Ovary obovate, papillose above; style long, slender; stigma 3-lobed; ovule single, basal. Fruit with membranous pericarp, rupturing irregularly in the middle or near the base.

Petrusson, L. & Thulin, M. (1996). Taxonomy and biogeography of Gymnocarpos (Caryophyllaceae). *Edinb. J. Bot.* 53 (in press). Chaudhri, M. N. (1968). A revision of the Paronychiinae. *Meded. Bot. Mus. Herb. Rijks Univ. Utrecht* 285: 1–427.

1.	Bracts small, stipule-like	**7. G. mahranus**
+	Bracts large, equalling or exceeding the flowers, or bracts leaf-like with stipules and lamina	2
2.	Bracts leaf-like, with small stipules and 2–4 mm long lamina	**1. G. decandrus**
+	Bracts large, scarious, only the lowermost ones sometimes with a vestigial to well developed lamina	3
3.	Leaves sessile	4
+	Leaves with 0.25–1.5 mm long petiole	5
4.	Sepals with long hairs at the base; bracts puberulous, white with a brown base only	**2. G. argenteus**
+	Sepals glabrous or practically so at the base; bracts glabrous, brown with a ± broad white margin	**3. G. dhofarensis**
5.	Leaves narrowly linear-oblanceolate; bracts brown throughout or with narrow white margin in upper part only	**4. G. bracteatus**
+	Leaves broadly linear-oblanceolate to almost circular in outline; bracts with narrow to broad white margin all around	6
6.	Bracts white with a brown base only	**5. G. rotundifolius**
+	Bracts mainly brown with a ± narrow white margin	**6. G. kuriensis**

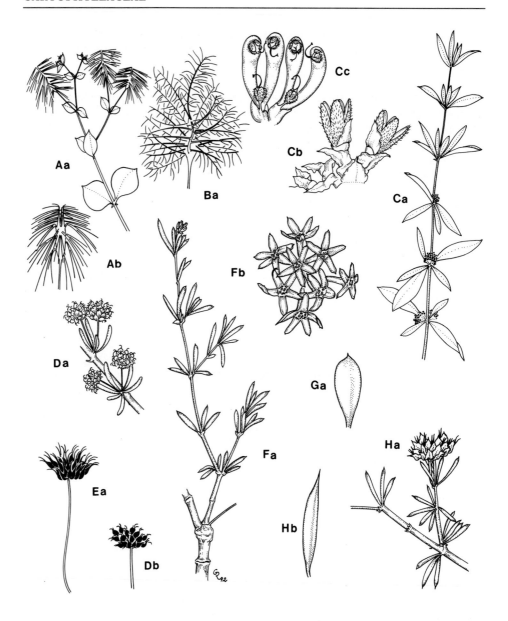

Fig. 33. **Caryophyllaceae**. A, *Cometes surattensis*: Aa, flowering shoot (×1); Ab, flowers (×2). B, *C. abyssinica*: Ba, flowers (×2). C, *Pollichia campestris*: Ca, flowering shoot (×1); Cb, part of inflorescence (×15); Cc, part of infructescence (×5). D, *Sphaerocoma aucheri*: Da, flowering shoot (×2); Db, inflorescence (×3). E, *S. hookeri*: Ea, inflorescence (×3). F, *Gymnocarpos decandrus*: Fa, flowering shoot (×1); Fb, inflorescence (×4). G, *G. kuriensis*: Ga, leaf (×5). H, *G. bracteatus*: Ha, flowering shoot (×1); Hb, leaf (×5).

1. G. decandrus Forssk. (1775 p.65). Syn.: *Trianthema fruticosa* Vahl (1790 p.32); *G. fruticosus* (Vahl) Pers., Syn. 1: 262 (1805). Illustr.: King & Kay (1984 p. 398).

Shrublet, up to 45 cm tall. Leaves sessile, linear, terete, 5–18 × 0.5–1.5(–2) mm, with mucro 0.2–0.6 mm long. Inflorescences dense, many-flowered; bracts leaf-like, stipulate, with 2–4 mm long lamina. Flowers 5–7 mm long, sessile; receptacle 2–3 mm long, papillose to densely pilose. Sepals 3–4.5 × 1 mm, pilose at the base; mucro 0.25–0.5 mm long, hairy at the base. Stamens with 1.4–2 mm long filaments; anthers c. 0.4 × 0.25 mm; staminodes linear-lanceolate, 1.25–1.5 mm long. Style 2–2.75 mm long. Fruit rupturing irregularly at the base. Seed 2–2.25 × 1.1 mm, ellipsoid, dark brown. **Map 214, Fig. 33.**

Dry rocky slopes, wadi-sides and stony deserts; on granite, volcanics and limestone; 0–1500m.

Saudi Arabia, Oman, UAE. N Africa, Canary Is., Palestine, Jordan, S Iran and Pakistan.

Also reported from Yemen (S) but the record has not been confirmed and seems unlikely.

2. G. argenteus Petruss. & Thulin in Edinb. J. Bot 53 (1996, in press). Illustr.: Petrusson & Thulin (1996). Type: Yemen (S), *Thulin, Eriksson, Gifri & Långström* 8330 (holo. UPS, iso. Aden Univ., E, K).

Shrublet, up to 40 cm tall. Leaves sessile, linear-oblanceolate, almost cylindrical, 5–20 × 1–1.4(–1.8) mm, with mucro (0.25–)0.4–0.8 mm long. Inflorescences dense, many-flowered; bracts broadly ovate to suborbicular, 5.5–6 × 5.5–6 mm, white, often brown at the base and/or along the midnerv, puberulous in lower part, denticulate, the lowermost ones sometimes with a ± well developed lamina. Flowers 4 × 1.5–2 mm, sessile; receptacle c. 1 mm long, pilose at the base with up to 0.8 mm long hairs. Sepals 2.5–2.75 × 1.1–1.5 mm, densely ciliate at the base and near the apex; mucro 0.5–0.8 mm long. Stamens with 1–1.2 mm long filaments; anthers (0.6–)1 × (0.25–)0.5 mm; staminodes narrowly triangular with concave sides, 0.6–0.8 mm long. Style c. 1.9 mm long. Fruits and seeds not seen. **Map 214.**

Open rocky places, also on gypsum; above 500m.

Yemen (S). Endemic.

3. G. dhofarensis Petruss. & Thulin in Edinb. J. Bot 53 (1996, in press). Illustr.: Petrusson & Thulin (1996). Type: Oman, *Miller* 2524 (holo. E, iso. K).

Shrublet, up to 30 cm high. Leaves sessile, linear-oblanceolate to fusiform or subcircular, 3–12.5 × 1.2–3 mm, with mucro (0.25–)0.5–0.75 mm long. Inflorescences dense; bracts ovate, 4–5 × 2.5 mm, ± brown with white ± broad margin, glabrous, entire, the lowermost ones sometimes with a ± well developed lamina. Flowers c. 5 × 2 mm, sessile; receptacle c. 1.5 mm long, usually ciliate at the base. Sepals 2.3–2.75 × 1–1.7 mm, glabrous or usually ± puberulous at the apex; mucro (0.35–)0.5–1 mm long. Stamens with c. 0.8 mm long filaments; anthers c. 0.6 × 0.4 mm; staminodes

0.5–0.7 mm long, narrowly triangular with concave sides. Style 1–1.9 mm long. Fruits and seed not seen. **Map 215.**

Open rocky slopes; 500–1300m.

Yemen (S), Oman. Endemic.

4. G. bracteatus (Balf. f.) Petruss. & Thulin in Edinb. J. Bot 53 (1996, in press). Syn.: *Lochia bracteata* Balf. f. (1884 p.409). Illustr.: Balfour (1888 t.84). Type: Socotra, *Balfour, Cockburn & Scott* 429 (lecto. E, iso. BM, K).

Shrublet, up to 35 cm tall. Leaves narrowly linear-oblanceolate, 7–13 × 1–1.5 mm, olive-green, with mucro 0.4–0.8 mm long; petiole 0.5–1(–1.5) mm long. Inflorescences dense, many-flowered; bracts broadly ovate to suborbicular, 4.5–5.75 × 5 mm, brown or streaked with brown throughout or with white margin in upper part only, the lowermost ones sometimes with a ± well developed lamina. Flowers 4–6 × 1.75–2 mm, sessile or the lower flower sometimes with 0.5–1 mm long pedicel; receptacle 1–2 mm long. Sepals 2–2.75 × 1.25 mm, sometimes slightly ciliate at the base; mucro 0.8–1 mm long. Stamens with 1.2–1.5 mm long filaments; anthers 0.35–0.6 × 0.25–0.3 mm; staminodes narrowly triangular with concave sides, 0.5–0.75 mm long. Style c. 1 mm long. Seed 1.75–2.3 × 0.8–1.4 mm, ellipsoid, dark brown. **Map 215, Fig. 33.**

Low shrubland on granitic slopes; 300–900m.

Socotra. Endemic.

5. G. rotundifolius Petruss. & Thulin in Edinb. J. Bot 53 (1996, in press). Illustr.: Petrusson & Thulin (1996). Type: Oman, *Miller* 6476 (holo. E, iso. UPS).

Shrublet, up to 25 cm high. Leaves elliptic-obovate to circular, 2–5 × 1.5–3.5 mm, with 0.1–0.4 mm long mucro; petiole 0.25–0.5 mm long. Inflorescences dense, many-flowered; bracts broadly ovate, 4–5 × 2.5–3 mm, white with brown base. Flowers 4.5–5.5 × 2–2.5 mm, sessile; receptacle c. 1.5 mm long, glabrous. Sepals c. 2.5 × 1 mm, ciliate at the base; mucro 0.4–0.6 mm long. Stamens with 0.8–1.2 mm long filaments; anthers (0.4–)0.8–1 × (0.25–)0.5 mm; staminodes narrowly triangular with concave sides, 0.8–1 mm long. Style 1.5–1.8 mm long. Fruits and seeds not seen. **Map 216.**

Open, desertic rocky ground and gravel plains; 0–100 m.

Oman. Endemic.

6. G. kuriensis (A.R. Smith) Petruss. & Thulin in Edinb. J. Bot 53 (1996, in press). Syn.: *Lochia bracteata* Balf. f. subsp. *abdulkuriana* Chaudhri, op. cit.: 60 (1968); *L. bracteata* Balf.f. subsp. *bracteata* forma *ciliata* Chaudhri, op. cit.: 60 (1968); *L. kuriensis* A.R. Smith in Hooker's Icon. Pl. 7(4): t. 3674 (1971). Illustr.: Radcliffe-Smith, loc. cit. (1971). Type: Yemen, Abd al-Kuri, *Forbes & Ogilvie-Grant* 84 (holo. E).

Shrublet, up to 30 cm tall. Leaves broadly linear-oblanceolate to subcircular in outline, 4–12 × 1.5–3(–5) mm, pale glaucous, with mucro 0.2–0.5 mm long; petiole 0.5–1 mm long. Inflorescences dense; bracts broadly ovate to suborbicular, 4–5.5 × 2.5–5 mm, ± chestnut-brown or streaked with brown at the base and middle part, apex and margin white, the lowermost ones sometimes with a ± well developed lamina. Flowers 4.5–5 × c. 2 mm, sessile; receptacle 1.5–2 mm long, sometimes ciliate at the base. Sepals c. 2.5 × 1–1.4 mm, occasionally minutely ciliate at the base; mucro 0.5–1 mm long. Stamens with 0.8–1.4 mm long filaments; anthers 0.5–0.6 × 0.3–0.4 mm; staminodes narrowly triangular with concave sides, 0.4–0.6 mm long. Style 1.2–1.75 mm long. Seed c. 2 × 1.25 mm, ovate, brown. **Map 216.**

Low sand-dunes, gravel plains and broken rocky slopes; 0–150m.

Socotra (Abd al-Kuri). Endemic.

7. G. mahranus Petruss. & Thulin in Edinb. J. Bot 53 (1996, in press). Illustr.: Petrusson & Thulin (1996). Type: Yemen, *Miller* 12131 (holo. E, iso. UPS).

Shrublet, up to at least 10 cm high. Leaves sessile, linear-oblanceolate, almost cylindrical, 4–8 × 1–1.5(–1.9) mm, with mucro 0.2–0.4 mm long. Inflorescences composed of small terminal dichasia with 3–7 flowers; bracts stipule-like, 0.6–1 mm long, brown with broad white margin. Flowers 4–6 × 1.5–2 mm, the middle flowers often with receptacle attenuate into c. 1.5 mm long pedicel, the other flowers sessile; receptacle 1–2 mm long. Sepals c. 2.5 × 1.5 mm, glabrous; mucro 0.1–0.2 mm long. Stamens with c. 0.6 mm long filaments; anthers c. 0.6 × 0.5 mm; staminodes triangular, c. 0.3 mm long. Style c. 1 mm long. Fruits and seeds not seen. **Map 217.**

Open rocky slopes; c. 550m.

Yemen (S). Endemic.
Known only from a single gathering.

4. SPHAEROCOMA T. Anderson

Small subshrubs. Leaves opposite or fascicled, narrowly linear to elliptic, fleshy, glabrous. Stipules very small. Flowers in pedunculate heads. Bracts shorter than the flowers, similar to the sepals. Sepals 5, free, unequal, ovate to ovate-oblong, with a broad scarious and fimbriate margin and a conspicuous deflexed apical awn, not hooded. Petals shorter than the sepals. Stamens 5. Style solitary, obscurely 2-lobed at the apex. Fruit one-seeded, indehiscent.

Chaudhri, M.N. (1968). A revision of the Paronychiinae. *Meded. Bot. Mus. Herb. Rijks Univ. Utrecht* 285: 30–34.

1.	Peduncles 12–27mm long; leaves linear, generally 10 × as long as wide; awns on fruiting heads up to 6.5mm long	**1. S. hookeri**
+	Peduncles 3–10mm long; leaves linear to elliptic, up to 7 × as long as wide but usually less; awns on fruiting heads up to 2mm long	**2. S. aucheri**

1. S. hookeri T. Anders. (1860 p.16, t.3). Syn.: *Psyllothamnus beevori* Oliv. in Hooker's Icon. Pl. 15: t.1499 (1885). Type: Yemen (S), *Thomson* (K).

Stems up to 60cm, glabrous or very rarely with a few hairs on the young growth; older wood grey with fascicular mounds at the nodes. Leaves linear, 7–21(–30) × 0.5–1(–1.5)mm, 2–13 in each fascicle. Stipules triangular, reddish brown. Inflorescence c.10mm across with numerous sterile flowers, spinescent, enlarging and becoming spherical in fruit; peduncles 12–27mm. Sepals ovate to ovate-oblong, 2.5–3 × 1–2.5mm; awns 0.4–3mm, recurved, increasing to 6.5mm in fruit. Petals white, oblong, $\frac{3}{4}$ as long as the sepals. Fruit ± globose, shorter than the sepals. **Map 217, Fig. 33.**

Dry, rocky slopes and semi-desert dwarf shrubland; 200–500m.

Yemen (S), Socotra. Somalia, Sudan and Egypt.

2. S. aucheri Boiss., Fl. orient. 1: 739 (1867). Illustr.: Cornes (1989 p.68); Western (1989 p.43).

Similar to *S. hookeri* but with at least the young shoots shortly lanate or scurfy; leaves linear to elliptic, 3.5–10 × 1–2(–3)mm, usually less than 7 in each fascicle; peduncles 3–10mm; sepal-awns less than 2mm in fruit. **Map 218, Fig. 33.**

In sandy areas, often near the coast; 0–100(–1700m).

Saudi Arabia, Yemen (S), Oman, UAE, Bahrain. Sudan, S Iran and S Pakistan.

This is typically a species of low lying coastal districts. The higher altitude given in brackets refers to two, apparently anomalous, records from Saudi Arabia (900m, *Zeller* 7504) and Oman (1700m, *Vesey-Fitzgerald* 13244/1).

5. PARONYCHIA (Tourn.) Miller

Annual or perennial herbs. Leaves opposite, sessile. Stipules scarious, silvery white, usually conspicuous. Flowers perigynous, in terminal and axillary clusters, often obscured by the large silvery-hyaline bracts. Sepals 5, equal or unequal, persistent. Petals absent. Stamens 5, rarely less; staminodes 5, filiform. Styles 2, ± free or fused above and then 2-lobed at the apex. Fruit one-seeded, indehiscent, membranous, rupturing irregularly at the base.

Chaudhri, M. N. (1968). A revision of the Paronychiinae. *Meded. Bot. Mus. Herb. Rijks Univ. Utrecht* 285: 1–427.

1.	Leaves mucronate; sepals with scarious margins, hooded and awned at the apex		**1. P. arabica**
+	Leaves not mucronate; sepals without scarious margins, not hooded or awned at the apex		2
2.	Sepals 2.5–3.5mm long; bracts suborbicular		**2. P. sinaica**
+	Sepals 4–7mm long; bracts ovate		**3. P. chlorothyrsa**

1. P. arabica (L.) DC. in Poir., Encycl. 5: 25 (1804). Syn. ? *Paronychia lenticulata* sensu Schwartz (1939) non (Forsskal) Aschers. Illustr.: Cornes (1989 p.62); Mandaville (1990 t.23); Phillips (1988 p.151); Western (1989 p.39).

Annual or perennial herb. Stems prostrate, 4–40cm, pilose to $\pm$ pubescent. Leaves linear-oblong to narrowly obovate, $3.5–18 \times 0.8–2.5$mm, mucronate, sparsely hairy. Stipules conspicuous, narrowly ovate, c.3×1mm. Bracts very conspicuous, ovate, c.3.5mm. Flowers 1.7–2.5mm, with antrorse hooked hairs near the base. Sepals equal, $\pm$ oblong, c.1.5mm, with broad scarious margins, hooded and awned at the apex. Staminodes c.$\frac{1}{4}$ as long as the sepals. Fruit oblong-ovoid, c.1.25×1mm, densely papillose in the upper parts. **Map 219, Fig. 34.**

Sandy and stony areas on the coast and inland, also as a field weed; 0–2150m.

Saudi Arabia, Yemen (N), Oman, UAE, Qatar, Bahrain, Kuwait. N Africa, S Iraq and S Iran.

A very variable species in which Chaudhri (op. cit.) recognises six subspecies and a number of varieties throughout its range based on characters of leaves, stipules and sepals. All Arabian material is referable to subsp. *breviseta* (Aschers. & Schweinf.) Chaudhri var. *breviseta*.

2. P. sinaica Fresen., Beitr. Fl. Aegypt., Mus. Senckenberg 1: 180 (1834).

Caespitose perennial herb, often mat-forming. Stems 4–8.5cm, densely pubescent. Leaves linear-oblong, $2–5 \times 0.5$-mm, the midrib and margin often thickened, antrorsely pubescent to hirtellous. Stipules narrowly ovate, usually shorter than the leaves. Bracts suborbicular, c.3×3.5mm, often rolled around and mostly exceeding and concealing the flowers. Flowers 2.5–3.5(–4)mm, densely adpressed-pubescent. Sepals unequal, 2–3mm, without broad scarious margins. Staminodes c.$\frac{1}{3}$ as long as the sepals. Fruit ovoid, c.1.4×1.25mm. **Map 220, Fig. 34.**

Rocky hillsides; 1000–1700m.

Saudi Arabia. Egypt, Sinai, the Negev and Jordan.

Rare in Arabia, known only from the extreme NW of the region.

3. P. chlorothyrsa Murbeck in Acta Univ. Lund. 33: 48, t.2/13–14 (1897). Illustr.: Collenette (1985 p.105).

Annual or perennial herb. Stems prostrate, rarely suberect, 3–8(–13)cm, retrorsely pubescent. Leaves linear-oblong to narrowly elliptic, slightly falcate, $3.5–8 \times 1–1.4$mm, antrorsely pubescent. Stipules $\pm$ ovate, c.3mm, somewhat acuminate. Bracts broadly ovate, c.4×4.5mm, often rolled around but subequal to or shorter than the flowers. Flowers c.4.5mm, antrorsely pubescent. Sepals unequal 3.5–4.3mm, without broad scarious margins. Staminodes c.$\frac{1}{8}$ as long as the sepals. Fruit oblong-ellipsoid, c.1.5×1mm, slightly narrowed at the apex. **Map 221, Fig. 34.**

Sandy soil, among granite boulders and rocky plateau areas; 750–2350m.

Saudi Arabia, Yemen (N). N Africa.

CARYOPHYLLACEAE

Fig. 34. Caryophyllaceae. A, *Sclerocephalus arabicus*: Aa, flowering shoot (×1); Ab, stipule (×3). B, *Pteranthus dichotomus*: Ba, flowering shoot (×1); Bb, inflorescence (×3). C, *Paronychia arabica*: Ca, habit (×1); Cb, flower (×13). D, *P. sinaica*: Da, flower (×13). E, *P. chlorothyrsa*: Ea, flower (×13). F, *Herniaria hemistemon*: Fa, calyx (×24). G, *H. mascatensis*: Ga, flowering shoot (×1); Gb, calyx (×24). H, *H. hirsuta*: Ha, calyx (×24). I, *Scleranthus orientalis*: Ia, habit (×1); Ib, inflorescence (×5).

6. PTERANTHUS Forsskal

Annual herb. Leaves opposite or fasciculate, fleshy, papillose-scabrid. Stipules scarious. Inflorescences clustered, glandular, each 3-flowered and borne on a broad flattened peduncle, the outer flowers sterile, the inner flower fertile; some bracts enlarging in fruit and becoming broadly spathulate with a long cusp and pink in colour. Sepals 4, hooded; the inner two membranous, ovate-lanceolate; the outer two linear-oblong, keeled. Petals absent. Stamens 4. Styles solitary, 2-lobed at the apex. Fruit 1-seeded, indehiscent.

1. P. dichotomus Forsskal (1775 p.36). Illustr.: Collenette (1985 p.106); Mandaville (1990 t.26).

Stems arching, 10–25cm, glabrous. Basal leaves spathulate, $9–13 \times 2–3$ mm; cauline linear to linear-spathulate, $9–20 \times c.1$mm; petioles absent. Stipules lanceolate, 3–4mm, toothed. Sepals 4–5mm. **Map 222, Fig. 34.**

Fallow fields, rocky ground etc; 0–900m.

Saudi Arabia, Kuwait. N Africa, SE Mediterranean region, Iraq, Iran and Pakistan.

7. SCLEROCEPHALUS Boiss.

Dwarf succulent annual herb. Leaves opposite or fasciculate, subulate, 5–15mm, spine-tipped. Stipules scarious. Inflorescences of compact terminal or axillary spherical heads, 4–7-flowered, subsessile or shortly pedunculate, becoming hardened in fruit; bracts like the leaves though more indurated in fruit, fused with the sepal bases. Sepals 5, linear-lanceolate with a hooded and spine-tipped apex. Petals absent. Stamens 5, borne on the margin of the cupular receptacle. Style solitary, 2–3-lobed at the apex. Fruit 1-seeded, indehiscent, ovoid, contained within the indurated receptacle.

1. S. arabicus Boiss., Diagn. pl. orient. sér. 1, 1 (3): 12 (1843). Syn.: *Paronychia sclerocephala* Decne. in Ann. Sci. Nat. Bot. sér. 2 (3): 262 (1835). Illustr.: Collenette (1985 p.107); Cornes (1989 p.64); Mandaville (1990 t.25); Phillips (1988 p.153); Western (1989 p.40). Type: Oman, *Aucher* 4513 (K).

Stems 1–10cm. Leaves 5–15mm, glabrous, sessile. Stipules lanceolate, c.2.5mm. Sepals 2.5–3 mm, glabrous above, densely lanate below. Fruit c.2.5mm long. **Map 223, Fig. 34.**

Rocky and sandy areas; 0–1100m.

Saudi Arabia, Oman, UAE, Qatar, Bahrain, Kuwait. N Africa, Sinai, Iraq and W Iran.

8. SCLERANTHUS L.

Prostrate or erect annual or perennial herbs. Leaves opposite, linear, the lower part with a scarious, often ciliate margin. Stipules absent or indistinct and connate to the leaf margins. Inflorescences of axillary or terminal cymes, remaining herbaceous in fruit; peduncles slender, up to 10mm; bracts leaf-like. Sepals 5, with hooded tips. Petals absent. Stamens (2–)5; staminodes present. Styles 2, free. Fruit 1-seeded, indehiscent, surrounded by the thickened wall of the perigynous zone of the flower, the complete flower falling at maturity.

1. S. orientalis Rössler in Phyton (Horn) 7: 207 (1957). Illustr.: Collenette (1985 p.107) as *S. annuus*.

Annual herb; stems prostrate or erect, 2–20cm. Leaves subulate, 0.5–5cm, connate, with a broad scarious margin at the base. Flowers sessile, in compact 5-flowered cymes. Sepals lanceolate, 3–5mm, usually slightly curved at the tip. Perigynous zone 3–5.5mm long in fruit, usually sparsely hairy, at least at first. **Map 224, Fig. 34.**

Damp, shaded gullies and bogs on granite; 2000–2100m.

Saudi Arabia. Ethiopia, Syria, Lebanon and Iran.

Intermediate between *S. annuus* L. (and possibly better treated as a subspecies of it), the range of which extends from Europe to W Turkey, and *S. uncinatus* Schur, from an area extending through S Europe and Turkey to N Iran. *S. annuus* is supposed to differ in the straight sepals and the often glabrous perigynous zone of the flower. *S. uncinatus* on the other hand has strongly incurved sepals and a more densely hairy perigynous zone in fruit.

9. HERNIARIA L.

Annual or perennial herbs. Stems prostrate to procumbent, brittle. Leaves opposite or apparently alternate on the flowering branches, elliptic to suborbicular, sessile or shortly petiolate. Stipules minute. Flowers very small, in 7–12-flowered leaf-opposed clusters, subsessile, slightly perigynous; bracts inconspicuous. Sepals 4 or 5, free, equal or unequal. Petals absent. Stamens 2–5; staminodes 4 or 5, filiform, sometimes absent. Stigmas 2, sessile or on a short style. Fruit one-seeded, indehiscent.

Williams, F. N. (1896). A systematic revision of the genus *Herniaria*. *Bull. Herb. Boiss.* 1 (4): 556–570. Chaudhri, M.N. (1968). A revision of the Paronychiinae. *Meded. Bot. Mus. Herb. Rijks Univ. Utrecht* 285: 1–427.

1. Sepals 4, unequal; outer 2 sepals broadly ovate, more than 1.5 × as long as the inner 2   **1. H. hemistemon**
+ Sepals 5, equal or unequal; outer 2 sepals narrowly ovate to oblong, not more than 1.3 × as long as the inner 3   2

2. Perennial; leaves suborbicular, distinctly petiolate; indumentum of short (c.0.2mm long) hairs **2. H. maskatensis**
+ Annual; leaves narrowly ovate to oblong, sessile; indumentum of long (0.4–0.5mm long) white hairs **3. H. hirsuta**

1. H. hemistemon J. Gay in Duchartre, Rev. Bot. 2: 371 (1847). Illustr.: Cornes (1989 p.62); Phillips (1988, p.150); Mandaville (1990 t.24); Western (1989 p.39).

Pubescent, mat-forming perennial (rarely annual or biennial) herb. Stems prostrate, 5–14(–21)cm, retrorsely pubescent below. Leaves oblong-elliptic, 2.5–5.7 × 1.5–2.6 (–3.1)mm, subsessile. Stipules and bracts similar, reddish brown. Flowers often with long hooked hairs in tufts near the base. Sepals 4, clearly unequal; outer pair broadly ovate, narrowed at the base, c.1.4 × 0.9mm, shortly ciliate at the apex; inner pair ovate-oblong, $\frac{1}{3}$ as long as the outer. Stamens 2; staminodes absent. Stigmas 2, sessile. Fruit ovoid-ellipsoid. **Map 225, Fig. 34.**

In depressions in flat sandy areas, often near the coast, locally common; 0–550m.

Saudi Arabia, UAE, Qatar, Bahrain, Kuwait. N Africa, Sinai, Palestine, Jordan, Syria, Iraq and S Iran

Of the 3 varieties recognized by Chaudhri (op. cit.), var. *hemistemon* is the only one recorded from the Arabian Peninsula. Var. *albostipulata* is endemic to S Iraq where it is found close to the Kuwait border. It is distinguished by its white stipules and bracts and might be expected to occur in Arabia.

2. H. maskatensis Bornm. in Mitth. Thuring. Bot. Vereins 6: 51 (1894). Type: Oman, *Bornmuller* 181 (B, G, JE, ?K).

Pubescent perennial herb, often dull purple-green. Stems prostrate, 2–8(–15)cm, retrorsely pubescent towards the base. Leaves broadly ovate to suborbicular, 1.6–4 × 1.5–3.7mm, with numerous sessile glands; petiole 0.5–1.5mm. Stipules often reddish brown. Flowers with a few hooked hairs near the base. Sepals 5, equal, ovate-oblong, c.0.6 × 0.45mm, with short stiff hairs. Stamens 5, staminodes triangular. Stigmas 2, subsessile. Fruit subglobose. **Map 226, Fig. 34.**

In sand, gravel and rocky places, often in gullies or channels; 20–1300m.

Yemen (S), Oman. Endemic.

3. H. hirsuta L., Sp. pl.: 218 (1753). Syn.: *H. cinerea* DC. in Lam., Fl. franç.: 375 (1815). Illustr.: Collenette (1985 p.104).

Hairy annual. Stems prostrate, 2.3–4.2(–20)cm, pubescent with stiff spreading hairs. Leaves narrowly elliptic to narrowly obovate (often obliquely so), sometimes attenuate at the base, 2.8–9 × 1–1.6mm, subsessile, covered with stiff white hairs, the older leaves glabrescent. Stipules whitish. Flowers with some hooked hairs towards the base. Sepals 5, ovate-oblong, c.1mm, equal to slightly unequal with the inner narrower than the outer, with long stiff white hairs. Staminodes filiform. Stamens 2–3(–5). Style short, 2-lobed. Fruit ovoid to subglobose. **Map 227, Fig. 34.**

Sandy areas and clay pans; 0–2150m.

Saudi Arabia, Oman, UAE, Bahrain. Eurasia (except the north) and N Africa.

10. HAYA Balf. f.

Perennial herbs. Leaves in whorls of 3–5, obovate to obovate-lanceolate. Stipules broadly lanceolate, cuspidate. Inflorescence a dense globose cluster of contracted 1–3-flowered spikes, with 6–10 flowers in total. Flowers subtended by 5 or more scarious bracts. Sepals 5. Stamens 5, alternating with minute staminodes. Ovary trigonous, with a solitary style and capitate stigma. Fruit a 1-seeded capsule, ovoid, dehiscing towards the base into three valves. Seeds ellipsoid.

A monotypic genus endemic to Socotra. There is a striking resemblance between this genus and *Polycarpaea*, especially *P. hayoides*, though the latter technically differs in its several-seeded capsules that dehisce from the apex.

1. H. obovata Balf. f. (1884 p.409). Type: Socotra, *Balfour, Cockburn & Scott* 250 (E). Much-branched perennial herb; stems prostrate 13–17cm. Leaves sessile, 6–18 × 4–10cm, with acuminate tips. Bracts scarious, ovate-lanceolate, c.4mm, cuspidate, reddish brown. Sepals oblong, white, c.1mm. **Map 228, Fig. 35.**

Ravines and hillsides, on limestone and granite, in drought-deciduous bushland; 200–900m.

Socotra. Endemic.

11. POLYCARPAEA Lam.

Annual or perennial herbs, sometimes woody-based or shrublets. Stems prostrate to erect. Leaves opposite, usually appearing whorled, linear to spathulate or suborbicular. Stipules scarious. Flowers sessile or shortly pedicellate, in terminal or axillary often spicate cymes. Bracts scarious. Sepals 5, free, unequal, the 3 inner longer than the 2 outer, not keeled, the midrib broad to narrow or rarely absent, the margins scarious, persistent. Petals broadly ovate to oblong, rarely with a short claw, shorter than the sepals. Stamens 5. Style solitary, 3-lobed at the apex. Fruit a many-seeded capsule, dehiscing by 3 valves, ± equalling or slightly longer than the sepals.

1.	Plant hairy	2
+	Plant glabrous	3
2.	Stiffly erect annual; sepals completely scarious, glabrous	**1. P. corymbosa**
+	Prostrate to ascending perennial; sepal midrib green, hairy	**2. P. repens**
3.	Petal differentiated into a limb and claw, the limb ovate to cordate, the claw short and linear	**13. P. robbairea**
+	Petal not differentiated into a limb and claw, narrowly ovate to linear	4
4.	Cauline leaves linear	5
+	Cauline leaves narrowly spathulate to suborbicular	8
5.	Woody-based perennial or shrublet	6
+	Delicate annual herb	7

Fig. 35. Caryophyllaceae. A, *Haya obovata*: Aa, flowering shoot (× 1); Ab, flower (× 12); Ac, fruit (× 15). B, *Polycarpaea hassalensis*: Ba, habit (× 1); Bb, flower (× 12); Bc, bract (× 15). C, *P. jazirensis*: Ca, habit (× 1); Cb, flower (× 12); Cc, bract (× 15). D, *P. hayoides*: Da, habit (× 1); Db, flower (× 12); Dc, bract (× 15); Dd, fruit (× 15).

6.	Woody-based perennial; stems tufted, thin and flexuous or becoming woody with elongated internodes, bearing whorls of leaves	**7. P. caespitosa**
+	Cushion-forming shrublet with densely tufted leaves; stems woody, bearing the overlapping remains of leaf bases	**8. P. pulvinata**
7.	Sepals 2–2.5mm long, ovate; cauline leaves $\frac{1}{3}$ as long as the internode or less	**3. P. spicata**
+	Sepals 3–4mm long, narrowly ovate; cauline leaves more than $\frac{1}{2}$ as long as the internode	**4. P. balfourii**
8.	Annual	9
+	Woody-based perennial	11
9.	Inflorescence sessile	**3. P. spicata**
+	Inflorescence pedunculate	10
10.	Stems up to 45cm long; sepals with a rounded tip	**5. P. paulayana**
+	Stems up to 6cm long; sepals with an acute tip	**6. hayoides**
11.	Cauline leaves spathulate to sub-orbicular, less than 2 × as long as broad	12
+	Cauline leaves narrowly spathulate, at least 3 × as long as broad	13
12.	Leaves glaucous, fleshy; basal leaves up to 10mm long; sepals c.2.5mm	**12. P. jazirensis**
+	Leaves bright green, not fleshy, basal leaves 15–40mm long; sepals c.2.5–3 mm	**11. P. haufensis**
13	Stems prostrate; sepals narrowly ovate, 3–4 × 0.8–1mm; peduncles swollen above	**9. P. kuriensis**
+	Stems erect; sepals ovate, 2.25–2.5 × 1–1.2mm; peduncles not swollen above	**10. P. hassalensis**

1. P. corymbosa (L.) Lam., Tabl. encycl. 2: 129 (1797).

Annual herb. Stems erect, 5–6.5(–30)cm, sparsely lanate. Leaves linear, 8–23 × 0.5–1.5mm. Stipules narrowly ovate, 3–6mm. Flowers in dense corymbose heads. Bracts ovate to narrowly ovate, 2–3mm, silvery-white. Sepals ovate, acuminate, 2–3mm, completely scarious, silvery-white or pink or pale brown. Petals c.$\frac{1}{2}$ as long as the sepals, erose at the apex. Style obscurely 3-lobed. Capsule elliptic. **Map 229.**

Socotra. Widely distributed in the tropics.

From our area only known from a single gathering made in the last century. This species has also been recorded, probably in error, from Aden.

2. P. repens (Forsskal) Aschers. & Schweinf. in Oesterr. Bot. Z. 39: 26 (1889). Syn.: *P. fragilis* Del. (1814 p.209) nom. illegit.; *Corrigiola repens* Forsskal (1775 p.207).

Perennial herb, somewhat woody at the base, shortly white-tomentose throughout. Stems decumbent to ± erect, 6–30cm. Leaves linear, rarely elliptic, 4–12 × 0.4–0.8(–2)mm, mucronate. Stipules ovate to narrowly ovate, c.5mm, acuminate. Flowers in ± dense terminal and axillary cymes. Bracts similar to the stipules. Sepals ovate, c.2mm; midrib broad, green. Petals ovate, c.1mm, acute. Capsule ovoid. **Map 230, Fig. 36.**

Sandy, gravelly and rocky areas, near the coast and inland; 0–2450m.

Saudi Arabia, Yemen (N & S), Socotra, Oman, UAE, Qatar, Bahrain, Kuwait. N Africa and Sudan eastwards to Pakistan.

Widespread and common in Arabia.

3. P. spicata Wight ex Arn. in Ann. Nat. Hist. sér. 1 (3): 91 (1839).

Glabrous annual herb. Stems erect, 2.5–16cm, di- or trichotomously branched above. Basal leaves rosette-forming; cauline leaves appearing whorled, narrowly linear or narrowly obovate to spathulate, 6–27 × 0.3–7mm, glaucous. Stipules ovate-oblong, c.1.5mm. Flowers in dense 2–4-spicate heads. Bracts ovate to ovate-oblong, c.2mm, white or brown. Sepals ovate to lanceolate, c.3mm; midrib on outer sepals broad or narrow, greenish brown. Petals oblong, c.1.3mm, the apex bidentate. Capsule narrowly ovoid. **Fig. 36.**

1.	Cauline leaves narrowly obovate to spathulate; sepal-midrib broad	var. **spicata**
+	Cauline leaves linear; sepal-midrib narrow	var. **capillaris**

var. **spicata.** Illustr.: Collenette (1985 p.106). **Map 231.**

On rocky slopes, open sandy and gravelly plains and grassland; 0–600m.

Saudi Arabia, Yemen (N & S), Socotra, Oman, UAE, Bahrain. Egypt, E Africa, S Iran, Pakistan and India.

var. **capillaris** Balf. f. (1884 p 403). Syntypes: Socotra, *Balfour* 211 (K); *Schweinfurth* 239 (K). **Map 232.**

Sea cliffs, amongst granite rocks and in deciduous shrubland; 10–550m.

Socotra. Endemic.

4. P. balfourii Briquet in Annuaire Conserv. Jard. Bot. Genève 13–14: 376 (1911). Syn.: *P. divaricata* Balf. f. (1882 p.502) non (Sol.) Poir. ex Steud. Type: Socotra, *Balfour, Cockburn & Scott* 684 (BM, E, K).

Glabrous annual herb. Stems erect, 6.5–26cm, di- or trichotomously branched above. Basal leaves rosette-forming, spathulate, 15–32 × 4–9mm; cauline leaves appearing whorled, linear to narrowly obovate, 7–32 × 0.5–2.5(–3)mm, mucronate. Stipules triangular, 1–1.5mm. Flowers in dense 2–4-spicate heads. Bracts narrowly ovate, c.2mm, white or brown. Sepals narrowly ovate, c.3.5(–4)mm, the midrib narrow, white suffused orange-brown. Petals linear-oblong, c.1–1.2mm, the apex bidentate or rarely tridentate. Capsule narrowly ovoid. **Map 233, Fig. 36.**

Gravelly plains, rocky slopes and wadi-sides in semi-desert shrubland; 15–500m.

Socotra. Endemic.

CARYOPHYLLACEAE

Fig. 36. Caryophyllaceae. A, *Xerotia arabica*: Aa, flowering shoot (× 1); Ab, part of inflorescence (× 5). B, *Polycarpaea robbairea*: Ba, habit (× 1); Bb, flower (× 8); Bc, petal (× 10). C, *P. caespitosa*: Ca, habit (× 1). D, *P. balfourii*: Da, habit (× 1). E, *P. paulayana*: Ea, flowering shoot (× 1). F, *P. spicata*: Fa, habit (× 1). G, *P. repens*: Ga, habit (× 1). H, *P. kuriensis*: Ha, habit (× 1); Hb, inflorescence (× 7).

5. P. paulayana Wagner in Kaiserl. Akad. Wiss. Wien, Math.-Naturwiss. Kl. Anz. 38 (3): 24 (1901). Illustr.: Vierhapper (1907 t.4/2). Syntypes: Socotra, 29 i 1899, *Paulay* s.n. (WU); 3 ii 1899, *Paulay* s.n. (WU).

Glabrous annual herb. Stems erect, prostrate or ascending, up to 45cm. Leaves appearing whorled, broadly spathulate, up to 60 × 15mm; petiole up to 40mm. Stipules minute with a narrow brown midrib. Flowers in dense 2–4-spicate heads; heads 2–10-flowered. Bracts scarious with a narrow brown midrib, c.$\frac{1}{2}$ as long as the sepals. Sepals ovate, longer than the capsules, the apex rounded. Petals probably equalling the capsule. **Map 234, Fig. 36.**

Coastal areas.

Socotra. Endemic.

Only known from the type specimens which have not been seen. The above account has been prepared from Vierhapper's description of the species (1907, p.31).

6. P. hayoides Chamberlain in Edinb. J. Bot. 51 (1): 5 (1994). Type: Socotra, *Miller et al.* M. 8235b (E, K).

Dwarf glabrous annual. Stems prostrate, 1–6cm. Leaves appearing whorled, spathulate, lamina 5–15 × 3–6mm; petiole 5–40mm. Stipules membranous, c.1mm, lanceolate with an aristate apex, margin sometimes fimbriate. Flowers in dense 2–4-spicate heads, on 5–30mm peduncles; spikes unbranched or branched, 2–5mm, 2–8-flowered. Bracts ovate-lanceolate to lanceolate, reddish-brown, with a scarious margin. Outer sepals ovate-lanceolate, acute, c.1.5mm; inner sepals lanceolate, acute, c.3mm. Petals minute, c.$\frac{1}{2}$ as long as the capsule. Capsule ovoid, c.3-seeded, dehiscing from the top into 3 valves. **Map 235, Fig. 35.**

Semi-desert shrubland on limestone, rocky wadi-sides and date gardens; 50–450m.

Socotra. Endemic.

Superficially resembling *Haya obovata* but differing in its 3-seeded capsule that opens from the top. In some respects this species links these two subfamilies of the Caryophyllaceae (see note under *Haya*).

7. P. caespitosa Balf. f. (1882 p.502). Type: Socotra, *Balfour, Cockburn & Scott* 683 (E, K, BM).

Woody-based, glabrous perennial. Stems erect, densely tufted, 6.5–23cm. Leaves tufted at the base, appearing whorled above, linear, 10–20(–30) × 0.4–1.4mm. Stipules triangular, minute. Flowers in dense 2–4-spicate heads. Bracts ovate to narrowly ovate, c.1.25–1.5mm, brown. Sepals narrowly ovate, c.3.2mm, with a narrow midrib. Petals c.$\frac{3}{4}$ as long as the sepals, $\pm$ entire to irregularly toothed at the apex, the margin minutely fimbriate towards the base. Capsules ovoid. **Map 236, Fig. 36.**

Large boulders, rocky wadi-sides and cliffs on limestone; 100–500m.

Socotra. Endemic.

8. P. pulvinata M.G. Gilbert in Kew Bull. 42(3): 703 (1987).

Glabrous shrublet forming low cushions up to 8cm across; stems woody, covered with old leaf bases. Leaves grey-green, densely tufted, needle-like, 5–8(–10) × c.0.4mm. Stipules narrowly lanceolate, up to 3(–4.5)mm long. Flowers in dense 2–4-spicate heads; peduncles up to 4cm long, about half-way bearing a pair of needle-like bracts. Bracts (subtending flowers) ovate-triangular, 1–1.5 × 1–2mm, scarious with a brown midrib. Sepals oblong-ovate, 3.5–5.5 × 1–1.5mm, with a narrow midrib. Petals $\frac{1}{2}$ to $\frac{2}{3}$ as long as the sepals, oblong-lanceolate. Capsule narrowly ovoid. **Map 236.**

Open rocky slopes on windswept mountain tops; 550m.

Yemen (S). Somalia.

In Arabia only known from a single gathering but likely to be found on other dry mountains in Mahra Governorate.

9. P. kuriensis Wagner in Anz. Kaiserl. Akad. Wiss. Wien, Math.-Naturwiss. Kl. 38, 3: 22 (1901). Illustr.: Vierhapper (1907 t.4).

Woody-based, glabrous perennial. Stems prostrate, forming dense mats, 6–15cm long. Leaves tufted below, apparently whorled above, narrowly spathulate, 15–40 × 2–6mm, the base long-attenuate, acute to obtuse at the tip. Stipules triangular, minute. Flowers in few-flowered heads; peduncles slightly swollen above. Bracts triangular-ovate, c.2 × 1mm. Sepals narrowly ovate, 3–4 × 0.8–1mm, with a narrow midrib. Petals $\frac{1}{2}$ to $\frac{3}{4}$ as long as the sepals, elliptic or obovate, the margin entire, acute or obtuse at the apex. Capsule ovoid. **Map 237, Fig. 36.**

Rocky slopes in semi-desert shrubland; 250–550m.

Socotra. Endemic.

10. P. hassalensis Chamberlain in Edinb. J. Bot. 51 (1): 53 (1994). Type: Abd al Kuri, *Miller* et al. 11399 (E).

Glabrous perennial with a woody rootstock. Stems erect, up to 5cm, branched from the base. Basal leaves densely tufted, narrowly spathulate, 10–17 × 3–4mm, with a shortly acuminate tip, the base long-attenuate; cauline leaves narrowly spathulate, 4–7 × 0.8–1.2mm. Stipules triangular-ovate, c.5 × 0.5mm, scarious with a brown midrib and a fimbriate margin. Flowers in dense 2–4-spicate heads; spikes 5–10mm; peduncle 12–20mm. Bracts broadly ovate, c.1.25 × 1mm, scarious with a narrow brown midrib. Sepals ovate, 2.25–2.5 × 1–1.2mm, reddish brown with a green midrib and narrow hyaline margin. Petals $\frac{3}{4}$ as long as the sepals, oblong-ovate, the margin fimbriate towards the base, entire above. Capsule broadly ovoid. **Map 238, Fig. 35.**

Rocky slopes in semi-desert shrubland; 150m.

Socotra (Abd al Kuri). Endemic.

11. P. haufensis A.G. Miller in Edinb. J. Bot. 53 (1996). Type: Yemen (S) *Miller* 12176 (E).

Glabrous perennial with a woody rootstock. Stems ascending to 12cm. Basal leaves tufted, spathulate, 2.5–5 × 0.3–0.8cm, the base long-attenuate, obtuse to rounded at the tip; cauline leaves appearing whorled, broadly spathulate, 3–15 × 3–5mm. Stipules triangular-ovate, c.1.5 × 0.5mm, scarious with a brown midrib and fimbriate margin. Flowers in dense 2–4-spicate heads; spikes 5–12mm; spike axis fleshy; peduncle 1.5–5cm. Bracts broadly ovate, c.1–2 × 1–2mm, scarious with a narrow brown midrib. Sepals oblong-ovate, 2.5–3.1 × 1–1.5mm, reddish brown with a green midrib and narrow hyaline margin. Petals c.2 × 0.75mm, the margin fimbriate towards the base, entire above. Capsule ovoid. **Map 238.**

On cliffs and large boulders near the sea; 0–20m.

Yemen (S). Endemic.

12. P. jazirensis R.A. Clement in Edinb. J. Bot. 51 (1): 53 (1994). Type: Oman, *Miller* 6470 (E).

Woody-based glabrous perennial, with a swollen rootstock. Stems erect, 5–9cm, di- to trichotomously branched above. Leaves tufted at the base, appearing whorled above, spathulate to suborbicular, 3–10 × 3–6mm, glaucous, slightly fleshy. Stipules broadly ovate, minute. Flowers in dense 2–4-spicate heads; spike axis fleshy. Bracts triangular, c.2mm, brown. Sepals ovate-oblong, c.2.5mm, with a broad midrib, reddish brown. Petals $\frac{3}{5}$ as long as the sepals, the apex ± acute, margin dentate towards the base. Capsule elliptic. **Map 239, Fig. 35.**

Sandy depressions and limestone cliffs; 150m.

Oman. Endemic.

13. P. robbairea (Kuntze) Greuter & Burdet in Willdenowia 12: 189 (1982). Syn.: *Robbairea delileana* Milne-Redhead in Kew Bull. 3: 452 (1948).

Glabrous, often glaucous, annual or perennial herb. Stems prostrate to ascending, 6–28cm. Leaves linear to narrowly obovate, rarely elliptic, 4–14(–20) × 0.5–3mm. Stipules triangular, 0.2–1.2mm. Flowers in many-flowered lax cymes; pedicels up to 2mm. Bracts ovate, c.1mm, white-scarious with a green midrib. Sepals ovate-elliptic, 1.5–2.5mm; midrib broad, green. Petals white or pink, with a broadly ovate to cordate limb narrowing abruptly into a short claw, ± equal to or slightly exceeding the sepals, the apex obtuse and ± entire. Capsule subglobose. **Map 240, Fig. 36.**

Sandy and gravelly areas, rocky slopes and as a weed of cultivated areas; 15–2300m.

Saudi Arabia, Yemen (N & S), Oman, UAE, Qatar, Kuwait. N Africa, Sudan and Palestine.
Widespread and common.

12. XEROTIA Oliv.

Dwarf glaucous shrub. Leaves succulent, opposite, glabrous, sessile. Stipules minute, scarious, broadly ovate-lanceolate, fimbriate. Inflorescence terminal, compound, articulated and easily fragmenting, the branches succulent, subtended by broadly ovate-lanceolate, scarious bracts, the ultimate branches cymose with 3–7 sessile flowers. Sepals 5, succulent. Petals 5, white. Stamens 5. Style solitary, obscurely three-lobed. Fruit a 3–6-seeded capsule, splitting into three coriaceous valves; seeds obliquely pyriform to ellipsoidal.

1. X. arabica Oliv. in Hooker's Icon. Pl. 24: t. 2359 (1895). Type: Yemen (S), *Lunt* 82 (E, K).

Stems erect, 10–60cm, sparsely villous. Leaves broadly ovate to orbicular, 3–5 × c.3mm, glabrous, sessile. Sepals c.1mm, blunt at the apex. Petals c.1mm. Fruit ovoid, usually slightly exserted from the calyx. **Map 241, Fig. 35.**

In sand amongst rocks and on sand-dunes in dwarf shrubland; 0–150m.

Oman, Yemen (S). Endemic.

13. POLYCARPON Loefl. ex L.

Low annual or perennial herbs. Leaves opposite or apparently in whorls of 4. Stipules scarious. Flowers in compound cymes. Bracts scarious. Sepals 5, free, keeled and hooded. Petals 5. Stamens 3–5. Style solitary, obscurely 3-lobed at the apex. Fruit a many-seeded capsule, dehiscing by 3 valves.

1.	Sepals acute, usually mucronate; leaves 1.5mm broad or more	**1. P. tetraphyllum**
+	Sepals obtuse, not mucronate; leaves up to 1.5mm broad	**2. P. succulentum**

1. P. tetraphyllum (L.) L., Syst. nat. ed. 10 (2): 881 (1759). Illustr.: Collenette (1985 p.106).

Glabrous annual or rarely perennial herb. Stems prostrate to procumbent, 2–7.5 (–12)cm. Leaves obovate to ± spathulate or elliptic, 3.5–14(–20) × 1.5–6(–9.2)mm, sometimes slightly succulent. Stipules ovate, c.2mm, acuminate. Inflorescence condensed to somewhat lax. Bracts similar to the stipules, fairly conspicuous. Sepals ovate, 1.5–2mm, keeled along the back, the margin scarious, hooded, acute and usually mucronate at the apex. Petals white, c.$\frac{1}{2}$ as long as the sepals. Stamens usually 3. Capsule ovoid, slightly shorter than the sepals. **Map 242, Fig. 37.**

Sandy and gravelly areas near the coast and inland, also as a weed of fields and by tracks; 15–2450m.

Saudi Arabia, Yemen (N), Oman, Qatar, Kuwait. W & C Europe, the Mediterranean area, Sinai, the Syrian Desert and N Iran.

2. P. succulentum (Del.) J. Gay in Rev. Bot. Bull. Mens. 2: 372 (1846).

Similar to *P. tetraphyllum* but the stems 1.5–4.5cm; leaves linear-oblong to narrowly spathulate, not more than 1.5mm broad, succulent; inflorescence condensed; sepals ovate-oblong, obtuse at the apex, not mucronate. **Map 243, Fig. 37.**

In sandy areas; 0–900m.

Saudi Arabia, Oman, Bahrain, Kuwait. Egypt, Sinai, Lebanon, Palestine and Iraq.

In Arabia distributed in sand areas along the coast and also widespread on sand-dunes in the north of the Peninsula.

P. prostratum (Forsskal) Aschers. & Schweinf. might be expected to occur in Arabia. It differs from the other species in its pilose indumentum.

14. LOEFLINGIA L.

Annual herbs. Leaves opposite, linear. Stipules scarious, fused with the leaf bases. Flowers in compound terminal and axillary cymes. Bracts leaf-like. Sepals 5, free, with 2 bristle-like lateral appendages. Petals 3 or 5. Stamens 3–5. Style solitary, obscurely 3-lobed. Fruit a many-seeded capsule, dehiscing by 3 valves.

1. L. hispanica L., Sp. pl.: 35 (1753).

Diffusely branched herb. Stems procumbent to erect, 3–13(–15)cm, glandular-hairy. Leaves linear, 4–19 × 0.6–0.9mm, with some glandular hairs, mucronulate. Stipules c.2mm long. Flowers ± sessile. Sepals ovate-lanceolate, 3–5mm, glandular-hairy, the margin scarious with a distinctive narrow linear appendage on each side, mucronulate. Petals white, much shorter than the sepals. Capsule ovoid, shorter than the sepals. **Map 244, Fig. 37.**

Gravel pans, sandy and silty places (including sand-dunes), fields and *Juniperus* woodland; 10–2150m.

Saudi Arabia, Bahrain, Kuwait. S Europe, N Africa, Turkey, the Syrian Desert and S Iran.

15. SPERGULA L.

Usually annual herbs. Leaves opposite with secondary fascicles in the axils, linear. Stipules scarious, free. Flowers in lax, terminal, compound cymes. Sepals 5, free, glabrous. Petals 5, white, entire, shorter than the sepals. Stamens usually 10. Styles 3 or 5, free. Fruit a many-seeded capsule, splitting into 3 or 5 valves. Seeds winged (in Arabia).

1. S. fallax (Lowe) Krause in Sturm, Deutschl. Fl. ed. 2, 5: 19 (1901). Syn.: *Spergularia fallax* Lowe in Bot. Misc. 8: 289 (1856). Illustr.: Collenette (1985 p.111); Mandaville (1990 t.21); Western (1989 p.42).

Glabrous annual herb; stems ascending, 5–40cm, branching from the base. Leaves

Fig. 37. Caryophyllaceae. A, *Polycarpon tetraphyllum*: Aa, habit (×1); Ab, calyx (×30). B, *P. succulentum*: Ba, calyx (×30). C, *Loeflingia hispanica*: Ca, habit (×1); Cb, sepal (×40). D, *Spergularia bocconei*: Da, habit (×1); Db, stipule (×10). E, *Spergula fallax*: Ea, habit (×1). F, *Minuartia filifolia*: Fa, habit (×1). G, *M. hybrida*: Ga, inflorescence (×1). H, *M. meyeri*: Ha, inflorescence (×1). I, *M. picta*: Ia, habit (×1); Ib, flower (×10).

linear, 10–35mm, sessile. Stipules scarious, broadly ovate-lanceolate, c.2mm. Flowers 5–10, in lax cymes; pedicels 5–17mm. Bracts scarious or with broad scarious margins, 1–1.5mm. Sepals lanceolate, 4–5mm, with a scarious margin. Petals ovate-lanceolate, c.4mm, white. Styles 3. Capsule subglobose, 4–5mm diam. **Map 245, Fig. 37.**

Sandy soil in wadis, by roads and as a weed of cultivated land; 0–2700m.

Saudi Arabia, Yemen (N), Socotra, Oman, UAE, Qatar, Bahrain, Kuwait. Canaries, N Africa, SW Asia and India.

16. SPERGULARIA (Pers.) J. & C. Presl

Annual, biennial or rarely perennial herbs, glandular-hairy mainly in the upper parts. Leaves opposite, sometimes with leaf-fascicles in the axils, linear, mucronate. Stipules scarious, white, united to surround the node. Flowers in lax, terminal, dichasial cymes. Sepals 5, free, glandular-hairy, margins scarious. Petals 5, shorter than or equalling the sepals, entire. Stamens (1–)2–5(–10). Styles 3, free. Fruit a many-seeded capsule, dehiscing by 3 valves. Seeds winged or not.

1. Mature seeds light brown, 0.6–0.7mm long, unwinged or a mixture of winged and unwinged; capsule usually 4–6mm long, exceeding the sepals; stipules on young shoots connate for $\frac{1}{3}$ to $\frac{1}{2}$ their length **1. S. marina**
+ Mature seeds blackish or if light brown, less than 0.5mm long, unwinged; capsule usually less than 4mm long, ± equalling the sepals; stipules on young shoots connate for less than $\frac{1}{3}$ their length **2**

2. Stipules 2mm or less; mature seeds black or brownish black; capsules subglobose **2. S. diandra**
+ Stipules more than 2mm; mature seeds light grey-brown; capsules ovoid **3. S. bocconei**

1. S. marina (L.) Bessler, Enum. pl. 97 (1822). Syn.: *Arenaria rubra* L. var. *marina* L., Sp. pl 423 (1753); *A. marina* (L.) Roth, Tent. fl. germ. 2(1): 482 (1789); *S. salina* J. & C. Presl, Fl. čech. 95 (1819); *S. marina* (L.) Griseb., Spic. fl. rumel. 1: 213 (1843). Illustr.: Western (1989 p.42).

Annual, biennial or sometimes perennial herb. Stems 7–26 cm, prostrate, procumbent or suberect. Leaves 7–30 × 0.5–1(–1.7)mm, slightly fleshy. Stipules ± triangular, 2–2.5 mm. Sepals 2.5–4(–4.5)mm. Petals pink, white near the base, not exceeding the sepals. Capsule ovoid, (3.5–)4–6mm long, usually clearly exceeding the sepals. Seeds 0.6–0.7mm, light brown, unwinged or a mixture of winged and unwinged. **Map 246.**

Weed of cultivated areas and waste ground; 0–2500m.

Saudi Arabia, Yemen (N), Oman, UAE, Bahrain, Kuwait.
Widespread in the N Hemisphere, introduced into the S Hemisphere.

2. S. diandra (Guss.) Boiss., Fl. orient. 1: 733 (1867). Illustr.: Mandaville (1990 t. 22).

Annual or biennial herb. Stems 4–15(–30)cm, ascending. Leaves 6–28 × 0.4–0.9mm, not usually fasciculate, not or slightly fleshy. Stipules ± triangular, c.1–1.5mm. Sepals 2–3mm. Petals pink to lilac, rarely white, ± equalling the sepals. Capsules subglobose, 2–3mm, ± equalling the sepals. Seeds 0.6–0.7mm, dark brown to black, unwinged. **Map 247.**

Usually on coastal dunes and saline areas; 0–1000m.

Saudi Arabia, Oman, UAE, Qatar, Bahrain, Kuwait. S Europe, N Africa, SW & C Asia.

3. S. bocconei (Scheele) Graebn. in Aschers. & Graebn., Syn. mitteleur. Fl. 5(1): 849 (1919). Illustr.: Cornes (1989 p.66).

Annual or biennial herb. Stems 8–18(–25)cm, ascending. Leaves linear, 12–35 × 0.5–0.8mm, not fasciculate, fleshy. Stipules triangular, c.2–2.5mm. Sepals (2–)2.5–3.5mm. Petals pink, white near the base, shorter than or equalling the sepals. Capsules ovoid, 2–3.5mm, shorter than or equalling the sepals. Seeds 0.4–0.5mm, greyish brown, unwinged. **Map 248, Fig. 37.**

Sandy places, often near irrigation or path-sides; 0–2900m.

Saudi Arabia, Yemen (N), Oman, Qatar, Bahrain. Mediterranean region; introduced elsewhere, particularly around ports.

Species doubtfully recorded

Two further species, *S. rubra* (L.) J. & C. Presl and *S. media* (L.) Presl have been reported as weeds in Arabia (Chaudhary 1983). No confirmed records are, however, known. *S. media* can be distinguished from all other recorded species by its all winged seeds. Records of *S. rubra* have probably been confused with *S. salina* or *S. bocconei*.

17. MINUARTIA L.

Annual or perennial herbs, sometimes somewhat woody at the base. Leaves opposite, linear to linear-filiform. Stipules absent. Flowers in terminal cymes. Sepals 5, free, with prominent nerves (rarely obscure) and scarious margins. Petals 5, rarely absent, white or pink. Stamens 10. Styles 3, free. Fruit a many-seeded capsule, dehiscing by 3 valves.

Mattfeld, J. *Repert. Spec. Nov. Regni Veg. Beih.* 15 (1922); McNeill, J. *Notes Roy. Bot. Gard. Edinburgh* 24: 311 (1963).

1.	Perennial herb, somewhat woody at the base, caespitose, often mat-forming	**1. M. filifolia**
+	Annual herbs, not caespitose	2
2.	Sepals ovate with a rounded apex; petals longer than the sepals	**2. M. picta**
+	Sepals linear to ovate with a narrowly acute apex; petals shorter than or subequal to the sepals	3

3. Sepals 5mm long or less; petals $\frac{1}{2}$ as long as the sepals or more

3. M. hybrida

+ Sepals 6mm long or more; petals $\frac{1}{3}$ as long as the sepals or less **4. M. meyeri**

1. M. filifolia (Forsskal) Mattfeld in Feddes Repert. Beih. 15: 93 (1922). Syn.: *Alsine* (sect. *Minuartia*) (Forsskal) Schweinf. (1896 p.175); *Arenaria filifolia* Forsskal (1775 p.211). Illustr.: Collenette (1985 p.104). Type: Yemen (N), *Forsskal* (C, BM).

Caespitose, somewhat woody-based, perennial. Stems decumbent to $\pm$ erect, 4–22(–25)cm, glabrous to glandular-hairy. Leaves linear-filiform, 5–20 × 0.3–0.5mm, often in dense axillary fascicles. Flowers in lax, 1–10(–14)-flowered cymes; pedicels 4–14mm, glabrous or with glandular hairs. Sepals narrowly ovate, 4.5–7mm, acuminate, prominently 3-nerved, glabrous or with a few sparse glandular-hairs at the base. Petals white, $\pm$ equal to the sepals. Capsule ovoid. **Map 249, Fig. 37.**

In crevices on rocky slopes and cliffs, frequently in *Juniperus* woodland; (1250–)2000–3000m.

Saudi Arabia, Yemen (N & S). NE and E tropical Africa.

2. M. picta (Sibth. & Sm.) Bornm. in Beih. Bot. Centralbl. 28, 2: 148 (1911). Illustr.: Collenette (1985 p.104).

Annual. Stems erect, 5.5–6(–12)cm, branched above, glabrous to glandular-hairy, leafy in the lower $\frac{1}{2}$ only. Leaves setaceous, 7–12(–20)mm, often in axillary fascicles. Flowers in lax, 5(–30)-flowered cymes; pedicels 2.5–8 (–15)mm. Sepals ovate, 1.5–2 (–2.5)mm, rounded at the apex, obscurely nerved. Petals pink, $1\frac{1}{2}$–2 × as long as the sepals. Capsule ovoid. **Map 250, Fig. 37.**

Rocky areas, usually in *Juniperus* woodland; 800–2800m.

Saudi Arabia. SW Asia.

3. M. hybrida (Vill.) Schischk., Fl. URSS 6: 488 (1936).

Annual. Stems erect, 4–17cm, usually glabrous. Leaves subulate to linear, 5–20 × 0.2–1mm, 3-nerved. Flowers in lax, many-flowered cymes. Sepals narrowly ovate to linear-ovate, acuminate, (2–)3.5–4mm, prominently 3-nerved, glandular-hairy, shorter than the pedicels. Petals white, $\frac{1}{2}$–$\frac{3}{4}$ as long as the sepals. Capsule cylindrical to narrowly ovoid, shorter or longer than the sepals. **Map 251, Fig. 37.**

Stony areas, open *Juniperus* woodland and fields; 1100–2800m.

Saudi Arabia, Oman. Mediterranean region and SW Asia.

Two subspecies have been recognized in SW Asia: subsp. *hybrida* with sepals more than 3.5 times as long as broad, petals $\frac{1}{2}$–$\frac{3}{4}$ as long as the sepals and cylindrical capsules; and subsp. *turcica* McNeill with sepals 3–3.5 times as long as, petals $\frac{3}{4}$ as long as or subequalling the sepals and the capsules narrowly ovoid.

4. M. meyeri (Boiss.) Bornm. in Beih. Bot. Centralbl. 27: 318 (1910).

Glandular-pubescent annual. Stems erect, 5.5–10(–15)cm, branched, often purplish. Leaves linear to linear-subulate, 10–21 × 0.6–1.5mm. Flowers in dense, many-flowered terminal and axillary cymes. Sepals narrowly ovate to linear-ovate, 5.5–8.5mm, acuminate, with 3 (to 5 at the base) prominent nerves, sometimes purplish. Petals white, less than $\frac{1}{2}$ as long as the sepals. Capsules narrowly ovoid, shorter than the sepals. **Map 252, Fig. 37.**

Rocky area in open *Juniperus* woodland; 1500–2150m.

Saudi Arabia, Oman. SW Asia.

18. HOLOSTEUM L.

Annual herbs. Leaves opposite, linear-lanceolate to oblong. Stipules absent. Flowers in terminal umbels. Sepals 5, free. Petals 5, white, entire to sinuate-dentate, slightly shorter to longer than the sepals. Stamens 3–5 or 10. Styles 3, free. Fruit a many-seeded capsule dehiscing by 6 revolute teeth. Seeds peltate, papillose.

1. H. glutinosum (Bieb.) Fisch. & C. Mey., Ind. sem. hort. petrop. 3: 39 (1839). Syn.: *H. umbellatum* L. var. *glutinosum* (Bieb.) Gay in Ann. Sci. Nat. Bot. sér. 3 (4): 33 (1845); *Arenaria glutinosa* Bieb., Fl. taur.-caucas. 1: 344 (1808). Illustr.: Collenette (1985 p.104) as *H. umbellatum* var. *glutinosum*; Fl. Iranica 163: t.26 (1988).

Annual herb. Stems upright, 5–20cm. Leaves linear-lanceolate to oblong, 10–25 × 2–6mm, glabrous or glandular-ciliate, contracted below into a short petiole. Umbels lax, 3–16-flowered; pedicels up to 30mm, often becoming deflexed, densely glandular-hairy. Sepals ovate-oblong, 3–4mm, sparsely glandular-hairy. Petals white, oblong-elliptic, 7–10mm, entire, glabrous at the base. Stamens 10. Capsule cylindrical, c.5mm long. **Map 253, Fig. 38.**

Silt pans, rock crevices and steep slopes in *Juniperus* woodland; 1350–2300m.

Saudi Arabia. SW Asia.

Closely allied to *H. umbellatum* L., a widespread species from Europe, N Africa and SW Asia, which differs in its glabrous inflorescence, petals only slightly longer than the sepals and in having only 3–5 stamens. A specimen, also from Saudi Arabia (*Collenette* 7717), is intermediate between these two species, with 3 stamens but a sparsely glandular inflorescence.

19. ARENARIA L.

Annual or perennial herbs. Leaves opposite, lanceolate to broadly ovate-lanceolate. Stipules absent. Flowers few, in terminal cymes. Sepals 5, free, lanceolate, broadening in fruit. Petals 5, white, entire or slightly emarginate, shorter than the sepals. Stamens 10. Styles 3, free. Fruit a many-seeded capsule, dehiscing by 6 teeth. Seeds reniform.

McNeill, J. (1963) *Notes Roy. Bot. Gard. Edinburgh* 24: 245–309.

19. Arenaria

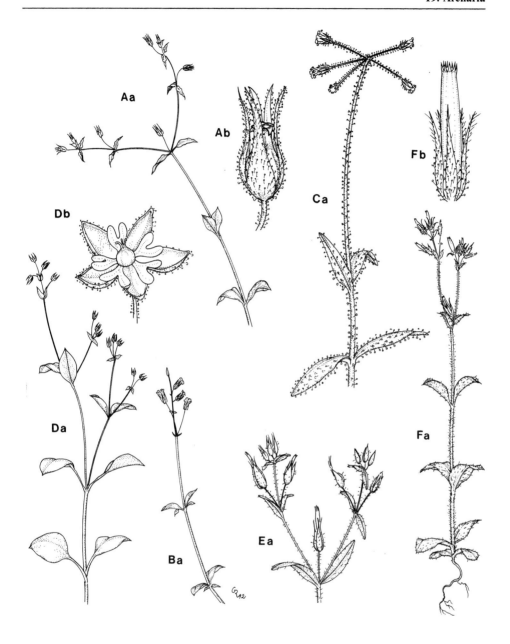

Fig. 38. Caryophyllaceae. A, *Arenaria foliacea*: Aa, flowering shoot (×1); Ab, flower (×10). B, *A. deflexa*: Ba, flowering shoot (×1). C, *Holosteum glutinosum*: Ca, flowering shoot (×1). D, *Stellaria media*: Da, flowering shoot (×1); Db, flower (×8). E, *Cerastium dichotomum*: Ea, inflorescence (×1). F, *C. glomeratum*: Fa, habit (×1); Fb, fruit and calyx (×6).

205

CARYOPHYLLACEAE

1.	Tufted perennial; petals c.5mm long	**1. A. deflexa**
+	Annual; petals 3–4mm long	**2. A. serpyllifolia** agg.

1. A. deflexa Decne, Florula Sinaica: 53 (1834).

Tufted perennial; stems 5–25cm. Leaves broadly ovate to elliptic, 3–10 × 3–5mm, acute, glandular-hairy; petioles absent or up to 3mm. Cymes lax, 4–6-flowered; pedicels 2–7mm. Bracts lanceolate, minute. Sepals 3.5–4.5mm, 3-nerved, glandular-pubescent. Petals c.5mm. Capsules c.4mm long, narrow. Seeds 5–6mm, tuberculate. **Map 254, Fig. 38.**

Shaded rocks in ravines; 1450m.

Saudi Arabia. Yugoslavia, Greece, Turkey, Lebanon, Palestine and Jordan.

Records of *R. graveolens* Schreber are almost certainly referable to this species.

2. A. serpyllifolia L. agg.

Annual herbs; stems 5–15cm. Leaves broadly ovate to ovate-lanceolate, 1–8mm long, acuminate, with a cuneate to rounded base, sparsely hairy, the hairs sometimes gland-tipped; petioles absent or up to 4mm. Cymes lax, 2–6-flowered; pedicels 3–10mm. Bracts minute or well-developed and leaf-like. Sepals 3–5mm. Petals narrowly lanceolate, 2–4mm. Capsules flask-shaped with curved sides, or conical and straight-sided. Seeds 4–7mm, black, sometimes papillose. **Map 255, Fig. 38.**

This is a complex of very closely allied taxa between which there is considerable overlap. Three of these have been recorded from the Arabian Peninsula and may be keyed out as follows:

1.	Bracts well-developed, leaf-like; lower leaves often petiolate, the lamina 4–8mm long	**2c. A. foliacea**
+	Bracts minute, narrowly lanceolate, not leaf-like; leaves sessile, the lamina 1–3mm long	2
2.	Capsule flask-shaped, with curved sides; seeds 0.55–0.7mm	**2a. A. serpyllifolia**
+	Capsule conical, with straight sides; seeds 0.4–0.55mm	**2b. A. leptoclados**

2a. A. serpyllifolia L., Sp. pl.: 423 (1753).

Recorded from Saudi Arabia and Yemen (N & S) but no specimens have been seen. Distributed widely across the temperate parts of Eurasia and N Africa and introduced in America and Australia.

2b. A. leptoclados (Reichenb.) Guss., Fl. sicul. syn. 2: 824 (1845). Syn.: *A. serpyllifolia* L. subsp. *leptoclados* (Reichenb.) Nyman, Consp. fl. eur.: 115 (1878).

Recorded from Saudi Arabia, Socotra and Oman but no specimens have been seen. External distribution as for *A. serpyllifolia*.

2c. A. foliacea Turrill in Kew Bull. 9: 415 (1954). Illustr.: Collenette (1985 p.101) as *A. ?serpyllifolia*; Fl. Trop. E. Afr.: 23 (1956).

Grassy banks, sandy soil and granite boulders; 400–3300m.

Saudi Arabia, Yemen (N), Oman. Tanzania, Somalia.

All the Arabian material seen differs from both *A. serpyllifolia* and *A. leptoclados* in its larger leaves and often leaf-like bracts, and on these characters is referable to *A. foliacea*. However, the frequently petiolate leaves distinguish at least some of the specimens from typical *A. foliacea*, and indeed the more extreme forms from the northern part of the range of the species (see the illustration in Collenette, 1985 p.101) are distinct from those from the south. There is, however, a complete overlap between the two extremes.

It is not clear whether all the records for this complex from the Arabian Peninsula are referable to *A. foliacea* but it seems likely that they are.

20. CERASTIUM L.

Annual glandular-pubescent herbs (in Arabia). Leaves lanceolate to elliptic or spathulate, sessile. Stipules absent. Flowers in lax or compact, few-flowered terminal cymes. Bracts herbaceous. Sepals (4–)5, free, herbaceous, with membranous margins. Petals (4–)5, sometimes absent in the lower flowers, white, shallowly or deeply bifid. Stamens 5–10. Styles (4–)5, free. Fruit a many-seeded capsule, dehiscing by 8–10 teeth, cylindrical, long-exserted from the calyx when ripe.

1. Sepals with long eglandular hairs at their tips; filaments glabrous; seeds 0.5–0.6mm diam. **1. C. glomeratum**
+ Sepals with glandular hairs at their tips; filaments hairy; seeds 1–1.4mm diam. **2. C. dichotomum**

1. C. glomeratum Thuill., Fl. env. Paris ed. 2: 226 (1799). Illustr.: Fl. Iranica 163: t.67 (1988).

Annual herb; stems 10–20(–45)cm, densely pilose with a mixture of glandular and eglandular hairs. Leaves spathulate to oblanceolate below, oblong to elliptic above, 0.5–2.5 × 0.3–1.2cm, with a rounded or sometimes acuminate apex, pilose with long hairs. Inflorescences compact, c.6-flowered, sometimes aggregated into dense heads; pedicels 1–3mm. Sepals 4–5mm, densely pilose with glandular and eglandular hairs, the hairs at the tips exclusively eglandular-pilose. Petals 4–5mm, deeply bifid (absent in the lower flowers). Stamens 5–8(–10), filaments glabrous. Capsule 5–10mm long, curved; seeds 0.5–0.6mm diam., pale brown, finely warted. **Map 256, Fig. 38.**

Montane grassland and roadsides; 1500–3200m.

Saudi Arabia, Yemen (N). A cosmopolitan weed.

Records from Yemen (N) of the closely allied and only doubtfully distinct *C. octandrum* A. Rich., an African species, probably refer to *C. glomeratum*.

2. C. dichotomum L., Sp. pl.: 438 (1753).

Annual herb; stems 10–20cm, densely villous, some of the hairs glandular. Leaves lanceolate, 1.3–2(–4) × 0.3–0.6cm, with an acute to rounded apex, villous with glandular hairs. Inflorescences of 3–6-flowered cymes that are dense in flower and more lax in fruit; pedicels 5–7mm. Calyx inflated or not in fruit; sepals 5–8mm, villous, some of the hairs glandular. Petals 4–6mm, shallowly bifid. Stamens 10, filaments ciliate at the base. Capsules 13–20mm long, usually straight; seeds 1–1.4mm diam., brown, strongly warted. **Map 257, Fig. 38.**

The two subspecies below can only be distinguished by the calyx when in fruit. It is not therefore always possible to assign flowering specimens to them.

1.	Calyx not inflated in fruit	subsp. **dichotomum**
+	Calyx inflated in fruit	subsp. **inflatum**

subsp. **dichotomum.** Illustr.: Fl. Iranica 163: t.70 (1988).

Rocky slopes, under *Juniperus*; 1700–2150m.

Saudi Arabia. S Europe, N Africa, SW and C Asia, widely introduced elsewhere.

subsp. **inflatum** (Link.) Cullen in Notes Roy. Bot. Gard. Edinburgh 27: 211 (1967). Syn.: *C. inflatum* Link in Desf., Tabl. écol. bot. ed. 3 (Cat. Hort. Paris) add.: 462 (1832). Illustr.: Collenette (1985 p.101); Fl. Iranica 163: t.71 (1988).

Rocky slopes; 1500–1650m.

Saudi Arabia. SW and C Asia, extending to Afghanistan and Pakistan.

It is noted that the Med. Checklist cites subsp. *inflatum* (Link.) Cullen as an invalid combination and that the literature citation given by Cullen is incorrect. This should be as given above. The status of the Additamenta of 1832 is uncertain and the type description has not been seen.

21. STELLARIA L.

Annual or perennial herbs. Leaves opposite, ovate-lanceolate, petiolate. Stipules absent. Flowers few, in lax dichasial cymes. Sepals 4–5, free. Petals 4–5, white, emarginate to deeply bifid. Stamens (3–)5–10. Styles usually 3, free. Fruit a many-seeded capsule, dehiscing by 6 valves, inflated below.

1.	Calyx (3.5–)4–5mm; petals usually present; seeds 0.8–1.4mm, red-brown; plants bright green	**1. S. media**
+	Calyx up to 3mm; petals often absent; seeds up to 0.8mm, pale brown; plants pale green	**2. S. pallida**

1. S. media (L.) Vill., Hist. pl. Dauphiné 3: 615 (1789). Illustr.: Fl. Iranica 163: t.32 (1988).

Annual herb, usually rather flaccid; stems ascending to prostrate, 10–30cm, with two lines of long hairs or glabrous. Leaves ovate-lanceolate, bright green, 1–3 × 0.8–1.7cm, the apex acute, glabrous; petioles absent or up to 1cm. Flowers 3–5, in lax axillary or terminal cymes; pedicels 7–20mm, sparsely villous. Sepals (3.5–)4–5mm, ovate, with a rounded tip, sparsely villous. Petals usually present, white, 2–3mm, deeply bifid. Capsule 6–7 mm long, slightly longer than the sepals. Seeds 0.8–1.4mm. **Map 258, Fig. 38.**

Weed of cultivation; 10–2400m.

Saudi Arabia, Yemen (N), Oman, Kuwait. A cosmopolitan weed.

2. S. pallida (Dumort.) Murb., Beitr. Fl. Südbosnien 158 (1891). Syn.: *S. media* L. subsp. *pallida* (Dumort.) Aschers. & Graeb., Fl. Nordostdeut. Flachl. 310 (1898). Illustr.: Fl. Iranica 163, 2: t.33 (1988).

Similar to *S. media* but the leaves pale green; calyx less than 3mm long; petals often absent; seeds up to 0.8mm. **Map 259.**

Weed of cultivation; 10–2300m.

Saudi Arabia, Yemen (N), Oman, Qatar, Kuwait. S Europe and SW Asia.

The differences between *S. media* and *S. pallida* are small and Arabian material can only be assigned with a degree of certainty. The widely used citation *S. pallida* (Dumort.) Piré (1863) was not validly published.

22. SILENE L.

Annual, biennial or perennial herbs, often woody below. Leaves various. Stipules absent. Inflorescence a monochasial (one-sided) or dichasial (two-sided) cyme, more rarely paniculate. Calyx tubular, 5-toothed, 10–60-ribbed, sometimes inflated in fruit. Petals 5, usually with a distinct lamina and claw, the base of the lamina often auriculate (with small lateral outgrowths); coronal scales usually present. Stamens 10. Styles 3 (–5), free. Fruit a many-seeded capsule, usually dehiscing by 6 teeth. Petals, stamens and ovaries borne on a stalk (the carpophore) that usually elongates in fruit.

Rohrbach, P. (1868). Monographie der Gattung Silene. Williams, F.N. (1894). Primary Subdivisions in the Genus Silene. *J. Linn. Soc., Bot.* 32: 1–196. 1896.

1.	Calyx with 15–30 ribs	2
+	Calyx with 10 ribs	5
2.	Perennial; calyx strongly inflated in fruit	**1. S. vulgaris**
+	Annual; calyx enlarged in fruit but not inflated	3

3.	Calyx 15–20-ribbed	**5. S. coniflora**
+	Calyx c.30-ribbed	4
4.	Calyx (16–)18–25(–29)mm long; calyx teeth 6–8mm long	**3. S. conoidea**
+	Calyx (10–)12–15(–16.5)mm long; calyx teeth 4–6mm long	**4. S. conica**
5.	Calyx 35–45mm long, entirely glabrous (teeth margins sometimes minutely scabrid); carpophore 25–30mm long	**2. S. macrosolen**
+	Calyx less than 30mm long, hairy at least on the teeth margins; carpophore less than 25mm long	6
6.	Perennials; leaves fleshy	**16. S. succulenta**
+	Annuals or perennials; leaves not fleshy	7
7.	Stems with eglandular hairs only above	8
+	Stems glabrous or with at least some glandular hairs above	11
8.	Petals usually not or only slightly exserted from the calyx, the limb 1.5–3mm long; carpophores 1–3mm long; annuals	**7. S. apetala**
+	Petals clearly exserted from the calyx, the limb 4–8mm long; carpophores 2–8mm long; annuals or perennials	9
9.	Annuals; inflorescences dichasial	**15. S. crassipes**
+	Annuals or perennials; inflorescences monochasial	10
10.	Perennials, sometimes flowering in the first year; leaves linear to broadly ovate	**8. S. burchellii** agg.
+	Annuals; leaves linear-elliptic	**12. S. colorata**
11.	Perennials; leaves of the basal rosettes broadly cuneate to spathulate, less than 2 × as long as broad	12
+	Annuals; leaves linear to obovate-cuneate, at least 4 × as long as broad	13
12.	Stems 10–45cm; lamina of basal leaves 15–30mm long, villous, glandular above	**9. S. yemensis**
+	Stems 65–100cm; lamina of basal leaves 5–12mm long, scabrid, eglandular	**10. S. corylina**
13.	Petals emarginate or entire; carpophores up to 1mm long	**6. S. gallica**
+	Petals bifid; carpophores 3–11mm long	14
14.	Pedicels and upper parts of the stems glandular-hairy	15
+	Pedicels and upper parts of the stems glabrous, viscid	18
15.	Calyx teeth c.1mm long; stems up to 11cm long	**17. S. hussonii**
+	Calyx teeth 2–3.5mm long; stems up to 45cm long	16
16.	Flowers in simple monochasial cymes	**11. S. arabica**
+	Flowers solitary or in dichasial cymes	17
17.	Fruiting pedicels erect, spreading or deflexed when mature; leaves 1–6(–10)mm wide	**13. S. villosa**
+	Fruiting pedicels stiffly erect when mature; leaves up to 19mm wide	**14. S. asirensis**

18.	Stems ascending to erect, 10–25cm long; calyx ribs raised in fruit **18. S. arenosa**
+	Stems erect, 10–100cm long; calyx ribs in the upper part of the calyx not raised when in fruit 19
19.	Lobes of the petal limbs linear **19. S. linearis**
+	Lobes of the petal limbs broadly elliptic **20. S. austroiranica**

1. S. vulgaris (Moench) Garcke, Fl. N. Mitt.-Deutschland ed. 9: 63 (1869). Syn.: *S. inflata* (Salisb.) Sm. (1800) nom. illegit. Illustr.: Collenette (1985 p.110); Fl. Palaest. 1: t. 101 (1966).

Glabrous to pubescent, glaucescent perennial; stems erect, 25–70cm. Leaves glabrous; basal leaves with elliptic lamina and long petioles; lower cauline leaves broadly ovate to narrowly lanceolate, 50–80(–100) × 7–25(–40)mm, sessile. Flowers in 6–25-flowered compound dichasial cymes; bracts and bracteoles scarious; pedicels 10–25(–40)mm, slender, glabrous. Calyx c.20-ribbed, 9–15mm, glabrous, strongly inflated and enclosing the capsule in fruit; calyx teeth broadly triangular, 3–4mm long. Corolla white, c.20mm long; limb c.9mm, divided almost to the base into two obovate lobes. Capsule ovoid, 8–10mm long; carpophore 2–4mm long, glabrous. **Map 260, Fig. 39.**

Roadsides, granite slopes and a field weed; 1800–2700m.

Saudi Arabia. Europe, N Africa and temperate Asia.

2. S. macrosolen Steud. ex A.Rich (1847 p.44). Syn.: ?*S. bupleuroides* L. subsp. *solenocalyx* (Boiss. & Huet) Melzh., Fl. Iranica 163, 2: 399 (1988). Illustr.: Collenette (1985 p.109).

Glaucous perennial; stems erect, 25–65cm, glabrous, viscid, arising from a woody base. Leaves linear, 25–50mm long, the basal up to c.4mm wide, the cauline 1–2mm wide, lamina glabrous, the margin scabrid, sessile. Flowers solitary or in 2–4(–6)-flowered lax dichasial cymes; pedicels 20–80mm, glabrous. Calyx 10-ribbed, narrowly cylindrical in flower, broadening above in fruit, (27–)30–45mm, glabrous; calyx teeth 2.5–4mm, similar or dissimilar, either all triangular-lanceolate with acute to cuspidate teeth or with blunt and cuspidate teeth alternating; margin entire or minutely scabrid, scarious. Corolla white to pink, c.25mm long; limb c.10mm, divided to the middle into two obovate lobes. Capsule c.12mm long; carpophore 25–30mm, glabrous. **Map 261, Fig. 39.**

Basalt rock faces, wadi-banks, steep river banks, and also a weed of disturbed ground; 1700–3000m.

Saudi Arabia, Yemen (N). E tropical Africa.

S. macrosolen is described as having calyx teeth of one type while in the closely allied *S. schimperiana* (from Sinai) they are described as dimorphic, alternating between those with blunt apices and those with acute to cuspidate apices. *S. schimperiana* may additionally be distinguished by its consistently smaller calyces (22–25mm long). Both

CARYOPHYLLACEAE

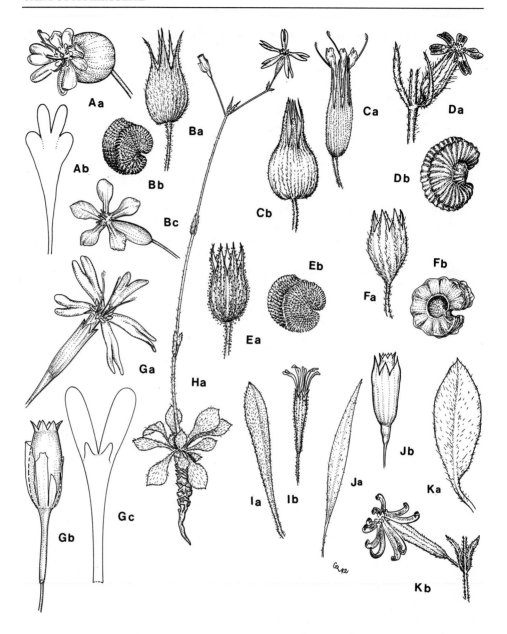

Fig. 39. Caryophyllaceae. A, *Silene vulgaris*: Aa, flower (×2.5); Ab, petal (×3). B, *S. conoidea*: Ba, fruiting calyx (×2.5); Bb, seed (×35); Bc, flower (×2.5). C, *S. conica*: Ca, flower (×2.5); Cb, fruiting calyx (×2.5). D, *S. gallica*: Da, flower (×2.5); Db, seed (×35). E, *S. coniflora*: Ea, fruiting calyx (×2.5); Eb, seed (×35). F, *S. apetala*: Fa, fruiting calyx (×2.5); Fb, seed (×35). G, *S. macrosolen*: Ga, flower (×2.5); Gb, fruiting calyx (×2.5); Gc, petal (×3). H, *S. yemensis*: Ha, habit (×0.6). I, *S. corylina*: Ia, leaf (×1.5); Ib, flower (×2.5). J, *S. burchelii*: Ja, leaf (×1.5); Jb, fruiting calyx (×2.5). K, *S. hochstetteri*: Ka, leaf (×1.5); Kb, flower (×2.5).

monomorphic and dimorphic calyx teeth do, however, occur consistently on plants with large calyces from Yemen (N) and probably also from E tropical Africa.

S. bupleuroides subsp. *solenocalyx*, from NE Turkey and NW Iran, is almost certainly synonymous with those forms of *S. macrosolen* in which the calyx teeth are dimorphic. However, the relationship between subsp. *solenocalyx* and subsp. *bupleuroides* is uncertain, especially as the material from the Arabian Peninsula does not always agree with the subspecies recognized by Melzheimer (loc. cit.).

3. S. conoidea L., Sp. pl.: 418 (1753). Illustr.: Collenette (1985 p.108); Fl. Iranica 163: t.240, 477/4–6 (1988).

Annual herb; stems 10–45(–60)cm, densely pilose, eglandular below, glandular above. Leaves oblong to lanceolate, 20–70 × 3–10mm, eglandular-pilose, usually sessile and clasping. Flowers in a 3–15(–30)-flowered dichasial cyme; pedicels 6–30mm, covered with a mixture of eglandular and glandular villous hairs. Calyx 30-ribbed, (16–)18–25(–29)mm, narrowly tubular-conical in flower, in fruit enlarged but not inflated, constricted above, the ribs predominantly glandular-hairy; calyx teeth subulate, 6–8mm. Corolla reddish-pink to lilac, 25–30mm long; limb 12–16mm, entire or shallowly 2-lobed. Capsule ovoid, (12–)13–16(–21)mm long; seeds reticulate-pectinate; carpophore 0–4mm long, pubescent. **Map 262, Fig. 39.**

Weed of cultivation; 0–2750m.

Saudi Arabia, Oman, Kuwait. Circumboreal.

4. S. conica L., Sp. pl.: 418 (1753). Illustr.: Fl. Iranica 163: t.239, 473/1–2 (1988).

Differs from *S. conoidea* in its shorter calyx, (10–)12–15 (–16.5)mm long, with teeth 4–5mm, and in its shorter capsule, usually 8–10mm long. Like *S. conoidea* the calyx is c.30-ribbed. **Fig. 39.**

Kuwait. Eurasia, extending to Afghanistan in the E.

The presence of this species in Arabia needs confirming. This record is based on an old literature record.

5. S. coniflora Otth in DC., Prodr. 1: 371 (1824). Illustr.: Collenette (1985 p.108); Fl. Iranica 163: t.241, 477/7–9 (1988).

Annual herb; stems erect, 8–30cm, densely covered with stout glandular and slender eglandular hairs. Leaves linear-lanceolate, 15–70 × 1.5–4mm, densely pilose, sessile, with a clasping base. Flowers in a 2–30-flowered dichasial cyme; pedicels usually present, c.5mm, villous with a mixture of eglandular and glandular hairs. Calyx 15–20-ribbed, 11–15.5(–17)mm, cylindrical in flower, narrowly ovoid in fruit, the ribs with a mixture of stout glandular and slender eglandular hairs; calyx teeth narrowly lanceolate, 3–6mm. Corolla white to deep pink, c.20mm long; petal limb 4–6mm, emarginate or shallowly 2-lobed. Capsule ovoid-oblong, 8–13mm long; seeds with a fine reticulate-pectinate pattern; carpophore absent or minute. **Map 263, Fig. 39.**

On gravel and sand in wadis; 700m.

Saudi Arabia. Spain, Portugal and SW and C Asia.

In Arabia only known from two gatherings in NW Saudi Arabia.

6. S. gallica L., Sp. pl. 1: 417 (1753). Illustr.: Fl. Palaest. 1: t.126 (1966); Collenette (1985 p.109) as var. *quinquevulnera*.

Annual herb; stems erect, 15–45cm, covered with short and long patent hairs, a few glandular in the upper part. Leaves obovate-spathulate, 40–70 × 7–15mm; upper cauline leaves sessile; lower cauline and basal leaves with petioles up to 15mm long, covered with long weak hairs, more sparsely so above. Flowers in a one-sided monochasial cyme, sessile or with pedicels up to 5mm, eglandular-villous. Calyx 10-ribbed, 7–8mm, with a dense mixture of glandular and short eglandular hairs, and with a few long eglandular hairs on the ribs; calyx teeth lanceolate, 2–3mm. Corolla white or pink to purple, sometimes with a purplish-red spot, 10–14mm; limb cuneate, 6mm, emarginate or entire. Capsule oblong-ovoid, 6–9mm long; seeds reticulate; carpophore absent or up to 1mm. **Map 264, Fig. 39.**

Damp sandy soil and fields; 2000–3000m.

Saudi Arabia, Yemen (N). Native in the Mediterranean Region, but widespread elsewhere as a weed of cultivated ground.

7. S. apetala Willd., Sp. pl. ed. 4, 2: 703 (1799). Illustr.: Collenette (1985 p.107); Fl. Iranica 163: t.246, 479/1–3 (1988).

Annual herb; stem erect to ascending, 8–60cm, minutely pubescent, eglandular. Leaves elliptic to oblong, (15–)25–40 × 2–6mm, sparsely pilose with tuberculate-based hairs, the margin ciliate; basal leaves with petioles 5–25mm; cauline leaves sessile, clasping at the base. Flowers in a lax, reduced 3–15(–25)-flowered dichasial cyme; pedicels sessile or 5(–15)mm, pubescent, eglandular. Calyx 10-ribbed, 8–10mm, cylindrical in flower, inflating and becoming broadly obovate in fruit, the ribs in particular coarsely hairy; calyx teeth narrowly triangular, 2–3mm, with scarious ciliate margins. Corolla white, greenish or pink, 6–8mm; petal limb 1.5–3mm, 2-lobed to the middle. Capsule 4–6.5mm long; seeds striate, minutely rugulose; carpophore 1–3mm, glabrous or minutely pubescent. **Map 265, Fig. 39.**

Rocky and gravelly ground, silty basins and wadis, abandoned fields, etc.; 600-2100m.

Saudi Arabia, Socotra, Oman. Mediterranean Region and SW Asia.

Field observations indicate that there are two forms of *S. apetala*. The first of these, with short petals scarcely longer than the calyx, equates with the type of the species. The second, with flowers up to 1 cm diam., and the petals long-exserted, may be an undescribed subspecies. It is not usually clear from herbarium material to which of these two forms individual specimens belong.

8. S. burchellii Otth in DC., Prodr. 1: 374 (1824) sensu lato.

Perennial, sometimes flowering in the first year; stems 5–40cm, when mature arising from a branched woody base, covered with short stiff hooked adpressed eglandular hairs. Leaves linear to broadly ovate, 15–30(–40) × 1.5–15(–20) mm, sparsely hispid, with a ciliate margin, the basal sometimes narrowed below into a c.5mm petiole. Flowers in 2–6-flowered monochasial cymes; pedicels absent or up to 20mm, pubescent, eglandular. Calyx c.10-nerved, 10–16mm, expanded above in fruit, stiffly pubescent; calyx teeth lanceolate, becoming triangular in fruit, 1.5–4mm, ciliate. Corolla white or cream to pink or reddish-purple, sometimes greenish below, 9–15mm; limb 5–7mm, emarginate or divided to the middle into two narrow lobes. Capsule ovoid, 7–8mm long; carpophore 2–7mm, glabrous. **Map 266, Fig. 39.**

The two taxa described below are part of a complex of species closely allied to *S. burchellii*. It is not however certain how these two taxa relate to the varieties described within the very polymorphic *S. burchellii*, a species described from S Africa but with a wide range throughout tropical East Africa. Furthermore, it appears that intermediates occur between *S. schweinfurthii* and *S. hochstetteri* in Arabia and the adjacent parts of East Africa. The whole complex is in need of thorough monographic revision throughout its extensive range.

S. burchellii in the strict sense can be distinguished from both the above mentioned by the presence of a tuberous rootstock; there is no evidence that any of the Arabian material exhibits this feature.

A specimen from NC Saudi Arabia (*Collenette* 6159) probably belongs to *S. hochstetteri* but differs in its broader leaves, the laminae of some of which are as long as broad.

  1. Basal leaves linear to narrowly elliptic, (4–)6–10 × as long as broad; calyx 10–13mm long; carpophore 2–3mm long   **8a. S. schweinfurthii**
  + Basal leaves elliptic to ovate, 2–2.5 × as long as broad; calyx 13–17mm long; carpophore up to 7mm long   **8b. S. hochstetteri**

8a. S. schweinfurthii Rohrb. in Bot. Zeitung 25: 82 (1867). Illustr.: Collenette (1985 p.110) as *S. yemensis*.

Rocky slopes, on granite and in sand pans; 650–3200m.

Saudi Arabia, Yemen (N), Oman. Sudan and Ethiopia.

8b. S. hochstetteri Rohrb., op. cit.: 81 (1867). **Fig. 39.**

Grassy slopes and rocky crevices; 700–2800m.

Saudi Arabia, Yemen (N), Oman. Ethiopia and Somalia.

9. S. yemensis Deflers (1889 p.112). Syn.: *S. engleri* Pax in Bot. Jahrb. Syst. 7: 586 (1893). Syntypes: Yemen (N), *Deflers* 304, 587 (P).

Densely villous perennial; stems 10–45cm, arising from a woody base, densely villous below, glandular-villous above. Basal leaves arranged in rosettes, the lamina broadly cuneate, 15–30 × 10–20mm, apiculate, villous, more densely so on the margins and midrib, glandular above, contracted below into a 10–40mm petiole; cauline leaves much reduced. Flowers solitary or in 2–5-flowered lax dichasial cymes; pedicels up to 20mm, glandular-villous. Calyx 10-ribbed, 13–15mm, slightly inflated in fruit, glandular-villous, especially on the ribs; calyx teeth triangular-lanceolate, c.2mm, acute. Corolla pink to brownish, c.20mm; limb c.7mm, divided in the top third into 2 linear lobes. Capsule 11–12mm long; carpophore 4–6mm, glabrous. **Map 267, Fig. 39.**

Rock ledges, gullies, open *Juniperus* woodland; 2200–2800m.

Saudi Arabia, Yemen (N). Ethiopia.

10. S. corylina D.F. Chamb. in Edinb. J. Bot. 51 (1): 56 (1994). Type: Saudi Arabia, *Collenette* 7788 (E).

Erect perennial herb with a thick woody base, forming clumps up to 35cm across; stems 65–100cm, shortly stipitate-glandular above, glabrous below. Basal leaves arranged in rosettes, spathulate, the lamina 5–12 × 3–5mm, scabrid, tapering below into a 7–15mm petiole; cauline leaves elliptic below, progressively more linear above, up to 15mm long. Flowers in lax dichasial cymes, 2–8(–12)-flowered; pedicels sessile or up to 17mm. Calyx 10-ribbed, 12–14mm, cylindrical in flower, stipitate-glandular; calyx teeth lanceolate, up to 2mm. Corolla greenish yellow with chestnut edges, c.20mm; limb 4–5mm, shallowly divided into rounded lobes. Capsule oblong, 4–6mm long; seeds 0.8–1mm, reniform, striate; carpophore 4–6mm, glabrous. **Map 268, Fig. 39.**

Amongst sandstone buttes; 1200m.

Saudi Arabia. Endemic.

This species is a member of Section *Brachypodae*, a group of species largely restricted to the E Mediterranean Region. It differs from all the presently known species in that section by its scabrid leaf indumentum. It is only known from one locality in NW Saudi Arabia where it is fairly common.

11. S. arabica Boiss., Fl. orient. 1: 593 (1967). Syn.: *S. affinis* Boiss., Diagn. pl. orient. sér. 2 (1): 72 (1853), non Godr. (1853). Illustr.: Collenette (1985 p 108); Mandaville (1990 t.18).

Erect annual herb; stems (6–)10–22cm, glandular-hairy, viscid. Leaves linear to linear-lanceolate, 15–35 × 1–3mm, sparsely glandular-hairy, sessile. Flowers in a simple monochasial cyme, sessile or with glandular-pubescent pedicels up to 12mm. Calyx 10-ribbed, 10–15mm, cylindrical in flower, becoming obovate and inflated in fruit, scabrous on the ribs; calyx teeth 2–2.5mm, ligulate, with broad scarious long-ciliate margins. Corolla white, sometimes with red lines on the petal reverse, c.15mm; limb 8–10mm, deeply divided into narrowly strap-shaped lobes with emarginate apices. Capsule oblong-ellipsoid, 5.5–9.5mm long; seeds ridged; carpophore 3–6.5mm, puberulent. **Map 269, Fig. 40.**

Silty soils in stony and sandy deserts and wadi-beds; 0–900m.

Saudi Arabia, Qatar, Kuwait. Iraq, Iran, Afghanistan and Pakistan.

Four varieties of this species are recognized in Palestine (Fl. Palaest. 1: 94 (1966)). Most of the Arabian material is referable to var. *arabica*, but a single specimen (*Collenette* 1581, from S. of Turayf in Saudi Arabia) belongs to var. *nabathaea* (Gomb. & A. Camus) Zohary, differing in its broader, obovate-lanceolate leaves, up to 10mm wide.

12. S. colorata Poiret subsp. **olivieriana** (Otth) Rohrb., Monogr. Sil. 116 (1868). Syn.: *S. olivieriana* Otth in DC., Prodr. 1: 373 (1824).

Annual; stems erect, 10–30cm, crispate-pubescent, eglandular. Basal leaves linear-elliptic, 20–35 × 3–5mm, crispate-pubescent, contracted below into a 10–15mm petiole; cauline leaves linear, sessile. Flowers in 4–6-flowered monochasial cymes; pedicels 0–5(–25)mm, when present pubescent, eglandular. Calyx 10-ribbed, 10–15mm, expanded above in fruit, pubescent on the ribs; calyx teeth 1.5–2(–4.5)mm, bluntly triangular. Corolla whitish to pink, 10–15mm; limb 5–7mm, deeply divided into 2 oblong-linear lobes. Capsule 7–8mm long; seeds c.1mm, auriculate, with undulate winged margins; carpophore 4–8mm, glabrous. **Map 270, Fig. 40.**

In silty and sandy runnels; 300–2400m.

Saudi Arabia, Yemen (N). Sinai, Turkey, Iraq, Syria, Jordan and Palestine.

All Arabian material examined is referable to subsp. *olivieriana* which differs from the type subspecies in its narrower leaves, often paler flowers and broader petals. This subspecies closely approaches *S. vivianii* (distributed from Libya to W Iran) which apparently differs in its less undulate seed margins. *S. colorata* subsp. *olivieriana* may be confused with *S. arabica*, but its characteristic seeds and eglandular indumentum readily distinguish it from that species.

13. S. villosa Forsskal, (1775 p.88). Illustr.: Phillips (1988 p.154); Western (1989 p.41).

Annual; stems 7–45cm, sparsely glandular-villous. Leaves linear to narrowly elliptic, 10–40 × 1–6(–10)mm, glandular-villous, sessile or with the lamina of the lower leaves contracted into a short petiole. Flowers solitary or in 2–6-flowered dichasial cymes; bracts large and leaf-like; pedicels 5–20mm, glandular-villous. Calyx 10-ribbed, 12–25mm, more markedly ribbed below, cylindrical, slightly inflated above in fruit, glandular-villous; calyx teeth 2–3.5mm, blunt. Corolla white to pink, 15–25mm; limb 5–10mm, divided to the middle into two narrow or broadly spathulate lobes. Capsule 10–11mm long; carpophore 5–11mm, glabrous. **Fig. 40.**

Two forms of this species are found in Arabia.

1. Calyx (10–)12–18mm long; capsule pendent when ripe, up to twice as long as the 5–7mm carpophore; stems decumbent or erect, 10–45cm long form **'A'**
+ Calyx 17–23(–25)mm; capsule erect, ± as long as the 10–11mm carpophore; stems erect or spreading, 7–18cm long form **'B'**

form **'A'**. Illustr.: Fl. Palaest. 1: t.119 (1966); Fl. Iranica 163: t.248, 479/7–9 (1988).

Stems decumbent or erect, 10–45cm. Calyx (10–)12–18mm. Petal limb 5–7mm,

CARYOPHYLLACEAE

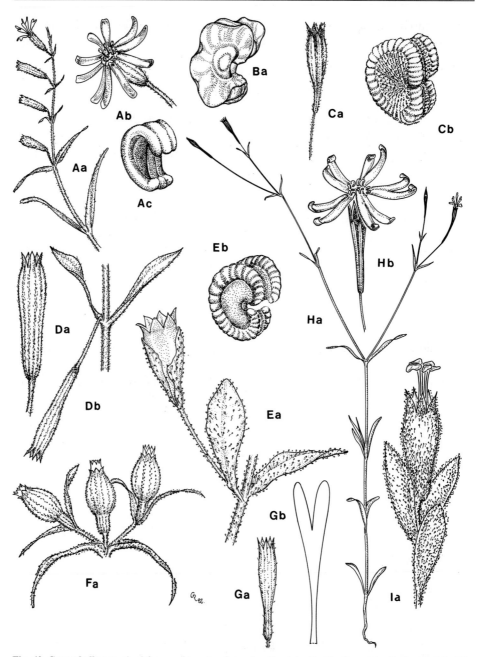

Fig. 40. **Caryophyllaceae**. A, *Silene arabica*: Aa, flowering shoot (×1); Ab, flower (×3); Ac, seed (×35). B, *S. colorata* subsp. *olivieriana*: Ba, seed (×35). C, *S. arenosa*: Ca, fruiting calyx (×3); Cb, seed (×35). D, *S. villosa*: Da, fruiting calyx - form 'B' (×3); Db, part of shoot and fruiting calyx - form 'A' (×3). E, *S. asirensis*: Ea, part of shoot showing fruiting calyx (×3.5); Eb, seed (×35). F, *S. crassipes*: Fa, fruiting calyces (×3). G, *S. hussonii*: Ga, fruiting calyx (×3); Gb, petal (×3). H, *S. linearis*: Ha, habit (×1); Hb, flower (×3). I, *S. succulenta*: Ia, tip of flowering shoot (×2.5).

divided into 2 narrow lobes. Capsule c. twice as long as the 5–7mm carpophore; seeds 0.8–1mm, hardly grooved, reticulate. **Map 271, Fig. 40.**

Rocky places and in sand and gravel; 550–2300m.

Saudi Arabia, Oman. Egypt, Palestine, Jordan and W Iran.

form **'B'**. Illustr.: Collenette (1985 p.110); Mandaville (1990 t.20).

Stems erect or spreading, 7–18cm. Calyx 17–23(–25)mm. Petal limb 10–12mm, divided into 2 broadly spathulate lobes. Capsule ± equalling the 9–11mm carpophore. **Map 272, Fig. 40.**

Sandy soil and sand-dunes; 0–1300m.

Saudi Arabia, Oman, UAE, Qatar, Bahrain, Kuwait. Jordan, ?Palestine, ?Egypt.

In the Arabian Peninsula forms 'A' & 'B' are perfectly distinct, and, while they do occur together in N Saudi Arabia, for the most part they do have disjunct distributions. The more conspicuously flowered form 'B' probably occurs in Lower Egypt from where *S. villosa* var. *villosa* was originally described. However, it is not certain whether it falls within the limits of that variety. Form 'A' may be referable to *S. villosa* var. *ismailitica* Schweinf. (in Aschers & Schweinf., Ill. Fl. Eg. Suppl. 748, 1889).

14. S. asirensis D.F. Chamb. in Edinb. J. Bot. 51 (1): 55 (1994). Type: Saudi Arabia, *Collenette* 7331 (E, K).

Erect annual; stems 30–50cm, glandular-villous. Leaves elliptic, loosely glandular-villous; basal leaves 40–50 × 10–17mm, with petioles up to c.12mm; cauline leaves progressively decreasing in size upwards, sessile. Flowers opening at night, in a (1–)5–10-flowered dichasial cyme; pedicels in flower c.5mm, in fruit stiffly erect, elongating to 25(–55)mm. Calyx 10-ribbed, 8–10mm, cylindrical, not much inflating in fruit, glandular-villous; calyx teeth c.2mm, ovate-lanceolate with a rounded tip. Corolla white c.12mm; limb 2–3mm, deeply divided into 2 oblong rounded lobes. Capsule elliptic, c.7mm long; seeds c.0.7mm, dark brown, deeply grooved, reticulate; carpophore c.2mm, glabrous. **Map 273, Fig. 40.**

Open *Acacia* bushland, rocky hillsides and in sand on basalt and granite; 1000–1300m.

Saudi Arabia. Endemic.

Probably allied to *S. villosa* but differing in its stiffly erect fruiting pedicels and smaller capsule.

15. S. crassipes Fenzl, Pug. pl. nov. Syr.: 8 (1842). Illustr.: Fl. Palaest. 1: t.115 (1966).

Annuals; stems erect, 15–30cm, scabrid, eglandular. Leaves scabrid; basal leaves with narrowly elliptic to elliptic lamina, 10–25 × 5–7mm, narrowed below into a 2–5mm petiole; cauline leaves linear to elliptic, 20–30 × 3–7mm, sessile. Flowers in 3–5-flowered simple or compound dichasial cymes, subsessile; bracts and bracteoles leaf-like. Calyx 10-nerved, 10–15mm, narrowly cylindrical in flower expanded above in fruit; calyx teeth triangular, c.1mm. Corolla pink to purple, 10–12 mm; limb cuneate,

3–5mm, entire or notched. Capsule ovoid, 7–11mm long; carpophore 2–4mm, glabrous. **Map 274, Fig. 40.**

Garden weed, probably introduced; 500m.

Saudi Arabia. E Mediterranean, from Turkey to Palestine.

The only Arabian record is of a garden weed.

16. S. succulenta Forsskal (1775 p.89). Illustr.: Fl. Palaest. 1: t.104 (1966).

Viscid, papillose-pubescent perennial; stems erect or ascending, 10–30cm, arising from a woody base. Leaves fleshy, broadly obovate to elliptic, 10–22 × 6-12mm, densely hispid, sessile. Flowers solitary or arranged in few-flowered leafy dichasial cymes; pedicels 0–5mm. Calyx 10-ribbed, 17–22mm, narrowly ellipsoid, not strongly inflated in fruit; calyx teeth, lanceolate, 4–5mm, acute. Corolla white or pink, c.25mm; limb up to c.15mm, divided to the middle into 2 oblong-spathulate lobes. Capsule oblong, c.8mm long; carpophore c.8mm, pubescent. **Fig. 40.**

Saudi Arabia. E Mediterranean Region.

No material has been seen of this species but it is very distinctive and so there is no reason to doubt the record (King & Kay op. cit.: 394).

17. S. hussonii Boiss., Diagn. pl. orient. sér. 1, 2 (8): 76 (1849). Illustr.: Fl. Palaest. 1: t.105 (1966).

Dwarf annual; stems 4–11cm, glandular-hairy, viscid. Leaves covered with spreading glandular hairs; basal leaves linear-elliptic, the lamina 8–10 × 2–3mm, petiole ± as long as the lamina; cauline leaves linear lanceolate, sessile. Flowers solitary or in lax 2–4-flowered dichasial cymes, sessile or with pedicels to c.5mm, sparsely glandular-villous. Calyx 11–12mm, minutely pubescent; calyx teeth c.1mm, linear, blunt. Corolla pale pink, 8–14mm; limb 4–5mm, deeply divided into two blunt lobes. Capsule not known; carpophore at least 4.5mm, glabrous. **Map 275, Fig. 40.**

Near a granite gully; 1800m.

Saudi Arabia. Egypt and Palestine.

In Arabia known only from a solitary specimen from NW Saudi Arabia.

18. S. arenosa C. Koch in Linnaea 15: 711 (1841). Syn.: *S. leyseroides* Boiss., Diagn. pl. orient. sér. 1, 1 (1): 41 (1843). Illustr.: Fl. Iranica 163: t. 255, 480 f.7–9 (1988).

Annual herbs; stems ascending to erect, 10–25cm, glandular-pubescent below, glabrescent above, viscid. Leaves linear-subulate, 15–25(–35) × 1–2.5mm, sparsely pubescent above, glabrescent beneath, the margin ciliate, sessile. Flowers in a lax compound dichasial cyme; inflorescence branches glabrous, 1–2.5cm long. Calyx 10-ribbed, 8–13(–15.5)mm, the ribs raised even in fruit, sparsely glandular-villous at first, sometimes glabrescent; calyx teeth lanceolate, 2–3.5mm, the margin ciliate. Corolla 8–10mm, white; limb 4.5–6mm, deeply divided into two linear lobes. Capsule 6–8mm long; seeds wrinkled and tuberculate; carpophore 4–6.5mm, pubescent. **Map 276, Fig. 40.**

Sandy sea shores, 0–10m.

Kuwait. Turkey, Iraq, Iran, Afghanistan and Pakistan.

19. S. linearis Decne in Ann. Sci. Nat. Bot. sér. 2, 3: 276 (1835). Illustr.: Collenette (1985 p.109); Mandaville (1990 t.19).

Annual glaucescent herb; stems erect, 10–40cm, minutely pubescent below, glabrescent above, viscid. Leaves linear to linear-elliptic, 15–25 × 1–2mm, pubescent, ± sessile. Flowers in a lax dichasial cyme; inflorescence branches glabrous, 1–3cm long. Calyx 10-ribbed, 10–13mm in flower, 15–20mm in fruit, the ribs in the upper part not raised in fruit, often lacking in the teeth, sparsely glandular-pubescent; calyx teeth 1–2mm, with a scarious and ciliolate margin. Corolla 8–12(–15)mm; limb 6–10mm, divided to the middle into 2 linear lobes. Capsule 6–8mm; seeds finely wrinkled; carpophore 5–8mm, scabrid. **Map 277, Fig. 40.**

Rocky slopes, ravines and wadis; 150–1550m.

Saudi Arabia, UAE. Palestine and Jordan.

Apparently intergrading with *S. austroiranica* (especially the plants from the United Arab Emirates).

20. S. austroiranica Rech. f., Aellen & Esfand. in Bot. Jahrb. Syst. 75: 349 (1951). Illustr.: Fl. Iranica 163: t.256, 480/10–12 (1988).

Annual herb; stems erect, 30–100cm, minutely glandular-hairy below, glabrous above, viscid. Leaves sparsely glandular-pubescent; basal and lower cauline leaves with linear to elliptic lamina, 12–15 × 1–4mm, tapering into a c.15mm petiole; upper cauline leaves linear, 10–15 × c.1mm, sessile. Flowers in a lax compound dichasial cyme; inflorescence branches glabrous, up to 5cm long. Calyx 10-ribbed, 12–15 mm, the ribs not raised in the upper part in fruit and often lacking in the teeth, glabrous; calyx teeth lanceolate, 2–3mm, the margin ciliate. Corolla 10–15mm; limb 4–8mm, divided to the middle into two broadly elliptic lobes. Capsule c.8mm long; seeds tuberculate; carpophore 6–10mm, pubescent. **Map 278.**

Gravel and rock crevices; 50–300m.

Saudi Arabia, Oman. Iraq, Iran, Pakistan.

Closely related to and possibly conspecific with *S. linearis*, with which the Arabian material appears to intergrade. Some forms of *S. austro-iranica* are difficult to distinguish from *S. linearis* on vegetative characters, though the relatively broad petal lobes serve as a more reliable character. It is not clear whether this species is distinct from *S. reinwardtii* Roth (1806), which is recorded from Jordan, Palestine, Syria, Lebanon and Turkey as well as from the Aegean.

Inadequately known taxa

S. flammulifolia Steud. ex A. Rich. var. **canescens** F.N. Williams in J. Linn. Soc., Bot. 32: 104 (1896).

This taxon is described from a specimen collected in Yemen (J. Schibam, nr. Manakhah). It is said to differ from the Ethiopian type variety in its incanous-pubescent indumentum, its scabrid-pubescent calyx nerves and its triangular-ovate, acute calyx teeth. From the description I suspect that the plant cited is a form of *S. burchellii* Otth. This record should therefore be treated with some caution.

S. aff. peduncularis Boiss., Diagn. pl. orient. sér. 1, 1 (1): 30 (1843). Illustr.: Collenette (1985 p.111) as *Silene* sp.

Bushy, twiggy herb up to 1m; stems glabrous, viscid, divaricately branched above. Upper cauline leaves glaucous, linear, 20–25 × c.2mm, glabrous, sessile. Flowers solitary or perhaps in dichasia, opening at night; peduncles up to 20mm. Calyx coriaceous, cylindrical in flower, enlarged above in fruit, 13–17(–20)mm; calyx teeth c.2mm, dimorphic, either rounded at the apex or acute, the margin scarious. Corolla 15mm across, creamy white. Anthers black. Capsule 5–6mm; carpophore 5–7 mm. **Map 279.**

Saudi Arabia.

The single specimen known (*Collenette* 4540, from NW Saudi Arabia), an illustration of which is cited above, is incomplete, lacking basal parts and good flowers. The short calyx indicates an affinity with *S. peduncularis* from W & NW Iran, a species that is allied to *S. macrosolen*. However, the present plant is more profusely branched and is taller than is usual in *S. peduncularis*.

23. VACCARIA Wolf

Glabrous annual herbs. Leaves opposite, ovate to lanceolate, sessile. Stipules absent. Flowers in lax paniculate corymbs; bracts absent. Calyx ovoid-pyramidate, inflated below, with 5 conspicuously winged veins, 5-toothed, the teeth lanceolate with scarious margins. Petals 5. Stamens 10. Styles 2. Fruit a many-seeded capsule dehiscing by 4 teeth. Seeds subglobose.

1. V. hispanica (Miller) Rauschert in Wiss. Z. Martin-Luther-Univ. Halle-Wittenberg, Math.-Naturwiss. Reihe 14: 496 (1965). Syn.: *Saponaria vaccaria* L., Sp. pl.: 409 (1753); *V. pyramidata* Medik., Philos. Bot. 1: 96 (1789).

Stems 30–50cm. Leaves 50–80 × 10–30mm. Calyx 12–16mm. Petals 20–25mm, pink, emarginate to entire. Fruit ovoid, papery. **Map 280, Fig. 41.**

Weed of cultivation; 10–2300m.

Saudi Arabia, Yemen (N), Qatar, ?Kuwait. A cosmopolitan weed.

24. GYPSOPHILA L.

Annual, biennial or perennial herbs, sometimes woody-based. Leaves opposite, linear to spathulate or elliptic. Stipules absent. Inflorescence a compound cyme, often paniculate. Bracts small. Calyx campanulate or narrowly obconical or tubular, 5-

toothed, hyaline between the nerves. Petals 5, white or pink, not usually differentiated into a limb and claw; coronal scales absent. Stamens 10. Styles 2. Fruit a many-seeded capsule, dehiscing by 4 valves. Seeds auriculate with a prominent radicle, tuberculate or ridged, rarely echinate.

Barkoudah, Y.I. (1962). A Revision of *Gypsophila, Bolanthus, Ankyropetalum* and *Phryna. Wentia* 9: 1–203.

1. Flowering calyx tubular, 2 × as long as wide; calyx teeth $\frac{1}{4}$ the length of the calyx   **2**
+ Flowering calyx campanulate or obconic, less than $1\frac{1}{2}$ × as long as wide; calyx teeth c.$\frac{1}{2}$ the length of the calyx or more   **3**

2. Petals with a line of thick hairs inside at the base of the limb; stem indumentum of short (c.0.3mm long) hairs   **1. G. bellidifolia**
+ Petals glabrous; stem indumentum of long (c.2mm) spreading hairs   **2. G. pilosa**

3. Stem internodes with bands of sessile glands, otherwise glabrous; calyx up to 2.5mm long   **3. G. viscosa**
+ Stem internodes without bands of sessile glands, glabrous or hairy; calyx 2–4mm long   **4**

4. Plants glabrous   **5**
+ Plants hairy, at least in the inflorescence   **6**

5. Cauline leaves spathulate, 4mm or more wide; most pedicels less than 5mm long   **4. G. montana**
+ Cauline leaves linear, 3mm or less wide; pedicels usually more than 10mm long   **6. G. capillaris**

6. Leaves spathulate; calyx narrowly obconical, up to 3.5mm long; seeds ridged or tuberculate   **4. G. montana**
+ Leaves oblong-elliptic to linear-oblong; calyx campanulate, 4mm or more long; seeds echinate   **5. G. umbricola**

1. G. bellidifolia Boiss., Diagn. pl. orient. sér. 1, 1 (1): 11 (1843). Syn.: *Saponaria barbata* Barkoudah, op. cit.: 180. Illustr.: Western (1989 p.38). Type: Oman, *Aucher-Eloy* 4263 (K)

Slightly viscid annual herb, sometimes apparently perennating, with a short indumentum of glandular hairs and rather coarse white eglandular hairs; stems erect, 5–30cm. Leaves often grouped near the base of the stem, spathulate to obovate or elliptic, 6–60 × 4–20mm, slightly fleshy. Inflorescence lax, with a narrow branching angle; pedicels 4–25mm. Calyx tubular, 3–4.5mm; teeth c.0.7mm. Petals white or pale pink to pale purple with white markings, ± cuneate, c.$1\frac{1}{2}$ × as long as the calyx, somewhat differentiated into a limb and claw, with a line of thick hairs on the inside at the base of the limb. Capsule elliptic, barely exceeding the calyx. Seeds c.0.7 × 0.5mm, obtusely ridged, furrowed along the back. **Map 281.**

Stony and rocky hillsides and wadis; 30–850m.

Oman, UAE. Endemic.

2. **G. pilosa** Huds. in Philos. Trans. 56: 252 (1767). Syn.: *G. porrigens* (L.) Boiss., Fl. orient. 1: 557 (1867).

Annual herb; stems erect (15–)30–45(–80)cm, glabrous below, with long spreading hairs above. Leaves narrowly ovate to linear-oblong, 30–90 × 7–21mm, acuminate, glabrous or villous. Inflorescence lax; pedicels capillary, 12–55mm, glabrous. Calyx ± tubular, 4.5–6mm, densely villous; teeth 1.5–2mm. Petals white to pink or pale mauve, linear-oblong, $1\frac{1}{2}$–2 × as long as the calyx, slightly contracted below the apex, glabrous. Capsule broadly ovoid, equal to or slightly exceeding the calyx. Seeds c.1.8 × 1.6mm, tuberculate. **Map 282, Fig. 41.**

A weed of cultivated fields; 0–10m.

Kuwait. S, C and SW Asia; adventive in Europe.

3. **G. viscosa** J.A. Murray in Commentat. Soc. Regiae Sci. Gott., Cl. Phys.: 9 (1783).

Glabrous, somewhat viscid annual herb; stems erect, 10–35cm, with bands of sessile glands on the internodes. Leaves narrowly ovate to ovate-oblong, 13–60 × 5–15mm, acuminate. Inflorescence lax, with a narrow branching angle; pedicels capillary, up to 15mm. Calyx broadly campanulate, 2–2.5mm; teeth ovate, 1–1.5mm. Petals white to pale pink, oblong, 2–3 × as long as the calyx. Capsule ± spherical, c.$1\frac{1}{2}$ × as long as the calyx. Seeds c.1.3 × 1.2mm, tuberculate. **Map 283, Fig. 41.**

Sand-dunes and rocky hills; 900–1250m.

Saudi Arabia, Kuwait. Turkey, Syria, Palestine, Jordan and Sinai.

King & Kay (op. cit.: 393) recorded *G. heteropoda* Freyn & Sint. from the Arabian Peninsula. The material on which this record was based is referable to *G. viscosa*.

4. **G. montana** Balf.f. (1882 p.501). Syn.: *Saponaria montana* (Balf.f.) Barkoudah, op. cit.: 183. Type: Socotra, *Balfour, Cockburn & Scott* 442 (E, K).

Woody-based perennial herb, glabrous or pubescent to extremely viscid-hairy; stems ± erect, 20–90cm, often whitish when young. Leaves spathulate, 9–30 × 4–10mm. Inflorescence paniculate, profusely branched, glabrous or viscid-hairy; pedicels usually less than 5mm. Calyx narrowly obconical, 2–3.5mm; teeth 1.2–1.6mm, narrowly acute. Petals white sometimes tinged pink, linear-oblong, 2.5–5.5mm. Capsule oblong-elliptic, ± equal to the calyx. Seeds c.0.8 × 0.9mm, ridged to obtusely tuberculate, with a shallow furrow along the back. **Map 284, Fig. 41.**

Open rocky slopes and gravelly wadi-beds; 30–1200m.

Yemen (N & S), Socotra, Oman. Somalia.

Balfour (1888) described three varieties (var. *montana*, var. *viscida* and var. *diffusa*) based on indumentum and inflorescence-form. However, because of numerous intermediates, we have been unable to recognise them in this account. Totally glandular-pubescent plants have been placed in subsp. *somalensis* (Franch.) M.G. Gilbert.

5. G. umbricola (J.R.I. Wood) R.A. Clement in Edinb. J. Bot. 51 (1): 57 (1994). Syn.: *Saponaria umbricola* J.R.I. Wood in Kew Bull. 39: 130 (1984). Type: Yemen (N), *J. Wood* 3458 (BM, E, K).

Woody-based perennial herb; stems decumbent, 7–35cm, glabrous or puberulent below, glandular-hairy above. Leaves oblong-elliptic, occasionally linear-oblong, 14–40 × 1.7–12mm, glabrous or sparsely glandular-hairy. Inflorescence lax, relatively few-flowered; pedicels 6–13mm. Calyx broadly campanulate, 3.2–4mm; teeth c.1.5mm. Petals white, cuneate, c.$1\frac{1}{4}$ × as long as the calyx. Capsule broadly ovoid, equal to or slightly exceeding the calyx. Seeds c.0.8 × 0.7mm, echinate. **Map 285, Fig. 41.**

On moist, partially shaded cliffs; 1600–2800m.

Saudi Arabia, Yemen (N). Endemic.

6. G. capillaris (Forsskal) C. Chr. in Dansk. Bot. Ark. 4 (3): 19 (1922).

Glabrous annual or biennial herb, often glaucous, or perennial with a woody rootstock. Stems erect, 7–75cm. Leaves linear to lanceolate below, linear above, 17–60 × 1–8mm, the margins papillose or not. Inflorescence a lax panicle; pedicels capillary, up to 33mm. Calyx campanulate-obconical, 2–3mm; teeth 1–1.6mm. Petals white with lilac to purple veins, oblong to narrowly obovate, c.2 × as long as the calyx. Capsule globose, ± equal to or slightly exceeding the calyx. Seeds 1–1.2 × 0.8–1mm; coat smooth or tuberculate. **Map 286, Fig. 41.**

Rocky and sandy places; 0–2400m.

Saudi Arabia, Kuwait. S Turkey, NE Egypt, Palestine, Jordan, Syria, Iraq and W Iran.

1. Seeds few per capsule; seed coat smooth; calyx persistent in fruit
 subsp. **capillaris**
+ Seeds 2–3(–8) per capsule; seed coat tuberculate; calyx deciduous
 subsp. **confusa**

subsp. **capillaris**. Syn.: *G. antari* Post & Beauv. in Dinsm., Pl. Post. & Dinsm. fasc. 1: 4 (1932); *Rokejeka capillaris* Forsskal (1775 p.90 & CXXIV).

Locally common and fairly widespread in the NE of the region.

subsp. **confusa** Zmarzty in Kew Bull. 48 (4): 694 (1994). Syn.: *G. arabica* Barkoudah in Wentia 9: 139 (1962); *G. obconica* Barkoudah in Wentia 9: 140 (1962).

Ripe fruit is required to distinguish between the two subspecies; no confirmed material of subsp. *confusa* has been seen from the Arabian Peninsula although it could be expected to occur. Barkoudah cites the following from Saudi Arabia: *Vesey-Fitzgerald* 13544 (HUJ, n.v.), *Burton* s.n., an immature specimen (K) & *Thesiger* s.n.

Fig. 41. Caryophyllaceae. A, *Velezia rigida*: Aa, flowering shoot (×1); Ab, flower (×4). B, *Vaccaria hispanica*: Ba, flowering shoot (×1); Bb, fruiting calyx (×2). C, *Gypsophila capillaris*: Ca, flowering shoot (×1); Cb, flower (×5). D, *G. bellidifolia*: Da, habit (×1); Db, petal (×10). E, *G. umbricola*: Ea, habit (×1); Eb, fruit (×5). F, *G. pilosa*: Fa, fruiting shoot (×1). G, *G. viscosa*: Ga, part of stem showing glandular hairs (×1); Gb, flower (×5). H, *G. montana*: Ha, base of shoot (×1); Hb, flower (×6).

(BM, n.v.). The identity of these specimens is uncertain. For a discussion of the subspecies of *G. capillaris* see Zmarzty in *op. cit.*: 683–697 (1994).

25. VELEZIA L.

Annual herbs. Stems dichotomously branched. Leaves opposite, linear. Stipules absent. Inflorescence a monochasial cyme or flowers solitary. Calyx narrowly cylindrical, 5-toothed. Petals 5, pink or reddish, differentiated into a limb and claw; coronal scales absent. Stamens 5 or 10. Styles 2. Fruit a few-seeded capsule, dehiscing by 4 teeth.

1. V. rigida L., Sp. pl.: 332 (1753). Illustr.: Collenette (1985 p.112).

Erect or ascending annual. Stems 4.5–22(–30)cm, rather stiff, glandular-hairy at least near the nodes. Leaves linear, (5–)10–20 × 0.7–1.2mm, with 3 prominent veins on the lower surface, tapering to a point, with a narrow scarious margin at the base which often joins around the nodes, with short antrorse hairs on the margin near the base. Pedicels stout, 1.4–3mm. Calyx 11–12mm, prominently veined, usually with some glandular hairs; teeth 1–1.5mm. Petals just exceeding the calyx, bifid. **Map 287, Fig. 41.**

Stony or gravelly areas in open *Acacia-Olea* woodland; 1850–2150m.

Saudi Arabia. Mediterranean region, SW and C Asia.

26. PETRORHAGIA (Ser.) Link

Annual herbs. Leaves opposite, narrow. Stipules absent. Inflorescence an open cyme. Calyx obconical, the tube 5-ribbed, hyaline between the ribs, 15-nerved (in Arabia), 5-toothed. Petals 5, without a distinct limb and claw, without coronal scales. Stamens 10. Styles 2. Fruit a many-seeded capsule, dehiscing by 4 teeth. Seeds peltate with a facial hilum.

Ball, P. W. & Heywood, V. H. (1964). A revision of the genus *Petrorhagia*. *Bull. Brit. Mus. (Nat. Hist.)* 3 (4): 119–172.

1. P. cretica (L.) Ball & Heywood, op. cit.: 142.

Glandular-hairy annual. Stems erect, 8–17cm, branched above. Leaves linear to linear-oblong, 10–26 × 1–2.5mm, ± glabrous. Flowers in a lax, open, 2–10-flowered cyme, ebracteate. Pedicels 9–20mm. Calyx obconical, 7–8(–10)mm; teeth acute, c.1.2mm long. Petals white, narrowly oblong or oblanceolate, entire ± equalling the calyx. Capsule equalling the calyx; seeds flattened, minutely papillose. **Map 288, Fig. 42.**

Sandy wadi-banks; c.1220m.

Saudi Arabia. Greece, Albania, Turkey, Palestine, the Syrian Desert and N Iran.
Known only from a single locality in N central Saudi Arabia.

27. DIANTHUS L.

Annual or perennial herbs, often with a woody base. Leaves opposite, linear to subulate. Stipules absent. Flowers solitary or in clusters of 2 or more. Calyx tubular, lacking differentiated intervals and commissural veins, 5-toothed, with 2–7 or more pairs of bracteoles (the epicalyx) at the base which sometimes intergrade with the leaf-like bracts. Petals 5, white to deep pink, clawed, subentire to dentate or fimbriate, sometimes also hairy on the upper surface (barbulate). Styles 2. Fruit a many-seeded capsule dehiscing by 4 teeth.

This genus includes a number of species that are grown as ornamental plants, including *D. caryophyllus* L. (the Carnation).

1.	Calyx at least 25mm long	2
+	Calyx up to 25mm long	4
2.	Petal limb subentire or obscurely toothed	**3. D. judaicus**
+	Petal limb dentate or fimbriate	3
3.	Cauline leaves imbricate; bracteoles at least half as long as the calyx	**6. D. longiglumis**
+	Cauline leaves not imbricate; bracteoles less than half as long as the calyx	**7. D. crinitus**
4.	Bracteoles 10–14; corolla fimbriate	**8. D. sinaicus**
+	Bracteoles 4–10; corolla dentate, sometimes deeply so	5
5.	Calyx minutely tuberculate-verruculose; annual; bracteoles $\frac{3}{4}$ as long to as long as the calyx	**1. D. cyri**
+	Calyx and bracteoles not tuberculate; perennials; bracteoles $\frac{1}{2}$ as long as the calyx	6
6.	Petals barbulate	**2. D. strictus**
+	Petals glabrous	7
7.	Lax plant, up to 30cm; flowers on 2–10cm peduncles long	**4. D. deserti**
+	Caespitose plant, up to 8cm; flowers sessile	**5. D. uniflorus**

Sect. Verruculosi Boiss., Fl. orient. 1: 479 (1867).

Annuals or perennials. Calyx minutely tuberculate-verruculose or smooth. Bracteoles 4–6. Petal limb dentate, glabrous or barbulate.

1. D. cyri Fischer & C. Meyer, Ind. sem. hort. petrop. 4: 34 (1837). Illustr.: Fl. Iranica 163, 2: t.376 (1988); Fl. Palaest. 1: t.147 (1966).

Annual herb; stems 20–40cm, glabrous. Cauline leaves linear, 20–50 × 1.5–4mm, sparsely pubescent, toothed below, grading into the bracts. Flowers solitary or in clusters of 2–3, with peduncles 15–30mm; bracteoles 4, the lower pair 8–15mm long, with a long stiff green arista. Calyx 11–15 × 3.5–5mm, minutely tuberculate-verruculose, coriaceous in fruit; teeth lanceolate, 5–7mm long, acute. Corolla rose-pink,

15–20mm long, limb 4–6mm long, narrowly ovate, dentate, sometimes hairy on the upper surface. **Map 289, Fig. 42.**

A weedy species of open ground in plantations; 0–300m.

Oman, UAE. SW Asia, from E Turkey to Afghanistan.

2. D. strictus Banks & Soland. subsp. **sublaevis** D.F. Chamb. in Edinb. J. Bot. 51 (1): 56 (1994). Illustr.: Collenette (1985 p.102) as *D. strictus*. Type: Saudi Arabia, *Collenette* 4899 (E).
Perennial herb; stems 12–75cm, arising from a woody base, velutinous to glabrescent. Leaves linear, the margins scabrid; basal leaves 15–30 × 1–2.5mm. Flowers solitary or in clusters of 2–3; peduncles 5–10cm. Bracteoles usually 4, ovate-lanceolate to lanceolate, 5–8mm, aristate, often tuberculate; margin broad, scarious and scabrid. Calyx 15–18mm, minutely tuberculate-verruculose; teeth lanceolate, up to 4mm long. Corolla pink with darker lines, c.20mm long; limb c.7mm long, deeply dentate, hairy on the upper surface. **Map 290, Fig. 42.**

Rocky ground; (550–)1000–2400m.

Saudi Arabia, Yemen (N). Endemic.

Material from NW & WC Saudi Arabia has velutinous stems and thus approaches the Palestinian var. *velutinus* (Boiss.) Eig. Some of the material has flowers in clusters of 2–3, so approaching var. *axilliflorus* Boiss. It therefore appears that much of the variation seen in the rest of *D. strictus* is represented in plants from Saudi Arabia. but they consistently differ in having calyces that are smooth or at most only slightly rough. It is probable that all records of *D. strictus* from the Arabian Peninsula belong to this subspecies.

Sect. Leiopetali Boiss., Fl. orient. 1: 479 (1867).
Perennials. Calyx smooth, not tuberculate. Bracteoles 4–10. Petal limb dentate, glabrous

3. D. judaicus Boiss., Diagn. pl. orient. sér. 1, 2 (8): 66 (1849). Illustr.: Fl. Palaest. 1: t.148 (1966).
Glaucous perennial herb; stems 20–35cm, arising from a woody base. Leaves tufted at the base of the plant and cauline; basal leaves setaceous, 30–40 × 1–3mm, glabrous. Flowers usually solitary; peduncles 3–9cm. Bracteoles 4, oblong-ovate, 12–18mm, acute to cuspidate, the cusp shorter than the lamina, sharply demarcated from the bracts. Calyx 30–35 × 4–7mm, smooth; teeth lanceolate, acute, glabrous. Corolla white to cream, flushed red on the back, 40–45mm long; limb narrowly oblong-obovate, 12–15mm long, subentire to obscurely toothed, glabrous. **Map 291, Fig. 42.**

Rocky deserts; 400–700m.

Saudi Arabia. Palestine, Egypt (Sinai) and SW Iraq.

Fig. 42. **Caryophyllaceae**. A, *Dianthus uniflorus*: Aa, habit (× 1); Ab, fruiting calyx (× 1.5); Ac, petal (× 2.5). B, *D. strictus*: Ba, petal (× 2.5); Bb, fruiting calyx (× 1.5). C, *D. sinaicus*: Ca, petal (× 2.5); Cb, fruiting calyx (× 1.5); D, *D. longiglumis*: Da, habit (× 0.6); Db, petal (× 2.5); Dc, fruiting calyx (× 1.5). E, *D. deserti*: Ea, petal (× 2.5); Eb, fruiting calyx (× 1.5). F, *D. judaicus*: Fa, petal (× 1.5); Fb, fruiting calyx (× 1.5); G, *D. cyri*: Ga, petal (× 2.5), Gb, fruiting calyx (× 1.5). H, *D. crinitus*: Ha, fruiting calyx (× 1.5); Hb, petal (× 2.5). I, *Petrorhagia cretica*: Ia, fruiting shoot (× 1); Ib, fruiting calyx (× 5).

4. D. deserti Kotschy in Sitzb. Akad. Wiss. Wien Sitzungber. Math.- Naturwiss. Kl. Abh. 1, 52: 262, t.7 (1866). Illustr.: ?Collenette (1985 p.102) as *D. uniflorus*. Type: Saudi Arabia, 1836–1838, *Anon.* (n.v.).

Perennial, usually dwarf herb; stems 7–30cm, arising from a woody base, glabrous. Leaves linear, glabrous, with a minutely scabrid margin; basal leaves 20–40 × 1–2mm; cauline leaves smaller. Flowers solitary; peduncles 2–10cm. Bracteoles 4, broadly ovate-lanceolate, 5–8mm, acuminate to aristate, glabrous, sharply demarcated from the bracts. Calyx 14–18mm, smooth, glabrous; teeth lanceolate, acute. Corolla deep pink, 17–22mm long; limb broadly elliptic, 4–6mm long, the margin subentire to dentate, glabrous. **Map 292, Fig. 42.**

Rocky and grassy places, especially on basalt, disturbed ground and *Juniperus* woodland; 2150–2800m.

Saudi Arabia, Yemen (S). Endemic.

A single specimen from Yemen (S) (*Lavranos* 8618 bis) differs in having 6 bracteoles with more markedly aristate tips that intergrade with the bracts. As the specimen has no basal parts and is geographically isolated from the main area of the species, I am uncertain of its status though it could represent a new subspecies of *D. deserti*.

5. D. uniflorus Forsskal (1775 p.CXI). Syn.: *D. pumilus* Vahl (1790 p.32). Type: Yemen (N), *Forsskal* (C).

Dwarf caespitose shrub; stems up to 8cm, arising from a woody base. Leaves stiff, linear, 5–10 × 1–2mm, margin scabrid. Flowers solitary, sessile. Bracteoles 8–10, broadly ovate-lanceolate, up to 4 × 2.5mm, the apex acute, grading into the bracts. Calyx c.12mm; teeth broadly ovate-lanceolate, with a fringed margin. Corolla 12–15mm long, pink; limb broad, c.6mm long, dentate, glabrous. **Map 293, Fig. 42.**

Montane grassland and open rocky slopes; 1700–3300m.

Yemen (N & S). Endemic.

6. D. longiglumis Del. in Ann. Sci. Nat. sér. 2, 20: 89 (1843).

Perennial herb; stems up to 30cm, arising from a woody base. Cauline leaves imbricate, the sheaths overlapping, stiff, 60–85 × c.4mm, margin more or less entire. Flowers either solitary with peduncles up to 15cm long or in clusters of 2–3 with peduncles 2–3cm long. Bracteoles ovate-lanceolate, 14–20mm, with a long aristate tip, glabrous but with a scabrid margin, usually clearly demarcated from the bracts. Calyx 36–40mm; teeth lanceolate, 15–20mm, glabrous or scabrid to minutely pubescent at the tip. Corolla pink, 45–50mm long; limb c.15mm long, broadly cuneate, coarsely dentate, glabrous. **Map 294, Fig. 42.**

Hanging from basaltic cliffs; 2100–2700m.

Yemen (N & S). Ethiopia.

A very distinctive species on account of its stiff, imbricated stem leaves.

Sect. Plumaria (Opiz) Aschers. & Graebn., Syn. mitteleur. Fl. 5 (2): 409 (1929).

Calyx smooth, not tuberculate. Bracteoles 4–14. Petal limb fimbriate, glabrous.

7. D. crinitus Sm. in Trans. Linn. Soc. London, Bot. 2: 300 (1794). Illustr.: Fl. Iranica 163, 2: t.411 (1988); Western (1989 p.37).

Perennial herb; stems 20–40cm, arising from a woody base, minutely scabrid, at least at first. Leaves linear, margins minutely scabrid; basal leaves 40–70 × 1–1.5mm. Flowers solitary; peduncles up to 20cm. Bracteoles 4–10, ovate-lanceolate to lanceolate, 7–12mm, acute to apiculate (sometimes aristate outside the Arabian Peninsula). Calyx 25–45mm; teeth lanceolate, up to 10mm, glabrous. Corolla white to pinkish, 35–55mm long; limb broadly obovate, c.10mm long, deeply fimbriate, glabrous. **Map 295, Fig. 42.**

Steep rocky slopes, grassland and *Juniperus* woodland; (450–)1000–3000m.

Oman, UAE. SE Europe and SW Asia from Turkey to Afghanistan and extending into C Asia.

A very variable species, originally described from the Caucasus. Five subspecies are recognized in Flora Iranica (op. cit.: 173). This species clearly requires a monographic revision over its whole range before the status of the Arabian material can be established.

8. D. sinaicus Boiss., Diagn. pl. orient. sér. 1, 1: 23 (1843). Illustr.: Collenette (1985 p.102); Fl. Palaest. 1: t.149 (1966).

Compact bushy glaucous shrub; stems 25–40cm, glabrous. Leaves mostly basal, subulate, 45–65 × 0.5mm, glabrous. Flowers mostly solitary, on 10–15cm peduncles, arranged on laxly branched leafless stems. Bracteoles 10–14, broadly ovate-lanceolate, 4–6mm, grading into the bracts. Calyx 20–25mm, pubescent; teeth lanceolate, 7–8mm. Corolla white to pale pink, 35–40mm long, limb narrowly oblong, 8–10mm long, deeply fimbriate, glabrous. **Map 296, Fig. 42.**

Rocky slopes on granite; 1650–2000m.

Saudi Arabia. Egypt, Palestine, Jordan.

Allied to the widespread SW Asian *D. orientalis* Adams which differs in usually having 4–8 bracteoles.

28. ACANTHOPHYLLUM C.A. Meyer

Small, spiny shrubs. Leaves opposite, needle-like, spiny. Stipules absent. Flowers clustered in shortly pedunculate globose heads; bracts (in ours) broadly ovate, apiculate, with broad, hyaline margins. Calyx cylindrical, 5-nerved, 5-toothed. Petals 5. Stamens 10. Ovary 4-ovulate; styles 2. Fruit 1-seeded by abortion, rupturing irregularly at the base.

1. A. sp. aff. bracteatum Boiss., Diagn. pl. orient. sér. 1, 1: 43 (1843).

Spiny, grey-green shrub to 50cm, shortly hirsute throughout. Leaves 1.5–6cm × 1–1.5mm, spine-tipped. Flower heads 8–15mm across; peduncles 5–15mm. Bracts c.8 × 6–7mm, the margins wavy. Calyx c.8mm. Petals narrowly spathulate, c.5mm long.

Steep gulley; 900m.

Saudi Arabia. Iraq, Iran and Pakistan.

In Arabia known only from a single gathering in the extreme north of Saudi Arabia.

Family 38. CHENOPODIACEAE

L. BOULOS

Herbs or shrubs, rarely small trees or climbers; stems often jointed and/or succulent. Leaves alternate, less often opposite, simple, often succulent, sometimes reduced and scale-like; stipules absent. Flowers small, often green, 1–many, axillary, usually regular, bisexual, sometimes unisexual on dioecious or monoecious plants; floral bracts usually 2. Calyx of (1–)5 free segments, these sometimes connate at the base or absent. Corolla absent. Stamens (1–)5, opposite the perianth-segments, often inserted on a hypogynous disc which sometimes shows distinct interstaminal lobes (staminodes). Ovary superior, of 2–3(–5) fused carpels, 1-locular; ovule 1, basal. Fruit usually a nutlet, often subtended by the persistent perianth or bracteoles; fruiting perianth often winged. Seeds with annular, curved or spiral embryo; perisperm starchy or absent; endosperm usually absent.

1. Stems jointed, leafless or with inconspicuous scale-like opposite leaves   2
+ Stems not jointed, leafy; leaves usually conspicuous, sometimes scale-like or succulent, usually alternate, rarely opposite   6

2. Halophytic succulents; fruiting perianth wingless   3
+ Desert shrubs or small trees; fruiting perianth winged   5

3. Flowers in short cone-like spikes; leaves succulent, scale-like, on small bud-like branches   **8. Halocnemum**
+ Flowers in elongated spikes; leaves absent or reduced to short extensions of the joints   4

4. Woody, much-branched perennials with many decumbent stems   **9. Arthrocnemum**
+ Herbaceous annuals with a single or few, erect stems   **10. Salicornia**

5. Seeds horizontal; perianth without fleece bundles within   **17. Haloxylon**
+ Seeds vertical; perianth with 2 dense fleece bundles within   **23. Anabasis**

CHENOPODIACEAE

6.	Branches short, spiny; leaves filiform or linear	**22. Noaea**
+	Branches not spiny, rarely spinescent (see *Salsola cyclophylla* & *S. spinescens*); leaves spiny or unarmed	7
7.	Leaves broad and flat	8
+	Leaves narrow, subulate, linear or filiform, globular or semiglobular, sessile or subsessile	11
8.	Fruit included in the bracteoles	9
+	Fruit not included in the bracteoles	10
9.	Plants dioecious; bracteoles spiny in fruit	**3. Spinacia**
+	Plants monoecious; bracteoles not spiny in fruit	**4. Atriplex**
10.	Fruits connate at the base in clusters of 2–4; fruiting perianth indurated	**2. Beta**
+	Fruits separate, single-seeded, the depressed-globular utricles with a membranous pericarp; fruiting perianth herbaceous	**1. Chenopodium**
11.	Leaves spine- (or bristle-)tipped	12
+	Leaves not spine-tipped	16
12.	Perianth segments winged in fruit	13
+	Perianth segments not winged in fruit	14
13.	Perennial shrub; only the leaves spine-tipped; spines (bristles) deciduous	**15. Agathophora**
+	Herbaceous annual; leaves and bracteoles spine-tipped; spines sharp, persistent	**18. Salsola** (*S. kali*)
14.	Leaves fleshy, ± terete	**13. Traganum**
+	Leaves rigid, not terete	15
15.	Leaves 0.5–1cm long, strongly clasping, with tufts of white woolly hairs in their axils	**14. Cornulaca**
+	Leaves 1–5cm long, conspicuously parallel-veined, slightly or not clasping, without tufts of hairs in their axils	**5. Agriophyllum**
16.	Perianth segments with conspicuous, membranous wings in fruit, the wings sometimes (in *Halothamnus*) becoming indurated	17
+	Perianth segments without conspicuous, membranous wings in fruit	20
17.	Leaves and branches all opposite; leaves cylindrical, thickening towards the tip; branches usually white-glossy; fruiting perianth wings unequal; staminodes 5, semi-orbicular, glandular-ciliate, fused with the bases of the filaments into a ring	**16. Seidlitzia**
+	Leaves alternate or if opposite then not with the above combination of characters	18

18.	Leaf bases distinctly constricted and soon becoming indurated; stamens alternating with staminode-like scales; fruiting perianth tube well-deveoloped with a conspicuous swollen base	**21. Lagenantha**
+	Leaf bases clasping the stem, never indurated; staminode-like appendages usually absent; fruiting perianth tube short or inconspicuous, without a swollen base	19
19.	Fruiting perianth membranous, or indurated at the base only; stigma narrow, not dilated or lobed	**18. Salsola**
+	Fruiting perianth indurated throughout, its base truncate, pentagonal; stigma flattened, ± lobed	**19. Halothamnus**
20.	Leaves perfoliate, subglobular, giving the stems a jointed appearance	**7. Halopeplis**
+	Leaves not perfoliate, the stems not appearing jointed	21
21.	Anthers with a yellow, petaloid appendage at the tip	**24. Halocharis**
+	Anthers without petaloid appendages	22
22.	Perianth lobes hairy or fleecy; leaves densely villous or hirsute or rarely adpressed hairy	**6. Bassia**
+	Perianth lobes glabrous or mealy; leaves glabrous or rarely the juvenile leaves hairy but then soon glabrescent, sometimes (in *Sevada*) with tufts of minute hairs in the leaf axils and the leaves rarely papillose	23
23.	Fruit orbicular with a fleshy, ± entire wing; herb	**12. Bienertia**
+	Fruit neither orbicular nor winged; shrubs or shrubby herbs	24
24.	Leaf bases distinctly constricted and soon becoming indurated; leaves with a tuft of minute hairs in the axils; stamens alternating with small, fleshy staminode-like appendages; perianth tube usually well-developed	**20. Sevada**
+	Leaf bases never indurated; leaves without a tuft of hairs in the axils; staminode-like appendages absent; perianth tube inconspicuous	**11. Suaeda**

1. CHENOPODIUM L.

Annual or short-lived perennial herbs or small shrubs (the latter not in Arabia); stems usually striated or angled. Leaves alternate, petiolate, lobed, dentate or entire, mealy or glandular and aromatic, sometimes malodorous. Inflorescence of small cymose clusters arranged in spike-like panicles. Flowers bisexual or female, sessile, bractless. Perianth segments 5 (rarely 2–4), green, free or variously united, persistent, not much altered or accrescent in fruit. Stamens 5 (rarely 1–4), free or connate at the base. Stigmas 2 (rarely 3–5), simple (rarely 2-lobed). Fruit depressed-globular, with a thin membranous pericarp. Seeds lenticular, horizontal, rarely vertical; embryo peripheral.

CHENOPODIACEAE

Boulos, L. (1991). A synopsis of *Chenopodium* L. Studies in the Chenopodiaceae of Arabia 4. *Kew Bull*. 46: 301–305.

1.	Plant strongly aromatic, with yellow- or amber-coloured glands or glandular hairs, not farinose	2
+	Plant not aromatic (smelling of decaying fish in *C. vulvaria*), glabrous or white-farinose, not glandular	6
2.	Flowers in dense axillary clusters; perianth segments keeled at least towards the tip, becoming thick and coriaceous in fruit	**1. C. carinatum**
+	Flowers and perianth segments not as above	3
3.	Inflorescence of dichasial cymes	**2. C. ambrosioides**
+	Inflorescence paniculate, of small sessile clusters, not dichasial cymes	4
4.	Glands sessile	**3. C. schraderianum**
+	Glands stalked	5
5.	Leaves up to 12.5 × 6cm; inflorescence conspicuously leafy almost to the top; stems leafy when the seeds mature; perianth segments connate at the base; seeds c.1mm diam.	**4. C. procerum**
+	Leaves not exceeding 5 × 2cm; inflorescence leafy mainly in the lower part, the upper part with much-reduced bracts; stems almost leafless when the seeds are mature; perianth segments free almost to the base; seeds c.0.7mm diam.	**5. C. botrys**
6.	Leaves entire, not toothed or lobed (though a lobe-like basal angle may be present); plant smelling strongly of decaying fish	**6. C. vulvaria**
+	Leaves toothed or lobed; plant not smelling of decaying fish	7
7.	Inflorescence leafy almost to the top	8
+	Inflorescence leafless in the upper part	10
8.	Leaves small, 0.3–1cm long, elliptic to lanceolate, sinuate or serrate, conspicuously glaucous or white-farinose beneath	**7. C. glaucum**
+	Leaves large, 1–5cm long and wide or more, broadly triangular or rhombic, coarsely or irregularly toothed, not glaucous or white-farinose beneath	9
9.	Seeds strongly keeled along the margin, 1.2–1.5mm diam.	**8. C. murale**
+	Seeds not keeled along the margin, 1.5–2mm diam.	**9. C. fasciculosum**
10.	Lower and median leaves as wide as long or slightly longer than wide, rarely exceeding 5cm in length	**10. C. opulifolium**
+	Lower and median leaves distinctly longer than wide, up to 10cm in length	11
11.	Pericarp persistent; seeds c.1.2mm diam.	**11. C. ficifolium**
+	Pericarp easily detached; seeds 1.2–1.6mm diam.	**12. C. album**

1. C. carinatum R. Br., Prodr.: 407 (1810).

Aromatic annual herb up to 40cm high; stems decumbent to erect, pilose, glandular, the glands sessile or stalked. Leaves ovate, 4–18 × 2–10mm, slightly lobed or coarsely serrate, the lower surface glandular on the veins. Inflorescence of dense axillary clusters of sessile flowers. Perianth segments 5, erect; stamens 0 or 1. Fruiting perianth thick, coriaceous, 1mm long, prominently keeled. Pericarp transparent, adherent to the seed; seeds vertical, lenticular, c.0.5mm diam. **Map 297, Fig. 43.**

Naturalized along roadsides and at the edges of cultivated ground; 1500m.

Saudi Arabia. Introduced from Australia.

2. C. ambrosioides L., Sp. pl.: 219 (1753). Illustr.: Fl. Libya 58: t.4 (1978); Collenette (1985 p.119).

Strongly aromatic annual or short-lived perennial herb, 20–80cm high, green; stems erect, much-branched, striate. Leaves 2–10 × 0.5–3.5cm, with yellowish glands on the lower surface; lower leaves well developed, elliptic-lanceolate, shortly petiolate, irregularly sinuate-dentate; upper leaves small and narrow. Inflorescence a much-branched panicle, with elongated spike-like clusters of sessile flowers. Perianth segments 3–5, connate at the base; stamens 4–5. Pericarp easily detached; seeds horizontal or vertical, deep reddish-brown, glossy, 0.6–0.9mm diam. **Map 298, Fig. 43.**

Waste ground, field edges, irrigation and drainage ditches and other moist ground; 1800–2400m.

Saudi Arabia, Yemen (N & S). Tropical and subtropical regions; a native of America now widely introduced and naturalized.

3. C. schraderianum Schultes, Syst. veg. 6: 260 (1820). Syn.: *Chenopodium foetidum* Schrader in Ges. Naturf. Freunde Berlin Mag. Neuesten Entdeck. Gesammten Naturk. 2: 79 (1808). Illustr.: Collenette (1985 p.119).

Strongly aromatic annual herb, 10–80cm high, green, pubescent to glabrescent; stems erect, simple or branched mainly near the base, striate. Leaves broadly or narrowly elliptic, 1–6 × 0.5–3cm, with sessile yellowish glands on the lower surface, pinnately divided with 3–5 pairs of lobes, these entire or with a few blunt teeth, obtuse and with a short mucro at the tip, the mid- and lateral veins conspicuous on the lower surface; lower leaves long-petiolate, the median short-petiolate and the upper sessile. Inflorescence of terminal and axillary cymes, these 0.3–1.5(–2.5)cm long, dichotomously branched. Perianth segments 5, with median crested keels; stamens 1–2. Pericarp easily detached; seeds horizontal, dark reddish-brown or blackish, glossy, c.0.75mm diam., slightly keeled. **Map 299, Fig. 43.**

Fields, waste ground and on the margins of cultivation; 1400–2800m.

Saudi Arabia, Yemen (N & S). Tropical E and SE Africa; introduced into E and C Europe.

4. C. procerum Hochst. ex Moq. in DC., Prodr. 13(2): 75 (1849).

Strongly aromatic annual herb, up to 1.5m high, dull greyish green, glandular and

CHENOPODIACEAE

Fig. 43. Chenopodiaceae. A–L, *Chenopodium* leaves and seeds. A, *C. ambrosioides*. B, *C. glaucum*. C, *C. ficifolium*. D, *C. murale*. E, *C. botrys*. F, *C. vulvaria*. G, *C. schraderianum*. H, *C. opulifolium*. I, *C. carinatum*. J, *C. fasciculosum* var. *muraliforme*. K, *C. procerum*. L, *C. album*. Leaves all × 0.6, seeds all × 10.

sparingly pubescent; stems erect, branched near the base, striate. Leaves elliptic or ovate-elliptic, 2–12.5 × 1–6cm, with stalked yellow glands on the lower surface, pubescent especially along the veins and margins, median and lower parts pinnately divided, upper part ± triangular, the midrib and veins conspicuous on the lower surface; basal leaves long-petiolate, becoming progressively smaller, less divided and shorter-petiolate towards the top. Inflorescence of terminal and axillary cymes, these 1.5–6cm long, dichotomously branched. Perianth segments 5, connate at the base, glandular especially along the keels, the margins membranous; stamens 1–2. Pericarp easily detached; seeds horizontal, blackish, glossy, c. 1mm diam., slightly keeled. **Map 300, Fig. 43.**

Cultivated ground and waste places; 1200–2600m.

Yemen (N & S). Tropical E Africa.

5. C. botrys L., Sp. pl.: 219 (1753). Illustr.: Fl. Libya 58: t.3 (1978).

Aromatic annual herb, 15–75cm high, green, glandular and sparingly pubescent; stems erect, much-branched, angled, almost leafless at maturity (most leaves falling before seed-maturity). Leaves ovate or elliptic, 1–5 × 0.5–2cm, with stalked yellow glands especially on the lower surface, pubescent especially along the veins and margins, irregularly pinnate, obtuse at the tip, mid- and lateral veins conspicuous on the lower surface; basal leaves long-petiolate, progressively smaller and short-petiolate to sessile towards the top; uppermost leaves reduced to small bracts. Inflorescence of terminal and axillary cymes, these 0.5–3cm long, dichotomously branched. Perianth segments 5, free almost to the base, mucronate at the tip; stamens 1–2 or absent. Pericarp not persistent; seeds mostly horizontal, keeled, blackish, glossy, c. 0.75mm diam. **Map 301, Fig. 43.**

Cultivated ground and roadsides; 1200–2300m.

Saudi Arabia, Yemen (N). Mediterranean region and Asia; introduced into C and E Europe and N America.

Frequently confused with *C. schraderianum* in Arabia.

6. C. vulvaria L., Sp. pl.: 220 (1753). Illustr.: Fl. Libya 58: t.5 (1978).

Annual herb, 10–60cm high, grey-farinose, smelling strongly of decaying fish; stems erect, ascending or procumbent, much-branched, ridged or angled. Leaves rhombic or deltoid-ovate, 1–2.5 × 0.5–2cm, grey-farinose especially on the lower surface, entire, acute or obtuse at the tip, mid-vein and 2 basal veins conspicuous on the upper and lower surfaces; petiole 0.3–1.5cm long. Inflorescence of terminal and axillary cymes, leafy, in short dense clusters at first, becoming looser and longer at seed-maturity. Perianth segments 5, united at the base, densely mealy, not keeled, acute at the tip; stamens 1–5, absent in female flowers. Seeds horizontal, black, glossy, 1–1.25mm diam., slightly keeled. **Map 302, Fig. 43.**

Fields and at the edge of cultivated ground; 1800–2300m.

Yemen (N), Oman. Europe, N Africa and Asia; introduced into N America and Australia.

7. C. glaucum L., Sp. pl.: 220 (1753).

Annual herb, 10–50cm high, glaucous, glabrous except on the lower surface of the leaves; stems erect, ascending or procumbent, much-branched especially at the base, angular, striated. Leaves elliptic, lanceolate or ovate, 1–5 × 0.3–1cm, green and glabrous on the upper surface, glaucous and farinose on the lower, the margins sinuate-serrate, rarely subentire, acute or obtuse at the tip, the mid-vein conspicuous on the lower and upper surfaces; basal and median leaves petiolate, the uppermost almost sessile. Inflorescence of terminal and axillary spike-like cymes, the terminal flowers often perfect. Perianth-segments 2–4, connate at the base, the margins broadly membranous; stamens 2–3. Pericarp thin, loosely attached to the seed; seeds vertical, occasionally horizontal in lateral flowers, reddish-brown, glossy, 0.8–1mm diam. **Map 303, Fig. 43.**

Moist waste ground and at the edges of cultivated fields; 0–600m.

Saudi Arabia, Bahrain, Kuwait. Europe, Mediterranean region and Asia.

8. C. murale L., Sp. pl.: 219 (1753). Illustr.: Cornes (1989 p.77); Fl. Qatar: 25 (1981); Fl. Kuwait 1: pl.161 (1985); Collenette (1985 p.119).

Annual herb, 10–60cm high, dark green, juvenile parts farinose, otherwise glabrous; stems erect or ascending, usually much-branched at the base, angular, striated. Leaves rather fleshy, ovate-rhombic or deltoid, less often narrowly elliptic-lanceolate, 1.5–10 × 1–7cm, coarsely dentate but not lobed, acute or obtuse at the tip, the mid-vein and basal lateral veins conspicuous on the lower surface; basal leaves long-petiolate, becoming progressively short-petiolate towards the top. Inflorescence of terminal and axillary leafy cymes, these divaricate, loose or compact. Perianth segments 5, connate to $\frac{1}{3}$ their length, papillose, keeled near the tip, the margins narrowly membranous; stamens 5. Pericarp firmly adherent to the seed; seeds horizontal, black, glossy, strongly keeled, 1.2–1.5mm diam. **Map 304, Fig. 43.**

Sandy desert soils, waste ground, gardens, orchards and fields; 0–2500m.

Saudi Arabia, Yemen (N & S), Socotra, Oman, UAE, Qatar, Bahrain, Kuwait. Virtually cosmopolitan.

9. C. fasciculosum Aellen var. **muraliforme** Aellen in Feddes Repert. Spec. Nov. Regni Veg. 24: 344 (1928).

Annual herb, 15–80cm high, juvenile parts farinose, soon becoming glabrous; stems erect or ascending, branched mainly at the base, grooved. Leaves broadly ovate to deltoid, 2–8 × 1.5–6cm, acute at the tip, the margins with irregular sharp teeth. Inflorescences terminal and lateral, of divaricately branched cymes, these 2–4cm long. Perianth segments 5, papillose, keeled near the tip, connate at the base; stamens 5. Pericarp easily detached from the seed; seeds vertical, black, shining, 1.5–2mm diam. **Map 305, Fig. 43.**

Cultivated and waste ground; 2000–2600 m.

Yemen (N). Kenya.

C. *fasciculosum* var. *fasciculosum* is known from Kenya, Tanzania and Ethiopia, while var. *muraliforme*, previously known only from Kenya (*Brenan* 1954), has recently been found in Yemen.

10. C. opulifolium Schrader ex W. Koch & Ziz, Cat. pl.: 6 (1814). Illustr.: Collenette (1985 p.119).

Annual or short-lived herbaceous perennial with a woody base, 30–200cm high, whitish-green, mealy, juvenile parts densely grey-mealy; stems slightly angled or terete, much-branched, striate. Lower and median leaves ovate-rhombic, 2–5 × 1.5–4cm, densely mealy on the lower surface, acute or obtuse at the tip, the margins few-toothed (but not distinctly 3-lobed in Arabia); petiole as long as or shorter than the blade; upper leaves smaller, almost entire, short-petiolate. Inflorescences terminal and lateral, densely grey-mealy, of dense or lax paniculate cymes. Perianth segments 5, connate at the base, papillose on the outer surface and margins; stamens 5. Pericarp somewhat persistent; seeds horizontal, black, glossy, 1.2–1.5mm diam. **Map 306, Fig. 43.**

Cultivated and waste ground; 0–2300 m.

Saudi Arabia, Yemen (N & S), Kuwait. Europe and Mediterranean region to Asia and tropical Africa; introduced into N America.

11. C. ficifolium Smith, Fl. brit. 1: 276 (1800).

Annual herb, 25–80cm high, whitish-green, mealy; stems erect or ascending, branched, angled, striate. Leaves 2–12 × 0.5–3.5cm, distinctly longer than wide, mealy on the lower surface, some distinctly 3-lobed with lateral lobes triangular and shorter than the median, obtuse at the tip, the margins sparsely and coarsely toothed; petiole 1–6cm long. Inflorescences terminal and lateral, of paniculate cymes, mealy, dense at first, becoming lax in late flowering and fruiting stages. Perianth segments 5, connate for $\frac{1}{2}$ their length, densely farinose, keeled towards the tip, membranous on margins; stamens usually 5. Pericarp firmly adherent to the seed; seeds horizontal, black, glossy, keeled, c. 1.2mm diam. **Map 307, Fig. 43.**

Cultivated fields and waste ground.

Saudi Arabia. Europe and Asia.

12. C. album L., Sp. pl.: 219 (1753). Illustr.: Collenette (1985 p.118).

Annual herb, 10–80(–120)cm high, grey-farinose; stems erect or ascending, usually much-branched, green or tinged with red, ridged or angled, striate. Leaves variable in shape and size, rhombic-ovate to narrowly lanceolate, 1–8 × 1–5cm, typically longer than wide, acute at the tip but often obtuse in the lower leaves, the margins dentate-sinuate to entire, mid-vein and 2 basal veins more conspicuous on the lower surface; petiole 0.5–3cm long. Inflorescences almost leafless, terminal and axillary, mealy, of compact, few-flowered, paniculate cymes. Perianth segments 5, free almost to the base, papillose, strongly keeled, membranous on the margins; stamens 5. Pericarp free; seeds horizontal, black, glossy, slightly keeled, 1.2–1.6mm diam. **Map 308, Fig. 43.**

Fields, field edges and moist ground; 0–2300 m.

Saudi Arabia, Yemen (N & S), Oman, Kuwait. Cosmopolitan.

Species incompletely known
C. triangulare Forsskal (1775 p.205).

2. BETA L.

Annual or perennial herbs; stems striate. Leaves alternate, entire, petiolate. Flowers connate at the base, in clusters of 2–4 in the leaf-axils, forming a leafy spicate inflorescence. Perianth segments 5; stamens 5; ovary fused to the base of the perianth and receptacle, the connate receptacles of adjacent flowers in each cluster falling as a unit in fruit; stigmas 2–3(–5). Fruit fleshy or indurated; seeds horizontal, lenticular or reniform, glossy.

Aellen, P. (1938) in *Ber. Schweiz. Bot. Ges.* 48: 470–484.

1. B. vulgaris L. subsp. **maritima** (L.) Arcang., Comp. fl. ital.: 593 (1882). Syn.: *Beta maritima* L., Sp. pl., ed. 2: 322 (1762). Illustr.: Fl. Libya 58: t.1 (1978); Collenette (1985 p.118)..

Annual or herbaceous perennial, up to 80cm high; stems erect or decumbent, branching at the base, green or reddish. Leaves 2–10 × 1–5cm, fleshy, glabrous; basal leaves in a rosette, ovate-cordate, long-petiolate; cauline leaves ovate-deltoid or rhombic, petiolate, uppermost bracteate and short-petiolate. Flowers in clusters of 2–4, in long, interrupted, leafy spikes; perianth segments 2–2.5mm long, green, fleshy, thick and indurated in fruit; stamens 5, on the rim of the glandular perigynous disc; ovary 3-carpellate, adherent to the perianth at the base. Seeds black, reticulate. **Map 309.**

Moist saline soils, at the edge of cultivated ground and along irrigation canals; 0–2600m.

Saudi Arabia, Yemen (N & S), Oman, UAE, Qatar, Bahrain, Kuwait. Atlantic Islands, W Europe, the Mediterranean region, W Asia and Sri Lanka.

3. SPINACIA L.

Dioecious annuals or perennials. Leaves hastate, pinnatifid or pinnatisect. Flowers in sessile clusters, the male forming spikes or panicles, the female axillary. Female flowers without perianths; floral bracts 2, fused around ovary and conspicuously accrescent in fruit, often each with a dorsal spine; style short or obscure; stigmas 4–5, filiform. Male flowers with 4(–5) perianth segments; stamens 4(–5), with slender filaments and

± rounded anthers. Fruit indehiscent, usually spiny, free or fused with the others of the cluster; seeds vertical.

1. S. oleracea L., Sp. pl.: 1027 (1753).

Annual or biennial herb, 15–40cm, glabrous, sparingly branched. Basal leaves deltoid-hastate, c.12 × 8cm, long-petiolate; upper leaves much smaller, lanceolate or narrowly hastate, the uppermost bract-like. Female flowers in clusters of 5–8; floral bracts with or without a dorsal spine; stigmas 1–3.5mm. Male flowers in panicles; perianth segments 4(–5), c.1.5mm long, oblong-ovate, obtuse; filaments 4(–5), c.2.5mm; anthers c.1mm. Fruits c. 3mm diam., not fused with others of the cluster. **Map 310.**

Saudi Arabia, Yemen (N). Cultivated as a pot-herb in most countries of the Arabian Peninsula; origin unknown, but possibly W Asia.

4. ATRIPLEX L.

Annuals, herbaceous perennials or shrubs, with mealy crust. Leaves mostly alternate. Flowers unisexual, on monoecious or dioecious plants, in terminal or axillary clusters or in spikes or panicles. Male flowers with 5 deeply lobed segments and without bracteoles; stamens 5. Female flowers without a perianth but with 2 bracteoles which enlarge and enclose the fruit. Pericarp membranous; seeds vertical, lenticular; embryo annular, with endosperm.

Aellen, P. (1939). *Bot. Jahrb. Syst.* 70: 1–66.

1.	Herbaceous annuals	2
+	Woody perennials	3
2.	Leaves sinuate-dentate or lobed	**8. A. tatarica**
+	Leaves entire, not lobed	**7. A. dimorphostegia**
3.	At least some leaves auriculate at the base	**1. A. farinosa**
+	Leaves not auriculate	4
4.	Flowers in axillary and terminal leafy inflorescences	5
+	Flowers in terminal, almost leafless inflorescences	6
5.	Leaves linear-lanceolate; fruiting bracteoles longer than wide and with a tooth on each side	**5. A. glauca**
+	Leaves triangular-deltoid; fruiting bracteoles as long as wide or wider than long, 3–5-lobed, each lobe 2–4-toothed	**6. A. leucoclada**
6.	Leaves slightly lobed or dentate, thin textured	**3. A. halimus**
+	Leaves entire or undulate, coriaceous	7
7.	Leaves slightly longer than wide to almost orbicular; petiole up to 4mm long; leaf tip retuse or rounded; fruiting bracteoles broadly ovate to cordate	**4. A. griffithii**
+	Leaves longer than wide, ovate to elliptic, sessile or subsessile; leaf tip obtuse to acute; fruiting bracteoles quadrangular	**2. A. coriacea**

1. A. farinosa Forsskal (1775 p.CXXIII) subsp. **farinosa**. Syn.: *A. hastata* sensu Forsskal (1775 p.175) non L. Illustr.: Collenette (1985 p.117). Type: Saudi Arabia, *Forsskal* (C).

Shrub or herb with a woody base, up to 1.2m, densely mealy on all parts, whitish-grey. Leaves petiolate, ovate-elliptic, $1.5–4 \times 0.8–2.2$cm, obtuse at the tip, entire, auriculate or cordate at the base. Flowers in terminal leafless panicles. Fruiting bracteoles variable in size and shape, $2–4 \times 2–3$mm, obconical to broadly elliptic, entire, connate at the base. **Map 311, Fig. 44.**

Coastal sandy soil; sea-level.

Saudi Arabia, Yemen (N & S), Socotra, Oman. Jordan, Egypt and NE tropical Africa.

The flowering (but not fruiting) specimen from the island of Abd-al-Kuri (*Radcliffe-Smith & Lavranos* 703(K)) is probably *A. farinosa* subsp. *farinosa* and not subsp. *keniensis* (Brenan) Friis & Gilbert; for a more definite determination fruits are needed as the fruiting bracteoles need to be examined.

2. A. coriacea Forsskal (1775 p.175). Syn.: *A. ocymifolium* Viv., Pl. aegypt. dec.: 23 (1831); *Obione coriacea* (Forsskal) Moq., Chenop. monogr. enum.: 71 (1840).

Shrub up to 1.5m. Leaves sessile or subsessile, ovate to elliptic, $0.5–2.8 \times 0.4–1.5$cm, obtuse or acute at the tip, entire, coriaceous, the midrib prominent on the lower surface. Flowers in terminal leafless panicles. Fruiting bracteoles quadrangular, 4×4mm, with 2–3 appendages on each side. **Map 312, Fig. 44.**

Coastal sand and waste ground; sea-level.

Saudi Arabia, Yemen (N & S). N Africa from Algeria to Sinai.

No verified material has been seen and the occurrence of this species in Arabia needs confirmation.

3. A. halimus L., Sp. pl.: 1052 (1753).

Shrub up to 1.5m; stems woody at the base, much-branched. Leaves ovate to triangular, $1.5–4 \times 0.5–2.5$cm, slightly lobed in the lower part, the small upper leaves entire; petioles 2–6mm long. Flowers staminate and pistillate, in mixed clusters in almost leafless terminal panicles. Fruiting bracteoles usually wider than long, reniform to broadly triangular-ovate, $2.5–4 \times 2.5–4.5$mm, connate at the base, the free upper part slightly dentate. **Map 313.**

Coastal and desert plains; 0–600m.

Saudi Arabia. Coasts of the Mediterranean region, Ethiopia and Tanzania.

Also cultivated in central Saudi Arabia, Yemen and Kuwait, and probably in other parts of Arabia, for its foliage which is used as fodder. The mapped record from central Saudi Arabia is of a cultivated plant; the plants are salt-tolerant.

4. A. griffithii Moq. subsp. **stocksii** (Boiss.) Boulos in Nord. J. Bot. 11: 310 (1991). Syn.: *Atriplex stocksii* Boiss., Diagn. pl. orient. sér. 2, 3(4): 73 (1859); *A. griffithii* var. *stocksii*

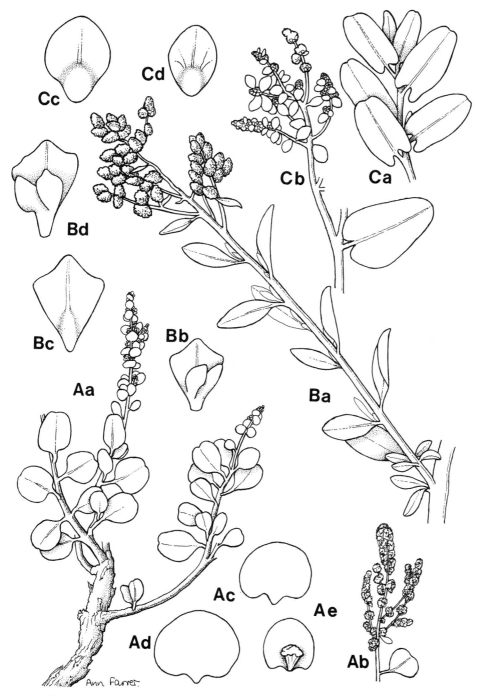

Fig. 44. Chenopodiaceae. A, *Atriplex griffithii* subsp. *stocksii*: Aa, habit (×1); Ab, inflorescence (×1); Ac–Ae, fruiting perianths (×4). B, *A. coriacea*: Ba, flowering branch (×1); Bb–Bd, fruiting bracteoles (×4). C, *Atriplex farinosa*: Ca, leaves (×1); Cb, flowering branch (×1); Cc–Cd, fruiting bracteoles (×4).

(Boiss.) Boiss., Fl. orient. 4: 916 (1879); *A. sokotranum* Vierh. in Oesterr. Bot. Z. 53: 481 (1903), excl. syn.; *A. stocksii* f. *sokotranum* (Vierh.) Vierh. (1907 p.18).

Subshrub up to 60cm; old stems woody at the base. Leaves broadly ovate to almost orbicular, 0.4–2 × 0.4–1.6cm, retuse or rounded at the tip, entire or undulate, not lobed or dentate, rounded to cuneate at the base, coriaceous, the midrib prominent on the lower surface; petiole 2–4mm. Flowers yellowish-green, in leafless panicles. Fruiting bracteoles broadly triangular-ovate to cordate, 3.5–5mm long and wide, reticulate on both sides, entire or with 2 small basal teeth. **Map 314, Fig. 44.**

Coastal sandy plains and sea cliffs; 0–300 m.

Yemen (S), Socotra, Oman. Somalia and Pakistan.

5. A. glauca L., Cent. pl. I: 34 (1755). Syn.: *Atriplex stylosa* Viv., Pl. aegypt. dec.: 23 (1831); *A. palaestina* Boiss., Diagn. pl. orient. sér. 1, 2(12): 96 (1853); *A. alexandrina* Boiss., Fl. orient. 4: 914 (1879); *A. crystallina* Boiss., op. cit.: 915 (1879). Illustr.: Fl. Libya 58: t.15 (1978).

Shrublet up to 30cm; stems erect to ascending, branched at the base, terete, whitish. Leaves sessile or subsessile, linear-lanceolate, 0.5–3 × 0.4–1 cm, acute at the tip, entire or sinuate. Flowers pistillate and staminate, in mixed clusters, forming terminal spikes or panicles; pistillate flowers in clusters only in the leaf axils. Fruiting bracteoles triangular to rhombic, 5–6 × 3–4.5mm, longer than wide, cuneate at the base, with 1 tooth on each side.

Wadi-beds and calcareous soils; 0–200 m.

? Saudi Arabia. Spain and Portugal; N Africa eastwards to Sinai and SW Asia.

No verified material of this species has been examined and its occurrence in Arabia needs confirmation.

6. A. leucoclada Boiss., Diagn. pl. orient. sér. 1, 2(12): 95 (1853). Syn.: *Obione leucoclada* (Boiss.) Ulbr. in Engler, Nat. Pflanzenfam. ed. 2, 16: 506 (1934). Illustr.: Mandaville (1990 pl.28); Fl. Qatar: pl.25 (1981); Fl. Kuwait 1: pls.162–3 (1985); Collenette (1985 p.119); Cornes (1989 p.73).

Perennial herb up to 80cm, basal part and lower branches woody; stems decumbent to erect, slender, branched mainly from the base. Leaves petiolate to sessile, triangular-deltoid to broadly ovate-cordate, 0.5–2.5 × 0.2–2cm, acute at the tip, entire, undulate or sinuate-dentate, hastate to almost rounded at the base. Flowers in axillary and terminal spikes or panicles, the axillary pistillate, the terminal staminate and pistillate. Fruiting bracteoles deltoid, quadrangular or campanulate, 3–5 × 2–5mm, lobed or toothed, connate at the base, free above.

Coastal plains and desert wadis on limestone, sandstone or volcanic soils; 0–2500m.

1. Leaves ovate-cordate, up to 1.5cm long, sessile or subsessile, entire or undulate; perianth in fruit triangular-deltoid with 5 apical lobes   var. **inamoena**
+ Leaves triangular-deltoid, up to 3.5cm long, petiolate, sinuate-dentate; perianth in fruit quadrangular to campanulate   var. **turcomanica**

var. **inamoena** (Aellen) Zoh., Fl. Palaest. 1: 147 (1966). Syn.: *Atriplex inamoena* Aellen in Bot. Jahrb. Syst. 70: 20 (1939). **Map. 315. Fig. 45.**

0–1675m.

Saudi Arabia, Yemen (S), Oman, UAE, Bahrain, Kuwait. Egypt to SW Asia.

var. **turcomanica** (Moq.) Zoh., Fl. Palaest. 1: 147 (1966). Syn.: *Atriplex laciniata* L. var. *turcomanica* Moq. in DC., Prodr. 13(2): 93 (1849); *A. leucoclada* subsp. *turcomanica* (Moq.) Aellen in Bot. Jahrb. Syst. 70: 22 (1939). **Map 316, Fig. 45.**

0–2500m.

Saudi Arabia, Yemen (N & S), UAE, Qatar, Bahrain, Kuwait. SW Asia eastwards to Afghanistan.

Much eaten by camels, especially in winter. A variable species, the above varieties being difficult to identify unless in fruit. Seasonal and intermediate forms are known.

7. A. dimorphostegia Karelin & Kir. in Bull. Soc. Nat. Moscou 15: 438 (1842). Illustr.: Mandaville (1990 pl.27); Fl. Kuwait 1: pl.164 (1985); Collenette (1985 p.115).

Annual 15–35cm; stems prostrate or ascending, much-branched especially from the base, glabrous, whitish. Leaves broadly ovate to deltoid, 1.5–5.5 × 1–4 cm, entire, green and almost glabrous on the upper surface, greyish-green and scurfy-mealy on the lower. Flowers in axillary clusters. Fruiting bracteoles of the upper flowers triangular-ovate, of the lower orbicular-cordate. **Map 317, Fig. 45.**

Sandy, slightly saline soils; 0–950 m.

Saudi Arabia, Kuwait. N Africa and SW Asia eastwards to Afghanistan and Pakistan.

8. A. tatarica L., Sp. pl.: 1053 (1753).

Annual up to 80cm; stems erect or ascending, much-branched. Leaves alternate above, opposite below, petiolate, broadly or narrowly triangular to lanceolate-hastate, 2–6(–8) × 1–3(–5)cm, sinuate-dentate or lobed. Flowers pistillate and staminate, in clusters forming leafless spikes or panicles, or the pistillate flowers solitary in the leaf axils. Fruiting bracteoles almost orbicular or oblong-rhombic, 5–7 × 2–4.5mm, dentate or entire, often with appendages on the dorsal side. **Map 318.**

Waste ground and roadsides.

Saudi Arabia, UAE. Mediterranean region, SW & C Asia.

CHENOPODIACEAE

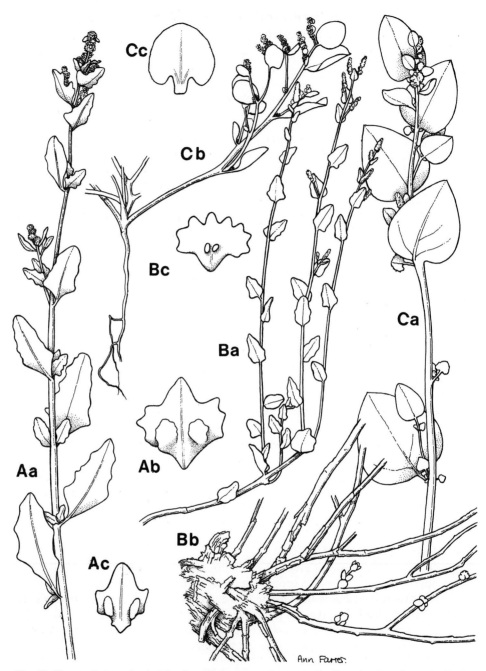

Fig. 45. Chenopodiaceae. A, *Atriplex leucoclada* var. *turcomanica*: Aa, flowering branch (×1); Ab–Ac, fruiting bracteoles (×3). B, *A. leucoclada* var. *inamoena*: Ba, flowering shoots (×1); Bb, basal branches (×1); Bc, fruiting bracteole (×3). C, *A. dimorphostegia*: Ca, flowering branch with large leaves (×1); Cb, flowering branch with small leaves (×1); Cc, fruiting bracteole (×3).

5. AGRIOPHYLLUM M. Bieb.

Spiny annual herbs, branching from the base. Leaves rigid, alternate, entire. Flowers solitary in spiny glomerules in the leaf axils; perianth of 1–5 segments, scarious; stamens 1–5. Fruit flattened, opening at maturity on the ventral and dorsal sides by means of 2 small scarious wings; seeds oval, compressed.

Iljin, M. (1936) in Komarov & Shishkin, *Fl. URSS* 6: 158–162.

1. A. minus Fischer & C. Meyer ex Ledeb., Fl. ross. 3(2): 755 (1851). Syn.: *A. montasirii* El-Gazzar in Mitt. Bot. Staatssamml. München 27: 16 (1988).

Spiny annual herb, 10–40cm, sparsely covered with branched hairs, juvenile parts more pubescent. Leaves linear-lanceolate, 1–5 × 0.1–0.6cm, spine-tipped, often attenuate towards the base, parallel-veined. Glomerules axillary, becoming densely spiny at maturity from recurved sharply pointed bracts. Fruit 4–5mm long with a 3-parted beak, the margin broadly winged, the lateral lobes toothed; seeds oval, 2 × 1mm. **Map 319, Fig. 46.**

Coastal and wind blown sandy soils; 30–150m.

Saudi Arabia, UAE, Qatar. Iraq and Iran to C Asia.

6. BASSIA All.

Syn.: *Chenolea* Thunb.; *Kochia* Roth; *Londesia* Fischer & C. Meyer; *Chenoleoides* (Ulbr.) Botsch.

Annual or perennial woody-based herbs. Leaves alternate, subsessile, thin or succulent, linear, entire, hairy. Flowers unisexual or bisexual, in the axils of leaf-like bracts, in leafy spikes or panicles; perianth membranous, of 5 segments connate above the middle and with inflexed lobes; stamens (3–)5; style short; stigmas 2–3, exserted. Fruiting perianth with horizontal wings, spines or lobes, or without appendages; fruit compressed, with a delicate membranous pericarp; seeds horizontal or vertical; embryo annular; endosperm abundant and mealy or absent

Scott, A.J. (1978). *Feddes Repert.* 89: 101–119.

1.	Fruiting perianth with 5 stellately-arranged spines	2
+	Fruiting perianth not spiny	3
2.	Spines of fruit 2.5–3.5mm long	**1. B. muricata**
+	Spines of fruit 1–1.5mm long, uncinate or circinnate at the tip	**2. B. hyssopifolia**
3.	Dwarf shrubs, up to 30cm; stems woody at the base	**6. B. arabica**
+	Annual (? or biennial) herbs, up to 150cm; stems herbaceous	4
4.	Flowers hidden in dense white fluffy hairs	**3. B. eriophora**
+	Flowers exposed	5

5.	Leaves 2–6(–8)cm long, glabrous or shortly pilose	**4. B. scoparia**
+	Leaves 0.5–1.5cm long, pubescent or pilose with long (up to 1.5mm) adpressed or spreading hairs mostly on the lower surface	**5. B. indica**

1. B. muricata (L.) Asch. in Schweinf., Beitr. Fl. Aethiop.: 187 (1867). Syn.: *Salsola muricata* L., Mant. 1: 54 (1767); *Kochia muricata* (L.) Schrader in Neues J. Bot. 3, 3, 4: 86 (1809); *Echinopsilon muricatus* (L.) Moq. in Ann. Sci. Nat. Paris, sér. 2, 2: 127 (1834). Illustr.: Boulos (1988 p.17); Collenette (1985 p. 117).

Annual up to 50cm high, often becoming stiff and rather woody at the base, densely villous; stems erect or decumbent, branched from the base. Leaves linear-lanceolate, 2–12(–20) × 1–2mm. Male and female flowers mixed in axillary clusters forming leafy spikes. Fruiting perianth segments connate at the indurated base forming a yellowish 5-armed star-shaped structure; arms needle-like, 2.5–3.5mm long; seeds 1mm diam., discoid, greyish, smooth. **Map 320, Fig. 47.**

Wadis, waste ground, roadsides and field margins; 0–2000m.

Saudi Arabia, Yemen (N), Oman, UAE, Qatar, Bahrain, Kuwait. N Africa and SW Asia eastwards to Iran.

2. B. hyssopifolia (Pallas) Kuntze, Revis. gen. pl. 2: 547 (1891). Syn.: *Salsola hyssopifolia* Pallas, Reise russ. Reich. 1: 491 (1771); *Kochia hyssopifolia* (Pallas) Roth, Neue Beytr. Bot.: 176 (1802); *Echinopsilon hyssopifolium* (Pallas) Moq. in Ann. Sci. Nat., sér. 2, 2: 127 (1834).

Annual, up to 1.5m; stems erect or ascending, branched from the base, rigid, whitish-yellow. Leaves flat, linear or oblong-linear, 5–15 × 1–2mm, acute, hirsute or villous. Bracts much smaller than the leaves, oblong, exceeding the flowers, fleecy at the base. Flowers bisexual and female, solitary or in clusters of 2–3, forming panicles; perianth 1–5mm diam., long-pilose, the segments connate at the base for $\frac{2}{3}$ their length, rounded, incurved and becoming accrescent and spiny in fruit; spines 1–1.5mm long, uncinate or circinnate at the tip; stigmas 2, filiform. Fruit 1.5mm diam.; seeds compressed, ovoid. **Map 321.**

Introduced and naturalized on roadsides, waste ground and sandy soils; 0–150m.

Saudi Arabia. SW Europe, Crimea, Caucasus and SW Asia.

3. B. eriophora (Schrader) Asch. in Schweinf., Beitr. Fl. Aethiop.: 187 (1867). Syn.: *Kochia eriophora* Schrader in Neues J. Bot. 3, 3, 4: 86, t.3 (1809); *K. latifolia* Fresen. in Mus. Senckenberg 1: 179 (1834); *Bassia latifolia* (Fresen.) Asch. & Schweinf. in Mém. Inst. Égypt. 2: 127 (1887). Illustr.: Boulos (1988 p.15); Mandaville (1990 pl.29); Collenette (1985 p. 117).

Annual, up to 60cm, villous in all parts; stems ascending to erect, much-branched from the base. Leaves oblong-linear to lanceolate or elliptic, 5–20 × 2–4mm, dark green, fleshy. Flowers hidden in dense fluffy hairs, forming dense leafy spikes. Fruiting perianth curved at the tip, the lobes with short appendages; seeds 1mm diam., discoid, brownish. **Map 322.**

Sandy soil in desert wadis, and waste places near cultivated ground; 0–950m.

Saudi Arabia, Qatar, Bahrain, Kuwait. Egypt and SW Asia eastwards to Pakistan.

4. B. scoparia (L.) A.J. Scott op. cit.: 108 (1978). Syn.: *Chenopodium scoparia* L., Sp. pl.: 221 (1753); *Kochia scoparia* (L.) Schrader in Neues J. Bot. 3, 3, 4: (1809).

Annual, up to 1.2m; stems erect, much-branched, whitish or often reddish, striate, the juvenile hairy. Leaves subsessile, flat, linear-lanceolate, 2–6(–8) × 0.1–0.5cm, acute, almost glabrous above, adpressed-hairy beneath. Flowers bisexual or unisexual, solitary or in pairs in the leaf axils, forming long leafy spikes. Fruiting perianth 3–4mm long, connate at the base, the free apical parts incurved; fruit 2–2.2mm diam.; seeds ovoid, compressed, brown to blackish. **Map 323.**

Cultivated in parks and gardens for its ornamental green foliage; also occurring as an escape.

Saudi Arabia, Kuwait. Europe, N Africa and temperate Asia.

5. B. indica (Wight) A.J. Scott op. cit.: 108 (1978). Syn.: *Kochia indica* Wight, Icon. pl. Ind. orient. t.1791 (1852); *Bassia joppensis* Bornm. & Dinsm. in Repert. Spec. Nov. Regni Veg. 17: 274 (1921); *K. scoparia* subsp. *indica* (Wight) Aellen in Hegi, Ill. Fl. Mitt.-Eur. 3(4): 711 (1961). Illustr.: Fl. Libya 58: t.9 (1978); Collenette (1985 p.122) as *Kochia indica*.

Annual, probably also biennial, up to 1.5m; stems erect, richly branched, the branches slender, spreading or ascending, pilose-pubescent. Leaves oblong-elliptic to narrowly linear, 5–15 × 1–5mm, with long (up to 1.5mm) soft adpressed and spreading hairs mostly on the lower surface; upper leaves reduced, densely white-hairy. Flowers solitary or 2–4 in sessile clusters forming leafy spikes; perianth white-hairy, the segments connate for up to $\frac{2}{3}$ their length, with short spreading wings in fruit; stamens exserted; anthers yellow; seeds not seen. **Map 324.**

Roadsides and waste ground; 0–150 m.

Saudi Arabia. N Africa and SW Asia eastwards to India; E Africa.

6. B. arabica (Boiss.) Maire & Weiller, Fl. Afr. nord 8: 54 (1962). Syn.: *Chenolea arabica* Boiss., Diagn. pl. orient. sér. 1, 2(12): 97 (1853); *Chenoleoides arabica* (Boiss.) Botsch. in Bot. Zhurn. (Moscow & Leningrad) 61: 1409 (1976).

Woody-based perennial herbs, up to 30cm; stems prostrate or decumbent, woolly-canescent. Leaves sessile, linear-oblong, 0.3–1 × 0.1–0.2cm, obtuse. Flowers bisexual, in axillary clusters forming dense leafy spikes; perianth 3mm long, densely woolly, with 5 hemispherical obtuse lobes; seeds horizontal; endosperm little or absent. **Map 325.**

Calcareous and sandy soils and desert wadis.

Saudi Arabia. N Africa and SW Asia.

7. HALOPEPLIS Bunge ex Ung.-Sternb.

Annual herbs or subshrubs; stems not jointed but appearing so because of the perfoliate leaves. Leaves mostly alternate, perfoliate, succulent. Flowers bisexual or female, in clusters of 3, often connate and adnate to the subtending bract, in spikes; perianth segments 3, not winged; stamens 1–2; ovary pear-shaped; stigmas 2. Fruit included; pericarp membranous; seeds ellipsoid; endosperm central, copious; embryo hook-shaped.

1. H. perfoliata (Forsskal) Bunge ex Asch. in Schweinf., Beitr. Fl. Aethiop.: 289 (1867). Syn.: *Salicornia perfoliata* Forsskal (1775 p.3). Illustr.: Fl. Qatar pl.27 (1981); Collenette (1985 p.121). Type: Saudi Arabia, *Forsskal* (C).

Subshrub, up to 50cm; stems richly branched, the older woody and leafless. Leaves perfoliate, $3–8 \times 4–10$mm, glabrous, juicy, green or reddish, threaded like beads around the younger stems and giving them a jointed appearance. Flowers red, fleshy, in alternate spikes; anthers yellow. Fruit 1-seeded; seed 0.8mm long. **Map 326, Fig. 47.**

Coastal salt-marshes, dunes and salt pans; 0–200 m.

Saudi Arabia, Yemen (N & S), Oman, UAE, Qatar, Bahrain. Sinai eastwards to S & SW Pakistan.

8. HALOCNEMUM M. Bieb.

Glabrous halophytic shrubs; old stems woody, richly branched; young stems jointed, succulent, with numerous opposite bud-like branches. Leaves opposite, succulent, decussate, the blades reduced to fleshy cups, connate at the base. Bracts opposite, free, deciduous. Flowers bisexual, in clusters, the clusters forming short, dense, lateral cone-like spikes; perianth segments 3, unequal, united at the base, inflexed at the tip; stamen 1, with flattened filament and ovoid anther; ovary ovoid; style short, thick; stigmas 2, filiform. Fruit ovoid, enclosed within the perianth; seeds vertical; endosperm absent; embryo arcuate.

1. H. strobilaceum (Pallas) M. Bieb., Fl. taur.-caucas. 3: 3 (1819). Syn.: *Salicornia strobilacea* Pallas, Reise russ. Reich. 1: 412 (1771); *S. cruciata* Forsskal (1775 p.2). Illustr.: Fl. Libya 58: t.19 (1978); Collenette (1985 p.120); Mandaville (1990 pl.30).

Shrub, up to 60cm, forming rounded hummocks up to 1.5m diam.; old stems intricate, with brownish bark; young stems much-branched, ascending to erect, $\pm$ straight. Leaves obovate, 1mm long, scarious on the margins. Flowers in clusters of 2–3 on short lateral and terminal branches; perianth segments broadly oblong, 1–1.25mm long, hyaline, truncate at the tip; stamens 2mm. Fruit 1.25mm diam., compressed-ovoid; seeds brownish, compressed-ovoid. **Map 327, Fig. 46.**

Coastal salt-marshes, low maritime dunes and adjacent saline plains and mud-flats; 0–600 m.

Mud-flats in tidal zones, often near mangrove vegetation; sea-level.

Saudi Arabia, Bahrain, Kuwait. Europe and the Mediterranean region eastwards to C Asia.

11. SUAEDA Forsskal ex Scop., *nom. conserv.*
Syn.: *Schanginia* C. Meyer.

Annual or perennial herbs, shrubs or rarely small trees, glabrous or slightly hairy on the juvenile parts. Leaves alternate, succulent, terete, subterete, subglobose, lenticular or flattened, sessile or short-petiolate, the base sometimes indurated or decurrent. Flowers bisexual or unisexual, solitary or in axillary or terminal sessile or subsessile glomerules, sometimes forming panicles; bracts 2–3, scarious, persistent, usually shorter than the perianth; perianth segments 5, herbaceous or succulent; stamens 5, inserted on the perianth segments, reduced to filaments (staminodes) in female flowers; ovary sessile, free or sometimes adnate to the perianth; male flowers with a rudimentary ovary; style short or absent; stigmas 2–3(–5). Fruiting perianth unchanged or becoming succulent or spongy; fruit free or adnate to the perianth; seeds horizontal or vertical; embryo spiral.

Boulos, L. (1991). Notes on *Suaeda* Forsskal ex Scop. Studies in the Chenopodiaceae of Arabia 2. *Kew Bull.* 46: 291–296.

1.	Small tree or shrub, up to 4m; conspicuous insect-galls frequently found on the stems, less often on the leaves; plant monoecious, the male flowers with a rudimentary ovary, the female with staminodes	**3. S. monoica**
+	Shrubs, up to 0.8(–1.5)m or annual or short-lived perennial herbs; insect-galls unknown on any part of the plant; flowers bisexual or bisexual and female	2
2.	Upper part of fruit conspicuously swollen and spongy, the lower part fused to the perianth at maturity	**1. S. aegyptiaca**
+	Fruit thin-walled, not spongy, enveloped by, but free from, the perianth at maturity	3
3.	Annuals; stems herbaceous, sometimes stiff and slightly woody at the base, usually drying greyish-green to brownish; anthers c.0.25mm	**2. S. maritima**
+	Shrubs; stems conspicuously woody, especially at the base, drying black; anthers 0.5–1mm	4
4.	Flowers in compact leafless terminal panicles, with strong pungent disagreeable musky smell (when fresh); leaves flattened or semi-terete; anthers c.0.5mm, subglobose	**5. S. moschata**
+	Flowers in leafy axillary and terminal spikes or loose panicles, with no particular smell; at least some of the leaves terete, or if flattened then on one side only; anthers 0.75–1mm, oblong	**4. S. vermiculata**

1. S. aegyptiaca (Hasselq.) Zoh. in J. Linn. Soc., Bot. 55: 635 (1957). Syn.: *Chenopodium aegyptiacum* Hasselq., Iter palaest.: 460 (1757); *Suaeda hortensis* Forsskal ex J.F. Gmelin, Syst. nat. ed. 1791, 2: 503 (1791); *S. baccata* Forsskal ex J.F. Gmelin, loc. cit.; *Schanginia baccata* (Forsskal ex J.F. Gmelin) Moq., Chenop. monogr. enum.: 119 (1840); *Schanginia hortensis* (Forsskal ex J.F. Gmelin) Moq., loc. cit.; *Suaeda maris-mortui* Post, Fl. Syria: 687 (1896); *Schanginia aegyptiaca* (Hasselq.) Aellen in Rech.f., Fl. Lowland Iraq: 195 (1964). Illustr.: Collenette (1985 p.127).

Succulent annual herb or short-lived perennial, up to 1m, often woody-based, glaucous, glabrous; stems erect or decumbent, terete, straw-yellow to white. Leaves dense on the vegetative branches, subterete to linear, 10–25 × 1–2.5mm, acute, fleshy. Flowers bisexual and female (the bisexual larger), sessile or short-pedicelled, in dense axillary clusters forming leafy spikes; bracts deltoid-ovate, c.1mm, with scarious margins; bisexual flowers with perianth segments c.1.75mm, deeply divided and incurved, greenish with scarious margins; perianth tube adnate to the ovary; stamens c.1.5mm; ovary tapering into a short cylindrical beak; stigmas 2–3(–4), c.1.5mm, filiform; female flowers with minute staminodes. Fruit c.1mm diam., thin-walled, immersed in the spongy receptacle; fruiting perianth lobes conspicuously swollen and spongy; seeds 1mm diam., reticulate, black and shining. **Map 330, Fig. 47.**

Sandy coastal and desert soils, borders of salt-marshes, areas of cultivation and waste ground; 0–2300m.

Saudi Arabia, Yemen (N & S), Oman, UAE, Qatar, Bahrain, Kuwait. Cyprus, N Africa and SW Asia eastwards to Iran; NE tropical Africa; naturalized in South Australia.

2. S. maritima (L.) Dumort., Fl. belg.: 22 (1827). Syn.: *Chenopodium maritimum* L., Sp. pl.: 221 (1753); *Suaeda prostrata* Pallas, Ill. pl.: 55, t.47 (1803).

Annual herb, 15–40(–60)cm, glabrous, glaucous, often purplish, drying greyish-green to brownish; stems erect to ascending, sometimes prostrate, terete to slightly angled. Leaves subterete to linear, 10–25 × 0.5–1mm, acute, fleshy, gradually shortening towards the tips of the branches. Flowers bisexual or rarely female, in axillary clusters of 2–5 or solitary, forming leafy spikes; bracts ovate to deltoid-ovate, c.0.5mm, scarious on the margins; perianth segments deltoid, c.0.8mm, free; ovary pyriform, slightly narrowed to a short beak; stigmas c.0.5mm, filiform; anthers c.0.25mm, subglobose. Fruit thin-walled; seeds c.1.5mm, compressed, horizontal, black or brownish-black, reticulate. **Map 331.**

Coastal mud-flats and salt-marshes; sea-level.

Saudi Arabia, Bahrain. Europe (excl. the NE and extreme N), Canary Is., N Africa, Cyprus, Caucasus, SW & C Asia eastwards to Siberia, Japan and Australasia; NE coasts of N America and Argentina.

3. S. monoica Forsskal ex J.F. Gmelin, Syst. nat. ed. 1791, 2: 503 (1791). Type: Yemen (N), *Forsskal* 180 (C).

Small glabrous tree or shrub, up to 4m high; trunk 8(–30)cm in girth at the base; stems much-branched, the branches erect or ascending, frequently with conspicuous insect galls. Leaves sessile or shortly petiolate, flattened on both sides, linear to linear-oblong, 10–40 × 1.5–3.5mm, obtuse to subacute. Flowers unisexual, greenish, solitary or in 2–6-flowered axillary clusters in loose leafy spikes; bracts deltoid-ovate, 0.75–1.25mm, membranous, ciliate. Male flowers c.2.75mm diam., developing first, with 5 stamens and a rudimentary ovary, the perianth segments fused in the basal $\frac{1}{3}$; stamens 1.75–2mm; anthers c.1mm, oblong. Female flowers much smaller, with perianth segments fused to the apex and with minute staminodes; ovary ovoid; stigmas 3–4, 1–1.25mm. Fruit 1.5–2mm, tightly enclosed within the membranous perianth; seeds 1.5–1.75mm diam., vertical, compressed, black, shining, smooth. **Map 332.**

Mainly on the coast at sea-level, less often in desert wadis, crater edges and sandy soils near salt-marshes; 0–1400m.

Saudi Arabia, Yemen (N & S), Socotra, Oman. Cape Verde Is. and Chad; eastern Africa from Mozambique northwards to Sudan; Egypt (incl. Sinai) to the southern part of the Dead Sea.

4. S. vermiculata Forsskal ex J.F. Gmelin, Syst. nat. ed. 1791, 2: 503 (1791). Syn.: *Suaeda fruticosa* Forsskal ex J.F. Gmelin, loc. cit.; *Salsola mollis* Desf., Fl. atlant. 1: 218 (1798); *Suaeda mollis* (Desf.) Del., (1813 p.57); *Suaeda paulayana* Vierh. in Oesterr. Bot. Z. 53: 481 (1903); *S. volkensii* C.B. Clarke in Dyer, Fl. Trop. Afr. 6(1): 92 (1909); *S. monodiana* Maire in Bull. Soc. Hist. Nat. Afrique N. 28: 377 (1937); *S. mesopotamica* Eig in Palestine J. Bot., Jerusalem Ser. 3: 127 (1945).

Shrub, 0.4–1(–1.8)m, glabrous or glabrescent; stems woody at the base, much-branched, the lower usually sprawling, the upper spreading, the young branches minutely puberulent. Leaves very shortly petiolate to sessile, succulent, obovate-oblong to subglobose, terete, or ± oblong to lenticular or curved and flattened on one side only, 2.5–20(–30) × 1–4mm, usually drying black. Flowers bisexual, usually in shortly pedunculate axillary 2–5-flowered clusters, sometimes solitary, in axillary and terminal leafy spikes or lax panicles; bracts deltoid-ovate, 0.75–1mm, scarious, denticulate; perianth segments c.1.25mm, fused at the base, the tips incurved, succulent; ovary free from the perianth except at the extreme base; stigmas 2–4(–5), c.1mm, linear; stamens 1.75–2mm; anthers 0.75–1mm, oblong. Fruit thin-walled; seeds vertical, black, glossy. **Map 333.**

Coastal and inland sandy plains, salt-marshes, desert wadis, volcanic soils and at the edge of cultivation; 0–1680m.

Saudi Arabia, Yemen (N & S), Socotra, Oman, UAE, Qatar, Bahrain, Kuwait. Cape Verde and Canary Is.; Senegal and Mauritania; N Africa and SW Asia eastwards to India and southwards to Kenya.

5. S. moschata A.J. Scott in Kew Bull. 36: 558 (1981). Type: Oman, *Radcliffe-Smith 5385* (K).

Shrub, 0.2–1m tall and up to 1m across; stems woody, especially at the base, much-branched, erect or ascending, young branches often reddish. Leaves succulent, sessile, oblong-ovate, 5–10 × 1–4mm, flattened or semiterete, obtuse or rounded at the tip.

CHENOPODIACEAE

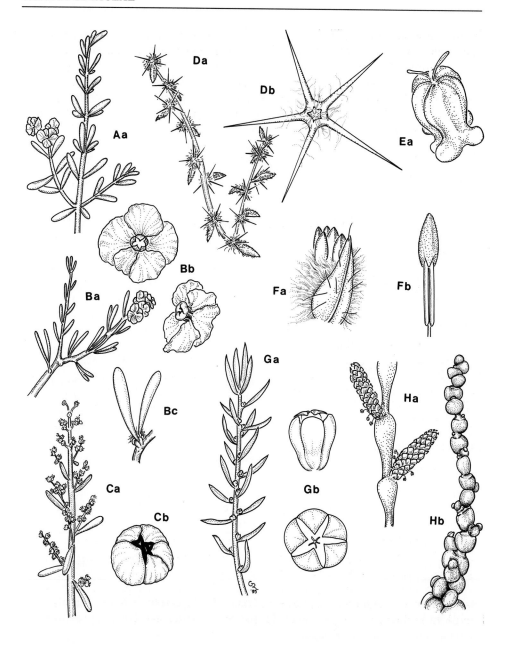

Fig. 47. Chenopodiaceae. A, *Seidlitzia rosmarinus*: Aa, fruiting branch (×1). B, *Lagenantha cycloptera*: Ba, fruiting branch (×1.5); Bb, fruit viewed from above and below (×10); Bc, leaves showing swollen base and tufts of hairs (×3). C, *Bienertia cycloptera*, Ca, flowering branch (×1); Cb, fruit (×10). D, *Bassia muricata*: Da, fruiting branch (×2); Db, fruit (×15). E, *Sevada schimperi*: Ea, fruit (×20). F, *Halocharis sulphurea*: Fa, flower (×10); Fb, stamen (×12). G, *Suaeda aegyptiaca*: Ga, flowering branch (×2); Gb, fruit, side and top views (×10). H, *Halopeplis perfoliata*: Ha, part of flowering branch (×3); Hb, sterile branch (×2).

Flowers bisexual, sessile, solitary or in clusters of 2–3, with a strong pungent disagreeable musky odour when fresh, in compact leafless terminal panicles; bracts 0.5–1mm, scarious; perianth segments c.1.5mm, fused for $\frac{1}{2}$ their length, yellow-green (drying black), obtuse, scarious on the margins; stamens c.1.5mm, with exserted filaments; anthers c.0.5mm, ± globose, yellow (drying black); ovary subglobose; stigmas 2–3, c.0.5mm long, papillose. Fruit with free pericarp; seeds 0.75–1mm, rounded. **Map 334.**

Sandy beaches, coastal dunes and wadi-beds; 0–600 m.

Oman. Endemic.

12. BIENERTIA Bunge ex Boiss.

Annual, halophytic glabrous herbs. Leaves succulent, alternate, deciduous, sessile. Flowers 5-merous, bisexual or female; perianth fleshy; stamens 5; stigmas 2–3. Fruiting perianth surrounded by a spongy discoid wing; fruit with a spongy utricle, adnate to the perianth; seeds horizontal.

1. B. cycloptera Bunge ex Boiss., Fl orient. 4: 945 (1879). Illustr.: Collenette (1985 p.118); Mandaville (1990 pl.35).

Stems much-branched, up to 60cm, erect or ascending. Leaves linear, 5–25 × 2–8mm, subterete, the base adnate to the stem. Flowers shortly pedunculate, in 5–10-flowered racemes, forming leafy panicles. Fruiting perianth 6–8mm diam., orbicular, with a single spongy, disc-shaped, horizontal, almost entire wing; seeds 1.5–2.5mm diam., rounded, crustaceous, granulate. **Map 335, Fig. 47.**

Coastal and inland salt-marshes, and sandy and gypseous soils; 0–100m.

Saudi Arabia, UAE, Kuwait. W Turkey, Iraq, Iran, Afghanistan and C Asia.

13. TRAGANUM Del.

Shrubs; stems cottony at the nodes. Leaves succulent, semiterete, alternate, sessile. Flowers bisexual, solitary or in axillary 2–3-flowered clusters; floral bracts 2; perianth segments 5, scarious; stamens 5, exserted, the filaments broad; anthers sagittate; staminodes absent or obscure; style bifid and with 2 subulate stigmas. Fruiting perianth thickened, indurated, with 2 horn-like teeth; fruit with a subglobose utricle, embedded in the perianth; pericarp free, membranous; seeds horizontal.

1. T. nudatum Del. (1814 p.204). Illustr.: Mandaville (1990 pl.36); Fl. Palaest. 1: t.236 (1966); non sensu Collenette (1985 p.129 = *Agathophora*).

Shrub, 30–80cm; stems much-branched, whitish, glabrous; old branches spinescent, with longitudinally cracked bark. Leaves forming dense clusters on the branches, ± oblong to narrowly triangular, 3–10 × 2–4mm, recurved, acute to rounded, clasping and slightly decurrent at the base, the axils with white cottony hairs. Flowers solitary or 2–3 in axillary clusters; perianth segments oblong-lanceolate, c.3mm. Fruiting perianth c.3.5mm, thickened, indurated and with dorsal horns. **Map 336, Fig. 46.**

Sandy and rocky soils; 0–1100m.

Saudi Arabia, UAE, Qatar, Kuwait. N Africa and SW Asia.

14. CORNULACA Del.

Shrubs or perennial herbs, rarely annuals, glabrous or scabrid with gland-like papillae. Leaves alternate, sessile, triangular, triangular-subulate or needle-like, aristate or spine-tipped, amplexicaul, often decurrent, the axils with white woolly hair-tufts. Flowers bisexual, in 1–8-flowered axillary glomerules, surrounded by dense tufts of white hair and subtended by a leaf-like bract and 2 bracteoles; perianth segments 5, hyaline and free at the base, becoming indurated and fused in fruit; stamens 5 with short filaments elongated after anthesis, monadelphous; ovary ovoid; stigmas 2. Seeds vertical; endosperm absent; embryo spirally coiled.

Aellen, P. (1950). *Verh. Naturf. Ges. Basel* 61: 158–166.

1.	Leaves 2–3.5cm long, developing conspicuous spines	**2. C. setifera**
+	Leaves 0.2–1cm long, spine-tipped or aristate	2
2.	Herbaceous annual or short-lived perennial becoming woody at the base	**1. C. aucheri**
+	Medium-sized, dwarf or sprawling shrubs	3
3.	Shrubs up to 80cm tall; stems usually with long internodes; leaves not decurrent	**3. C. monacantha**
+	Dwarf or sprawling shrubs up to 35cm tall; stems with short internodes; leaves decurrent	4
4.	Dwarf shrub up to 20cm tall; clasping leaf-base c.2mm long, narrowly triangular-subulate, conspicuously white-margined	**4. C. amblyacantha**
+	Sprawling shrub up to 35cm tall and 2m wide; clasping leaf-base 4–6mm long; free part of leaf broadly triangular to triangular-ovate, 2–3mm long, leathery, not white-margined	**5. C. ehrenbergii**

1. C. aucheri Moq., Chenop. monogr. enum.: 163 (1840). Syn.: *Cornulaca leucacantha* Charif & Aellen in Verh. Naturf. Ges. Basel 61: 161 (1950).

Annual or perennial herbs up to 50cm; stems much-branched, rigid, becoming woody at the base in perennial specimens. Leaves acicular to triangular-subulate, 4–10mm long, straight or recurved, decurrent; leaf axils in juvenile specimens with tufts of long white hair extending beyond the leaf, the tufts becoming much shorter in adult plants. Flowers 1–8 in glomerules, surrounded by tufts of short white hair; bracts triangular subulate-aristate, 6–8mm long; bracteoles almost as long as or shorter than the bract; perianth segments spathulate, 2mm long, obtuse; filaments fused into a tube, 1.5mm long when fully developed; anthers 1mm. Fruiting perianth 2–3mm; seeds 1.5mm diam., compressed, rounded, yellow. **Map 337, Fig 48.**

Sandy soils, plains and wadis; 0–300m.

Saudi Arabia, Oman, Qatar, Bahrain, Kuwait. Iraq, Iran, SW Pakistan (Baluchistan) and SW Afghanistan.

2. C. setifera (DC.) Moq. in DC., Prodr. 13(2): 218 (1849). Syn.: *Astragalus setiferus* DC., Prodr. 2: 296 (1825); *Cornulaca tragacanthoides* Moq., Chenop. monogr. enum.: 163 (1840). Illustr.: Collenette (1985 p.120) as *Cornula*ca sp.

Shrub up to 35cm; stems branched mainly from the base, whitish, striate. Leaves subulate, rigid and straight, 2–3.5cm long, aristate and yellowish at the tip, the margins scarious, the base expanded and clasping the stem, glabrescent, pale green becoming whitish, the axils with dense white hairs. Flowers 2–3 in glomerules, surrounded by tufts of short white hairs; bracts subulate-aristate, 8–12mm; bracteoles similar to but slightly shorter than the bracts; perianth segments lanceolate-spathulate, 2–2.5mm, acute; filaments fused into a tube 1.5mm, the free tips 1mm long; anthers 0.75mm; stigmas filiform, 0.5mm. Fruit not seen. **Map 338, Fig 48.**

Drifted shallow dunes over limestone plateaux; 750–900m.

Saudi Arabia. The Syrian Desert and southern Iraq.

3. C. monacantha Del. (1814 p.206). Syn.: *Cornulaca arabica* Botsch. in Kew Bull. 23: 439 (1969). Illustr.: Collenette (1985 p.120); Mandaville (1990 pl.51).

Shrub up to 80cm; stems woody, richly branched, usually elongate with long internodes, sometimes intricate, the old stems with cracking bark. Leaves triangular or triangular-subulate, 3–10mm long, subulate or acicular, with a broadened rounded base. Flowers 2–5 in glomerules, surrounded by dense tufts of white hairs up to 5mm long; glomerules well separated, rarely congested; bracts triangular-subulate or triangular-aristate, up to 8mm, recurved; bracteoles shorter, straight; perianth segments narrowly spathulate, 2.5–3mm; filaments 2–3mm, fused in the basal $\frac{1}{4}-\frac{1}{3}$, with papillose appendages between the bases of the free tips; anthers 1.5mm; ovary ovoid. Fruiting perianth pyriform, 3–4mm; seeds 1.25mm diam., compressed, rounded, yellowish. **Map 339, Fig 48.**

Coastal and inland sandy and gravelly plains, dunes and wadis; 0–750m.

Saudi Arabia, Yemen (S), Oman, UAE, Qatar, Kuwait. W tropical and N Africa; SW Asia eastwards to Pakistan.

A variable species occupying a vast geographical area from the Sahara to Pakistan. It may grow under extremely arid conditions and is usually the last survivor in areas suffering prolonged periods of drought. Some forms are almost spineless, others have short internodes, dense leaves and congested glomerules.

4. C. amblyacantha Bunge in Mem. Acad. Imp. Sci. St. Petersbourg 4: 88 (1862).

Low shrub up to 20cm, often cushion-like; stems much-branched, thick at the base,

CHENOPODIACEAE

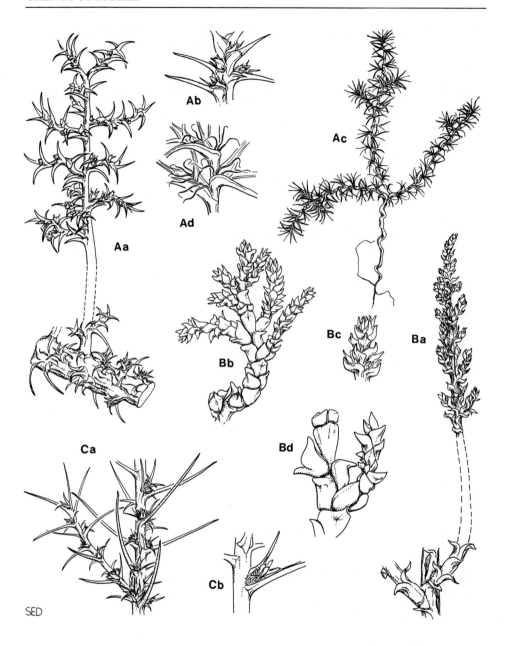

Fig. 48. Chenopodiaceae. A, *Cornulaca aucheri*: Aa, branch of adult plant (× 1); Ab, leaves of adult plant (× 2); Ac, juvenile plant (× 1.5); Ad, leaves of juvenile plant (× 2). B, *C. ehrenbergii*: Ba, leafy branch (× 1); Bb–Bd, leaves (× 2). C, *C. setifera*: Ca, branch (× 1); Cb leaves (× 2).

whitish. Leaves closely packed or overlapping on the branches, triangular-subulate, 5–6.5 × 3–4mm, the base c.2mm long, broadened, strongly clasping, the free part c.4mm long, strongly keeled and with conspicuous white margins, the spiny tip 1–1.5mm, thorny, acuminate. Flowers 3–5 in inconspicuous axillary glomerules; bracteoles shorter and narrower than the leaves; perianth segments narrowly linear, 1.5–2mm, spathulate, hyaline; anthers 0.8mm, ovate-elliptic; style elongated; stigmas 2, thread-like. Fruit not seen. **Map 340, Fig 48.**

Coastal and inland wadis with calcareous boulders; 0–1050m.

Yemen (S), Oman. Iran.

5. C. ehrenbergii Asch. in Schweinf., Beitr. Fl. Aethiop.: 184 (1867).

Sprawling shrub up to 35cm tall and 2m across; stems richly branched, the older with flaking bark. Leaves broadly triangular-ovate, 6–9mm long, leathery, decurrent, the base 4–6mm long, clasping and completely concealing the young stems, the free part 2–3mm long, its acute spiny tip c.1mm. Flowers solitary; perianth segments linear-oblong, 2–2.5mm, membranous, obtuse, pilose at the base; filaments 1.75–2mm; style 1.5mm; stigmas short, subulate. Fruit not seen. **Map 341.**

Coastal dunes; sea-level.

Saudi Arabia, Yemen (S). Egypt, Sudan, Ethiopia and Somalia.

15. AGATHOPHORA (Fenzl) Bunge
Syn.: *Halogeton* C. Meyer

Subshrubs, richly branched from the base; young branches white. Leaves alternate, fleshy, almost globular to cylindrical or clavate, spine-tipped (spine sometimes caducous on the older leaves); leaf axil with a thick tuft of long hair and a few reduced leaves. Flowers sessile in glomerules, 2-bracteate; perianth segments 5, laterally compressed, free, membranous; stamens 5; stigmas 2. Fruiting perianth with horizontal membranous wings at or just above the middle; fruit ovoid, with persistent style; pericarp membranous; seeds vertical; embryo spiral.

Botschantzev, V.P. (1977). *Bot. Zhurn.* (*Moscow & Leningrad*) 62: 1447–1451.

Botschantzev recognizes five species of *Agathophora* which are treated here as one, extending from N Africa to Arabia.

1. A. alopecuroides (Del.) Fenzl ex Bunge in Mém. Acad. Imp. Sci. Saint Pétersbourg, sér. 7, 4(11): 92 (1862). Syn.: *Salsola alopecuroides* Del. (1814 p.200); *Halogeton alopecuroides* (Del.) Moq., Chenop. monogr. enum.: 161 (1840); *Anabasis alopecuroides* (Del.) Moq. in DC., Prodr. 13(2): 210 (1849); *Traganum undatum* [*nudatum*] sensu Collenette (1985 p.129). Illustr.: Collenette (1985 p.129) as *T. undatum*; Mandaville (1990 pl.49).

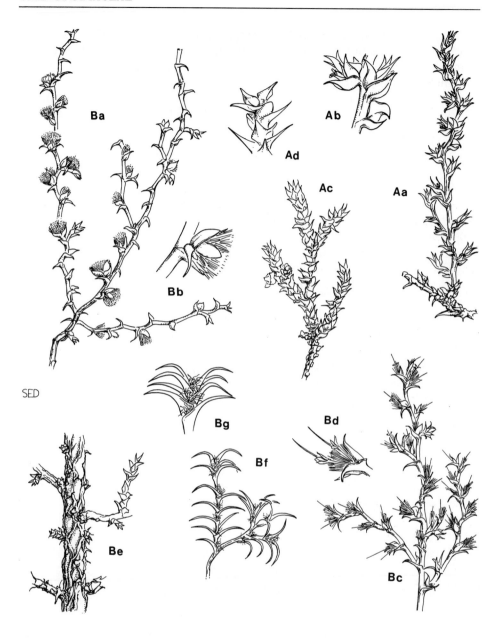

Fig. 49. Chenopodiaceae. A, *Cornulaca amblyacantha*: Aa, leafy branch (×1); Ab, leaves (both from *Sanadiki* s.n.) (×2); Ac, leafy branch (×1); Ad, leaves (both from *Boulos* et al. 16897) (×2). B, *C. monacantha*: Ba, leafy branch (×1); Bb, leaves (both from *Collenette* 6058) (×2); Bc, leafy branch (×1); Bd, leaves (both from *Guest* et al. 16057B) (×2); Be, old stem (×1); Bf, seedling (×1); Bg, leaves from seedling (×2).

Subshrub, 40–80cm. Leaves 13–18mm long, spine-tipped. Flowers in glomerules of 3–10(–15) in leaf axils of the upper branches; bracts ovate, fleshy, spine-tipped, with a tuft of hair in the axil. Fruiting perianth with 3(–5) wings, three of equal size, the other two, if developed, much smaller. **Fig. 46.**

Sandy and gravelly wadis, escarpments and alluvial desert plains; 1000–1700 m.

1.	Branches and leaves glabrous, not papillose-hispid	var. **alopecuroides**
+	Branches and leaves papillose-hispid	var. **papillosa**

var. **alopecuroides**. Syn.: *Salsola postii* Eig in Palestine J. Bot., Jerusalem Ser. 3(3): 131, t.4 (1945); *Aellenia postii* (Eig) Aellen in Mouterde, Fl. Djebel Druze: 87 (1953); *Agathophora postii* (Eig) Botsch. in Bot. Zhurn. (Moscow & Leningrad) 62: 1449 (1977); *A. galalensis* Botsch., loc. cit.: 1450.

Branches and leaves glabrous, not papillose-hispid; upper part of leaves around glomerules terete, twice as long as wide. **Map 342.**

Saudi Arabia. N Africa eastwards to Pakistan.

var. **papillosa** (Maire) Boulos in Kew Bull. 47: 284 (1992). Syn.: *Halogeton alopecuroides* (Del.) Moq. var. *papillosa* Maire in Bull. Soc. Hist. Nat. Afrique N. 34: 190 (1943); *Agathophora iraqensis* Botsch. in Bot. Zhurn. (Moscow & Leningrad) 62: 1451 (1977); *A. algeriensis* Botsch., op. cit.: 1452.

Branches and leaves papillose-hispid, not glabrous; upper part of leaves around glomerules terete, scarcely as long as wide. **Map 343.**

Saudi Arabia, Kuwait. N Africa eastwards to Pakistan.

16. SEIDLITZIA Bunge ex Boiss.

Shrubs or annual herbs, glabrous, the branches opposite. Leaves opposite, cylindrical, succulent, thickening towards the tip. Flowers bisexual, solitary or in few-flowered axillary clusters; floral bracts 2; perianth segments 5, shortly fused at the base; stamens 5, exserted; staminodes 5, semi-orbicular, glandular-ciliate, fused with the bases of the filaments into a disc; style short; stigmas 2. Fruit with a depressed utricle; fruiting perianth spoon-shaped below the wings, the wings subequal; seeds horizontal; embryo spiral.

1. S. rosmarinus Bunge ex Boiss., Fl. orient. 4: 951 (1879). Syn.: *Suaeda rosmarinus* Ehrenb. ex Boiss., loc. cit., pro syn.; *Salsola rosmarinus* (Bunge ex Boiss.) Solms-Laub. in Bot. Zeitung, 2 Abt. 59: 171 (1901). Illustr.: Fl. Palaest. 1: t.243 (1966); Collenette (1985 p.126); Mandaville (1990 pl.37). Type: Arabia, *Ehrenberg* 123 (K).

Shrub, 40–120cm; stems much-branched, the younger white, glossy. Leaves sessile, cylindrical, 4–35 × 1–3mm, thickening towards the rounded or subacute tip and tapering towards the clasping and decurrent base, often curved, with a tuft of white hairs in the axil. Flowers axillary, in 2–5-flowered clusters; perianth segments obtuse. Fruit 1–1.2cm diam. including the wings; seeds horizontal. **Map 344, Fig. 47.**

Coastal and desert sandy and calcareous soils; 0–850m.

Saudi Arabia, Socotra, Oman, Qatar, Bahrain, Kuwait. Egypt and SW Asia eastwards to Afghanistan and Tadzhikistan.

17. HALOXYLON Bunge
Syn.: *Hammada* Iljin

Suffruticose herbs (but not in Arabia), shrubs or small trees; stems cylindrical, jointed. Leaves opposite, connate, scale-like or rudimentary. Flowers solitary, axillary, small, bisexual or male; bracteoles 2; perianth segments 5, membranous, free or connate at the base; fruiting perianth developing horizontal wings near the middle; stamens 5, the filaments united at the base on a hypogynous disc; staminodes 5, connate with the filaments; styles 2–3, short, thick; stigmas 2–5, filiform. Fruit included in the perianth; seeds horizontal; embryo spiral.

1. Small tree up to 3.5m; staminodes thin, glabrous; stigmas 5; wings of fruiting perianth ± equal **1. H. persicum**
+ Shrub up to 60cm; staminodes thick, papillose-glandular; stigmas 2; wings of fruiting perianth unequal **2. H. salicornicum**

1. H. persicum Bunge in Nouv. Mém. Soc. Imp. Naturalistes Moscou 12: 189 (1860). Syn.: *Haloxylon ammodendron* sensu Collenette (1985) non (Cam.) Bunge. Illustr.: Collenette (1985 p.121) as *Haloxylon ammodendron*.

Shrub or small tree up to 3.5m; trunk 10–25cm diam.; stems richly branched, glabrous, the younger herbaceous, green, slender. Leaves scale-like, triangular, 0.5–1.25mm long, acute, connate at the base into a cup, with a tuft of short hairs in the axil. Flowers on short slender spicate lateral branches; stamens exserted; staminodes ovate, membranous, thin, glabrous (not glandular), exceeding the ovary; stigmas 5, sessile. Fruiting perianth with wings c.8mm diam., ± equal, entire. **Map 345.**

Desert wadis and plains with drifted sand; 150–1250 m.

Saudi Arabia, Oman, UAE. Egypt and SW Asia eastwards to Pakistan and C Asia.

2. H. salicornicum (Moq.) Bunge ex Boiss., Fl. orient. 4: 949 (1879). Syn.: *Caroxylon salicornicum* Moq. in DC., Prodr. 13(2): 174 (1849); *Hammada salicornica* (Moq.) Iljin in Bot. Zhurn. (Moscow & Leningrad) 33: 583 (1948); *H. elegans* (Bunge) Botsch. in Novosti Sist. Vyssh. Rast. 1: 362 (1964). Illustr.: Fl. Qatar pl.29 (1981); Collenette (1985 p.121) as *Hammada salicornica*; Mandaville (1990 pls.45–6).

Shrub up to 60cm; stems richly branched from the base, the older thicker with yellowish cracking bark. Leaves scale-like, shortly triangular, connate at the base into a cup, membranous on the margins, densely woolly in the axils. Flowers in dense, slender spikes; bracteoles ovate; staminodes linear-ovate, papillose at the tip; stigmas 2, club-shaped, papillose on the inner side. Fruiting perianth with wings 6–8mm diam., ovate-orbicular, unequal, overlapping, greenish, pink or pale brown; seeds 1.25mm diam. **Map 346, Fig. 46.**

Gravelly and sandy deserts, wadis; 0–1250 m.

Saudi Arabia, Oman, UAE, Qatar, Bahrain, Kuwait. Egypt and SW Asia.

Often one of the most abundant species in N and central Arabia, being dominant over vast sandy plains. It is a very variable species with several distinct forms.

18. SALSOLA L.

Shrubs, subshrubs or annuals or perennial herbs. Leaves small, often succulent, alternate or rarely opposite, sessile, entire, the base clasping or not. Flowers bisexual or less often unisexual, solitary in the leaf axils or in glomerules forming loose or dense spikes, each subtended by 2 bracts; perianth segments 5, rarely 4, scarious, usually developing a prominent transverse wing in fruit; stamens 5, rarely 4, the anthers sometimes with appendages; staminodes rarely present; stigmas 2(–3); style present or absent. Fruit with a disc-shaped to subglobose utricle included within the persistent fruiting perianth, this obconical with a narrow base, usually not indurated, rarely with a narrow indurated rim around the pedicel; seeds horizontal, subglobose; testa membranous; embryo spiral.

Botschantzev, V. P. (1969). *Bot. Zhurn. (Moscow & Leningrad)* 54: 989–1000; Chaudhary, S. & Akram, M. (1986). *Proc. Saudi Biol. Soc.* 9: 57–89; Freitag, H. (1989). *Flora* 183: 158–170.

1.	Annuals, becoming ± woody in fruit	2
+	Shrubs or sub-shrubs, the branches woody at least at the base	5
2.	Plants glabrescent or scabridulous; leaves mucronate-spiny	**1. S. kali**
+	Plants villous or papillose-mealy; leaves not mucronate-spiny	3
3.	Plants smelling of rotten fish when bruised; young and adult stems and leaves villous; hairs rough, straight and perpendicular to the stem; fruiting perianth forming a pubescent cone above the wings	**2. S. volkensii**
+	Plants with no particular smell when bruised; mature parts mealy or papillose-mealy, the young stems and leaves pilose to villous; fruiting perianth adpressed above the wings, scurfy	4
4.	Floral bracts 3–3.5mm, usually persistent; fruiting perianth (including wings) 8–11mm diam.; leaves 1–2cm long	**3. S. jordanicola**
+	Floral bracts 1.2–2mm, usually deciduous in fruit; fruiting perianth (including wings) 4–8mm diam.; leaves 0.2–1cm long	**4. S. inermis**

5.	Leaves entirely glabrous, terete, clavate or obpyriform	6
+	Leaves hairy, not terete, clavate or obpyriform	7
6.	Leaves clavate, obpyriform or subglobular, 2–5(–8)mm long	**5. S. drummondii**
+	Leaves linear, 5–25mm long, often curved	**6. S. schweinfurthii**
7.	Leaves in 2 opposite ranks, overlapping and concealing the young branches	8
+	Leaves not as above	9
8.	Leaves 4–9 × 1–2mm, silvery pubescent, but glabrescent with age	**7. S. rubescens**
+	Leaves 1.5 × 1.5mm, densely covered with adpressed scurfy scale-like hairs	**8. S. omanensis**
9.	Fruiting bracts strongly hooded, as high as or exceeding the wings; fruiting perianth obscure, even at maturity	**9. S. arabica**
+	Fruiting bracts not or faintly hooded, seldom reaching the wings; fruiting perianth conspicuous at maturity	10
10.	Branching divaricate, the older lateral branches spinescent; leaves densely crowded into small cone-like knots	11
+	Branching usually not divaricate, the branches not spinescent; leaves not crowded into cone-like knots (sometimes loosely so in *S. imbricata*)	12
11.	Indumentum on young leaves and shoots of simple sessile hairs	**10. S. cyclophylla**
+	Indumentum of stalked, medifixed hairs	**11. S. spinescens**
12.	Leaves opposite; flowers unisexual (many) with 4 stamens and 4 staminodes, and bisexual (few) with 5 stamens	**12. S. tetrandra**
+	Leaves alternate; all flowers bisexual with 5 stamens	13
13.	Sprawling shrub smelling of rotten fish when bruised; leaves of mature plants ovate-orbicular, 1–2.5mm long, succulent; flowers in short dense spikes	**13. S. imbricata**
+	Erect sub-shrubs with no particular smell; leaves linear-subulate, up to 2cm long, not succulent; flowers in axillary white-cottony glomerules or in loosely congested spikes	14
14.	Leaves up to 2cm long, shortly tomentose; flowers in paniculate spikes, buried in white-cottony axillary glomerules; fruiting perianth not developing wings	**14. S. lachnantha**
+	Leaves up to 1.2cm long, with yellowish denticulate hairs; flowers in loosely congested spikes, not in white-cottony glomerules; fruiting perianth developing conspicuous wings	**15. S. villosa**

1. S. kali L., Sp. pl.: 222 (1753). Illustr.: Fl. Palaest. 1: t.246 (1966); Fl. Libya 58: t.28 (1978). Illustr.: Collenette (1985 p.124).

Succulent annual herb, 25–80cm, glabrous to scabridulous or hispid-puberulent; stems ascending, angular. Leaves opposite below, subopposite or alternate above, linear-subulate, 5–25 × 2–3mm, fleshy, semiterete, mucronate-spiny, slightly clasping at the base. Flowers 1–3 in the leaf axils and forming loose leafy spikes; bracts slightly exceeding the perianth, spine-tipped; perianth segments 3–4mm, membranous; stamens (4–)5; stigmas 2–3. Fruiting perianth 0.5–1cm diam. (including wings); seeds turbinate-subglobose. **Map 347.**

Waste ground; 1400–2000m.

Saudi Arabia, Yemen (N). Widespread, especially on seashores in the northern hemisphere.

In Yemen (N) a common field weed on the high plateau near Sana'a.

2. S. volkensii Asch. & Schweinf., Ill. fl. Égypte: 130 (1887). Illustr.: Fl. Palaest. 1: t.250 (1966). Illustr.: Collenette (1985 p.125).

Succulent annual herb, 15–60cm, smelling of rotten fish when bruised, bluish-green, becoming brownish-yellow on drying, villous with rough, straight hairs perpendicular to the stem; stems erect or ascending, much-branched. Leaves alternate, linear-subulate, 3–8 × 1.5–2.5mm, succulent, semiterete. Flowers solitary in the leaf axils, forming loose or dense spikes; bracts suborbicular, 2 × 1.5mm, succulent; perianth segments connivent, oblong-lanceolate to ovate, 3 × 1.5mm, acute, villous. Fruiting perianth 6–9mm diam. (including wings), forming a pubescent cone above the wings, these imbricate. **Map 348.**

Waste ground and sandy desert soils; 500–700m.

Saudi Arabia. Egypt, Palestine, Jordan and Iraq.

3. S. jordanicola Eig in Palestine J. Bot., Jerusalem Ser. 3: 130 (1945). Illustr.: Fl. Palaest. 1: t. 248 (1966).

Annual, 10–40cm, papillose-mealy, the young parts villous, glabrescent with age; stems ascending to erect, much-branched. Leaves alternate, linear, 10–20 × 1.5–2mm, papillose or villous, soon deciduous. Flowers solitary in the leaf axils, forming spikes; bracts orbicular, 3–3.5mm, with white margins, forming cavities 3–4mm in diameter after the fruit has been shed; perianth segments ovate, 2–3 × 1–2mm. Fruiting perianth 8–11mm diam. (including wings), adpressed and scurfy-mealy or scurfy-pubescent above the wings. **Map 349, Fig 50.**

Sandy coastal soils; 0–50m.

Saudi Arabia, Kuwait. Palestine, Jordan, Syria, Iraq and Iran.

4. S. inermis Forsskal (1775 p.57), non sensu Collenette (1985 = *S. schweinfurthii*). Illustr.: Fl. Palaest. 1: t.249 (1966).

Sprawling annual, 10–30cm, sometimes with a tough woody base, the mature parts densely papillose-mealy, the younger parts pilose to villous; stems slender and wiry,

CHENOPODIACEAE

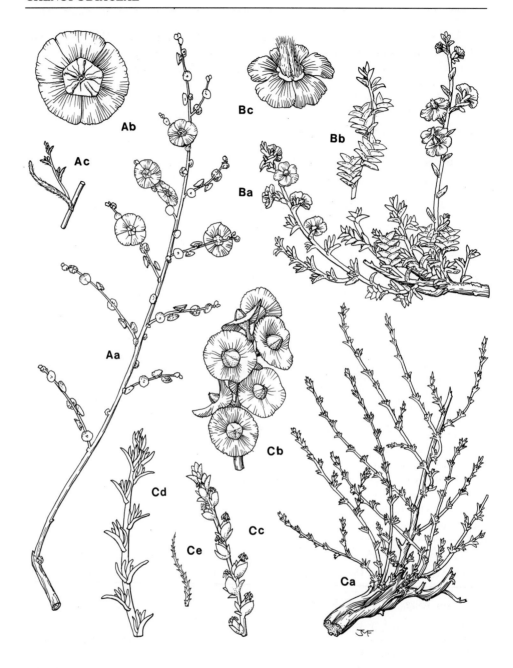

Fig. 50. Chenopodiaceae. A, *Salsola jordanicola*: Aa, fruiting branch (×1); Ab, fruit (×3); Ac, basal leaves (×1). B, *S. rubescens*: Ba, habit (×1); Bb, vegetative branch (×1); Bc, fruit (×3). C, *S. villosa*: Ca, habit (×1); Cb, fruits (×2); Cc, flowering branch (×2); Cd, sterile branch (×2); Ce, detail of hair (×20).

much-branched from the base. Leaves alternate, linear-subulate, 2–10 × 1mm, clasping at the base. Flowers solitary, in loose or dense spikes; bracts suborbicular, 1.5–2mm, caducous in fruit; perianth segments mealy. Fruiting perianth 4–8mm diam. (including wings), the wings connivent.

Sandy desert soils.

Saudi Arabia. Egypt and SW Asia.

5. S. drummondii Ulbr. in Engl. & Prantl, Nat. Pflanzenfam. 2 Aufl. 16C: 565 (1934). Syn.: *Salsola obpyrifolia* Botsch. & Akhani in Bot. Zhurn. (Moscow & Leningrad) 74: 1665 (1989). Illustr.: Mandaville (1990 pl.42).

Shrub, 30–75cm, glabrous, the younger branches greyish-white, the older brownish and knotty. Leaves mostly alternate, sometimes subopposite, clavate, obpyriform or subglobular, 2–5(–8) × 2–3mm, succulent, terete, sessile, rounded at the tip. Flowers in leafy terminal panicles; bracts leaf-like, succulent; perianth segments not connivent above the wings, exposing the ovary. Fruiting perianth 5–7mm diam. (including wings), the wings subequal, straw-yellow to pinkish. **Map 350, Fig 50.**

Sandy wadis and saline depressions near the coast; 0–150m.

Saudi Arabia, Oman, UAE, Qatar. S Iran, Pakistan and India (Punjab).

6. S. schweinfurthii Solms-Laub. in Bot. Zeit. 59: 173 (1901). Syn.: *Darniella schweinfurthii* (Solms-Laub.) Brullo in Webbia 38: 313 (1984); *Salsola inermis* sensu Collenette (1985) non Forsskal. Illustr.: Fl. Palaest. 1: t.254 (1966); Collenette (1985 p.124).

Shrub, 20–60cm, glabrous with ascending branches, the younger branches smooth and white, the older brownish with longitudinally-fissured bark. Leaves opposite, subopposite or alternate, linear-terete, 5–25 × 2–3mm, sessile, straight or curved, apiculate at the tip with a caducous bristle, slightly decurrent at the base, with tuft of short woolly hairs in the axils. Flowers 2–5, in the upper leaf axils, forming loose or dense spikes; bracts suborbicular, succulent; perianth segments with white margins; staminodes present but not conspicuous. Fruiting perianth 3.5–7.5mm diam. (including wings), the wings subequal. **Map 351.**

Limestone rocks, sandstone and salty sand overlying fossil coral; 0–1350m.

Saudi Arabia, Oman. Egypt, Palestine and Jordan.

7. S. rubescens Franchet, Sert. somal.: 60 (1882). Syn.: *Salsola hadramautica* Baker in Bull. Misc. Inform. 1894: 340 (1894); *S. leucophylla* Baker, loc. cit.

Shrub, 25–50cm, densely adpressed silvery-pubescent, the older parts glabrescent. Leaves in 2 opposite ranks, overlapping and concealing the young branches, separated by internodes on the older branches, ± cylindrical, 4–9 × 1–2mm, sessile, rounded at the tip, decurrent at the base. Flowers solitary in the leaf axils, forming leafy spikes; bracts leaf-like; perianth segments 2–3.5 × 2mm, unequal, pilose at the tip, reddish; stamens 5. Fruiting perianth 4–5mm diam. (including wings), densely pilose in the upper part. **Map 352, Fig 50.**

Limestone rocks and sandy soils; 0–1350m.

Yemen (S), Oman, UAE. Somalia.

8. **S. omanensis** Boulos in Kew Bull. 46: 297, t.1 (1991). Type: Oman, *Miller* 6409 (E, K, KTUH, ON).

Shrub up to 50cm, divaricately branched; older branches woody with grey-brown bark; younger lateral branches short, ± cylindrical, densely covered with leaves. Leaves adnate to and concealing the stem, subopposite, ovoid-triangular, 1.5 × 1.5mm, grey-green, sessile, rigid, thick at the base, densely covered with adpressed scurfy scale-like hairs on the outer concave side and margins. Flowers few, at the tips of young branches, sessile, ovoid, bisexual; floral-bracts 2, similar to the leaves but thinner at the base; perianth segments 5, 3 × 1–1.5mm, free, boat-like, acute at the tip, scarious on the margins; stamens 5; anthers sagittate. Fruiting perianth connivent, developing a dorsal thinly membranous scarlet wing, 5.5–6mm diam.; seeds discoid, 1.5mm diam. **Map 353, Fig 51.**

Sandy coastal plains and limestone cliffs; 50–300m.

Oman. Endemic.

9. **S. arabica** Botsch. in Bot. Zhurn. (Moscow & Leningrad) 60 (4): 499 (1975). Illustr.: Mandaville (1990 pls.38–9). Type: Saudi Arabia, *Mandaville* 2954 (BM, LE).

Shrub, 20–60cm; stems much-branched, canescent. Leaves lanceolate to broadly triangular, 2–5 × 1.5–2mm, ± triquetrous, pubescent, glabrescent with age. Flowers in dense lateral spikes; bracts hooded, pubescent. Fruiting bracts becoming swollen and strongly hooded, reaching or exceeding the wings. Fruiting perianth 3–3.5mm diam. (including wings), the wings often obscure and not well developed even at maturity. **Map 354.**

Rocky and sandy wadis; 350–600m.

Saudi Arabia. Endemic.

10. **S. cyclophylla** Baker in Bull. Misc. Inform. 1894: 340 (1894). Illustr.: Collenette (1985 p.126) as *Salsola* sp. no. 2480; Mandaville (1990 pl.40). Type: Yemen (S), *Lunt* 53 (BM).

Shrub, 20–60cm, conspicuously woody with cracked whitish-grey bark at the base, the older lateral branches spinescent. Leaves crowded into knots on the main and lateral branches, suborbicular to broadly triangular, 1.5–2 × 1.5mm, densely adpressed-tomentose with simple smooth silvery hairs. Flowers in densely contracted lateral spikes; bracts 1–1.5 × 1.2–1.5mm; perianth segments connivent, 2mm, densely hairy above and below the wings. Fruiting perianth 3–5mm diam. (including wings), the segments densely pubescent and projecting above the wings. **Map 355, Fig 52.**

Sand-dunes, rocky slopes and outcrops; limestone and sandstone plateaux; 50–900m.

18. Salsola

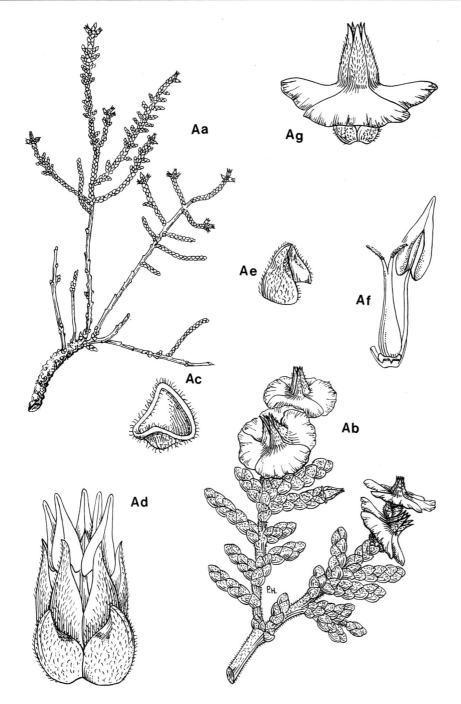

Fig. 51. Chenopodiaceae. A, *Salsola omanensis*: Aa, flowering branch (× 1); Ab, fruiting branch (× 3); Ac, leaf (× 15); Ad, flower (× 15); Ae, floral bract (× 10); Af, ovary and 1 stamen (× 15); Ag, fruit (× 6).

CHENOPODIACEAE

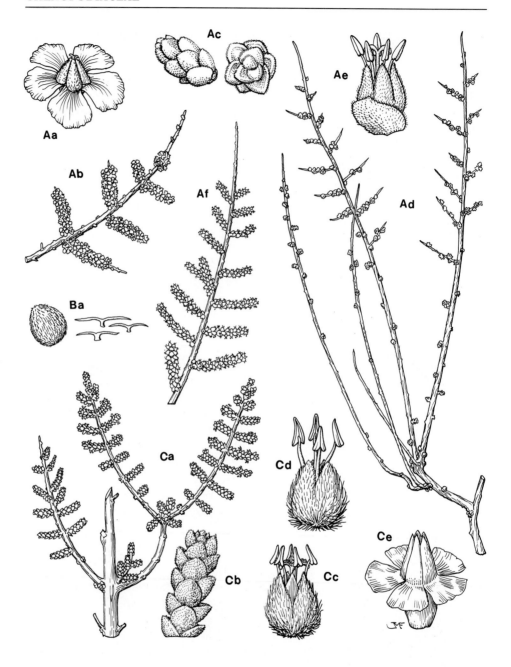

Fig. 52. Chenopodiaceae. A, *Salsola cyclophylla*: Aa, fruit (×6); Ab, fruiting branch (×1); Ac, leaves (×8); Ad, sterile branch (×1); Ae, flower (×8); Af, flowering branch (×1). B, *S. spinescens*: Ba, bract (×20) and magnified hairs. C, *S. tetrandra*: Ca, flowering branch (×1); Cb, leaves (×6); Cc, bisexual flower (×8); Cd, male flower (×8); Ce, fruit (×12).

Saudi Arabia, Yemen (S), Oman, UAE, Qatar, Bahrain, Kuwait. Egypt to Iraq; Sudan and Ethiopia.

11. S. spinescens Moq. in DC., Prodr. 13(2): 179 (1849). Syn.: *Salsola forskalii* Schweinf. (1896 p.160); *S. congesta* N.E. Br. in Bull. Misc. Inform. 1909: 50 (1909); *S. aethiopica* Botsch. in Bot. Zhurn. (Moscow & Leningrad) 60: 498 (1975); *S. imbricata* sensu Collenette (1985). Illustr.: Collenette (1985 p.124). Type: Saudi Arabia, *Botta* (P).

Rigid shrub, 25–150cm, the older stems and branches strongly spinescent; young shoots bud-like, minute, formed of overlapping leaves. Leaves ovate, 1.5–2 × 1.5mm, rounded at the tip, scarious on the margins, densely covered with adpressed medifixed hairs when young. Flowers solitary, 1.5–2.5mm diam.; bracts leaf-like; perianth segments 1.5–2.5mm. Fruiting perianth 3.5–5mm diam. (including wings), the wings pinkish white. **Map 356, Fig 52.**

Sandy and rocky ground; 0–2100m.

Saudi Arabia, Yemen (N & S), Socotra, Oman. Sudan, Djibouti and Ethiopia.

12. S. tetrandra Forsskal (1775 p.58). Syn.: *Halogeton tetrandrus* (Forsskal) Moq., Chenop. monogr. enum.: 160 (1840); *Salsola baryosma* sensu Collenette (1985); non Collenette (1985 p.124 = *Agathophora alopecuroides*). Illustr.: Collenette (1985 p.123) as *Salsola baryosma*.

Shrub, 20–50cm, greyish-white, villous-tomentose; older stems woody, much-branched, the young shoots tetragonous-cylindrical. Leaves opposite, imbricate, broadly ovate-triangular, c.2 × 2mm, succulent, villous, with scarious margins and hairy axils. Flowers unisexual and bisexual, mixed in short spikes; unisexual flowers many, with 4 perianth segments and 4 stamens alternating with 4 obscure staminodes; bisexual flowers few, with 5 perianth segments and 5 stamens. Fruiting perianth 3.5–5mm diam. (including wings), connivent and forming a short cone above the wings, the wings much-reduced or almost absent and white to pinkish. **Map 357, Fig 52.**

Saline and sandy soils; volcanic and limestone outcrops; 600–900m.

Saudi Arabia. Canary Is., N Africa, Palestine and Jordan.

13. S. imbricata Forsskal (1775 pp. XCVII, CVIII, 57). Syn.: *Chenopodium baryosmon* Roemer & Schultes, Syst. veg. 6: 269 (1820); *Salsola foetida* Del. ex Sprengel, Syst. veg. 1: 925 (1824) non Vest (1820); *Caroxylon imbricatum* (Forsskal) Moq. in DC., Prodr. 13(2): 177 (1849); *Salsola baryosma* (Roemer & Schultes) Dandy in Andrews, Fl. Pl. Sudan 1: 111 (1950); non Collenette (1985 p.123 = *Salsola spinescens*). Illustr.: Fl. Palaest. 1: t.256 (1966); Collenette (1985 p.122) as *S.* sp. aff. *alopecuroides*. Neotype: Yemen (N), *J.R.I. Wood* 1184 (K, E).

Multiform sprawling shrub, 30–80cm tall, sometimes up to 2m across, greyish green, smelling of rotten fish when bruised, irregularly much-branched, the younger plants with reddish stems and linear hairy leaves up to 1.5cm long. Mature leaves alternate, imbricate, ovate-orbicular, 1–2.5 × c.1.5mm, succulent, with narrow scarious margins. Flowers solitary, in short dense spikes, often forming panicles; bracts imbricate, 1.5–2.5 × c.2mm; perianth segments connivent, c.1.5 × 1.25mm, obtuse. Fruiting perianth 3.5–6mm diam. (including wings). **Map 358.**

Sandy and saline soils bordering salt-marshes, rocky wadis and at the edge of cultivated fields; 0–2100m.

Saudi Arabia, Yemen (N & S), Oman, UAE, Qatar, Bahrain, Kuwait. N Africa, tropical NE Africa, Kenya, eastwards to India, and northwards to Syria. For a discussion of the identity of *S. imbricata* see Boulos in Kew Bull. 46(1): 137–140.

14. **S. lachnantha** (Botsch.) Botsch. in Novosti Sist. Vyssh. Rast. 17: 124 (1980). Syn.: *Salsola tomentosa* (Moq.) Spach subsp. *lachnantha* Botsch. in Bot. Zhurn. (Moscow & Leningrad) 53: 1448 (1968).

Subshrub, 20–50cm; stems erect and spreading, the older woody with light brown fissured bark; younger stems herbaceous, straw-yellow to whitish and shortly tomentose. Leaves alternate, linear, 0.3–2cm, subterete, obtuse, slightly clasping at the base, shortly tomentose. Flowers 2–4 in axillary white-cottony glomerules, forming much-branched, often dense panicles; bracts green, succulent, buried in the dense white tomentum; perianth segments membranous, hyaline, hairy. Fruiting perianth not developing wings. **Map 359.**

Sandy and gravelly hillsides and wadis; 300–600m.

Saudi Arabia. Iraq and Iran.

15. **S. villosa** Schultes, Syst. veg. 6: 232 (1820). Syn.: *Salsola vermiculata* sensu Mandaville (1990); *S. vermiculata* L. var. *villosa* (Schultes) Moq., Chenop. monogr. enum.: 141 (1840); *S. vermiculata* L. subsp. *villosa* (Schultes) Eig in Palestine J. Bot., Jerusalem Ser. 3: 132 (1945); *S. delileana* Botsch. in Novosti Sist. Vyssh. Rast. 1: 371 (1964); *S. mandavillei* Botsch. in Bot. Zhurn. (Moscow & Leningrad) 60: 502 (1975); *S. chaudharyi* Botsch., op. cit. 65: 687 (1984). Illustr.: Mandaville (1990 pl.41) as *S. vermiculata*.

Multiform shrub, 25–75cm, woody and much-branched at the base; young branches herbaceous, yellowish-villous, becoming indurated and glabrous with age. Lower leaves alternate, narrowly triangular to linear-subulate, 0.4–1.2cm long, slightly clasping at the base, villous with denticulate hairs; upper leaves closely imbricate, triangular, 1–4mm long, adpressed pubescent. Flowers solitary, in loose or congested spikes; bracts suborbicular, with scarious margins, keeled; perianth segments connivent, pubescent, with scarious margins. Fruiting perianth with straw-yellow wings, these often with purplish to brownish veins. **Map 360, Fig 50.**

Open sandy deserts, gravelly wadis and soil overlying limestone and volcanic rocks; 450–1800m.

Saudi Arabia. SW Sahara eastwards to SW Asia and western India.

19. HALOTHAMNUS Jaub. & Spach
Syn: *Aellenia* Ulbr.; *Salsola* L. sect. *Sephragidanthus* Iljin

Shrubs, subshrubs (in Arabia) or annual herbs; stems flattened (but not in our species), angular or terete, young branches alternate. Leaves adpressed or reflexed. Flowers solitary (but not in Arabia) or well-spaced in panicles; bracts and bracteoles leaf- or scale-like, not indurated; perianth segments 5, ovate or lanceolate, scarious on the margins; stamens 5 with linear filaments and anthers divided almost to the middle; style thickened below, distinct from the ovary; stigmas flattened, ± lobed. Fruiting perianth developing horizontal wings above, the lower part indurated with truncate-pentagonal base and thickened margins, star-like and 5-rayed with 5 grooves between the rays when viewed from beneath; seeds horizontal; embryo spiral.

Botschantzev, V.P. (1981). *Novosti Sist. Vyssh. Rast.* 18: 146–176. Kothe-Heinrich, G. (1993). *Bibliotheca Botanica* 143: 1–176.

1. Leaves lanceolate-subulate, up to 4.5cm long, decurrent, axillary hairs absent; branches angular **3. H. lancifolius**
+ Leaves triangular or oblong-linear, up to 2cm long, not decurrent, axils with tufts of hair; branches ± terete **2**

2. Branches stiff, spinescent; leaves triangular, 1–2(–5)mm long; flowering perianth 2–3.5mm long; wings of fruiting perianth 5–6(–8)mm diam. **1. H. bottae**
+ Branches brittle, not spinescent; leaves on young branches oblong-linear, 0.5–2cm long; flowering perianth 4–5.5mm long; wings of fruiting perianth 10–14mm diam. **2. H. iraqensis**

1. H. bottae Jaub. & Spach, Ill. pl. orient. 2: 50, t.136 (1845). Syn.: *Caroxylon bottae* (Jaub. & Spach) Moq. in DC., Prodr. 13(2): 178 (1849); *Salsola bottae* (Jaub. & Spach) Boiss., Fl. orient. 4: 964 (1879). Illustr.: Collenette (1985 p.123) as *Salsola bottae*.

Shrub up to 50cm; stems richly branched, the older with longitudinally cracking and peeling cortex; branches rigid, spinescent. Leaves triangular, 1–2(–5) × 1–1.2mm, acute, with axillary tufts of hair. Bracts triangular, 1.2–1.5 × 1–1.4mm; bracteoles broadly ovate, 1.5–2 × 1.5–2.2mm, mucronate. Perianth segments triangular-ovate, 2–3.5mm, with scarious margins; filaments 0.4mm, broadened at the base; anthers 2mm; stigmas rounded and toothed at the tip. Fruiting perianth developing wings below the middle, these 5–6(–8)mm diam. **Map 361, Fig 53.**

Sandy desert plains and wadis, roadsides and waste ground; 20–2600 m.

Saudi Arabia, Yemen (N & S), Oman, UAE. Endemic.

Some specimens, particularly from the area around Aden, dry black; these are treated as a distinct subspecies: subsp. *niger* Kothe-Heinrich in Flora & Vegetatio Mundi 9: 46 (1991).

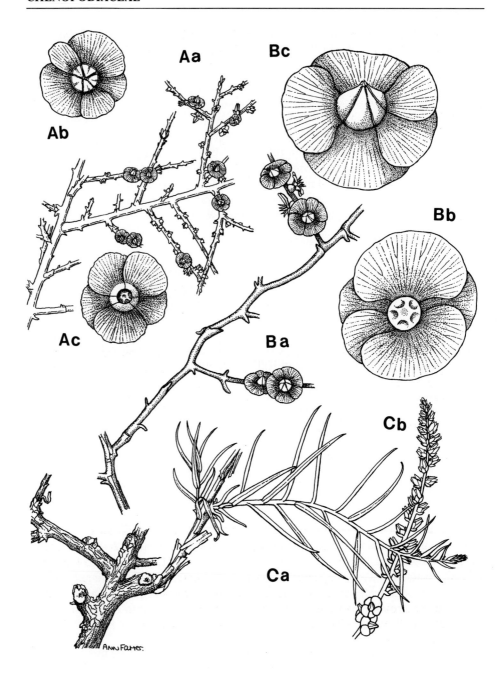

Fig. 53. Chenopodiaceae. A, *Halothamnus bottae*: Aa, fruiting branch (×1); Ab, upper view of fruit (×3); Ac, lower view of fruit (×3). B, *H. iraqensis*: Ba, fruiting branch (×1); Bb, lower view of fruit (×3); Bc, upper view of fruit (×3). C, *H. lancifolius*: Ca, habit (×1); Cb, flowering and fruiting branch (×1).

2. H. iraqensis Botsch. in Novosti Sist. Vyssh. Rast. 18: 152 (1981). Syn.: *Halothamnus iraqensis* var. *hispidulus* Botsch., op. cit.: 153; *Aellenia subaphylla* sensu Collenette (1985) non (C.A.Mey.) Aellen. Illustr.: Collenette (1985 p.115) as *Aellenia subaphylla*; Mandaville (1990 pls.43–4).

Subshrub up to 60cm; stems richly branched, the older branches woody; younger herbaceous, brittle, not spinescent. Leaves on young branches oblong-linear, $5–20 \times 1–2$ mm, with axillary tufts of short hair. Bracts broadly linear to broadly ovate, $1–1.5 \times 2–3.5$mm, acute; bracteoles 2–2.5mm, as long as wide, acute. Perianth segments 4–5.5mm; filaments 0.75mm, broadened at the base; anthers 3.5mm; stigmas rounded and toothed at the tip. Fruiting perianth with wings 10–14mm diam. **Map 362, Fig 53.**

Sandy plains, wadis and sandstone hills; 0–900m.

Saudi Arabia, Kuwait. Syria, Iraq and southern Iran.

3. H. lancifolius (Boiss.) Kothe-Heinrich op. cit.: 88 (1993). Syn.: *Caroxylon lancifolium* Boiss., Diagn. pl. orient. sér. 1, 2(12): 98 (1853); *Salsola lancifolia* (Boiss.) Boiss., Fl. orient. 4: 958 (1879); *Aellenia lancifolia* (Boiss.) Ulbr. in Engl. & Prantl, Nat. Pflanzenfam., ed. 2, 16C: 567 (1934).

Perennial herb up to 60cm; stems woody at the base, erect, the branches spreading or ascending, angular. Leaves lanceolate-subulate, $10–45 \times 2–4$mm, succulent, decurrent and slightly clasping at the base, without axillary hairs. Bracts linear-lanceolate, $5–20 \times 1.5–2.5$mm, persistent after anthesis; bracteoles broadly triangular, c.1mm, as long as broad, acute. Perianth segments 4–6mm. Fruiting perianth with wings up to 14mm diam. **Map 363, Fig 53.**

Sandy desert plains.

Saudi Arabia. Egypt, Palestine, Jordan, Syria and Iraq.

20. SEVADA Moq.

Low shrubs. Leaves alternate or opposite, often in axillary clusters, linear, succulent, terete, with a shortly indurated base, glaucous. Flowers bisexual and female, in dense axillary and terminal spikes or panicles; bracteoles small, scarious. Perianth segments 5, fused into a short tube, the lobes incurved; stamens 5, with exserted anthers, alternating with small staminodal scales; ovary free; style tapered; stigmas 2–3, short, recurved. Fruit ovoid with acute beak; seeds vertical.

Botschantzev, V.P. (1975). *Kew Bull.* 30: 367–370; Boulos, L., Friis, I. & Gilbert, M.G. (1991). *Nord. J. Bot.* 11: 313–315.

1. S. schimperi Moq. in DC., Prodr. 13(2): 154 (1849). Syn.: *Suaeda schimperi* (Moq.) Martelli in Nuovo Giorn. Bot. Ital. 20: 367 (1888); *S. vermiculata* Forsskal ex J.F. Gmelin var. *puberula* C.B. Clarke in Dyer, Fl. Trop. Afr. 6(1): 92 (1909); *Salsola longifolia* sensu Blatter (1923 p.412), pro parte non Forsskal. Illustr.: Collenette (1985 p.128) as *Suaeda schimperi*. Type: Saudi Arabia, *Schimper* 867 (G, E, LE, BM).

Shrub up to 40cm, glabrous or minutely papillose; stems woody, densely branching, the branches spreading to ascending, the older stems with fissured bark. Leaves opposite or alternate, 2–15 × 1–2mm, indurated, obtuse, constricted at the base, yellowish brown, with an axillary tuft of short white hairs. Flowers in clusters of 3–10, forming short axillary spikes or panicles; perianth segments subglobose, c.1mm; stamens c.1.5mm, anthers c.0.5mm, yellow. Fruiting perianth c.2mm, the tube longitudinally ribbed, slightly constricted towards the expanded base; fruit ovoid, with an acute beak, but none seen in maturity. **Map 364, Fig. 47.**

Sandy coastal plains and adjacent hills; 0–700 m.

Saudi Arabia, Yemen (S), Socotra, Oman. Egypt and E tropical Africa.

Some specimens from Oman, possessing puberulent leaves, resemble material from the Red Sea coastal region of Sudan. These may be found, in due course, to represent a distinct variety.

21. LAGENANTHA Chiov.
Syn.: *Choriptera* Botsch.; *Gyroptera* Botsch.

Closely related to *Sevada* Moq., but the mature perianth develops a prominent transverse wing around the top of the tube; the wing entire or divided into lobes with radial veins.

Botschantzev, V.P. (1975). *Kew Bull.* 30: 367–370; Boulos, L., Friis, I & Gilbert, M. (1991). *Nord. J. Bot.* 11: 313–315.

1. L. cycloptera (Stapf) M.G. Gilbert & Friis in Nord. J. Bot. 11: 315 (1991). Syn.: *Salsola cycloptera* Stapf in Forbes, Nat. Hist. Socotra: 526 (mid 1903); *S. semhahensis* Vierh. Osterr. bot. Zeitschr.: 434 (Nov. 1903); *Choriptera semhahensis* (Vierh.) Botsch. in Bot. Zhurn. (Moscow & Leningrad) 52: 806, t.3 (1967); *Gyroptera cycloptera* (Stapf) Botsch. op. cit.: 809. Type: Abd al Kuri, *Forbes* (K).

Low glabrous shrub, 20–40cm; stems erect or spreading, whitish, the older with cracking brownish bark. Leaves opposite or alternate, sometimes in small clusters, subglobose to narrowly cylindrical, 3–20 × 1–3mm, succulent, usually curved, rounded at the tip, slightly decurrent at the base and attenuate into a short indurated petiole, with a tuft of short white hair in the axils. Flowers solitary, sessile in the leaf axils, forming short lateral spikes; bracts minute, succulent, with scarious margins; perianth segments c.1mm, inflexed, the tube constricted above the swollen base; stamens exserted. Fruiting perianth accrescent, the wings 5–7mm diam., whitish to straw-yellow; fruit with a distinct thickened rim; seeds horizontal. **Map 365, Fig. 47.**

Sandy coastal plain; sea-level.

Socotra. Somalia.

22. NOAEA Moq.

Shrubs or annual herbs; stems rigid, branched, the branches thorny. Leaves alternate, sessile. Flowers bisexual, solitary, with 2 floral bracts; perianth segments 5, shortly fused at the base; stamens 5, on a fleshy lobed disc; anthers narrowly sagittate; staminodes absent; ovary ovoid; style elongated; stigmas 2, recurved. Fruiting perianth winged, membranous; fruit with a membranous utricle, included in but not adherent to the perianth segments; seeds vertical, compressed, orbicular; embryo spiral.

1. N. mucronata (Forsskal) Asch. & Schweinf., Ill. fl. Égypte 2: 131 (1887). Syn.: *Salsola mucronata* Forsskal (1775 p.56); *Noaea spinosissima* (L.f.) Moq. in DC., Prodr. 13(2): 209 (1849). Illustr.: Fl. Palaest. 1: t.257 (1966); Collenette (1985 p.122).

Sprawling shrub, 20–100cm; stems much-branched, the older with greyish fissured bark; younger branches glabrous, rigid, terminating in a sharp spine. Leaves narrowly linear, 3–12 × 1–2mm, semiterete, mucronate, slightly decurrent at the base, caducous. Flowers axillary, in loose or dense spikes; floral bracts ovate-deltoid, 3–4 × 1.5–2mm, apiculate, with a narrow scarious margin; perianth segments c.4mm. Fruiting perianth winged, the wings 5–7mm diam., membranous, white or reddish; seeds c.2mm diam., discoid; testa membranous. **Map 366, Fig. 46.**

Rocky slopes and sandy soils; 1200–2300m.

Saudi Arabia. N Africa and the E Mediterranean region eastwards to C Asia.

23. ANABASIS L.

Perennial herbs, subshrubs or shrubs; stems articulate, often succulent. Leaves opposite, fleshy, connate and amplexicaul, often reduced to scales. Flowers bisexual and female, axillary, solitary or rarely in clusters, subtended by 2 bracteoles; perianth segments 5, free, 3 or 5 of them developing a transverse wing on the back in fruit; stamens 5; staminodes 5, shorter than and alternating with the stamens; style short; stigmas 2. Fruit with a membranous or fleshy utricle included within the winged perianth; seeds vertical, lenticular; endosperm absent; embryo spiral.

1. Leaves cylindrical, up to 15mm long, succulent   2
+ Leaves minute, not succulent   3

2. Lateral branches short, densely covered by overlapping leaves; leaves 2–3(–5)mm long, without a terminal bristle   **3. A. ehrenbergii**
+ Lateral branches well-developed, sparsely covered by leaves; leaves (3–)5–15mm long, with a terminal bristle (caducous on old leaves)   **4. setifera**

3. Shrub, 50–80cm tall; leaves reduced to a short 2-lobed cupule; anthers obovate-cordate; wings of the fruiting perianth 6–8mm diam.   **1. A. articulata**
+ Dwarf shrub, 15–30cm tall; leaves minute, rounded, scarcely 1mm long,

divergent from the stem; anthers cylidrical-elliptic; wings of the fruiting perianth c.8mm diam. **2. A. lachnantha**

1. A. articulata (Forsskal) Moq. in DC., Prodr. 13(2): 212 (1849), emend. Asch. & Schweinf., Ill. Fl. Égypte 2: 128 (1887). Syn.: *Salsola articulata* Forsskal (1775 p.55), non Cav. (1794).

Shrub up to 80cm; stems with split bark; branches opposite, brittle. Leaves reduced to a short 2-lobed cupule, villous within. Flowers solitary, opposite, in spikes at the ends of branches, 4–5mm diam.; perianth segments membranous; stamens exserted; anthers ciliolate; staminodes thick, papillose; ovary papillose towards the short stigmas. Fruiting perianth membranous, with 5 subequal wings, 6–8mm diam., dull pink. **Map 367.**

Rocky and sandy soils, desert wadis; 0–1400m.

Saudi Arabia, Bahrain. S Spain and N Africa eastwards to Palestine and Jordan.

2. A. lachnantha Aellen & Rech.f. in Anz. Oesterr. Akad. Wiss. math.-naturw. Kl. 1961: 26 (1961). Illustr.: Mandaville (1990 pl.47).

Dwarf shrub, 15–30cm; stems richly branched from the base, divaricate, numerous, straw-yellow. Leaves minute, rounded, scarcely 1mm, divergent from the stem. Flowers solitary, axillary, minute, in short spikes at the tips of branches; anthers cylindrical-elliptic, shortly ciliate; ovary laterally flattened. Fruiting perianth c.8mm diam. (including wings), the wings ovate-orbicular, sinuate, pinkish. **Map 368.**

Shallow silty soil over limestone; 100–600m.

Saudi Arabia, Kuwait. Iraq and S Iran.

3. A. ehrenbergii Schweinf. ex Boiss., Fl. orient. 4: 970 (1879). Illustr.: Collenette (1985 p.115).

Spreading or prostrate shrub; stems up to 75cm, often rooting at the nodes, with split bark. Young leaves cylindrical, 3–5 × 1mm; mature leaves overlapping on short lateral branches, obovoid, 2–3 × 1–1.5mm, obtuse. Flowers in axillary 1.5–3cm spikes; bracteoles fleshy; perianth segments oblong, 2.5mm, scarious, rounded at the tip; anthers exserted, acuminate, bright crimson; stigmas purple. Fruit not seen. **Map 369.**

Seashores and low dunes; sea-level.

Saudi Arabia, Yemen (S). Sudan, Ethiopia and Somalia.

4. A. setifera Moq., Chenop. monogr. enum.: 164 (1840). Syn.: *Hammada scoparia* sensu Collenette (1985) non (Pomel) Iljin. Illustr.: Fl. Qatar pl.24 (1981); Collenette (1985 p.115); Mandaville (1990 pl.48).

Subshrub up to 50cm; stems richly branched, the older woody with split bark; younger stems fleshy, 4-angled, papillose. Leaves cylindrical or clavate, 0.5–1.5cm, usually spreading, with a terminal caducous bristle. Flowers 3–5, in axillary clusters on the upper parts of the stems; perianth segments elliptic, 1.2mm; stamens 1mm;

staminodes minute, fimbrillate; ovary papillose. Fruiting perianth 5mm diam., the wings unequal, membranous, yellowish white. glossy. **Map 370, Fig. 46.**

Wadi-beds, edges of salt-marshes, often on volcanic soils; 0–1000 m.

Saudi Arabia, Oman, UAE, Qatar, Bahrain, Kuwait. Egypt eastwards to Pakistan.

24. HALOCHARIS Moq.

Annual herbs. Leaves alternate, sessile, somewhat fleshy, linear, entire, hairy. Flowers bisexual, in short and dense spikes in the axils of leaf-like bracts; perianth membranous, of 5 free segments; stamens 5, each bearing a conspicuous petaloid anther-appendage; style long; stigmas 2. Fruiting perianth unchanged in fruit; seeds vertical; embryo spiral.

1. H. sulphurea (Moq). Moq. in DC., Prodr. 13(2): 201 (1849).

Annual herb, densely hairy throughout; stems erect or prostrate, much-branched, up to 40cm. Leaves 10–20mm, obtuse. Bracts densely hirsute. Anther-appendages yellow, oblong, 1–2 × 0.5–1mm. **Fig. 47.**

Rocky desert; c.350m.

Saudi Arabia. Syria, Iraq, Iran and Afghanistan.

Recorded only once from Arabia. Readily distinguished by its characteristic petaloid, yellow anther-appendages.

Family 39. AMARANTHACEAE

A.G. MILLER

Hermaphrodite or more rarely monoecious or dioecious herbs, subshrubs or shrubs. Leaves opposite or alternate, simple, usually entire; stipules absent. Flowers unisexual or bisexual, in heads, spikes or panicles, usually subtended by a membranous or rigid bract and 2 bracteoles; fertile flowers sometimes subtended by 1–2 variously modified sterile flowers. Perianth of 4–5 tepals; tepals free or fused at the base, membranous or firm. Stamens 5, rarely 1 or 2, opposite the tepals; anthers 1- or 2-celled; filaments usually fused into a basal cup or tube, alternating with pseudostaminodes or not. Ovary superior, 1-celled; ovules 1–several with basal placentation; styles solitary; stigmas capitate or 2—3 and filiform. Fruit a capsule with circumscissile dehiscence or rupturing irregularly, a nutlet or rarely a berry; seeds usually black or brown and shining.

1.	Leaves predominantly opposite	2
+	Leaves predominantly alternate	7

AMARANTHACEAE

2.	Fertile flowers subtended by 2 sterile flowers which are modified into rigid, hooked spines		**4. Pupalia**
+	Flowers all fertile, not subtended by modified sterile flowers		3
3.	Bracteoles strongly keeled and crested at least along the upper part of the midrib		**10. Gomphrena**
+	Bracteoles not keeled or crested		4
4.	Stamens 1–2; flowers minute, tepals c.1.25mm		**6. Nothosaerva**
+	Stamens 5; flowers with tepals 2mm or longer		5
5.	Flowers in sessile axillary clusters		**9. Alternanthera**
+	Flowers in lax or compact spikes		6
6.	Fruits deflexed; pseudostaminodes present		**8. Achyranthes**
+	Fruits never deflexed; pseudostaminodes absent		**7. Psilotrichum**
7.	Fertile flowers subtended by modified sterile flowers		8
+	Flowers all fertile, modified sterile flowers absent		9
8.	Annual herb; fertile flowers subtended by 2 antler-shaped sterile flowers		**3. Digera**
+	Shrub; fertile flowers subtended by sterile flowers modified into plumose filiform appendages		**11. Saltia**
9.	Tepals glabrous		10
+	Tepals lanate or villous		11
10.	Fruits 2–10-seeded; flowers bisexual		**1. Celosia**
+	Fruits 1-seeded; flowers unisexual		**2. Amaranthus**
11.	Perennial herbs or subshrubs; stamens 5; pseudostaminodes present		**5. Aerva**
+	Annual herbs; stamens 1–2; pseudostaminodes absent		**6. Nothosaerva**

1. CELOSIA L.

Annual or perennial herbs, sometimes woody-based. Leaves alternate. Flowers bisexual, in spikes or spike-like thyrses. Tepals 5, free. Stamens 5, united below into a short tube. Pseudostaminodes absent. Style slender or ± absent; stigmas capitate or 2–3-lobed. Capsule 2–10-seeded, circumscissile.

Townsend, C. C. (1975) in *Hooker's Icon. Pl.*, 38 (2): 1–123.

1.	Inflorescence a dense, many-flowered terminal spike; tepals 6–10mm	**4. C. argentea**
+	Inflorescence a spike-like thyrse, lateral cymes distant; tepals 1.75–2.75mm	2
2.	Top of fruit swelling and becoming spongy	**3. C. anthelminthica**
+	Top of fruit not swollen and spongy	3

3. Tepals with a dark brown midrib, acute or obtuse, not mucronate
1. C. polystachia
+ Tepals with a pale midrib, mucronate
2. C. trigyna

1. C. polystachia (Forsskal) C. C. Townsend in Hooker's Icon. Pl. 38 (2): 23 t.3728 (1975). Syn.: *Achyranthes polystachia* Forsskal (1775 p.48); *Celosia populifolia* Moq. in DC., Prodr. 13(2): 239 (1849). Illustr.: Collenette (1985 p.43). Type: Yemen (N), *Forsskal* (BM).

Sprawling perennial herb, often somewhat woody, glabrous or glabrescent; stems up to 3m. Leaves ovate, 1–6(–8) × 0.5–3.5(–4.5)cm, acute or acuminate. Flowers in terminal and axillary elongate thyrses comprising distant 1–8-flowered cymes and often forming an open terminal panicle. Tepals narrowly elliptic-oblong, c.2mm, acute or obtuse, with a dark brown midrib and membranous margin. Style short; stigmas 2–3, longer than style. **Map 371, Fig. 54.**

Acacia-Commiphora bushland on rocky slopes and on wadi-sides; 300–650m.

Saudi Arabia, Yemen (N & S). Sudan, Ethiopia and Somalia.

2. C. trigyna L., Mant. pl. 2: 212 (1771). Syn.: *Celosia trigyna* L. var. *fasciculiflora* Fenzl ex Moq. in DC., Prodr. 13 (2): 241 (1849); *C. caudata* Vahl (1790 p.21); *Achyranthes decumbens* Forsskal (1775 p.47); *A. paniculata* Forsskal (1775 p.48). Illustr.: Collenette (1985 p.43). Type: Yemen (N), *Forsskal* (C).

Similar to *C. polystachia* but an erect annual; tepals 1.75–2.75mm, shortly mucronate, the midrib colourless. **Map 372.**

Disturbed ground by roadsides and cultivated areas; in woodland and in shade by water; 200–1850m.

Saudi Arabia, Yemen (N), Oman. Widespread in tropical Africa.

3. C. anthelminthica Asch. in Schweinf., Beitr. Fl. Aethiop.: 176 (1867). Illustr.: Fl Trop. E. Afr.: 10 (1985).

Similar to *C. trigyna* but tepals strongly concave; stigmas ± sessile, 2–3-lobed; fruit swelling and becoming spongy at the apex. **Map 373.**

No habitat details available.

Yemen (S). Rwanda, Ethiopia, Tanzania, Kenya and Uganda.
Very rare in Arabia, only collected once in the last century.

4. C. argentea L., Sp. pl.: 205 (1753). Illustr.: Fl. Trop. E. Afr.: 18 (1985).

Erect annual herb up to 2m, glabrous throughout. Leaves narrowly elliptic to lanceolate, 2–15 × 0.5–3cm, acute. Inflorescence a dense, many-flowered, terminal spike, 2–20 × 1.5–2cm, silvery. Tepals lanceolate, 6–10mm, hyaline. Style filiform; stigma ± capitate. **Map 374, Fig 54.**

Weed of cultivation.

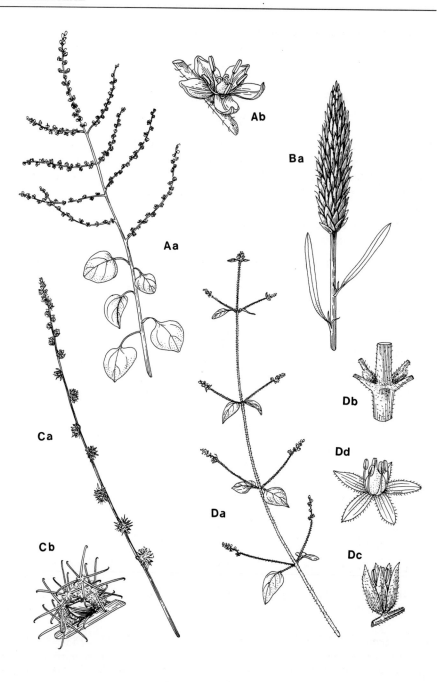

Fig. 54. Amaranthaceae. A, *Celosia polystachia*: Aa, flowering shoot (×0.6); Ab, flower (×8). B, *C. argentea*: Ba, inflorescence (×0.6). C, *Pupalia lappacea*: Ca, flowering and fruiting shoot (×0.6); Cb, part of infructescence (×3). D, *Psilotrichum gnaphalobryum*: Da, flowering shoot (×0.6); Db, node (×4); Dc, flower (×8); Dd, opened flower (×8).

Yemen (N & S). Pantropical weed.

Rare in Arabia. No Arabian material seen, description based on African material.

2. AMARANTHUS L.

Monoecious or dioecious annual herbs, glabrous or thinly pubescent. Leaves alternate, long-petiolate. Flowers unisexual, arranged in sessile clusters in the axils of the leaves or sometimes also in leafless terminal spikes or panicles; bracts and bracteoles scarious with a green midrib. Tepals 3–5, free, membranous. Stamens 0–5. Stigmas 2–3; ovule solitary. Fruit 1-seeded, dehiscing by a circumscissile lid, rupturing irregularly or indehiscent. Seeds black, shiny.

1.	Plants with paired spines in the axils of the leaves	**3. A. spinosus**
+	Plants without spines	2
2.	Fruits ± woody, arranged in stellately divergent clusters in the axils of the leaves	**8. A. sparganiocephalus**
+	Fruits thin-walled, never in stellately divergent clusters	3
3.	Flowers in axillary clusters only, the stems thus appearing leafy ± to the tips	4
+	Flowers in spikes forming a terminal panicle as well as in axillary spikes or clusters	7
4.	Bracteoles spinescent, about twice as long as the flowers; leaves spinescent	**5. A. albus**
+	Bracteoles acute, mucronate, shortly aristate or with hair-like aristae, not spinescent, equalling or just exceeding the flowers; leaves acute, obtuse or emarginate, never spinescent	5
5.	Bracts and bracteoles produced into long, fine, hair-like awns	**6. A. tricolor**
+	Bracts and bracteoles acute, mucronate or shortly aristate	6
6.	Leaves acute, obtuse or at most retuse; capsule strongly wrinkled	**7. A. graecizans**
+	Leaves broadly emarginate; capsule smooth or somewhat wrinkled	**9. A. lividus**
7.	Terminal spike of inflorescence long and pendulous; tepals of female flowers spathulate	**1. A. caudatus**
+	Terminal spike of inflorescence erect or somewhat nodding; tepals of female flowers narrowly oblong to narrowly spathulate	8
8.	Capsule opening by a circumscissile lid; female tepals 3–5	9
+	Capsule indehiscent or rupturing irregularly; female tepals 3(–4)	11

9. Bracts and bracteoles produced into fine hair-like awns; female flowers with 3(–4) tepals   **6. A. tricolor**
+ Bracts and bracteoles acute, mucronate or shortly awned; female flowers with 5 tepals   10

10. Male flowers intermixed with female throughout the spikes; lid of capsule smooth or somewhat wrinkled below neck   **2. A. hybridus**
+ Male flowers ± restricted to the tips of the spikes, rarely intermixed with female below; lid of capsule strongly wrinkled near the line of dehiscence   **4. A. dubius**

11. Capsule smooth or somewhat wrinkled on drying, distinctly longer than the tepals   **9. A. lividus**
+ Capsule strongly wrinkled, shorter than or scarcely as long as the tepals   **10. A. viridus**

1. A. caudatus L., Sp. pl.: 990 (1753). Illustr.: Fl. Trop. E. Afr.: t.4/4 (1985).

Erect herb up to 1.5m, frequently reddish tinged throughout. Leaf lamina broadly ovate to rhombic-ovate, 3–15 × 1–8cm, obtuse or subacute, mucronulate. Flowers in terminal and axillary spikes, the male and female intermixed throughout; terminal spike long and pendulous, up to 30cm. Bracts and bracteoles ovate, produced into a rigid arista. Female tepals 5, broadly obovate to spathulate, 2.5–3.5mm, blunt or emarginate. Stigmas 3. Capsule ovoid-globose, 2–2.5mm, dehiscent, the lid smooth or furrowed below. **Map 375.**

Cultivated as a garden ornamental, sometimes naturalized; 150–1000m.

Yemen (N & S). Pantropical.

2. A. hybridus L., Sp. pl.: 990 (1753).

Erect or ascending herb up to 100(–200)cm, frequently reddish tinged throughout. Leaf lamina ovate to rhombic-ovate, 2–12(–20) × 1–8cm, acute to obtuse, sometimes mucronate. Flowers in terminal and axillary spikes, the male and female intermixed throughout. Bracts and bracteoles narrowly ovate, produced into a stout arista. Female tepals 5, narrowly oblong to narrowly oblong-obovate, 1.5–3(–3.5)mm, acute or shortly aristate. Stigmas (2–)3. Capsule subglobose to ovoid, 2–3mm, dehiscent, smooth or wrinkled below neck.

1. Longer bracteoles of the female flowers twice as long as the tepals; stigma base swollen in fruit   subsp. **hybridus**
+ Longer bracteoles of the female flowers 1.5× as long as the tepals; stigma base not swollen in fruit   subsp. **cruentus**

subsp. **hybridus**. Syn.: *A. chlorostachys* Willd., Hist. Amaranth.: 34, t.x/19 (1790); *A. hypochondriacus* L., Sp. pl.: 991 (1753). Illustr.: Fl. Trop. E. Afr.: t.4/7 (1985). **Map 376, Fig. 55.**

Weed of gardens and waste places; 10–2500m.

Saudi Arabia, Yemen (N & S), Oman, UAE, Qatar. Cosmopolitan.

subsp. **cruentus** (L.) Thell., Fl. adv. Montpellier: 205 (1912). Syn: *A. cruentus* L., Syst. Nat. ed. 10, 2: 1269 (1759); *A. paniculatus* L., Sp. pl., ed. 2: 1406 (1763). Illustr.: Fl. Trop. E. Afr.: t.4/9 (1985). **Map 377**.

A weed of gardens and waste-places.

?Saudi Arabia, Yemen (?N & S). Tropical and subtropical regions of the world.
Recorded from Saudi Arabia but no authenticated material has been seen.

3. A. spinosus L., Sp. pl.: 991 (1753). Illustr.: Fl. Trop. E. Afr.: t.4/10 (1985); Collenette (1985 p.42).
Erect herb up to 60(–150)cm, sometimes reddish tinged; stems with pairs of spines in the leaf axils. Leaf lamina ovate or rhombic-ovate to oblong-ovate, 1.5–8(–12) × 1–4(–6)cm, subacute or obtuse to retuse. Flowers in terminal and axillary spikes, the lower spikes entirely female, the upper spikes female below and male in the upper $\frac{1}{4}-\frac{2}{3}$. Bracts and bracteoles ovate, produced into a reddish arista. Female tepals 5, narrowly oblong to oblong-obovate, 1.5–2.5mm, acute or obtuse, mucronulate. Stigmas (2–)3. Capsule ovoid, c.1.5mm, with inflated beak, dehiscent or rarely indehiscent, the lid wrinkled below the neck. **Map 378, Fig. 55**.

Weed of gardens and irrigated land, and by water points; 50–2000m.

Saudi Arabia, Yemen (N & S), UAE. Pantropical weed.

4. A. dubius Thell., Fl. adv. Montpellier: 203 (1912). Illustr.: Fl. Trop. E. Afr.: t.4/11 (1985).
Erect herb up to 40(–100)cm. Leaf lamina ovate to rhombic-ovate, 1.5–5(–8) × 0.5–2(–5)cm, blunt or retuse. Flowers in terminal and axillary spikes, the male found mainly at the tips of spikes or rarely below. Bracts and bracteoles ovate, produced into a reddish arista. Female tepals 5, narrowly oblong to spathulate-oblong, 1.5–2.75mm, acute or obtuse, mucronulate. Stigmas 3. Capsule ovoid, 1.5–1.75mm, with an inflated beak, dehiscent or rarely indehiscent, the lid wrinkled near the line of dehiscence. **Map 379, Fig. 55**.

Weed of irrigated date gardens; 340m.

Oman. Tropical regions of the world.

5. A. albus L., Syst. Nat. ed. 10, 2: 1268 (1759).
Erect or procumbent herb up to 50cm. Leaf lamina spathulate or oblong, 5–20 (–30) × 2–10(–15)mm, spinescent. Flowers in axillary clusters, the male and female intermixed. Bracts and bracteoles narrowly ovate, spine-tipped. Female tepals 3, narrowly oblong-elliptic, c.1.5mm, acuminate. Fruit subglobose, c.1.5mm, circumscissile, strongly wrinkled. **Map 380, Fig. 55**.

AMARANTHACEAE

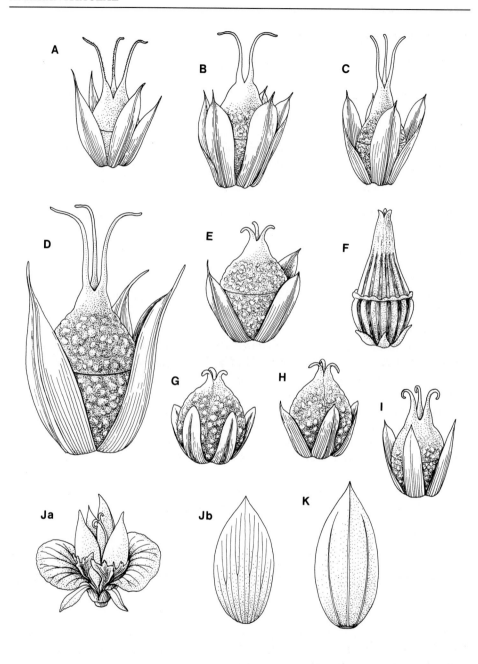

Fig. 55. Amaranthaceae. A–I. *Amaranthus* fruits (×20): A, *A. hybridus*; B, *A. spinosus*; C, *A. dubius*; D, *A. tricolor*; E, *A. graecizans*; F, *A. sparganiocephalus*; G, *A. lividus*; H, *A. viridis*; I, *A. albus*. J, *Digera muricata* subsp. *muricata*: Ja, perianth and sterile flowers (×8); Jb, outer tepal (×15). K, *D. muricata* subsp. *trinervis* var. *trinervis*: outer tepal (×15).

Weed of waste ground; 760m.

Saudi Arabia. Temperate regions of the world.

6. A. tricolor L., Sp. pl.: 989 (1753). Syn.: *A. gangeticus* L., Sp. pl., ed. 2: 1403 (1763); *A. mangostanus* L., Cent. pl. I: 32 (1755); *A. tristis* L., Sp. pl.: 989 (1753).

Erect or ascending herb up to 125cm. Leaf lamina ovate or rhombic-ovate to oblong-ovate, very variable in size, acute or obtuse to emarginate. Flowers in terminal spikes and globose axillary clusters, the male and female intermixed throughout. Bracts and bracteoles broadly ovate, produced into a long, fine, flexuous awn. Female tepals 3, oblong-elliptic to oblong-obovate, 3–5mm, produced into long, pale-tipped awns. Stigmas 3. Capsule ovoid, 2.25–2.75mm, dehiscent, obscurely wrinkled. **Map 381, Fig. 55.**

Cultivated.

Saudi Arabia, Yemen (S). Widespread in the tropics.

Cultivated in Yemen (S) according to Schwartz (1939) but no material seen.

7. A. graecizans L., Sp. pl.: 990 (1753). Syn.: *A. blitum* sensu Balfour (1888) non L.; *A. polygamus* sensu Blatter (1919–36) non L. Illustr.: Western (1989 p.54).

Erect, ascending or prostrate herb up to 45(–70)cm. Leaf lamina linear-lanceolate, ovate or rhombic-ovate to obovate, 5–45(–55) × 1.5–25(–30)mm, acute to obtuse or retuse. Flowers in axillary clusters, the male and female intermixed throughout but the male commonest in the upper clusters. Bracts and bracteoles narrowly ovate-oblong, produced into an arista. Female tepals 3, narrowly oblong to narrowly ovate, 1.25–2mm, gradually to abruptly mucronate or shortly aristate. Stigmas 3. Capsule subglobose to ovoid, 1.5–2mm, dehiscent or not, usually strongly wrinkled. **Fig. 55.**

1. Tepals aristate, leaves linear to rhombic-spathulate   subsp. **thellungianus**
+ Tepals mucronate; leaves linear or oblong to broadly obovate   2

2. Leaves linear-lanceolate to oblong; usually 2.5 × as long as broad or longer
    subsp. **graecizans**
+ Leaves elliptic to broadly ovate, up to 2.5 × as long as broad.
    subsp. **silvestris**

subsp. **graecizans**. Syn.: *A. angustifolius* Lam. subsp. *aschersonianus* Thell. in Asch. & P. Graebner, Syn. mitteleur. Fl. 5, 1: 300 (1919). *A. angustifolius* Lam. var. *graecizans* (L.) Thell. in Asch., op. cit.: 306. **Map 382.**

Weed of disturbed ground, gardens, irrigated land and in sandy wadi-beds; 0—2500m.

Saudi Arabia, Yemen (N & S), Oman, UAE, Qatar, Bahrain, Kuwait. Warmer temperate and tropical regions of the Old World.

subsp. **silvestris** (Villars) Heukels, Geill. schoolfl. Nederl., ed. 11: 170 (1934). Illustr.: Fl. Trop. E. Afr.: t.4/4 (1985). **Map 383.**

Stony plains, wadi-beds and a weed of gardens and irrigated land; 0–2500m.

Saudi Arabia, Yemen (N & S), Oman. Warmer temperate and tropical regions of the Old World.

subsp. **thellungianus** (Nevski) Gusev in Bot. Zhurn. (Moscow & Leningrad) 57 (5): 462 (1972). Illustr.: Fl. Trop. E. Afr.: t.4/4 (1985). **Map 383.**

Dry rocky gullies; 0–50m.

Socotra, Oman. Tropical Africa and India.

8. A. sparganiocephalus Thell. in Asch. & P. Graebner, Syn. mitteleur. Fl. 5 (1): 312 (1914). Illustr.: Fl. Trop. E. Afr.: 33 (1985).

Erect herb up to 45(–60)cm. Leaf lamina ovate to oblong-ovate, 1–4 × 0.5–3cm, obtuse to broadly retuse. Flowers in compact axillary clusters, the male and female intermixed throughout. Bracts and bracteoles small, oblong-ovate, acute. Female tepals 3, elliptic-oblong to oblong-obovate, 0.75–1.25mm, apiculate. Stigmas 2. Fruiting clusters rigidly stellate; capsule somewhat woody, ovoid, conical above, 2.75–4mm, dehiscent, longitudinally ridged. **Map 384, Fig 55.**

Fields, roadsides and by water-points; 0–1450m.

Yemen (S), Socotra, Oman. E and NE tropical Africa.

9. A. lividus L., Sp. pl.: 990 (1753). Syn.: *A. blitum* L., Sp. pl.: 990 (1753); *A. oleraceus* L., Sp. pl. ed. 2: 1403 (1763).

Erect, ascending or prostrate herb up to 90cm. Leaf lamina ovate to rhombic-ovate, 1–10 × 0.5–6cm, broadly emarginate. Flowers in terminal spikes and axillary clusters, the male and female intermixed throughout. Bracts and bracteoles narrowly ovate, produced into a short mucro. Female tepals 3, narrowly ovate to oblong-ovate, 0.75–2mm, subacute and mucronate or obtuse. Stigmas 2–3. Capsule subglobose, 1.25–2.5mm, indehiscent, smooth or slightly wrinkled on drying. **Map 385, Fig. 55.**

Weed of gardens, cultivated fields and waste places.

Saudi Arabia, Yemen (N). Warmer regions of the World.

Rare in Arabia, no authenticated material seen. Recorded from Yemen (N) by Forsskal (1755).

10. A. viridis L., Sp. pl., ed. 2: 1405 (1763). Syn.: *A. gracilis* Desf., Tabl. école bot. ed. 1: 43 (1804). Illustr.: Fl. Trop. E. Afr.: t.4/8 (1985).

Erect or ascending herbs up to 60(–100)cm, sometimes reddish tinged. Leaf lamina ovate to triangular-ovate, 2–7 × 1.5–5cm, acute to obtuse or retuse. Flowers in terminal spikes and axillary clusters, the male and female intermixed throughout. Bracts and bracteoles narrowly ovate, produced into short awns. Female tepals 3(–4), narrowly oblong to narrowly obovate, 1.25–1.75mm, minutely mucronulate. Stigmas 2–3. Capsule subglobose, 1.25–1.5mm, indehiscent, strongly wrinkled throughout. **Map 386, Fig. 55.**

Weed of gardens, cultivated fields, irrigated land and waste places; 0–2200m.

Saudi Arabia, Yemen (N & S), Oman, UAE, Qatar, Bahrain, Kuwait. Cosmopolitan weed.

Species doubtfully recorded

A. retroflexus L., a species mainly of the temperate regions, has been recorded, probably in error, from Yemen (N). It is similar to *A. hybridus* but the perianth segments of the female flowers are spathulate with obtuse or emarginate tips, the stems are densely pubescent and the inflorescences are terminal and paniculate.

3. DIGERA Forsskal

Annual herbs. Leaves alternate. Flowers bisexual, in long axillary spikes, each subtended by 2 antler-shaped sterile flowers. Tepals (4–)5, the 2 outer stiff and strongly nerved, the inner membranous. Stamens (4–)5. Pseudostaminodes absent. Style filiform; stigma bilobed. Fruit a nutlet, enclosed by the persistent perianth.

1. D. muricata (L.) Mart. in Nova Acta Phys.-Med. Acad. Caes. Leop.-Carol. Nat. Cur. 13(1): 285 (1826). Syn.: *D. arvensis* Forsskal (1775 p.65). Illustr.: Fl. Trop. E. Afr.: 38 (1985); Fl. Pakistan 71: 22 (1974).

Stems erect, up to 45(–75)cm, glabrous or thinly pubescent. Leaves ovate to oblong-elliptic, 2–7 × 1–4cm, acute; petioles 1.5–5cm. Flowers pink, in 5–20cm spikes; sterile flowers accrescent, up to 1.5mm in fruit, appressed. Outer tepals ovate, 3–4mm, 3–12-nerved; inner tepals slightly shorter, blunt or erose. Fruit subglobose, c.2mm. **Map 387, Fig. 55.**

1.	Outer tepals 7–12-nerved	subsp. **muricata**
+	Outer tepals 3(–5)-nerved (subsp. *trinervis*)	2
2.	Leaves glabrous	var. **trinervis**
+	Leaves hairy beneath	var. **patentipilosa**

subsp. **muricata**. Syn.: *Desmochaeta alternifolia* (L.) DC., Cat. pl. horti monsp.: 103 (1813); *Digera alternifolia* (L.) Aschers. in Schweinf., Beitr. Fl. Aeth.: 180 (1867). Illustr.: Collenette (1985 p.43).

Weed of cultivation, wadi-sides and drought-deciduous woodland; 10–1200m.

Saudi Arabia, Yemen (N & S), Socotra, Oman. Kenya, Somalia, Ethiopia, Madagascar, Malaysia, Afghanistan, Indonesia, India, Pakistan and Sri Lanka.

subsp. **trinervis** C. C. Townsend in Kew Bull. 28: 141 (1973).
var. **trinervis**

?Socotra. Sudan, Ethiopia, Uganda, Kenya and Tanzania.

Recorded from Socotra by Townsend in Fl. Trop. E. Afr. (1985 p.37) but no authenticated material has been seen.

var. **patentipilosa** C. C. Townsend in Kew Bull. 28: 142 (1973).

Drought-deciduous shrubland with *Croton socotranus* and *Jatropha unicostata*; 50m.

Socotra. Kenya.

4. PUPALIA A.L. Juss.

Annual or perennial herbs or subshrubs. Leaves opposite. Flowers in distant clusters along a terminal spike; clusters bracteate, consisting of 3 bisexual flowers, the outer pair subtended by two modified sterile flowers; sterile flowers consisting of a number of strongly accrescent sharply hooked spines. Tepals 5. Stamens 5, fused at base into a cup; pseudostaminodes absent. Style slender; stigma capitate. Fruit 1-seeded, irregularly rupturing, shed with the hooks of the sterile flowers making a highly adhesive burr.

Townsend, C. C. (1979). A survey of *Pupalia* Juss. *Kew Bull*. 34: 131–142.

1.	Tepals oblong-ovate, less than 6mm; style up to 2(–3)mm		**1. P. lappacea**
+	Tepals oblong-lanceolate, (6–)7–9mm; style (2.75–)3–4.5mm		2
2.	Subshrub; leaves triangular-ovate with blunt tips; tepals densely lanate-tomentose		**3. P. robecchii**
+	Perennial herb; leaves narrowly to broadly ovate, with acuminate tips; tepals pilose with forward pointing hairs		**2. P. grandiflora**

1. P. lappacea (L.) A. L. Juss. in Ann. Mus. Natl. Hist. Nat. Paris 2: 132 (1803). Illustr.: Fl. Trop. E. Afr.: 76 (1985); Collenette (1985 p.44).

Annual or perennial herb; stems erect or sprawling, up to 1.5m. Leaves ovate to elliptic or circular, 1–6(–10) × 1–3(–5)cm, acute or acuminate, glabrescent to tomentose or sericeous; petioles up to 2cm. Spike up to 30cm. Tepals oblong-ovate, 4–5(–6)mm, lanate on back. Style 1–2(–3)mm. Spines of sterile flowers up to 6mm, the burr 7–13(–18)mm diam. **Map 388, Fig. 54.**

Rocky and stony slopes, gravel plains and wadi-sides; disturbed and cultivated ground; 0–2000m.

Saudi Arabia, Yemen (N & S), Socotra, Oman. Throughout the Old World tropics.

All Arabian material is referable to var. *velutina* (Moq.) Hook.f., Fl. Brit. Ind. 4: 724 (1885), characterized by its lanate tepals and sterile flowers with straw-coloured spines.

2. P. grandiflora Peter in Feddes Repert. Beih. 40 (2), Descr.: 22 (1932). Illustr.: Fl. Trop. E. Afr.: 76 (1985).

Similar to *P. lappacea* but flowers much larger; tepals narrowly oblong-lanceolate, (6–)7–8mm; style (2.75–)3–3.5mm. **Map 389.**

In a shaded gully; 1600m.

Yemen (N). NE, E and C tropical Africa.

In Arabia only known from a single specimen from J. Milhan in Yemen (N).

3. P. robecchii Lopr. in Bot. Jahrb. Syst. 27: 55 (1899).

Similar to *P. lappacea* but a subshrub; leaves triangular-ovate; flowers much larger, tepals oblong-lanceolate, 7–9mm; style 3–4.5mm. **Map 390.**

No habitat details available.

Yemen (S). Somalia.

In Arabia only collected once from Ras Fartak in Yemen (S).

5. AERVA Forsskal

Perennial herbs or subshrubs. Leaves alternate, rarely opposite. Flowers unisexual or bisexual, in terminal and axillary spikes which often form a terminal panicle. Tepals 5–6. Stamens 4 or 5, united below, alternating with oblong or triangular pseudo-staminodes. Stigma bilobed or ± capitate. Capsule 1-seeded, thin-walled, rupturing irregularly.

1.	Stems and leaves glabrous except for tufts of hairs in the leaf axils	**4. A. microphylla**
+	Stems and leaves lanate or tomentose	2
2.	Leaves glabrous above, shortly tomentose beneath, margins revolute	**3. A. revoluta**
+	Leaves lanate or tomentose above, margins not revolute	3
3.	Flowers in 3(–7)-flowered clusters along interrupted spikes; tepals 6; stamens 4; stigma capitate, sessile	**5. A. artemisioides**
+	Flowers in dense cylindrical or globular spikes; tepals 5; stamens 4; stigma bilobed, on a short style	4
4.	Flowers unisexual; outer 2 tepals acute or obtuse; inner 3 tepals with a narrow midrib and broad membranous margin; inflorescence usually leafless; spikes 1–15cm long	**1. A. javanica**
+	Flowers usually bisexual; outer 2 tepals mucronate; inner 3 tepals with a broad midrib and narrow membranous margin; inflorescence usually leafy; spikes 0.5–1(–2)cm long	**2. A. lanata**

1. A. javanica (Burm.f.) Schultes, Syst. veg. 5: 565 (1819). Syn.: *A. persica* (Burm.f.) Merrill in Philipp. J. Sci. 19: 348 (1921); *A. tomentosa* Forsskal (1775 pp. CXXII &

170). Illustr.: Fl. Pakistan 71: 28 (1974); Fl. Qatar pl.31 (1981); Collenette (1985 p.41); Mandaville (1990 pl.52); Western (1989 p.53).

Dioecious, perennial herb, often woody-based; stems erect, up to 1m, densely whitish or yellowish tomentose. Leaves very variable from linear-oblanceolate to broadly elliptic, 1–10 × 0.5–3cm, densely tomentose. Flowers unisexual, in cylindrical spikes; spikes 10–150 × 3–8mm (male spikes narrower), forming a ± leafless panicle. Tepals 5, obovate, 1.5–3mm, margins membranous, densely lanate dorsally. Female flowers: outer 2 tepals acute or obtuse; inner 3 tepals with a narrow, greenish midrib; style distinct, 0.5–0.8mm; stigmas filiform. Male flowers smaller; stamens 5; pseudostaminodes narrowly triangular; ovary rudimentary. **Map 391/392, Fig. 56.**

Habitat very variable including coastal sand, rocky slopes, sandy and stony plains and deserts and around cultivation; 0–2600m.

Saudi Arabia, Yemen (N & S), Socotra, Oman, UAE, Qatar, Bahrain. Widespread in the drier parts of the tropics and subtropics of the Old World from Africa eastwards to Burma.

A very common, variable and widespread species. Two varieties are found in Arabia. Var. *bovei* Webb differs from the commoner var. *javanica* in having narrower, linear-oblong leaves and more slender and often interrupted spikes. Male plants are apparently much rarer than the female in Arabia.

2. A. lanata (L.) A.L. Juss. in Ann. Mus. Natl. Hist. Nat. 2: 131 (1803). Syn.: *Achyranthes villosa* Forsskal (1775 p.48); *Amaranthus lanatus* A.L. Juss. var. *rotundifolia* Moq. in DC., Prodr. 13(2): 304 (1849); *A. lanatus* var. *viridis*, loc. cit. Illustr.: Collenette (1985 p.41).

Perennial herb, often woody-based; stems prostrate to erect, up to 60cm, densely lanate. Leaves obovate or broadly elliptic to circular, 8–50 × 2–28mm, rounded or acute, thinly to densely lanate. Flowers usually bisexual, in globular to cylindrical spikes; spikes sessile, 5–15(–20) × c.3mm, solitary or clustered in the leaf axils, usually forming a narrow leafy panicle. Tepals 5, ovate to elliptic, (0.75–)1.25–2mm, densely lanate dorsally; outer 2 shortly mucronate, membranous; inner 3 narrower, midrib broad and green, the margin membranous; stamens 5; pseudostaminodes narrowly triangular. Style and stigma lobes short. **Map 393, Fig 56.**

Habitats very variable from wadi-beds and cultivated ground to stony plains and rocky slopes; 50–2600m.

Saudi Arabia, Yemen (N & S), Socotra. Widespread in drier parts of the tropics and subtropics of the Old World from Africa eastwards to Malaysia and the Philippines.

An extremely variable species in which the infraspecific taxa recognized by various authors are impossible to distinguish in Arabia; see Townsend in Kew Bull. 29: 46 (1974).

3. A. revoluta Balf. f. (1883 p.92). Illustr.: Hooker's Icon. Pl., 37: t.3695 (1971). Syntypes: Socotra, *Balfour* 478 (BM) & *Schweinfurth* 558 (K).

Subshrub; stems erect, up to 50cm, shortly creamy-tomentose. Leaves obovate, 10–25 × 6–15mm, obtuse, the margins revolute, glabrous or glabrescent above, densely

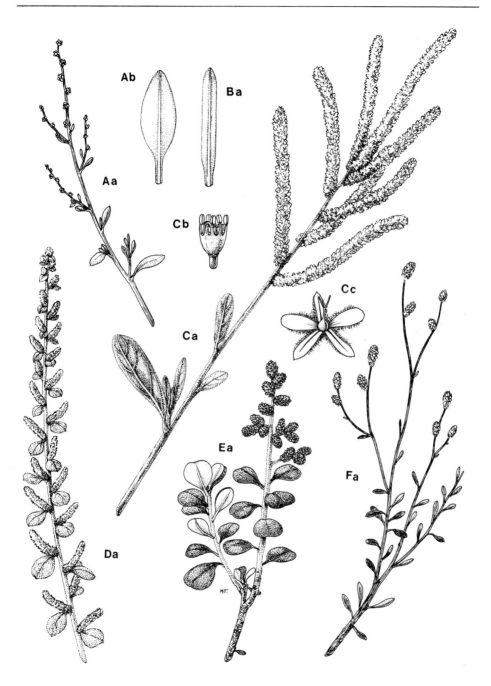

Fig. 56. **Amaranthaceae**. A, *Aerva artemisioides* subsp. *batharitica*: Aa, flowering branch (×0.6); Ab, leaf (×3). B, *A. artemisioides* subsp. *artemisioides*: Ba, leaf (×3). C, *A. javanica* var. *javanica*: Ca, flowering branch (×0.6); Cb, male flower (×12); Cc, female flower (×6). D, *A. lanata*: Da, flowering branch (×0.6); E, *A. revoluta*: Ea, flowering branch (×0.6); F, *A. microphylla*: Fa, flowering branch (×0.6).

tomentose beneath. Flowers bisexual, in short, sessile spikes; spikes, 5–15 × 3–5mm, usually forming a narrow leafless panicle. Tepals 5, ovate, 2–2.5mm, acute, the margins membranous, dorsally lanate. Stamens 5; pseudostaminodes triangular. Style short; stigma bifid. **Map 394, Fig. 56.**

Rock crevices and boulder strewn wadis in semi-deciduous shrubland; apparently restricted to granite; 200–950m.

Socotra. Endemic.

4. A. microphylla Moq. in DC., Prodr. 13(2): 301 (1849). Illustr.: Balf. f. (1888 t.85). Type: Socotra, *Nimmo* s.n. (K).

Subshrub up to 60cm; stems erect or decumbent, glabrous. Leaves fleshy, obovate, 5–15 × 2–7mm, obtuse, glabrous, lanate in the axils. Flowers bisexual, in short globular or cylindrical spikes; spikes 5–12 × 3–5mm, solitary or clustered at the ends of branches. Tepals 5, elliptic-oblong to obovate, the margins broadly membranous, dorsally lanate. Stamens 5; pseudostaminodes oblong or triangular. Style short; stigma bifid. **Map 395, Fig. 56.**

On cliffs and in rock crevices; 10–900m.

Socotra. Endemic.

Low plants with broader, relatively smaller leaves have been placed in var. *humilis* Vierh. (1907 p.25); this, however, seems to be merely a grazed and stunted form.

5. A. artemisioides Vierh. & Schwartz in Schwartz (1939 p.43). Syntypes: Yemen (S), *Hein* s.n. (WU) & *Paulay* s.n. (WU).

Subshrub; stems erect, up to 1m, densely tomentose. Leaves linear-oblong to linear-oblanceolate or elliptic to ovate, 4–40 × 2–12mm, tomentose. Flowers bisexual, in 3(or 7)-flowered clusters arranged along interrupted spikes and forming open leafless panicles. Tepals 6, oblong-elliptic to broadly obovate, 1–1.5mm, densely tomentose dorsally; 2 outer, rounded, the membranous margin broad; 4 inner acute, the membranous margin narrow. Stamens 4; pseudostaminodes triangular or oblong. Stigma sessile, capitate. **Map 396, Fig. 56.**

1. Leaves linear-oblong to linear-oblanceolate, 10–40 × 2–5mm, long-attenuate at the base into a petiole or ± sessile   subsp. **artemisioides**
+ Leaves elliptic to ovate, 4–25 × 5–12mm, base cuneate, petiole short
          subsp. **batharitica**

subsp. **artemisioides**

Dry rocky slopes and wadi-sides; 450–650m.

Yemen (S). Endemic.

subsp. **batharitica** A.G. Miller in Edinb. J. Bot. 51 (1): 35 (1994).

Dry cliffs and rocky slopes on limestone; 20–300m.

Oman. Endemic.

6. NOTHOSAERVA Wight

Annual herb. Leaves opposite or alternate. Flowers minute, bisexual, in dense spikes; spikes sessile, solitary or usually clustered, axillary or on short axillary shoots. Tepals 3–4(–5), hyaline. Stamens 1 or 2, with filiform filaments; pseudostaminodes absent. Style short with a capitate stigma. Capsule 1-seeded, thin-walled, rupturing irregularly.

1. N. brachiata (L.) Wight, Ic. Pl. Ind. Or. 6: 1 (1853).

Slender herb to 15(–45)cm, glabrous or thinly hairy. Leaves narrowly elliptic to elliptic-oblong or ovate, 8–50 × 3–15(–20)mm, obtuse or subacute at the tip, entire, attenuate at the base. Spikes 3–5(–15) × c.2mm. Bracts hyaline, c.0.5mm; bracteoles minute, hyaline. Tepals broadly ovate, c.1.25mm, acute, villous externally, white and with a green midrib in the basal two-thirds. Capsule falling with the persistent perianth.

Locally common in areas liable to periodic flooding; 10m.

Saudi Arabia. Tropical Africa, India, Sri Lanka and Burma.

7. PSILOTRICHUM Blume

Perennial herbs. Leaves opposite, rarely absent. Flowers bisexual, in lax or compact spikes, often forming open panicles. Tepals 5, free. Stamens 5, shortly fused at the base; pseudostaminodes absent. Style slender; stigma capitate. Capsule 1-seeded, thin-walled, rupturing irregularly.

1. Stems aphyllous; flowers c.5mm long, in short, compact, up to 6-flowered spikes **4. P. aphyllum**
+ Stems leafy; flowers 2–3.5mm long, in lax spikes, forming large, open panicles 2

2. Leaves linear, glabrous **3. P. virgatum**
+ Leaves ovate, appressed-hairy or sericeous 3

3. Flowers 3–3.5mm long; nodes glabrous or with a few hairs; young leaves appressed-hairy, never silvery-sericeous beneath **1. P. gnaphalobryum**
+ Flowers 2–2.5mm long; nodes with tufts of long fine hairs; young leaves at least silvery-sericeous beneath **2. P. sericeum**

1. P. gnaphalobryum (Hochst.) Schinz in Vierteljahrsschr. Naturf. Ges. Zürich 57: 550 (1912). Syn.: *Achyranthes cordata* Hochst. & Steud. nom. nud; *Psilotrichum cordatum* Moq. in DC., Prodr. 13 (2): 280 (1849). Illustr.: Collenette (1985 p.43). Type: Saudi Arabia, *Schimper* 785 (K).

Perennial herb; stems erect or scrambling, much-branched, up to 2m, glabrescent, with tufted sericeous hairs at the nodes. Leaves ovate, 10–40 × 3–20mm, becoming linear above, softly appressed-hairy. Flowers in lax spikes forming open panicles, the ultimate branches zig-zagging; bracts c.1mm. Tepals green; outer oblong-ovate, 3–3.5mm, 3-nerved; inner with broad membranous margins. **Map 397, Fig. 54.**

Dry stony hills and rocky slopes in *Acacia-Commiphora* bushland; 0–2000m.

Saudi Arabia, Yemen (N & S). Egypt, Sudan, Ethiopia, Somalia and Kenya.

2. P. sericeum (Roxb.) Dalz. in Dalz. & A. Gibson, Bombay fl.: 216 (1861). Syn.: *Psilostachys sericea* (Roxb.) Hook.f. in Benth. & Hook.f., Gen. Pl., 3: 32 (1880). Illustr.: Fl. Trop. E. Afr.: 93 (1985).

Similar to *P. gnaphalobryum* but the nodes with a ring of long (c.5mm) fine hairs; at least young leaves silvery-sericeous beneath; flowers smaller, 2–2.5mm long. **Map 398.**

Amongst boulders and on sandy soil etc. near the sea; 0–15m.

Socotra. Kenya, Tanzania, Somalia and India.

3. P. virgatum C.C. Townsend in Kew Bull. 35: 377 (1980).

Perennial herb, glabrous throughout; stems virgate, 50–60cm. Leaves glaucous, linear, 10–60 × 0.25–2mm. Flowers in lax spikes forming open panicles, the ultimate branches not zig-zagging; bracts c.0.5–0.75mm. Tepals whitish green; outer narrowly oblong, c.2.5mm, 3-nerved; inner with membranous margins. **Map 399.**

Rocky slopes in open *Acacia-Commiphora* bushland; 50–200m.

Yemen (S), Oman. Somalia.

4. P. aphyllum C. C. Townsend in Kew Bull. 35: 134 (1980). Type: Socotra, *G. Popov GP/SO 314* (BM, EAH).

Aphyllous perennial herb; stems intricately-branched, appressed-hairy or glabrescent. Leaves, when present, aristate-subulate, c.4mm long. Flowers in a short, compact, up to 6-flowered spike. Outer tepals oblong-lanceolate, several-nerved, with narrow membranous margins; inner tepals with broad membranous margins. **Map 400.**

Hanging from rock crevices in a narrow limestone gully; 15m.

Socotra. Endemic.

A rare plant known only from a single gathering.

8. ACHYRANTHES L.

Perennial herbs. Leaves opposite, petiolate. Flowers bisexual, becoming deflexed and appressed to the axis, in elongated terminal and axillary spikes; bracts and bracteoles spine-tipped. Tepals 5, free. Stamens 5, filaments fused into a short basal cup, alternating with oblong pseudostaminodes. Style filiform, stigma capitate. Fruit 1-seeded, indehiscent, falling with the bracteoles and perianth.

1. A. aspera L., Sp. pl.: 204 (1753). Illustr.: Fl. Trop. E. Afr.: 103 (1985); Collenette (1985 p.40).

Erect or sprawling herb up to 1.5(–2)m. Leaves ovate to elliptic or broadly elliptic to broadly obovate, 2–8 × 0.8–4.5cm, acuminate to apiculate or acute to rounded, glabrescent to densely appressed-pubescent; petiole up to 2cm. Spikes up to 30cm. Flowers erect and congested at first, becoming deflexed and distant. Tepals whitish, greenish or pink, lanceolate, 3–7mm, acute. Pseudostaminodes bearing a fimbriate scale or rarely this obsolete. Fruit oblong-ovoid, 1–3mm. **Fig. 56.**

1. Leaves broadly elliptic to broadly obovate, rounded or apiculate; flowers 3.5–5mm   var. **aspera**
+ Leaves ovate to elliptic, acuminate; flowers 3–7mm   2

2. Robust plant; flowers (4.5–)5–7mm; leaves bluntly to shortly acuminate with a variable indumentum from glabrescent to tomentose but never green above and silvery canescent beneath   var. **pubescens**
+ Slender plant; flowers 3(–4.5)mm; leaves long acuminate, green above and silvery canescent beneath, at least when young   var. **sicula**

var. **aspera**. **Map 401.**

A common weed of cultivated land, also on rocky slopes, wadi-sides etc.; 0–3000m.

Saudi Arabia, Yemen (N & S), Socotra, Oman, UAE. Widespread in tropical and warm temperate parts of the world.

var. **pubescens** (Moq.) C.C. Townsend in Kew Bull. 29: 473 (1974). **Map 402.**

Habitat and external distribution as above.

Saudi Arabia, Yemen (N & S), Oman.

var. **sicula** L., Sp. pl.: 204 (1753). Syn.: *A. aspera* L. var. *argentea* (Lam.) Boiss., Fl. orient. 4: 994 (1879). **Map 403.**

Habitat and external distribution as above.

Saudi Arabia, Yemen (N & S), Socotra, Oman.

Recorded from N Oman by Schwartz (1939) but not seen there recently.

Species incompletely known

A. **capitata** Forsskal (1775 p.48).

9. ALTERNANTHERA Forsskal

Annual or perennial herbs. Leaves opposite. Flowers bisexual, in sessile axillary clusters (in Arabia). Tepals 5, free, all similar or strongly dissimilar, sometimes spine-tipped. Stamens 5, sometimes some without anthers; filaments united into a basal cup and alternating with pseudostaminodes. Style short; stigma capitate. Fruit 1-seeded, indehiscent.

1. Prostrate, mat-forming herb; bracts and outer tepals spine-tipped; tepals very dissimilar in form; leaves broadly obovate, the tips rounded or apiculate **1. A. pungens**
+ Erect or ascending herbs; bracts and outer tepals acute, mucronate or shortly aristate, unarmed; tepals similar in form, subequal; leaves elliptic to oblanceolate, the tips acute   2

2. Outer tepals 1.5–2.5mm, glabrous, 1-nerved   **2. A. sessilis**
+ Outer tepals 3.5–4mm, whitish-pilose on back, 3-nerved   **3. A. tenella var. bettzickiana**

1. A. pungens Kunth in Humb., Bonpl. & Kunth, Nov. gen. sp. 2: 206 (1818). Syn.: *Alternanthera repens* (L.) Link, Enum. hort. berol. alt. 1: 154 (1821).; Illustr.: Fl Pakistan 71: 40 (1974); Collenette (1985 p.41).

Prostrate, mat-forming perennial herb; stems up to 50cm, densely villous, often glabrescent with age. Leaves broadly ovate to broadly obovate, 1–3.5 × 0.3–2mm, the tip rounded or apiculate, mucronate, narrowing below into a short petiole. Bracts and bracteoles spine-tipped. Tepals dissimilar: outer pair lanceolate, c.5mm, spine-tipped; inner (adaxial) tepal oblong, c.3mm, dentate at apex; inner (lateral) pair c.2mm, bearing tufts of barbed hairs at base. Stamens 5, all bearing anthers. Fruit ± circular, c.2mm diam., compressed, apex rounded to emarginate. **Map 404, Fig. 57.**

Wadi-beds, and on tracks and roadsides, apparently preferring bare, well-trodden ground; 350–2200m.

Saudi Arabia, Yemen (N & S). A widespread weed of the tropics and subtropics.

2. A. sessilis (L.) DC., Cat. pl. horti monsp.: 77 (1813). Illustr.: Fl. Pakistan 71: 40 (1974); Collenette (1985 p.42).

Erect or ascending annual or perennial herb; stems up to 45cm, glabrous or with 2 lines of hair. Leaves elliptic to oblanceolate, 1.5–7 × 0.5–2cm, the tip acute, the base attenuate, ± sessile. Bracts and bracteoles mucronate. Tepals all similar, lanceolate, 1.5–2.5mm, acute, mucronate, glabrous. Stamens 5, 2 without anthers. Fruit obcordate, 2–2.5mm diam., strongly compressed, the tip emarginate. **Map 405, Fig. 57.**

Weed of cultivation, frequently by water; 150–1150m.

Saudi Arabia, Yemen (N). A widespread weed of the tropics and subtropics.

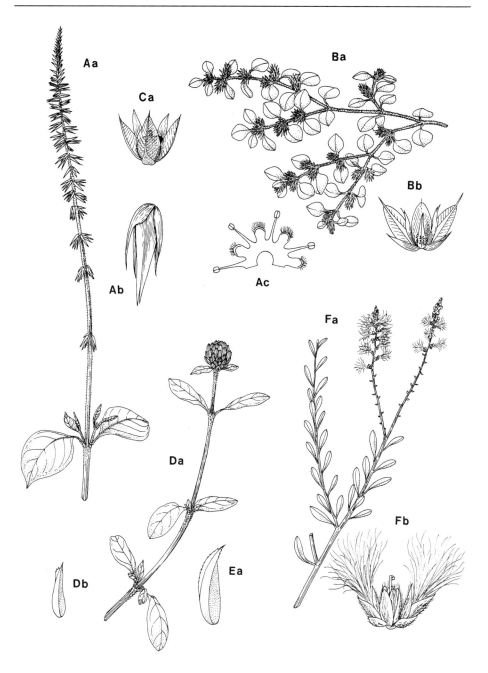

Fig. 57. Amaranthaceae. A, *Achyranthes aspera* var. *pubescens*: Aa, flowering branch (×0.6); Ab, flower (×6); Ac, androecium (×6). B, *Alternanthera pungens*: Ba, habit (×0.6); Bb, flower (×6). C, *A. sessilis*: Ca, flower (×6). D, *Gomphrena celosioides*: Da, flowering branch (×0.6); Db, bracteole (×4). E, *G. globosa*: Ea, bracteole (×4). F, *Saltia papposa*: Fa, fruiting branch (×0.6); Fb, part of inflorescence showing sterile flowers with plumose appendages (×4).

3. A. tenella Colla var. **bettzickiana** Veldk. in Taxon 27: 313 (1978).

Erect or ascending perennial herb; stems up to 45cm, villous at first but soon glabrescent. Leaves often purplish, elliptic to oblanceolate, 1–2.5 × 0.5–1.5cm, acute or acuminate, attenuate below into a short petiole. Bracts and bracteoles shortly aristate. Tepals all similar, lanceolate, 3.5–4mm, shortly aristate, the outer 2 whitish-pilose on back. Stamens 5, all bearing anthers. Fruit not seen. **Map 406.**

Saudi Arabia. Possibly a native of S America; now widely cultivated as an ornamental. A rare ?escape from cultivation, only collected once in Arabia.

10. GOMPHRENA L.

Annual herbs. Leaves opposite. Flowers bisexual, in globose or spicate terminal heads; bracteoles keeled and strongly laterally compressed, crested along the back on the midrib. Tepals 5, free; outer 3 flat, lanate at the base; inner 2 lanate to the tip. Stamens 5, filaments united into a tube; tube toothed at the apex. Style long, stigmas 2. Fruit 1-seeded, indehiscent, compressed.

1. Bracteoles with a conspicuous dorsal crest extending from the base to the tip; flower heads c.2cm diam. **1. G. globosa**
+ Bracteoles with a small dorsal crest towards the tip; flower heads c.1.25cm diam. **2. G. celosioides**

1. G. globosa L., Sp. pl.: 224 (1753).

Erect annual up to 50cm. Leaves elliptic-oblong to oblanceolate, 3–10(–15) × 1–3 (–6)cm, acute, attenuate into a short petiole below, thinly pilose with long appressed hairs. Flower-head globose or rarely ovoid, c.2cm diam., pinkish or purple. Bracts triangular-ovate, 3–5mm; bracteoles c.10mm, mucronate; crest conspicuous, irregularly serrulate, running ± from the base to the tip. Tepals 6–6.5mm. Fruit oblong-ovoid, c.2mm. **Map 407, Fig 57.**

Cultivated as an ornamental; 800–1200m.

Saudi Arabia, Yemen (N & S). A native of tropical America now widely cultivated in the warmer parts of the world.

2. G. celosioides Mart. in Nova Acta Phys.-Med. Acad. Caes. Leop.-Carol. Nat. Cur. 13 (1): 301 (1826). Illustr.: Fl. Trop. E. Afr.: 128 (1985).

Prostrate, decumbent or erect perennial herb up to 15(–30)cm. Leaves oblong to oblong-elliptic, 1–3(–8) × 0.5–1(–1.8)cm, obtuse to subacute, attenuate into a short petiole below, glabrous to densely lanate particularly beneath. Flower-head globose, becoming cylindrical in fruit, 1–7 × c.1.25cm, white. Bracts triangular-ovate, 2–4mm; bracteoles 5–6mm, mucronate; crest small, irregularly serrulate, confined to the upper third. Tepals 4.5–5mm. Fruit ovoid, compressed, c.1.5mm. **Map 408, Fig. 57.**

Track-sides; 500–700m.

Socotra. A native of S America now widely distributed throughout the tropics and subtropics.

11. SALTIA R. Br.

Shrubs. Leaves alternate or fascicled, somewhat fleshy. Flowers in terminal spikes, subtended by 2 bi-bracteolate modified sterile flowers. Tepals 5, free. Stamens 5, united into a basal cup; pseudostaminodes absent. Style slender; stigma capitate. Sterile flowers modified into plumose filiform appendages which greatly elongate in fruit. Fruit 1-seeded, indehiscent.

Endemic genus

1. S. papposa (Forsskal) Moq. in DC., Prodr., 13(2): 325 (1849). Syn.: *Achyranthes papposa* Forsskal (1775 p.48). Type: Yemen (N), *Forsskal* (C).

Shrub up to 1.5(–3.5)m; stems puberulous becoming glabrous. Leaves linear to narrowly obovate, 5–25 × 1–4mm, obtuse, attenuate below, subsessile, glabrous. Spikes 1–5cm, sometimes forming a terminal panicle. Bracts and bracteoles broadly ovate, cuspidate, 1–3mm long. Tepals all similar, oblong-ovate, 3.25–4 × c.1mm, thinly to densely sericeous on the back. Plumose appendages of sterile flowers inconspicuous at first, elongating to 15mm in fruit. Fruit oblong-ovoid. **Map 409, Fig. 57.**

Dry rocky slopes in *Acacia-Commiphora* bushland; 50–800m.

Yemen (N & S). Endemic.

Also very doubtfully recorded from J. Sidr near Jeddah in Saudi Arabia. Townsend (Kew Bull. 28: 143 (1973)) points out that the monotypic genus *Psilodgera* Susseng (Mitt. Bot. Staatssamml. München 1: 109 (1952)) belongs in *Saltia*. It was based on a specimen supposedly collected in Tanzania but which almost certainly came from Aden.

Family 40. CACTACEAE

A. G. MILLER

Succulent perennials of diverse habit. In Arabia stems cylindric; branches jointed, with flattened segments which are usually leafless and bear spines and bristles arising from cushion-like 'areoles'. Leaves small, soon deciduous. Flowers bisexual, solitary, sessile, actinomorphic. Perianth segments many, in several whorls, petaloid, free or ± fused below. Stamens many, inserted at the base of the perianth. Ovary inferior, 1-celled; ovules numerous on parietal placentas; style single with several stigmatic lobes. Fruit a spiny berry. Seeds numerous, encased in hard white arils.

CACTACEAE

OPUNTIA Miller

Much-branched shrubs with flattened, jointed branches; stem-segments obovate to elliptic, flattened; areoles bearing barbed bristles (glochids) and one or more stouter spines. Leaves small, subulate to cylindric. Flowers large and showy, yellow or orange; perianth segments spreading. Fruit a spiny berry.

Ellenberg, H. (1989). *Opuntia dillenii* als problematischer Neophyt im Nordjemen. *Flora* 182: 3–12.

1. Stem-segments elliptic to narrowly obovate; areoles with no spines or occasionally a few thinnish greyish spines present; fruits ellipsoidal, straight, usually orange-yellow when ripe **1. O. ficus-indica**
+ Stem-segments obovate to broadly obovate; areoles bearing 2–5 coarse yellow spines; fruits subglobose or obovoid, curved below, purple-red when ripe **2. O. dillenii**

1. O. ficus-indica (L.) Miller, Gard. dict. ed. 8, no. 2 (1768).

Erect or sprawling shrub up to 3(–5)m. Stem-segments dull green, elliptic to narrowly obovate, 30–40cm long × 15–20cm wide and 1–1.5cm thick; areoles with spines absent or occasionally bearing one or more greyish spines, the spines up to 1.5cm long; glochids deciduous. Leaves subulate, 3–4mm long. Flowers yellow or orange, 5–8cm in diam. Fruits ellipsoid, 5–8 × 3–6cm, usually orange-yellow when ripe. **Map 410.**

Naturalized on rocky hills and around villages; (600–)1500–3000m.

Saudi Arabia, Yemen (N), ? Oman. Probably a native of Mexico, now widely cultivated and naturalized in the Old World.

Planted as a hedge plant and for its edible fruit (the Prickly Pear). Very common on rocky slopes around villages on the SW Escarpment mountains and the High Plateau. Ghazanfar (1992) records a species of *Opuntia*, cultivated near villages in the J. Akhdar range of northern Oman, which is possibly referable to this species.

2. O. dillenii (Ker Gawl.) Haw., Suppl. pl. succ.: 79 (1819). Syn.: *O. stricta* (Haw.) Haw. var. *dillenii* (Ker Gawl.) L. Benson in J. Cact. Succ. Soc. Amer. 41: 126 (1969).

Similar to *O. ficus-indica* but the stem-segments obovate and somewhat glaucous; areoles bearing 2–5 coarse, yellow spines; flowers bright yellow; fruit subglobose or obovoid, curved below, c.5cm long, purple-red when ripe. **Map 411.**

Naturalized on rocky hills and around villages; 200–1200m.

Saudi Arabia, Yemen (N). Native of America, now widely cultivated and naturalized in the New World.

John Wood (pers. comm.) observes that the fruits are not particularly palatable and that in Yemen *O. dillenii* has been spread mainly by baboons. It is therefore common on isolated hillsides as well as by villages. It is considered a serious pest in some areas (Ellenberg op. cit.).

Family 41. MAGNOLIACEAE

A.G. MILLER

Trees or shrubs. Leaves alternate, simple, entire, coriaceous. Stipules large, deciduous, leaving prominent scars. Flowers solitary, axillary, bisexual, actinomorphic. Perianth segments 9–12, free, in whorls of 3, petaloid or the outer sepaloid. Stamens numerous, spirally arranged. Carpels numerous, free, spirally arranged on an elongated receptacle, unilocular, 2–many-seeded. Fruit a spike of follicles, dehiscing by longitudinal sutures; seeds 1 or more, suspended by thread-like funicles.

MICHELIA L.

Description as for the family.

1. M. champaca L., Sp. pl.: 536 (1753). Illustr.: Fl. Pakistan 64: 2 (1974).

Tree up to 7(–20)m. Leaves ovate-elliptic, 10–20(–25) × 4–6(–10)cm, acute to acuminate, the base obtuse or cuneate, pubescent becoming glabrous; petiole 1.5–2(–3)cm. Flowers yellow or orange, 5–7cm across, fragrant. Fruiting receptacle 4–10cm long; follicles obpyriform, 0.5–2cm across, glabrous or sparsely hairy, verrucose. **Map 412.**

Coffee plantations; 800m.

Yemen (N). Asia from Pakistan to Indonesia.

A rare plant in Arabia, known only from coffee plantations on J. Raymah in the outer escarpment mountains. Introduced.

Family 42. ANNONACEAE

A.G. MILLER

Trees or shrubs. Leaves simple, alternate. Stipules absent. Flowers bisexual, actinomorphic, usually 3-merous, solitary, clustered or cymose. Sepals 3, free. Petals 6, in 2 whorls. Stamens numerous, spirally arranged. Ovary superior; carpels 1–numerous, free; ovules 1–numerous; styles free; stigmas free or united. Fruiting carpels 1–several, either united to become a many-celled fruit or separate and sessile or stalked; seeds 1–several in each carpel.

A large tropical family with no genera native in Arabia. The family description is based on those genera likely to be cultivated in Arabia.

Polyalthia longifolia (Sonn.) Thw. is grown as an ornamental in Bahrain. It is a small tree with large (20–30 × 4–6cm) lanceolate leaves, greenish yellow flowers and fruits of several, stalked, 1-seeded carpels.

ANNONA L.

Small trees or shrubs. Petals 6, in 2 whorls, the inner whorl scale-like or absent. Carpels numerous, 1-seeded, ± united to form a 1-locular ovary with parietal placentation; styles clavate. Fruit a large fleshy syncarp formed from the fusion of the carpels and receptacle.

1. A. squamosa L., Sp. pl.: 537 (1753).

Shrub or small tree up to 4(–6)m tall. Leaves elliptic-oblong, 6–10 × 2–4cm, obtuse or subacute, glabrous above, pubescent or glabrescent beneath; petioles 4–10(–15)mm. Petals green or yellowish-brown, narrowly oblong, c.10(–25)mm long. Fruit ± globose, c.5cm diam., the surface covered with rounded lumps, green becoming blackish when mature, the seeds surrounded by white edible pulp. **Map 413.**

Cultivated and naturalized in drought-deciduous woodland; 400–1450m.

Saudi Arabia, Yemen (N & S). A native of the West Indies, widely cultivated in the tropics for its edible fruit.

The "Sweetsop" or "Custard Apple" is cultivated and now widely naturalized in the outer escarpment mountains of Yemen. Apparently "wild" trees have also been collected on J. Fayfa in the extreme SW of Saudi Arabia.

Schwartz (1939) reports *A. muricata* L. cultivated at Lahedj in Yemen. This differs in having a softly spiny fruit.

Family 43. LAURACEAE

D. R. McKEAN

Trees, shrubs or (in Arabia) parasitic twiners. Leaves opposite or alternate, simple, leathery and evergreen or (in Arabia) reduced to minute scales; stipules absent. Flowers small, bisexual, actinomorphic, 3-merous, in axillary inflorescences. Perianth of (4–)6 segments. Stamens (3–)12–18, the innermost often reduced to staminodes; anthers continuous with the filaments, 2(–4)-celled, opening by 2(–4) flaps from the base upwards. Ovary superior, 1-celled with a single pendulous ovule; style terminal, simple; stigma small. Fruit a 1-seeded drupe.

This large tropical family is represented in Arabia by a single genus which is sometimes segregated into a separate family, the Cassythaceae. The family description is based on the Arabian genus.

CASSYTHA L.

Stems twining, thread-like, adhering to the host by haustoria. Flowers white, in lax spikes; bracteoles 3. Perianth segments 6, in two whorls, the outer 3 resembling the bracteoles, the inner 3 longer. Stamens 12–18, in several dissimilar whorls of 3, the two outermost whorls with petaloid filaments and dehiscing inwards, the third whorl dehiscing outwards, the innermost whorls reduced to staminodes. Ovary globose with a short style and capitate stigma. Fruit enclosed in the enlarged perianth tube and crowned by the persistent perianth segments.

1. C. filiformis L., Sp. pl.: 1: 35 (1753). Syn.: *Volutella aphylla* Forsskal (1775 p.84). Illustr.: Collenette (1985 p.284).

Stems bright yellow, glabrous to tomentose. Leaf-scales ovate to subulate, 1–2mm. Inflorescence 2.5–5cm, 3- to 10-flowered. Flowers up to 2.2mm across; bracteoles triangular, c.1 × 1mm, ciliolate. Outer perianth segments similar to the bracteoles; inner perianth segments broadly ovate, c.2 × 1.75mm, rounded. Fruit white, spherical, 4–6mm diam. **Map 414.**

Acacia-Commiphora bushland and valley forest; 300–1200m.

Saudi Arabia, Yemen (N). A widely distributed plant throughout the tropics of Africa and Asia.

Parasitic and often completely enveloping trees and bushes in a web of thread-like stems. Recorded on a variety of species including: *Mimusops laurifolia*, *Ficus* spp., *Salvadora persica* etc.

Family 44. RANUNCULACEAE

A.G. MILLER & J.A. NYBERG

Annual or perennial herbs, rarely woody climbers (*Clematis*). Leaves alternate or rarely opposite (*Clematis*), usually deeply divided or compound; stipules minute or absent. Flowers actinomorphic or zygomorphic, solitary, racemose or paniculate. Perianths uniseriate or biseriate. Sepals 3–5(–8), free, often petaloid. Petals 5–8, rarely more, sometimes absent, free or united, often nectar-secreting. Stamens numerous, spirally arranged. Carpels numerous, rarely 1–5, free or rarely fused; ovules 1–numerous, marginal. Fruit a cluster of achenes or follicles.

1.	Woody climbers; leaves opposite	**1. Clematis**	
+	Annual or perennial herbs, never climbing; leaves alternate		2
2.	Flowers zygomorphic, spurred	**6. Delphinium**	
+	Flowers actinomorphic, without spurs		3
3.	Fruit a group of several fused follicles; petals 2-lobed	**5. Nigella**	
+	Fruit a group of free achenes; petals absent or entire		4
4.	Petals absent; sepals 4, often petaloid; perennial herb up to 1m	**2. Thalictrum**	
+	Petals present; sepals 5; annual or perennial herbs up to 50cm		5

5.	Petals without a basal nectary pouch; leaves pinnatisectly divided into linear segments; never aquatic **3. Adonis**
+	Petals with a basal nectary pouch; leaves palmately divided, if with linear segments then a submerged aquatic **4. Ranunculus**

1. CLEMATIS L.

Woody climbers. Leaves opposite, pinnately or bi-pinnately divided, the leaflets petiolulate. Flowers actinomorphic, solitary or in few- to many-flowered panicles. Sepals 4, sometimes petaloid. Petals absent. Carpels many, usually uniovulate. Fruit a cluster of achenes bearing persistent plumose styles.

1.	Leaves bipinnate, usually with more than 10 leaflets **4. C. orientalis**
+	Leaves pinnate with 3–7 leaflets 2
2.	Leaves rusty-tomentose beneath; lateral leaflets asymmetric; sepals 3.5–5cm, erect at anthesis **3. C. longicauda**
+	Leaves sometimes tomentose but never rust-coloured beneath; lateral leaflets symmetric; sepals 0.8–15(–30)mm, spreading at anthesis 3
3.	Leaflets often trilobed, the margins crenate-serrate to the base and the tip, tomentose or rarely thinly pubescent beneath **1. C. hirsuta**
+	Leaflets simple, the margins crenate-serrate at the middle but entire at the tip and base, glabrous or thinly pubescent beneath **2. C. simensis**

1. C. hirsuta Guillemin & Perr., Fl. Seneg. tent. 1: 1 (1831). Syn.: *C. incisodentata* A. Rich. (1847–8 p. 2); *C. wightiana* Wallich, Numer. List n.4674 (1831). Illustr.: Collenette (1985 p.416) as *C. incisodentata*.

Stems up to 4m. Leaves with 3, 5 or 7 leaflets; leaflets often 3-lobed, ovate to broadly ovate, 3–9 × 2–8cm, the margin crenate-serrate, sparsely pubescent or glabrous above, tomentose or more rarely thinly pubescent beneath. Inflorescence many-flowered. Sepals white or cream, c.8–15(–30) × 5–6mm, tomentose. Achenes ellipsoid, 2.5–3.5 × 2–2.5mm, the persistent style 2–3.5cm. **Map 415, Fig. 58.**

On terrace-walls and hedges, in bushland by water-courses; 700–2000(–2900)m.

Saudi Arabia, Yemen (N). Tropical Africa and India

2. C. simensis Fresen., Beitr. Fl. Abyssin. 2: 267 (1837). Illustr.: Collenette (1985 p.416).

Stems up to 2(–10)m. Leaves with 5 or sometimes 3 leaflets; leaflets simple, ovate, 4.5–12 × 2.5–5.5cm, tip acuminate, the margin crenate-serrate at the middle and entire at the base and tip, glabrescent or sometimes pubescent beneath. Inflorescence many-flowered. Sepals white or cream, c.8–10 × 3–5mm, tomentose. Achenes ellipsoid, c.3 × 2.5mm, the persistent style up to 4cm. **Map 416, Fig. 58.**

On terrace walls and hedges, in bushland by watercourses; 1700–1900m.

Saudi Arabia, Yemen (N & S). Tropical Africa.

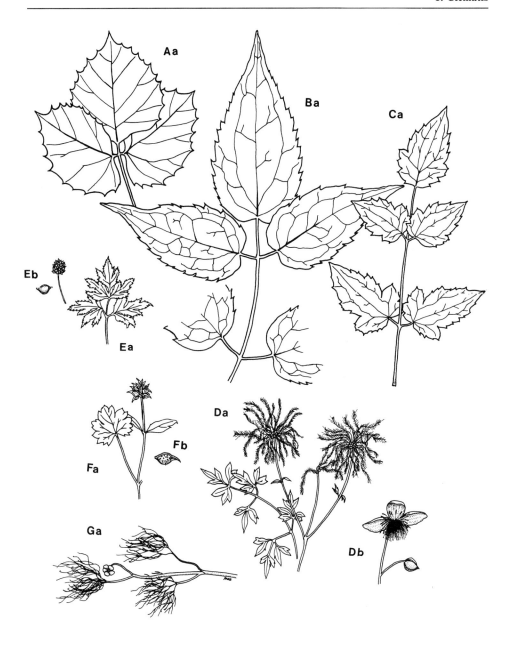

Fig. 58. Ranunculaceae. A, *Clematis longicauda*: Aa, leaf. B, *C. simensis*: Ba, leaf. C, *C. hirsuta*: Ca, leaf. D, *C. orientalis*: Da, leaf and fruits; Db, flower and bud. E, *Ranunculus multifidus*: Ea, leaf; Eb, fruiting head and achene. F, *R. muricatus*: Fa, leaf and fruiting head; Fb, achene. G, *R. rionii*: Ga, flowering shoot. All × 0.4 except achenes × 3.

3. C. longicauda Steud. ex A. Rich. (1847 p.2).

Stems up to 8m. Leaves with 3, rarely 5 leaflets; leaflets simple or sometimes 3-lobed, ovate to transversely ovate, 4–9 × 3.5–7cm, the lateral leaflets asymmetric, the margin dentate, glabrescent above rusty-tomentose beneath. Inflorescence few-flowered. Sepals yellowish brown, 3–5 × 1.5cm, tomentose. Achenes ellipsoid, c.4 × 2mm, the persistent style up to 7.5cm. **Map 417, Fig. 58.**

1950m.

Yemen (N). Ethiopia

Known only from a single sterile specimen (J. Nasira near Hajjah, *Müller-Hohenstein & Deil* 701). Flowering material is required so that its presence in Arabia can be confirmed. The fruiting and flowering measurements given above are taken from Ethiopian plants.

4. C. orientalis L., Sp. pl.: 543 (1753).

Stems up to 6m. Leaves bipinnate with more than 10 leaflets; leaflets lanceolate, 15–35 × 5–10mm, entire or with 1–2 teeth, glabrescent. Flowers solitary or in few-flowered panicles. Sepals yellowish, 6–18 × 4–10mm. Achenes ellipsoid, 3–3.5 × 2–2.5mm, the persistent style 3–3.5cm. **Map 418, Fig. 58.**

Scrambling over rocks in *Juniperus* woodland and by villages; 1900–3000m.

Oman. SE Europe, SW Asia, NW India and C Asia.

A very variable and widespread species in the Middle East. The Arabian plants differ from plants elsewhere in the range of the species by their more divided leaflets.

2. THALICTRUM L.

Perennial herbs. Leaves alternate, pinnately divided, the bases sheathing, the leaflets petiolulate. Flowers actinomorphic, small, in panicles. Sepals 4, usually soon deciduous. Petals absent. Carpels numerous, uniovulate. Fruit a cluster of free achenes, with short persistent styles.

1. T. minus L., Sp. pl.: 546 (1753).

Glabrous or glandular-hairy herb; stems erect, up to 1m. Leaves triangular in outline, up to 30cm long; ultimate segments oblong to obtriangular, 3–7-lobed or toothed, glabrous or sparsely glandular-pubescent beneath. Flowers in lax panicles. Sepals greenish yellow, c.4mm. Stamens c.7mm. Achenes sessile, ellipsoid, ribbed, c.3mm long, glabrous or sparsely glandular. **Map 419, Fig. 59.**

Cliffs and rocky gullies; 1600–3000m.

Yemen (N & S). Europe, NW Africa, Ethiopia, South Africa, SW Asia and Siberia.

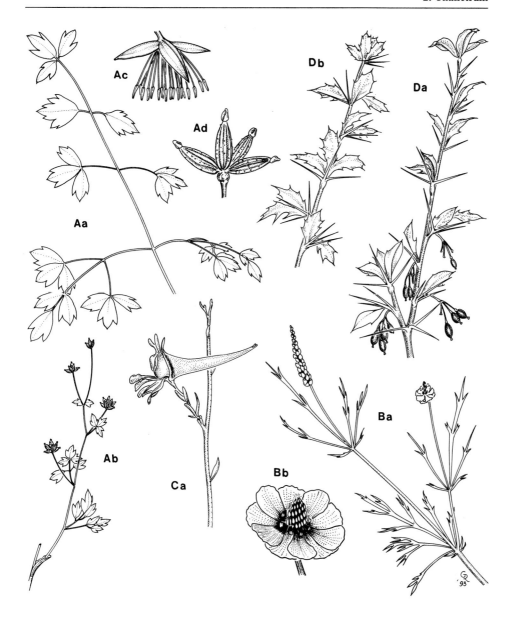

Fig. 59. Ranunculaceae. A, *Thalictrum minus*: Aa, leaf (× 1); Ab, infructescence (× 1); Ac, flower (× 4); Ad, fruit (× 10). B, *Adonis dentata*: Ba, flowering and fruiting shoot (× 1); Bb, flower (× 3). C, *Delphinium sheilae*: Ca, flower (× 3). **Berberidaceae.** D, *Berberis hostii*: Da, fruiting branch (× 1); Db, sterile branch (× 1).

3. ADONIS L.

Annual or perennial herbs. Leaves alternate, 2–3-pinnate with linear segments. Flowers actinomorphic, solitary at the ends of branches. Sepals 5, membranous. Petals 5–20, showy. Carpels numerous, uniovulate. Fruit an elongated head of beaked achenes.

1. A. dentata Del. (1813 p.143). Illustr.: Collenette (1985 p.415).

Glabrous annual herb; stems erect, up to 20cm. Leaf-segments linear. Sepals 3–5mm. Petals orange or yellow, 5–10 × 3–5mm. Anthers violet-black. Fruiting head up to 1.5cm long; achenes wrinkled, c.2 × 1mm. **Map 420, Fig. 59.**

Sandy and gravelly plains; 885m.

Saudi Arabia, Yemen (N), Kuwait. Cyprus, N Africa and SW Asia.

4. RANUNCULUS L.

Terrestrial or aquatic, annual or perennial herbs. Leaves alternate, simple or compound, with or without stipules. Flowers actinomorphic, solitary or in panicles. Sepals 5 (in Arabia). Petals yellow or white, 5 or more, bearing a basal nectary pit which is often covered by a scale. Carpels few to many, uniovulate. Fruit a group of free, beaked achenes.

1.	Aquatic plants; leaves submerged and divided into linear segments; petals white with a yellow base	2
+	Terrestrial plants; leaves with broad laminas; petals yellow	3
2.	Fruiting peduncle shorter than the subtending leaf; petals 3–5mm long	**4. R. rionii**
+	Fruiting peduncle longer than the subtending leaf; petals 6–10mm long	**5. R. sphaerospermus**
3.	Perennial; achenes 1.5–2mm long, including the beak	**3. R. multifidus**
+	Annuals; achenes 4–8mm long, including the beak	4
4.	Fruiting peduncle equalling or shorter than the subtending leaf; lower leaves suborbicular to broadly ovate, entire and crenate-dentate or 3-lobed into broad segments	**1. R. muricatus**
+	Fruiting peduncle much longer than the subtending leaf; lower leaves ovate, 3-lobed, the lobes themselves divided into narrow segments	**2. R. cornutus**

Subgenus **Ranunculus**

Terrestrial herbs with leaves various but never submerged and divided into linear segments; petals yellow or red; achenes smooth, sculptured or tuberculate, rarely transversely ridged, usually distinctly beaked.

1. R. muricatus L., Sp. pl.: 555 (1753). Illustr.: Collenette (1985 p.417); Mandaville (1990 pl. 11).

Terrestrial annual herb, 5–30cm, glabrous or sparsely pilose. Lower leaves simple or 3-lobed, subcircular to transversely ovate, 0.5–7.5 × 0.5–10cm, the margin coarsely crenate-dentate, the base cordate or truncate; petiole 1–13cm; upper leaves with narrower segments and more shortly petiolate. Flowers on a peduncle which is equal to or shorter than the subtending leaf. Sepals 5, reflexed. Petals yellow, 3–6(–9)mm. Achenes narrowly elliptic, 6–8 × 3–4mm, compressed, faces bearing short spines, margin grooved, with a hooked beak at the tip. **Map 421, Fig. 58.**

Weed of irrigated areas; 300–2000m.

Saudi Arabia, Oman, UAE. S Europe, SW Asia, N Africa.

2. R. cornutus DC., Syst. nat. 1: 300 (1817).
Similar to *R. muricatus* but the lower leaves ovate, 3-lobed with the lobes themselves divided into narrow segments; peduncles much exceeding the subtending leaves. **Map 422.**

Weed of irrigated areas; 10–100m.

Saudi Arabia. SE Europe and SW Asia.
A rare weed of oases in E Saudi Arabia.

3. R. multifidus Forsskal (1775 p.102). Syn.: *R. forskoehlii* DC., Syst. nat. 1: 303 (1817). Illustr.: Collenette (1985 p.417) Type: Yemen (N), *Forsskal* (C).
Terrestrial perennial herb, 10–50cm, pilose. Leaves triangular, deeply pinnately-dissected into lanceolate segments, 3–9 × 4–10cm, the segments serrate; petioles 1–10cm. Sepals 5, reflexed. Petals yellow, 3–7mm. Achenes obovate, c.1.5 × 1.5–2mm, compressed, faces smooth or with scattered tubercles, with a hooked beak at the tip. **Map 423, Fig. 58.**

Wet and irrigated areas, 1800–2800m.

Saudi Arabia, Yemen (N). Tropical and South Africa

Subgenus **Batrachium** (DC.) A. Gray
Aquatic herbs with submerged leaves finely dissected into linear segments; petals white with a yellow base; achenes transversely ridged with a short beak.

4. R. rionii Lagger in Flora 31: 49 (1848). Syn.: *R. trichophyllus* Chaix var. *rionii* (Lagger) Rikli in Schinz & R. Keller, Fl. Schweiz: 193 (1900).
Aquatic, annual or perennial herb, glabrous or sparsely pilose. Leaves all submerged, up to 6cm long, subsessile or shortly petiolate, finely divided into linear segments. Fruiting peduncle shorter than the subtending leaf. Sepals 5. Petals white with a yellow base, 3–5mm. Achenes elliptic, 0.75–1mm, somewhat compressed, faces wrinkled, shortly beaked at the tip. **Map 424, Fig. 58.**

Submerged in pools and slow-running water; 2100–2900m.

Saudi Arabia, Yemen (N). N Temperate regions, SW Asia and southern Africa.

Literature records (e.g. Schwartz p.59) of *R. aquatilis* L. from Arabia are probably referable to this species. *R. aquatilis* differs in its larger achenes (1.5–2mm) and petals (up to 10mm).

5. R. sphaerospermus Boiss. & Blanche, Diagn. pl. orient. sér. 2, (3)5: 6 (1856). Illustr.: Collenette (1985 p.417) as *R. trichophyllus*.

Similar to *R. rionii* but differing in the fruiting peduncles exceeding the subtending leaves in fruit and the longer (6–10mm) petals. **Map 425.**

Pools in slow-running water; 2550–2900m.

Saudi Arabia. SE Europe, SW Asia and the Himalayas.

Known only from two pools in the Asir mountains of Saudi Arabia. The determination of the Arabian material as *R. sphaerospermus* is provisional and therefore the presence of this species in Arabia requires confirmation. The material examined (*Collenette* 1170 and 4669) has the relatively long petals and elongating peduncles which distinguish it from the related species, *R. rionii* and *R. trichophyllus*. However, in the Arabian material the achenes are c.1.5mm long, whereas in non-Arabian material they are usually less than 1mm. The species of subgenus *Batrachium* are notoriously difficult and further collecting in Arabia is needed.

5. NIGELLA L.

Annual herbs. Leaves alternate, 1–3-pinnatisect into ± linear segments. Flowers solitary at the ends of branches. Sepals 5, petaloid (in Arabia). Petals 5(–8), nectariferous, with a bent claw and two lobes, the inner lobe simple, the outer lobe bifid. Stamens numerous. Carpels 2–10, free or partly fused. Fruit of several follicles, many-seeded, free or partly fused to form a capsule and crowned by the persistent styles.

1. N. sativa L., Sp. pl.: 534 (1753).

Erect herb up to 50cm, pubescent or glandular-hirsute. Leaf-segments linear-lanceolate. Sepals whitish, petaloid, ovate, c.7–14 × 6–8mm, shortly clawed. Petals much smaller than the sepals. Anthers mucronate. Follicles united to the apex. **Map 426.**

Cultivated in the highlands; 300–2300m.

Saudi Arabia, Yemen (N & S). Native of SW Asia, widely cultivated elsewhere for its seeds. SW Asia, N Africa and S Europe.

Cultivated as a culinary herb and for medicinal purposes. Records of *N. arvensis* L. (distinguished by the follicles which are free at the apices) probably refer to this species.

6. DELPHINIUM L.
(including *Consolida* (DC.) S.F. Gray)

Annual or perennial herbs. Leaves alternate, palmately divided or laciniate. Flowers zygomorphic, in racemes. Sepals 5, petaloid, the 2 uppermost fused towards the base and forming a spur. Petals 4, free or united, the uppermost produced into a nectar-secreting spur within the sepal spur. Stamens numerous. Ovary of 1–9 carpels. Fruit of 1–5 free or partly fused follicles.

1. Flowers violet; petals united into a 3-lobed limb; follicles solitary **3. D. orientale**
+ Flowers bright blue or pale blue suffused with purple; petals free; follicles 2–5 2

2. Annual; stems retrorsely pubescent at the base **1. D. sheilae**
+ Perennial; stems viscid-pubescent throughout **2. D. penicillatum**

Subgenus **Delphinium**
Annual or perennial. Petals free. Carpels 2–5.

1. D. sheilae Kit Tan in Notes Roy. Bot. Gard. Edinburgh 42: 17 (1984). Illustr.: Collenette (1985 p. 417). Type: Saudi Arabia, *Collenette* 4531 (E, K).

Annual; stem erect, up to 85cm, retrorsely pubescent at the base. Basal leaves soon withering, apparently up to c.8cm, long-petiolate, laciniate at the tip; cauline leaves entire or tripartite, oblanceolate to lanceolate, up to 4cm. Flowers in lax racemes, bright blue or pale blue flushed with purple, c.2cm long; spur 1.5cm long. Fruit glabrous, 6–7mm. **Map 427, Fig. 59.**

N facing cliffs on granite and on open basaltic hillsides; 1450–1600m.

Saudi Arabia. Endemic.
Known only from two gatherings in NW Saudi Arabia.

2. D. penicillatum Boiss. in Ann. Sci. Nat. Bot. sér. 2, 16: 369 (1841). Type: Oman, *Aucher-Eloy* 4034 (K, P)

Perennial; stem erect, up to 30cm, viscid-pubescent throughout. Leaves suborbicular; blade 1.5–2cm, tripartite; segments linear, bipinnatifid; petioles 3–6cm. Flowers in dense racemes, pale blue, c.1.5cm long; spur c.1cm long. **Map 428.**

Oman. Iran.
In Arabia known only from a single gathering made near Muscat in the last century, apparently now extinct there.

Subgenus **Consolida** (DC.) Huth.
Annual. Petals united into a 3–5-lobed or subentire limb. Carpel solitary.

3. D. orientale J. Gay in Des Moul., Cat. rais. pl. Dordogne: 12 (1839). Syn.: *D. ajacis* L., Sp. pl.: 531 (1753), nomen confusum; *Consolida orientalis* (J. Gay) Schröd. in Abhandl. Zool.-Bot. Ges. Wien 4(5): 62 (1909).

Annual; stem erect, up to 10(–70)cm, crisped-hairy, with glandular hairs above. Leaves pinnately ternate with linear segments. Flowers in dense racemes, violet, c.2cm; spur 5–10mm, shorter than the sepals; bracteoles reaching to at least the base of the flowers. Follicles pubescent. **Map 429.**

In thickets of *Ziziphus nummularia*.

Saudi Arabia. N Africa, Mediterranean region, SW & C Asia.

In Arabia known only from a single gathering from C Saudi Arabia.

Family 45. BERBERIDACEAE

D.F. CHAMBERLAIN

Spiny shrubs (in Arabia). Leaves alternate, simple (in Arabia), borne on short shoots. Stipules usually absent. Flowers yellow, actinomorphic, bisexual, usually 3-merous, in clusters, racemes or panicles. Sepals usually 6, in 2 whorls, free. Petals similar to the sepals, with basal nectaries or nectariferous scales. Stamens 6, opposite the petals, the anthers dehiscing by valves. Ovary superior, of 1 carpel with few to many basal or marginal ovules; style short or absent; stigma rounded. Fruit a berry.

BERBERIS L.

Description as for the family.

1. B. holstii Engler in Abh. Königl. Akad. Wiss. Berlin 1894: 64 (1894). Syn.: *B. aristata* sensu Oliv, Fl. Trop. Afr. 1: 51 (1868) non DC.; *B. petitiana* C.K. Schneider in Bull. Herb. Boissier sér. 2, 5: 455 (1905); ? *B. forskaliana* C.K. Schneider loc. cit.: 456.

Shrub up to 3m; young stems grooved, purplish red, glabrous; spines numerous, stout, 3-fid. Leaves thick, obovate, 7–18 × 3–8mm, usually entire, occasionally with 1–3 pairs of spine-tipped teeth, glabrous, sessile. Flowers yellow, 6–12, in clusters or short simple or compound racemes. Petals and inner sepals c.6mm. Berries oblong-ellipsoid, 7–8 × 4–5mm, dark red when ripe, pruinose; style 0.5–0.8mm. **Map 430, Fig. 59.**

Cliffs and grassland in Yemen; *Juniperus* woodland in Oman; (1400–)2000–3200m.

Yemen (N), Oman. Tropical E Africa

Berberis forskaliana is doubtfully distinct from *B. holstii*. The type specimen (*Schweinfurth* 1682) lacks both flowers and fruits but differs from *B. holstii* in its generally more spiny leaves (with 3–4 pairs of spines). There are two further gatherings (*Miller & King* 5329 and *Chaudhary* s.n.) from Yemen with spiny leaves; both are also sterile and may belong to the same taxon.

Family 46. MENISPERMACEAE

A.G. MILLER

Dioecious, twining or trailing lianes or erect shrubs, sometimes bearing spiny cladodes. Leaves alternate, petiolate, sometimes peltate, simple, entire or lobed. Stipules absent. Flowers unisexual, actinomorphic, in sessile clusters or pedunculate panicles or in dense clusters on the cladodes. Male flowers: sepals free, 3–15, in 2 or 4–5 whorls; petals free, 2–4 or 6, in 2 whorls; stamens 2–9, free or united into a central column. Female flowers: sepals and petals similar to those of the male flowers; staminodes present or not; carpels 1 or 3–6, free; ovules 2 (one soon aborting). Fruits drupaceous, bearing the scar of the style subterminally or near the base due to excentric growth; seeds surrounded by a bony endocarp, horseshoe-shaped.

1.	Leaves peltate	**3. Stephania**
+	Leaves not peltate	2
2.	Leaves ovate or oblong-ovate to obovate or absent, the base cuneate to rounded; carpels 3–6; twining or trailing lianes or erect shrubs bearing spiny cladodes	**1. Cocculus**
+	Leaves broadly ovate-triangular, the base cordate; carpels 3; twining lianes	**2. Tinospora**

1. COCCULUS DC.
Cebatha Forsskal (1775); *Laeba* Forsskal (1775).

Lianes or shrubs with branches bearing spiny cladodes. Leaves entire or lobed. Flowers in clusters. Sepals 6–15, in 2 or 4–5 whorls. Petals 6, auriculate at the base and clasping the filaments. Male flowers with 6–9 stamens. Female flowers with 3–6 carpels. Drupes globose, with a style-scar near the base, laterally compressed and ribbed on the lateral surfaces.

1.	Shrub bearing spine-tipped cladodes; leaves soon deciduous	**3. C. balfourii**
+	Twining or trailing liane; leaves persistent	2
2.	Branches usually white-pubescent; leaves usually glabrescent; fruits dark red, the septum surrounding the cavity at the centre of the endocarp not perforated	**1. C. pendulus**
+	Branches yellowish-tomentose, the bark brown; leaves usually tomentose; fruits black with a blue bloom, the septum surrounding the cavity at the centre of the endocarp perforated	**2. C. hirsutus**

1. C. pendulus (J. Forster) Diels in Engl. & Prantl, Nat. Pflanzenfam. 4, 94: 237 (1910). Syn.: *C. cebatha* DC., Syst. nat. 1: 527 (1817); *C. laeba* DC. op. cit.: 529. Illustr.: Fl. Pakistan 74: p.6 (1974); Collenette (1985 p.367).

Twining or trailing liane; stems up to 10m; young branches whitish pubescent; trunk white. Leaves ovate to oblong-ovate or deltoid, sometimes 1–2-lobed at the base, 0.5–5 × 0.3–2.5cm, obtuse and mucronulate or emarginate, the margin entire, the base cuneate or rounded, thinly pubescent, often becoming glabrous, whitish green; petiole 0.1–1cm. Flowers greenish yellow; clusters axillary or on leafless branches. Sepals 6, in 2 whorls: the outer ovate, 0.7–1 × 0.4–0.6mm; the inner ± circular, 1.5–2 × 1.5mm. Petals emarginate, 1–2mm. Carpels 6. Fruits 4–6mm long, dark red, the septum surrounding the cavity at the centre of the endocarp not perforated. **Map 431, Fig. 60.**

Dry rocky slopes, sandy plains, wadi-beds and cliffs in open semi-desert bushland; 5–1900m.

Saudi Arabia, Yemen (N & S), Socotra, Oman, UAE. Qatar. N Africa, tropical E Africa southwards to Kenya, S Iran, Pakistan & India.

2. C. hirsutus (L.) Theob. in Mason, Burmah (ed. Theob.) 2: 657 (1883). Syn.: *Cebatha villosa* (Lam.) C. Christ. (1922 p.37); *Cocculus villosus* DC., Syst. nat. 1: 525 (1817). Illustr.: Fl. Pakistan 74: 6 (1974); Fl. Trop. E. Afr.: 11 (1956).

Similar to *C. pendulus* but generally more densely hairy with yellowish tomentose branches and leaves; bark rough, brown; leaves oblong-ovate, up to 4 × 9cm; fruits black with a blue bloom, the septum surrounding the centre of the endocarp perforated. **Map 432, Fig. 60.**

Open *Acacia* bushland, on cliffs and terraces; 500–1550m.

Saudi Arabia, Yemen (N). Tropical Africa and tropical Asia to S China.

3. C. balfourii Schweinf. ex Balf. f. (1882 p.500); Illustr.: Balf.f. (1888 t.1). Type: Socotra, *Balfour* 439 (lecto. E, K).

Densely branched, ± aphyllous shrub; stems up to 1m, puberulous, bearing cladodes; cladodes flattened or ± terete, spine-tipped, 3–9 × 0.1–1cm. Leaves soon deciduous, ovate to obovate, 1–2 × 0.5–1cm, retuse or apiculate, entire, the base cuneate, shortly petiolate. Flowers cream, in sessile clusters on the cladodes or on the stem below the cladodes. Sepals 12–15, in 4–5 whorls, reducing in size outwards; inner sepals broadly obovate, c.1.5 × 1.25mm. Petals acute, c.1.25mm. Carpels 3. Fruits c.4mm, bright red. **Map 433, Fig. 60.**

On the mainland on rocky slopes (usually on limestone) in *Acacia-Commiphora* bushland; on Socotra in low *Cephalocroton socotranus-Rhus thyrsiflora* bushland, on granite; 50–1600m (above 460m on Socotra).

1. Cocculus

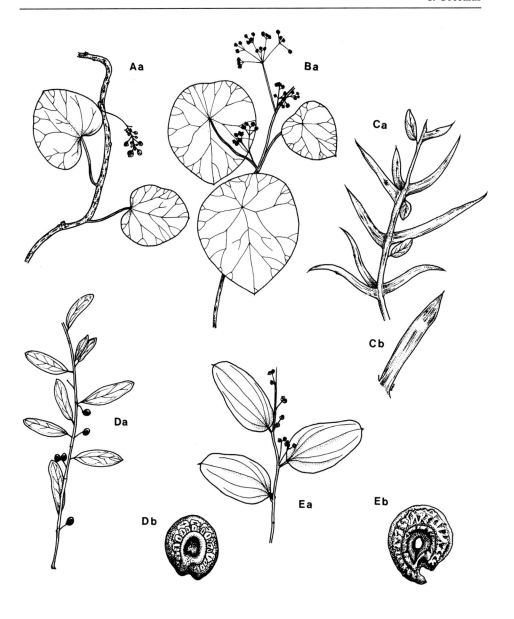

Fig. 60. Menispermaceae. A, *Tinospora bakis*: Aa, flowering branch (×½). B, *Stephania abyssinica*: Ba, flowering branch (×½). C, *Cocculus balfourii*: Ca, flowering branch (×½); Cb, cladode (×1). D, *C. pendulus*: Da, fruiting branch (×½); Db, fruit (×5). E, *C. hirsutus*: Ea, flowering branch (×½); Eb, fruit (×5).

Yemen (S), Socotra, Oman. Endemic.

For a discussion of this species see Forman, L.L. (1980). *Kew Bull.* 35: 379–381.

2. TINOSPORA Miers

Twining lianes. Leaves simple, entire. Flowers in raceme-like panicles. Sepals 6, in 2 whorls. Petals 6, fleshy with inrolled margins. Male flowers: stamens 6. Female flowers: carpels 3. Drupes ovoid, with the style-scar subterminal and with a rugose endocarp.

1. T. bakis (A. Rich.) Miers in Hook., Niger Fl.: 215 (1849).

Stems succulent, glabrous, up to 5m. Leaves broadly ovate-triangular, $2–3 \times 3–5$cm, with a rounded to shortly acuminate tip and cordate base, glabrous or puberulous beneath; petiole c. 0.8mm. Inflorescence 2.5(–12)cm long. Flowers greenish yellow; outer sepals ovate, c.1×0.5mm; inner sepals obovate c.4×2mm; petals c. 1.8×1mm. Drupes 6×4mm. **Map 434, Fig. 60.**

Twining on shrubs in sandy wadis; c.150m.

Saudi Arabia, ?Yemen (N). Sahel region and tropical E Africa.

In Arabia very rare, known only from a single wadi in the extreme SW of Saudi Arabia. J. Wood (pers. comm.) also collected a sterile specimen of *Tinospora* (at about 300m between Bait Faqih and Mansuriah in Yemen) which may be referable to this species.

3. STEPHANIA Lour.

Twining lianes. Leaves simple, entire, peltate. Flowers in pedunculate, umbel-like panicles. Male flowers: sepals 6–8; petals 3–4; stamens 2–6, fused into a central column with the anthers in a horizontal ring. Female flowers: sepals 3–6; petals 2–4; carpel solitary, 3-lobed at the apex. Drupe subglobose, with a style-scar near the base, the endocarp with 3 rows of small prickles or tubercles.

1. S. abyssinica (Dillon & A. Rich.) Walp., Repert. bot. syst. 1: 96 (1842).

Scrambling or climbing glabrous herb. Leaves broadly ovate to suborbicular, $2–9 \times 2–8$cm, with a rounded to obtuse tip and rounded base; petioles 3–10cm. Panicles solitary or in clusters of 2–4; peduncles 4–10cm. Flowers yellow-green. Sepals ovate to obovate, 1.2–2.5mm. Petals broadly ovate to suborbicular, 0.8–1.2mm. Drupe 5–8mm. **Map 435, Fig. 60.**

Scrambling over shrubs and walls; c.2200m.

Yemen (N). Tropical and southern Africa.

Family 47. NYMPHAEACEAE

D.R. McKEAN

Aquatic perennials with stout rhizomes. Leaves floating or sometimes submerged, alternate, peltate, cordate, long-petiolate. Stipules present. Flowers actinomorphic, bisexual, floating, solitary, usually large and showy, long-pedunculate. Sepals 4, free. Petals numerous, gradually passing into the stamens. Stamens numerous, the outer with petaloid filaments. Ovary semi-inferior, many-locular; carpels numerous, sunk into the receptacle, each with an inwardly-curved stylar appendage and a radiating stigmatic ridge; ovules numerous in each carpel; placentation parietal. Fruit a spongy capsule, ripening under water; seeds floating, arillate.

A widespread tropical family extending into the temperate zones and containing, in the broad sense, 6 genera. Only one, very doubtfully native, species is found in Arabia.

NYMPHAEA L.

Description as for the family.

1. N. nouchali Burm.f. var. **caerulea** (Savigny) Verdc. in Fl. Trop. E. Afr.: 7 (1989). Illustr.: loc. cit.: 8.

Leaves ± round, 15–40cm diam., entire or slightly undulate, slightly peltate; petiole up to 1.5m. Flowers 6–20cm diam. Sepals 3–8cm. Petals pinkish blue, 12–24, oblong-lanceolate, 3–8cm, acute. Stamens bright yellow. Carpels 14–21. Fruit depressed-globose, 2–4cm diam. **Map 436.**

In pools; 1700m.

Yemen (N). Throughout most of Africa.

Only known from a single pool, near Turbah, in Yemen. Probably introduced.

Family 48. CERATOPHYLLACEAE

A.G. MILLER

Aquatic perennial herbs, usually free-floating, monoecious. Leaves in whorls of 6–10, usually rigid and brittle, dichotomously branched into linear segments; ultimate segments minutely toothed. Flowers minute, sessile or shortly pedicellate, axillary, the male and female at different nodes. Perianth of 6–15 linear or strap-shaped lobes, united at the base; lobes truncate, with 2 teeth and a central reddish projection at the apex. Male flowers with numerous stamens; stamens spirally arranged; immature anthers resembling the perianth lobes, later swollen; filaments short or absent. Female flowers with a superior, 1-celled, uniovulate ovary; style straight; stigma subulate.

Fruit a nut, terminated by the persistent style and sometimes with 2 prominent spines at the base.

Wilmot-Dear, M. (1985). *Ceratophyllum* revised - a study in fruit and leaf variation. *Kew Bull.* 40: 243–271.

CERATOPHYLLUM L.

Description as for the family.

1. Leaves robust, 1–2(–4) × forked, with many, prominent marginal teeth; fruit with a pair of basal spines **1. C. demersum**
+ Leaves delicate, 3–4 × forked, with few and inconspicuous marginal teeth; fruit lacking basal spines **2. C. submersum**

1. C. demersum L., Sp. pl.: 992 (1753).

Submerged perennial herb up to 2m. Leaves 1–4cm, 1–2(–4) × forked, the margins usually with many prominent teeth. Perianth lobes 0.5–1.3 × 0.2–0.4mm. Male flowers up to 2.5(–3.5) diam., numerous, 1–3 per whorl. Female flowers less numerous than the male, 1 per whorl. Fruit (var. *demersum*) ovoid or ellipsoid, 4–5mm long (excluding spines), compressed, smooth, with a long apical and 2 basal spines, base rarely unarmed; spines 2–5mm. **Map 437, Fig. 61.**

In ponds, cisterns and sluggish to fast-moving streams and canals; according to Wilmot-Dear (op. cit.) apparently not found in conditions of high alkalinity; 150–1600m.

Saudi Arabia, Yemen (N), Oman. Cosmopolitan.

All Arabian material is apparently referable to var. *demersum* which has fruits with a smooth surface and long basal spines. However, a single specimen (*Collenette* 6310) from Saudi Arabia is rather problematic. On some pieces of this specimen the leaves are consistently 2 × forked, as in typical *C. demerusm*, whilst on others they are less robust, inconspicuously few-toothed and 3(–4) × forked suggesting *C. submersum*. The only fruit (found on a piece with less robust leaves) is too immature for certain identification. This specimen may represent a mixed gathering although Wilmot-Dear (pers. comm.) considers this would be very unusual as there are no records of the two species ever found growing together. For a discussion and key to the infraspecific taxa of *C. demersum* and *C. submersum* see Wilmot-Dear (op. cit.).

2. C. submersum L., Sp. pl., ed. 2: 1409 (1763).

Similar to *C. demersum* but usually a more delicate plant; leaves 3–4 × forked, with the marginal teeth few and inconspicuous; fruit (var. *submersum*) ellipsoid, the surface usually prominently warty-papillose, the apical spine usually less than 1mm, lacking basal spines. **Map 437, Fig. 61.**

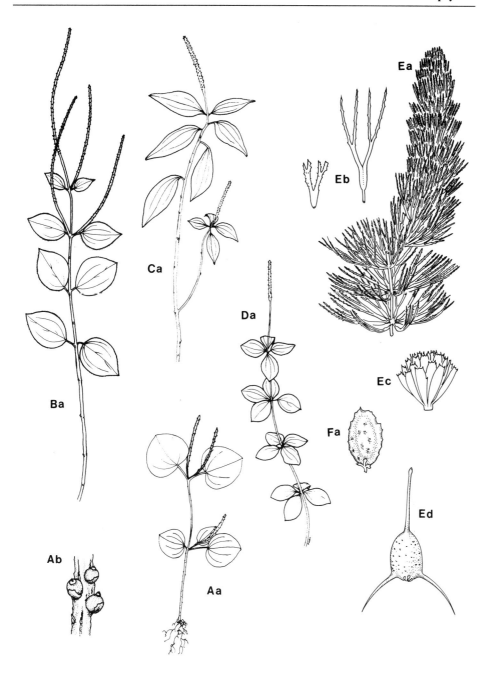

Fig. 61. Piperaceae. A, *Peperomia pellucida*: Aa, habit (×1); Ab, detail of flowering spike (×10). B, *P. blanda* var. *leptostachya*: Ba, habit (×1). C, *P. abyssinica*: Ca, habit (×1). D, *P. tetraphylla*: Da, habit (×1). **Ceratophyllaceae.** E. *Ceratophyllum demersum*: Ea, habit (×1); Eb, leaves (×1.5); Ec, male flower (×10); Ed, fruit (×3). F, *C. submersum*: Fa, fruit (×30).

Rooted in mud in slow-moving stream; 600m; in Africa (Wilmot-Dear op. cit.) it is found in pools and swamps, often in alkaline conditions, and is tolerant of a narrower range of conditions than *C. demersum*.

Saudi Arabia, Yemen (N?), Oman. Europe, Africa, Asia and N America (distribution of var. *submersum*).

This species is noted from Yemen by Wilmot-Dear (op. cit.) but no material is cited and the record must be considered uncertain. The Arabian material is apparently referable to var. *submersum*, the fruit of which is warty-papillose, has a short apical spine and lacks basal spines.

Family 49. PIPERACEAE

A.G. MILLER

Succulent annual or perennial herbs, often creeping. Leaves simple, alternate, opposite or whorled, entire, palmately veined at the base. Stipules absent. Flowers bisexual, in terminal and leaf-opposed spikes, minute, much reduced, borne in the axils of peltate bracts. Perianth absent; stamens 2; anther thecae 2, usually confluent; ovary superior, 1-celled, 1-ovulate; stigma sessile and brush-like. Fruit a drupe.

Düll, R. (1973). Die Peperomia-Arten Afrikas. *Bot. Jahrb. Syst.* 93: 56–129.

1. PEPEROMIA Ruiz & Pavón

Description as for the family

1. Delicate annual; leaves thin, semi-translucent, cordate at the base
 3. P. pellucida
+ Perennials; leaves thick, not translucent, rounded or cuneate at the base 2

2. Leaves alternate **4. P. abyssinica**
+ Leaves opposite or whorled 3

3. Prostrate herb, often rooting at the nodes; leaves glabrous or glabrescent, mostly whorled, up to 12mm long **1. P. tetraphylla**
+ Erect or ascending herb; leaves sparsely pubescent, mostly opposite, more than 15mm long **2. P. blanda**

1. P. tetraphylla (J. Forster) Hook.f. & Arn., Bot. Beechey Voy.: 97 (1841). Syn.: *P. reflexa* (L.f.) A. Dietrich, Sp. pl. ed. 6, 1: 180 (1831). Illustr.: Düll op. cit.: 74.

Prostrate herb; stems ribbed, often rooting at the nodes. Leaves opposite or mostly in whorls of 3–4, obovate or broadly elliptic to suborbicular, 7–12 × 5–8mm, with a rounded tip and rounded or broadly cuneate base, glabrous or glabrescent; petiole 1–3mm. Spikes solitary, terminal, 10–20(–30) × c.1.5mm, pubescent. Drupes ellipsoid, c.1mm. **Map 438, Fig. 61.**

In shade between large boulders in low shrubland; 900–1500m.

Socotra. Throughout the tropics.

2. P. blanda (Jacq.) Kunth in Humb., Bonpl. & Kunth, Nov. gen. sp. 1: 67 (1815). Syn.: *P. arabica* Decne ex Miq. Syst. Piperac. 1: 121 (1843); *P. goudotii* sensu Balfour (1888 p.260) non Miq.

Erect or ascending herb; up to 30cm, finely tomentose. Leaves mainly opposite, a few in whorls of 3 above, obovate, 1.5–5 × 1–3cm, with a shortly acuminate or rounded tip and rounded or narrowly cuneate base, sparsely pubescent; petiole 0.5–1.5cm. Spikes solitary or in groups of 2–5, in the axils of the upper leaves and terminal, 30–70 × c.1.5mm, glabrous. Drupe cylindrical, c.1mm. **Map 439, Fig. 61.**

Rock faces in mist-affected areas; 1400–3000m (mainland); in shade amongst large boulders in low shrubland; 600–1100m (Socotra).

Yemen (N), Socotra. New World tropics, tropical Africa and Sri Lanka.

All Arabian material belongs to var. *leptostachya* (Hook. & Arn.) Düll.

3. P. pellucida (L.) Kunth in Humb., Bonpl. & Kunth, Nov. gen. sp. 1: 64 (1815). Illustr.: Düll op. cit.: 73, f. 1.

Delicate annual herb; stems erect or prostrate, up to 10cm, glabrous throughout. Leaves pale green and semi-translucent, alternate, broadly ovate to suborbicular or transversely ovate, 5–18 × 5–23mm, with a rounded or acute tip and cordate base; petiole 2–7mm. Spikes solitary, terminal and leaf-opposed, 10–20(–50) × c.1mm. Drupe ellipsoid, c.0.75mm. **Map 440, Fig. 61.**

In shade on wet rocks and banks; 200–1500m.

Yemen (N & S), Oman. Throughout the tropics.

4. P. abyssinica Miq. in Bot. Misc. 4: 419 (1845).

Creeping glabrous herb; stems succulent, rooting at the nodes and leafless at the base. Leaves alternate, ovate to elliptic or obovate, 8–30 × 5–17mm, with an acute or obtuse tip and rounded or cuneate base; petiole 1–4mm. Spikes solitary, terminal and leaf-opposed, (20–)40–60(–80) × 1–2mm. Drupe ellipsoid, c.1mm. **Map 441, Fig. 61.**

Rock faces in mist-affected areas; 2600–3000m.

Yemen (N). Central and NE tropical Africa.

Family 50. ARISTOLOCHIACEAE

A. G. MILLER

Perennial rhizomatous herbs. Leaves alternate, simple, entire; stipules absent. Flowers axillary, solitary, zygomorphic, bisexual. Perianth tubular, curved, with a limb at the mouth, inflated at the base, foetid. Stamens 6, adnate to style and forming a column. Ovary inferior, 6-locular; ovules numerous; placentation axile; style short, columnar, with 3 or 6 stigmatic lobes. Fruit a many-seeded, 6-valved capsule.

ARISTOLOCHIA L.

Description as for the family.

1.	Leaves broadly ovate-triangular with a cordate base	**1. A. bracteolata**
+	Leaves linear-triangular with a rounded or auriculate base	**2. A. rigida**

1. A. bracteolata Lam., Encycl. 1: 258 (1783). Syn.: *A. bracteata* Retz., Observ. Bot. 5: 29 (1788); *A. sempervirens* sensu Forsskal (1775) non L.; *A. maurorum* L. var. *latifolia* sensu Blatter (1919-36 p.419) non Boiss. Illustr.: Collenette (1985 p.52); Western (1989 p.55).

Sprawling herb; stems up to 1m, glabrous. Leaves broadly ovate-triangular to $\pm$ circular, 1.5–7 × 2–7cm, rounded at tip, the margin minutely crisped, cordate at the base; petiole 1–3cm. Pedicel bearing a small, leaf-like bract which is 0.5–1mm long. Perianth tube yellowish green suffused with red, tubular, 1.5–2cm long, 5–8mm across at mouth, with a globose (c.5mm across) swelling at the base; limb dark reddish brown, narrowly oblong, 1.5–3 × 0.6–0.8cm. Capsule obpyriform or cylindrical, 1.5–2.5 × 1–1.25cm. **Map 442, Fig. 62.**

Wadi-sides, gardens and cultivated areas on volcanic shingle and sandy soils; 20–1500m.

Saudi Arabia, Yemen (N & S), Oman, UAE. Tropical Africa, India and Pakistan.

2. A. rigida Duchartre in DC., Prodr. 15(1): 495 (1864).

Erect perennial herb, whitish green. Leaves linear-triangular, 2–5 × 0.2–0.6cm, acute, the margin scabrid, rounded or auriculate at the base, sessile. Pedicel bearing an ovate bract which is broader than the leaves. Perianth yellow and purple-black, funnel-shaped, 3–5cm long, 2–3cm across at mouth, with a globose swelling (1–1.5cm across) at the base; limb lip-like, recurved. Fruit not known. **Map 443, Fig. 62.**

On sand dunes; 50–150m.

Yemen (S). Somalia, ?Kenya.

Balfour (1888) recorded a sterile *Aristolochia* on Socotra; the specimen was never determined to species and cannot now be traced.

Family 51. HYDNORACEAE

L.J. MUSSELMAN

Subterranean parasitic herbs, lacking leaves, often with massive root systems spreading laterally from the host; roots terete, 4–5-angled or sometimes flattened, covered

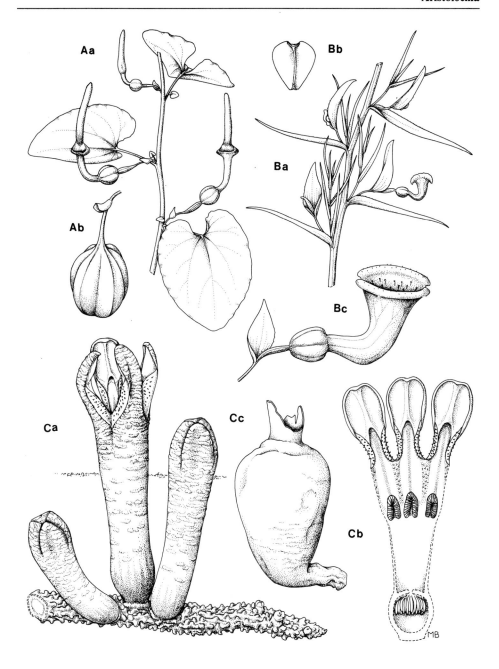

Fig. 62. Aristolochiaceae. A, *Aristolochia bracteolata*: Aa, flowering branch (×1); Ab, fruit (×1). B, *A. rigida*: Ba, part of flowering branch (×1); Bb, seed (×2); Bc, flower (×1). **Hydnoraceae.** C, *Hydnora johannis*: Ca, habit (×0.5); Cb, section through flower (×0.5); Cc, fruit (×0.5).

with warty outgrowths. Flowers solitary, appearing above ground after rain, tubular below, 3–5-lobed above; lobes valvate, fleshy, patent and resting on the soil or remaining connate at the tips and the flower then opening by lateral slits. Anthers sessile, inserted on the perianth tube. Ovary inferior, 1-locular, with numerous infolded placentas and many ovules; stigma sessile, cushion-shaped, grooved. Fruit globose, ripening below ground, many-seeded

Harms, H. (1935). Hydnoraceae in Engler & Prantl, Nat. Pflanzenfam., ed. 2, 16b: 282–295; Musselman, L.J. (1984). Some parasitic angiosperms of Sudan: Hydnoraceae, Orobanchaceae and Cuscuta (Convolvulaceae). *Notes Roy. Bot. Gard. Edinburgh.* 42: 21–38; Musselman, L.J. & Visser, J.H. (1989). Taxonomy and natural history of *Hydnora* (Hydnoraceae). *Aliso* 12 (2): 317–326.

HYDNORA Thunb.

Description as for the family.

1. H. johannis Beccari in Nuovo Giorn. Bot. Ital., 3: 5 (1871). Syn.: *H. abyssinica* A.Br. ex Decne in Bull. Soc. Bot. France 20: 76 (1873). Illustr.: Collenette (1985 p.260); Miller (1988 p.149) as *H.* aff. *africana*.

Roots fleshy, up to 5cm across. Flowers (3–)4(–5)-merous, foetid, 5–25cm long; perianth tube 3–4cm diam.; lobes hooded at the tips, 6–8cm × 1.5–2cm, bright red, pale orange or pink within, densely hairy at base. Fruit fleshy, 10–15cm diam. **Map 444, Fig. 62.**

In sandy wadis; parasitic on *Acacia* and ?*Tamarix* spp; 100–1500m.

Saudi Arabia, Yemen (N & S), Oman. Tropical east Africa.

Family 52. OCHNACEAE

D.R. MCKEAN

Trees or shrubs. Leaves usually alternate, simple, the margins serrate to ciliate or rarely entire, petiolate. Stipules deciduous. Flowers bisexual, actinomorphic, either solitary or paired (in Arabia) or in panicles, racemes or umbellate clusters; pedicels articulated. Sepals (4–)5, free, enlarging in fruit. Petals 5(–10), contorted in bud. Stamens 20–numerous; anthers dehiscing by terminal pores (in Arabia). Ovary (in Arabia) superior, of (3–)5(–15) carpels; carpels free at the base, 1-ovulate, with gynobasic styles which are united almost to the apex. Fruit (in Arabia) a cluster of 1-seeded drupes inserted on the enlarged red receptacle.

OCHNA Schreb.

Description as for the family.

1. O. inermis (Forsskal) Schweinf. in Atti Congr. Bot. Int. Genova (1892): 335 (1893). Syn.: *Euonymus inermis* Forsskal (1775 p.204); *O. parvifolia* Vahl (1790 p.33). Illustr.: Collenette (1985 p.377). Type: Yemen (N), *Forsskal* (C).

Deciduous shrub or small tree, up to 1–2(–6)m, glabrous throughout; trunk with smooth whitish bark. Leaves alternate or subopposite, dark glossy green, elliptic to narrowly oblong or rarely subcircular, 2–6 × 2–3cm, obtuse or rounded, serrulate, rounded to cuneate at the base; petioles 1–2.5mm. Flowers fragrant; pedicels 1–3cm. Sepals elliptic, c.5mm long, becoming red and enlarging to 10–16mm in fruit. Petals yellow, obovate to subcircular, c.10 × 5mm, clawed. Ovary 5-lobed; carpels 5. Drupes c.10 × 6mm, black, oblong-ovoid. **Map 445, Fig. 12.**

Acacia-Commiphora bushland; (700–)1000–2000m.

Saudi Arabia, Yemen (N). Tropical Africa from Ethiopia to northern Transvaal.

Restricted in Arabia to the SW escarpment mountains where it is locally common in Yemen but rare in Saudi Arabia. John Wood (pers. comm.) mentions a small-leaved form of *O. inermis* from Yemen which favours drier, *Acacia-Grewia* bushland.

Family 53. GUTTIFERAE

N.K.B. ROBSON

Herbs, shrubs or trees. Leaves opposite, simple, often glandular with translucent dots and streaks and sometimes with black gland-dots. Stipules absent. Flowers bisexual, actinomorphic, solitary or in terminal cymes or panicles. Sepals 5, free, imbricate, often glandular. Petals 5, free, contorted in bud. Stamens numerous in 3 or 5 bundles. Ovary superior, 1–5-celled; ovules numerous; styles 3 or 5, free or fused. Fruit a many-seeded, 3- or 5-valved capsule.

Robson, N.K.B. (1985, 1993). Studies in the genus Hypericum L. *Bull. Brit. Mus. (Nat. Hist.) Bot.* 12: 163–325 (1985); 23 (2): 67–70 (1993).

HYPERICUM L.

Description as for the family.

1.	Stems and leaves pubescent or puberulous	2
+	Stems and leaves glabrous	4
2.	Low shrub; leaves glabrous above; black glands absent	**6. H. fieriense**
+	Erect to prostrate perennial herbs; leaves with hairs on both sides; black glands present	3

3.	Stems erect, not rooting; leaves triangular-lanceolate to narrowly oblong; flowers c.25mm diam.	**10. H. collenettiae**
+	Stems ascending to prostrate, sometimes rooting; leaves narrowly oblong to elliptic or oblanceolate; flowers 9–13mm diam.	**11. H. sinaicum**
4.	Erect perennial herbs	5
+	Small trees or shrubs	6
5.	Sepals and bracts fringed with black, glandular cilia; bracts auriculate	**9. H. annulatum**
+	Sepals and bracts without black glands; bracts not auriculate	**12. H. perforatum**
6.	Styles 3, free, outcurving or spreading	7
+	Styles 5, wholly or partly coherent	9
7.	Leaves 25–75mm long, lanceolate to ovate, with densely reticulate venation between the main lateral veins; stamen bundles 5, deciduous	**5. H. hircinum**
+	Leaves 8–25mm long, oblong-elliptic to obovate, with obscure venation between the main lateral veins; stamen bundles 3, persistent	8
8.	Flowers 1(2), terminal and in the axils of older leaves; sepals ensiform; capsule valves vittate; habit erect	**8. H. scopulorum**
+	Flowers 7–13 in terminal subumbellate inflorescence; sepals elliptic or oblong-oblanceolate to oblong; capsule valves verrucose; habit spreading or straggling	**7. H. tortuosum**
9.	Leaves with black submarginal gland-dots; styles completely coherent	**4. H. quartinianum**
+	Leaves with pale (pellucid) submarginal gland-dots; styles partly coherent below	10
10.	Sepals with gland-dots; flowers solitary; main lateral veins joining to form an uninterrupted submarginal vein	**1. H. revolutum**
+	Sepals without gland-dots; flowers in 1–5(–10)-flowered terminal cymes; main lateral veins not joining to form an uninterrupted submarginal vein	11
11.	Leaves narrowly elliptic or narrowly lanceolate, acute; leaf venation parallel (with 3–7 main veins) towards the base, becoming pinnate along the midrib towards the tip	**2. H. balfourii**
+	Leaves elliptic to obovate or subcircular, acute to rounded; leaf venation mainly pinnate with 1 or 2 pairs of basal lateral veins extending around the margin towards the tip	**3. H. socotranum**

1. H. revolutum Vahl (1790 p.66). Syn.: *H. lanceolatum* auctt. non Lam. (1797). Illustr.: Webbia 22: 239 (1967); Collenette (1985 p.261). Type: Yemen (N), *Forsskal* (C).

Glabrous shrub or small tree up to 3m. Leaves sessile, lanceolate to narrowly oblong-elliptic, 8–15 × 2.5–4mm, acute, reflexed-auriculate at the base, black glands absent; main lateral veins joining to form an uninterrupted submarginal vein. Flowers

3.5–5.5cm across, solitary. Sepals ovate to broadly ovate, 7–9 × 4–8mm, acute to obtuse, with marginal black gland-dots. Petals bright yellow to yellow-orange (15–)20–30 × 5–25mm, persistent. Stamen bundles 5. Styles 5, 6–10mm, united below. Capsule 5-valved, ovoid to subglobose, 9–12 × 8–10mm. **Map 446, Fig. 63.**

Montane shrubland and grassland; (2200–)2500–3500m; flowering in early summer.

Saudi Arabia, Yemen (N). Tropical and South Africa

H. revolutum comprises two subspecies which differ mainly in leaf venation and style length. Subsp. *keniense* (Schweinf.) N. Robson is restricted to the mountains of E Africa whereas subsp. *revolutum* is more widespread, occurring in SW Arabia and throughout the mountains of tropical and southern Africa.

2. H. balfourii N. Robson, op. cit.: 191 (1985). Syn.: *H. mysorense* sensu Balf. f. (1888) non Heyne. Illustr.: Webbia 22: 263 (1967) as *H. mysorense*. Type: Socotra, *Balfour, Cockburn & Scott* 606 (A, BM, E, K).

Glabrous shrub or small tree up to 3m. Leaves sessile, narrowly elliptic to narrowly lanceolate, 10–45 × 4–12mm, acute, cuneate to subauriculate at the base, black glands absent; venation parallel (with 3–7 main veins) towards the base, becoming pinnate along the midrib towards the tip. Flowers 4–7cm across, in 1–3(–10)-flowered terminal cymes. Sepals ovate to oblong, 5–9 × 3–5.5mm, acute or obtuse, black glands absent. Petals golden yellow, 20–35 × 10–20mm, deciduous. Stamen bundles 5. Styles 5, 5–8mm, united below. Capsule 5-valved, ovoid, 13–15 × 7–9mm. **Map 447, Fig. 63.**

Low montane shrubland and dwarf *Hypericum-Crotonopsis* shrubland; (120–)600–1100m.

Socotra. Endemic.

Closely related to *H. mysorense*, a species from southern India and Sri Lanka. Normally found above about 600m, the lower altitudes are atypical.

3. H. socotranum Good in J. Bot. 65: 334 (1927).

Glabrous shrub or small tree up to 4.5m. Leaves sessile, elliptic to obovate or subcircular, 12–30 × 5–22mm, acute to rounded, cuneate at the base, black glands absent; venation mainly pinnate with 1–2 pairs of basal lateral veins extending around the margin towards the tip. Flowers (2.5–)4–7.5cm across, in 1–5-flowered terminal cymes. Sepals ovate to oblong, 4.5–6.5 × 2.5–3mm, apiculate or obtuse to rounded, black glands absent. Petals golden yellow, 15–40 × 9–24mm, deciduous. Stamen bundles 5. Styles 5, 5–8mm, united below. Capsule 5-valved, ovoid, 12–13 × 6–7mm. **Map 448, Fig. 63.**

Socotra. Endemic.

GUTTIFERAE

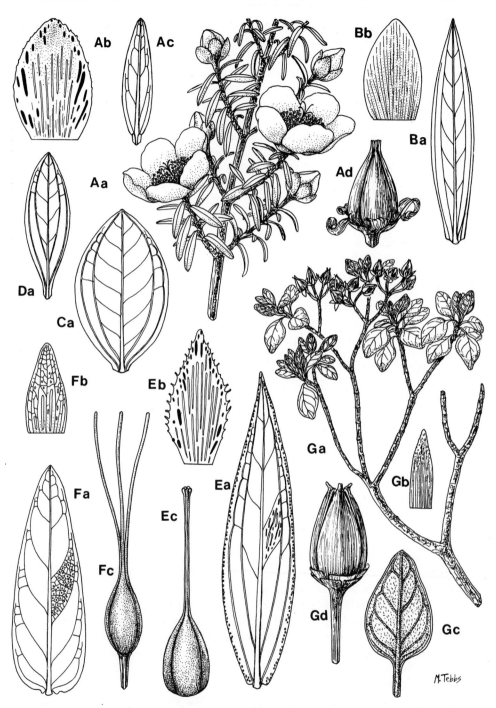

Fig. 63. Hypericaceae. A, *Hypericum revolutum*: Aa flowering branch (×½); Ab, sepal (×3); Ac, leaf (×1½); Ad, capsule (×1½). B, *H. balfourii*: Ba, leaf (×1½); Bb sepal (×3). C, *H. socotranum subsp. socotranum*: Ca, leaf (×1½). D, *H. socotranum subsp. smithii*: Da, leaf (×1½). E, *H. quartinianum*: Ea, leaf (×1½); Eb, sepal (×3); Ec, capsule (×3). F, *H. hircinum*: Fa, leaf (×1½); Fb, sepal (×3); Fc, capsule (×3) G, *H. fierense*: Ga, flowering branch (×½); Gb, sepal (×3); Gc, leaf (×1½); Gd, capsule (×3).

1. Leaves elliptic to oblanceolate, 5–10mm wide, acute to obtuse; pedicels 1–6mm; flowers solitary   subsp. **socotranum**
+ Leaves broadly elliptic to subcircular, 8–22mm wide, apiculate to rounded; pedicels 5–25mm; flowers 1–5   subsp. **smithii**

subsp. **socotranum**. Syn.: *H. lanceolatum* sensu Balf. f. (1888) non Lam. Illustr.: Webbia 22: 259 (1967); Robson, op. cit.: 193 (1985). Type: Socotra, *Balfour, Cockburn & Scott* 246 (BM, E, K, LE, P).

Limestone escarpments and rock crevices in the western part of the island on; 210m.

subsp. **smithii** N. Robson, op. cit.: 194 (1985). Illustr.: Robson, op. cit.: 194. Type: Socotra, *Smith & Lavranos* 272 (BM, K).

Limestone escarpments and rock crevices in the central and eastern part of the island; (120–)400–650m.

4. H. quartinianum A. Rich. (1847 p.97). Illustr.: Robson, op. cit.: 195 (1985); Webbia 22: 255 (1967).

Glabrous shrub up to 1m. Leaves sessile, lanceolate to oblong-elliptic, $20–100 \times 5–35$mm, acute, cuneate to subcordate at the base, with marginal black gland-dots; venation pinnate with only 2–3 pairs of lateral veins towards the base. Flowers 3.5–7cm across, in few-flowered terminal cymes. Sepals ovate to lanceolate, $5–15 \times 2–5$mm, acute, with submarginal black gland-dots. Petals bright yellow, $20–40 \times 10–25$mm, tardily deciduous. Stamen bundles 5. Styles 5, 8–13mm, completely united. Capsule 5-valved, ovoid, $10–16 \times 5–10$mm. **Map 449, Fig. 63.**

Humid gullies; 2000m.

Yemen (N). Eastern tropical Africa from Ethiopia and Somalia south to Zambia and Moçambique.

A rare plant of the outer escarpment mountains in Yemen.

5. H. hircinum L., Sp. pl.: 784 (1753). Illustr.: Robson, op. cit.: 308 (1985); Fl. Palaest. 1: t.324 (1966); Collenette (1985 p.261).

Glabrous shrub up to 1.5m. Leaves sessile, lanceolate to ovate, $25–60(–75) \times 10–25(–35)$mm, acute or obtuse, rounded or cordate at the base, black glands absent; venation pinnate with a prominent reticulum between the main lateral veins. Flowers 3–4cm across, in few-flowered terminal cymes. Sepals lanceolate, $4–9 \times 1.5–3$mm, acute, black glands absent. Petals bright yellow, $(11–)15–20 \times (4–)8–9$mm, deciduous. Stamen bundles 5. Styles 3, (10–)15–25mm, free. Capsule 3-valved, ellipsoid, $8–14 \times 4–7$mm. **Map 450, Fig. 63.**

In damp sand by streams; 1800–2250m.

Saudi Arabia. Mediterranean region; widely cultivated and naturalized elsewhere.

The Arabian plants are all referable to subsp. *majus* (Aiton) N. Robson.

6. H. fieriense N. Robson, loc. cit.: 68 (1993). Type: Socotra, *Smith & Lavranos* 475 (K).

Low shrub up to 1m with stems densely to sparsely fawn-puberulous and leaves densely whitish-puberulous beneath. Leaves with petiole 4–6mm, triangular-ovate to oblong-ovate, 10–17 × 7–12mm, subobtuse to rounded, broadly cuneate to truncate at the base, black glands absent; venation pinnate, obscure between the main lateral veins. Flowers c.1cm across, in subumbellate 3–5-flowered terminal cymes. Sepals linear-lanceolate, 4–5 × c.1.5mm, acute, black glands absent. Petals not seen. Stamen bundles 3. Styles 3, free, not seen complete. Capsule 3-valved, pyramidal-ovoid, 6–7.5 × 4.5–5mm. **Map 451, Fig. 63.**

Low *Hypericum-Crotonopsis* shrubland among trees of *Dracaena cinnabari*; 1350m.

Socotra. Endemic

7. H. tortuosum Balf. f., (1882 p.502). Illustr.: Balfour (1888 t. 4B); Webbia 22: 271 (1967). Type: Socotra, *Schweinfurth* 757 (E)

Spreading or straggling glabrous shrub, up to 50cm. Leaves shortly petiolate, becoming sessile above, obovate to oblong-elliptic, 8–25 × 4–20mm, obtuse, cuneate at the base, sometimes auriculate, black glands absent; venation pinnate, obscure between the main lateral veins. Flowers 1–1.5cm across, in subumbellate, many-flowered, terminal cymes. Sepals elliptic or oblong-oblanceolate to oblong, 3.5–6 × 1–2mm, acute or truncate, black glands absent. Petals yellow, 5–7.5 × 2–2.5mm, persistent. Stamen bundles 3. Styles 3, free, 2.5–3mm. Capsule 3-valved, subglobose, c. 4.5 × 4mm, the valves verrucose. **Map 452, Fig. 64.**

Cliffs on granite and limestone and in cushion vegetation on granite pinnacles; 600–1150m.

Socotra. Endemic.

8. H. scopulorum Balf. f. (1882 p.502). Illustr.: Balfour (1888 t. 4A); Webbia 22: 270 (1967). Type: Socotra, *Schweinfurth* 756 (E).

Erect, glabrous shrub up to 120cm. Leaves sessile, obovate to oblong-elliptic. 10–24 × 4–11mm, obtuse, subauriculate at the base, black glands absent; venation pinnate, obscure between the main lateral veins. Flowers 1–1.5cm across, solitary in the axils of the uppermost leaves. Sepals ensiform, 5–7.5 × 0.5–1.2mm, acute, black glands absent. Petals golden-yellow to golden-orange, 5–11 × 1.5–2mm, persistent. Stamen bundles 3. Styles 3, free, c. 1.5mm. Capsule 3-valved, subglobose, c. 5 × 4mm, the valves vittate. **Map 453, Fig. 64.**

Common in dwarf *Hypericum-Crotonopsis* shrubland on the higher slopes of the granite mountains and less commonly on limestone cliffs; 750–1000m.

Socotra. Endemic.

9. H. annulatum Moris, Stirp. sard. elench. 1: 9 (1827). Illustr.: Webbia 22: 275 (1967); Collenette (1985 p.261).

Erect, glabrous, perennial herb up to 40cm. Leaves sessile, oblong-ovate to oblong-elliptic, 15–40 × 0.5–15mm, acute or obtuse, rounded at the base, with marginal black

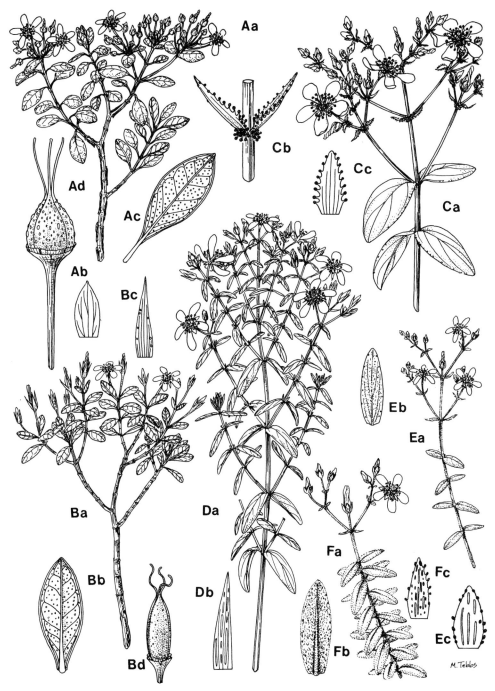

Fig. 64. Hypericaceae. A, *Hypericum tortuosum*: Aa, flowering branch; Ab, sepal; Ac, leaf; Ad, capsule. B, *H. scopulorum*: Ba, flowering branch; Bb, leaf; Bc, sepal; Bd, capsule. C, *H. annulatum*: Ca, flowering branch; Cb, bracts; Cc, sepal. D, *H. perforatum*: Da, flowering branch; Db, sepal. E, *H. sinaicum*: Ea, flowering branch; Eb, leaf; Ec, sepal. F, *H. collenettiae*: Fa, flowering branch; Fb, leaf; Fc, sepal. Flowering branches all × $\frac{1}{2}$, sepals and capsules all × 3, leaves all × 2.

gland-dots. Flowers 2–3cm across, in many-flowered pyramidal cymes. Bracts with dense black glandular-ciliate auricles. Sepals lanceolate, 3.5–4.5 × 0.8–1.5mm, acute, with black gland-dots and black glandular-ciliate margins. Petals yellow, c. 15 × 3–4mm, persistent. Stamen bundles 3. Styles 3, 6–7mm, free. Capsule 3-valved, ovoid-pyramidal, c. 4 × 2.5mm. **Map 454, Fig. 64.**

Shady banks, amongst grass and shrubs; 1050–1750m.

Saudi Arabia. External distribution of subsp. *intermedium*: Sudan and northern Ethiopia.

A species with an interesting disjunct distribution between southern Europe and tropical E Africa with the Arabian population, which is somewhat intermediate geographically and morphologically, referable to subsp. *intermedium* (Steud. ex A. Rich.) N. Robson, op. cit.: 69 (1993).

10. H. collenettiae N. Robson, op. cit.: 69 (1993). Illustr.: Collenette (1985 p.262) as *H.* sp. aff. *sinaicum*. Type: Saudi Arabia, *Collenette* 3752 (BM, K).

Erect woody-based herb up to 20cm; stem and leaves white-puberulous. Leaves sessile, markedly tetrastichous, narrowly oblong to triangular-lanceolate, 7.5–13 × 1.5–4mm, obtuse, rounded to subcordate at the base, with submarginal black gland-dots only. Flowers c.2.5cm across, in (2–)5–11-flowered terminal cymes. Sepals narrowly lanceolate, 3.5–4 × 0.8–1.2mm, acute to subaristate, with black gland-dots, the margins with irregular black glandular cilia and prominent sessile black glands. Petals golden yellow, 9–10 × c.4.5mm, persistent. Stamen bundles 3. Styles 3, c.4mm, free. Capsule 3-valved, ovoid-cylindric, 4.5–5 × 3–4mm. **Map 455, Fig. 64.**

In shady rock crevices; 1800m.

Saudi Arabia. Endemic.

Known only from two localities in the mountains of the Asir. It is intermediate, both morphologically and geographically, between *H. somaliense* N. Robson (N Somalia) and *H. sinaicum*.

11. H. sinaicum Hochst. & Steud. ex Boiss., Fl. orient. 1: 808 (1867). Illustr.: Täckholm (1974 t.42B).

Ascending woody-based herb up to 15cm; stems and leaves white-tomentose. Leaves sessile, ovate to elliptic, 5–17 × 3–8mm, obtuse to acute, subcordate at the base, with scattered black gland-dots. Flowers 1–1.5cm across, in few to several-flowered terminal cymes. Sepals ovate, 3–4.5 × 1–1.25mm, acute, with black gland-dots, the margins with stalked black glands. Petals yellow, 6–8 × 3–4mm, persistent. Stamen bundles 3. Styles 3, 4–6mm, free. Capsule 3-valved, ovoid, c. 5 × 3mm. **Map 456, Fig. 64.**

In rock crevices by water; 1200–1500m.

Saudi Arabia. Sinai.

Outside the Sinai Peninsula known only from around J. Lawz in the extreme NW of Saudi Arabia.

12. H. perforatum L., Sp. pl.: 785 (1753).

Erect, glabrous, perennial herb up to 50cm. Leaves sessile, ± linear to narrowly oblong or narrowly lanceolate, 5–30 × 2–7mm, acute or obtuse, cuneate or ± rounded at the base, with marginal black gland-dots. Flowers 1–2cm across, in many-flowered pyramidal cymes. Bracts non-auriculate, with scattered black gland-dots. Sepals linear to lanceolate, 5–9 × 1–1.25mm, entire or with a few non-glandular cilia on the margins. Petals yellow, c. 12 × 5mm, persistent. Stamen bundles 3. Styles 3, 4–5mm, free. Capsule 3-valved, ovoid to pyramidal, 5–8 × 3–4.5mm. **Map 457, Fig. 64.**

Banks bordering fields; 2070m.

Saudi Arabia. Europe, N Africa, SW and C Asia, Siberia and W China.

Family 54. PAPAVERACEAE

D.R. MCKEAN

Annual or perennial herbs, usually with milky or coloured latex. Leaves mainly alternate, rarely opposite or in a basal rosette, often lobed or deeply dissected. Stipules absent. Flowers actinomorphic, bisexual, solitary or in cymes or racemes. Sepals 2(–3), free, caducous. Petals 4–6, free, usually soon caducous. Stamens 4–6 or numerous; filaments free. Ovary superior, 1-celled or apparently 2–20-celled because of deeply intrusive placentas; placentation parietal. Fruit a many-seeded capsule opening by valves or pores or a lomentum (*Hypecoum*).

1.	Fruit a capsule dehiscing by terminal pores	**4. Papaver**	
+	Fruit a capsule dehiscing by longitudinal valves or a lomentum		2
2.	Leaves armed with prickles	**1. Argemone**	
+	Leaves unarmed		3
3.	Petals yellow, in 2 dissimilar pairs, the inner pair lobed; fruit a lomentum	**5. Hypecoum**	
+	Petals red or purple, all similar, never lobed; fruit a linear capsule		4
4.	Flowers red or yellow; fruit 10–20cm long	**2. Glaucium**	
+	Flowers purple; fruit up to 4.5cm long	**3. Roemeria**	

1. ARGEMONE L.

Annual or short-lived perennial herbs, armed throughout with sharp prickles; latex white or yellow. Leaves sinuate to pinnatifid, glaucous, with conspicuous white veins, margins spiny, amplexicaul at the base. Flowers showy, solitary in the upper leaf axils. Sepals 2–3, spine-tipped. Petals 4–6, yellow. Stamens numerous. Ovary 1-

locular, with 3–5 placentas. Stigma sessile, star-shaped. Fruit a many-seeded capsule, splitting by 3–6 valves at the top; seeds black.

1. Petals golden yellow; buds spherical; the top of the capsule obscured by the stigmatic arms **1. A. mexicana**
+ Petals pale whitish yellow; buds oblong; the top of the capsule visible between the stigmatic arms **2. A. ochroleuca**

1. A. mexicana L., Sp. pl.: 508 (1753). Illustr.: Collenette (1985 p.388).

Stems erect, up to 1m. Leaves somewhat fleshy, pinnatifid, elliptic-oblong, up to 12 × 5cm, the margins and the veins beneath armed with sharp prickles. Flowers sessile. Petals golden yellow, obovate, 2–3.5 × 1.5–2.5cm. Capsule oblong, 3–4 × 1.5–2cm, armed with prickles, the top surface obscured by the stigmatic arms. **Map 458, Fig. 65.**

Fields and dry waste places; 0–1850m.

Saudi Arabia, Yemen (N & S), Socotra, Oman, Bahrain. A pantropical weed, originally a native of the New World.

2. A. ochroleuca Sweet, Brit. fl. gard. 3, t.232 (1829). Illustr.: Collenette (1985 p.389).

Similar to *A. mexicana* but the leaves more pointed and less fleshy; petals pale yellow; the top surface of the capsule visible between the stigmatic arms. **Map 459.**

A weed of cultivation; 400–1850m.

Saudi Arabia, Yemen (N). A pantropical weed, originally a native of C America.

Often confused with *A. mexicana* and consequently possibly under-recorded in Arabia.

2. GLAUCIUM Miller

Annual, biennial or perennial herbs, pubescent; latex yellow. Leaves fleshy, lobed or dissected, sometimes in a basal rosette, glaucous. Flowers showy, solitary. Sepals 2. Petals 4, red or yellow. Stamens numerous, the filaments filiform. Ovary 2-locular. Stigma 2-lobed. Fruit a linear capsule, 2-horned at the apex, 2-valved, many-seeded, the valves dehiscing from the apex almost to the base.

1. Stem-leaves mainly with 5 or more pairs of lobes, sparsely villous; peduncles shorter than the subtending leaves; annual or biennial **1. G. corniculatum**
+ Stem-leaves mainly with c.3 pairs of lobes, densely villous at least initially; peduncles longer than the subtending leaves; perennial **2. G. arabicum**

1. G. corniculatum (L) J. H. Rudolph, Fl. jen. pl. 13 (1781). Illustr.: Fl. Palaest. 2: t.339 (1966).

Annual or biennial herb up to 50cm. Basal leaves petiolate, oblong, c.10–20 × 6cm, pinnatipartite with narrowly oblong coarsely-toothed lobes, sparsely villous; stem-leaves amplexicaul, with 5 or more pairs of lobes. Flowers usually red with a dark base, sometimes orange or yellow, 2.5–5cm across; peduncles shorter than or equalling the subtending leaf. Sepals c.3cm, glabrous or sparsely hirsute. Capsule up to 10–20cm long. **Map 460.**

Field margins and rocky deserts; 1750–2150m.

Saudi Arabia, Kuwait. SE Europe, the Mediterranean region, N Africa, SW Asia to W Iran, C Asia.

2. G. arabicum Fresen., Mus. Senckenberg. 1: 174 (1834); Illustr.: Fl. Palaest. 2: t.340 (1966); Collenette (1985 p.389).

Similar to *G. corniculatum* but a generally much more densely hairy perennial herb; basal leaves usually forming a large basal rosette, with the terminal lobes much larger than the lateral segments; stem-leaves smaller, with c.3 pairs of lobes; flowers yellow or orange-red with a dark base, the peduncles much exceeding the subtending leaves. **Map 461, Fig. 65.**

Wadi beds and dry rocky slopes in deserts; c.650m.

Saudi Arabia. Egypt, Sinai and Palestine.

Very similar to the preceding species but generally a much showier plant with a branched inflorescence and mainly basal leaves. *G. corniculatum* is generally a bushier and more leafy plant. Both species can have a black blotch at the base of the petals.

3. ROEMERIA Medic.

Annual herbs, with foetid yellow latex. Leaves deeply 2–3-pinnatisect into narrow segments. Flowers solitary. Sepals 2. Petals 4, red or violet. Stamens numerous. Ovary 1-locular. Stigma globose, 3–4-lobed. Fruit a linear-cylindrical capsule, 3–4-valved, many-seeded, the valves dehiscing from the apex almost to the base, unilocular.

Kadereit, J.W. (1987). The taxonomy, distribution and variability of the genus *Roemeria* Medic. *Flora* 179: 135–153.

1. R. hybrida (L.) DC. subsp. **dodecandra** (Forsskal) E.A. Durande & Barratte, Fl. libyc. prodr.: 6 (1910). Syn.: *Chelidonium dodecandrum* Forsskal (1775 p.100). *R. dodecandra* (Forsskal) Stapf in Akad. Wiss. Wien, Math.-Naturwiss. Kl., Denkschr. 51: 295 (1886); *R. hybrida* subsp. *hybrida* sensu Fl. Kuwait (1985). Illustr.: Collenette (1985 p.392); Mandaville (1990 pl.12); Fl. Kuwait 1: pls. 76–78 (1985) as subsp. *hybrida*.

Erect or ascending annual herb up to 5–50cm, hispid throughout and sometimes also thinly pilose, rarely glabrous. Leaves 2-pinnatisect (except in depauperate forms), 3–12 × 4–5cm, the segments linear to ovate-oblong or triangular, the apex terminated by a straight bristle, the lower and basal long-petiolate. Sepals oblong, c.10 × 3mm. Petals violet-purple, obovate, 10–20mm long. Capsule 2–4.5cm long, covered with setae, rarely glabrous. **Map 462, Fig. 66.**

Rocky and silty deserts, and a weed of cultivated and waste places; 100–1550m.

Saudi Arabia, Kuwait. N Africa & SW Asia.

All Arabian material is referable to subsp. *dodecandra* which differs from subsp. *hybrida* in its shorter capsules (up to 4.5cm), smaller seeds (up to 0.8mm long) and the capsules entirely covered by setae. For a discussion of the distribution and variability of *R. hybrida* see Kadereit (op. cit.). Two specimens (*Collenette* 6535 and 9064) from Saudi Arabia are totally glabrous and may be referable to *R. latiloba* (Hausskn. & Bornm.) Fedde.

4. PAPAVER L.

Annual or biennial herbs with milky latex (in Arabia). Leaves variously dissected or lobed. Flowers solitary on long peduncles. Sepals 2. Petals 4–6, red, violet, white or yellow. Stamens numerous. Ovary 1-celled but often appearing 3–20-celled because of intrusive placentas. Stigmas 4–20, radiating from the centre of a sessile lobed disc. Fruit a many-seeded, globose or ellipsoid capsule dehiscing from pores beneath the stigmatic disc.

Kadereit, J.W. (1988). A revision of *Papaver* L. section *Rhoeadium* Spach. *Notes Roy. Bot. Gard. Edinburgh.* 45: 225–286.

1.	Stem-leaves amplexicaul	2
+	Stem-leaves sessile, but not amplexicaul	4
2.	Petals lilac or white with a purple blotch at the base; capsules 3–7 × 4–5cm; leaves simple, serrate-dentate	**1. P. somniferum**
+	Petals red; capsules up to 3 × 2cm, leaves usually deeply dissected	3
3.	Capsule globose, rounded at the base, not or obscurely ribbed	**2. P. glaucum**
+	Capsule obconical, tapered to the base, ribbed	**3. P. decaisnei**
4.	Capsule beset with ascending setae	**8. P. hybridum**
+	Capsule glabrous	5
5.	Capsule subglobose, rounded at the base	**5. P. rhoeas**
+	Capsule obconical or oblong-elliptic, tapered to the base, usually at least 2 × as long as broad	6
6.	Stigmatic rays distinctly ridged, the sinuses between them extending to over half the radius of the disc	**4. P. macrostomum**
+	Stigmatic rays not ridged, the sinuses between them very shallow, inconspicuous	7
7.	Capsule with a pronounced conical apex or umbo; terminal leaf-lobes much broader than the laterals	**6. P. umbonatum**
+	Capsule flat-topped or slightly conical; terminal leaf-lobes about as broad as the laterals	**7. P. dubium** subsp. **laevigatum**

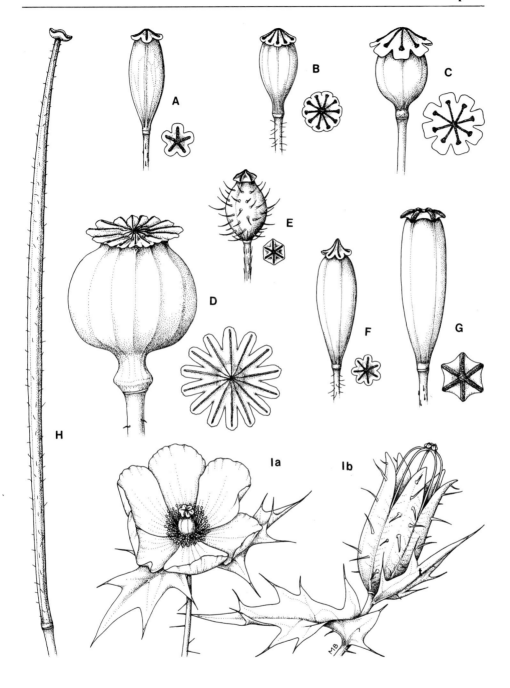

Fig. 65. **Papaveraceae**. A–G, *Papaver* fruits (all × 1): A, *P. dubium*; B, *P. rhoeas*; C, *P. glaucum*; D, *P. somniferum*; E, *P. hybridum*; F, *P. umbonatum*; G, *P. macrostomum*. H, *Glaucium arabicum*: fruit (× 1). I, *Argemone mexicana*: Ia, flower (× 1); Ib, dehisced fruit (× 1).

PAPAVERACEAE

1. P. somniferum L., Sp. pl.: 508 (1753).

Erect glaucous annual, simple or branched, up to 30–100cm, usually glabrous, rarely sparsely setose. Leaves ovate-oblong, undulate, irregularly serrate-dentate; stem-leaves amplexicaul. Flowers large and showy; peduncles glabrous or setose. Petals orbicular, lilac to white, usually with a dark purple blotch at the base. Filaments clavate. Capsule subglobose, 3–7 × 4–5cm, glabrous; stigmatic disc mainly flat or slightly conical. **Map 463, Fig. 65.**

Weed of waste places, rubbish dumps, etc; 10–2000m.

Saudi Arabia, Yemen (N), ?Kuwait. A widespread weed in the warm temperate regions of the world.

Rarely (?still) cultivated in the SW escarpment mountains. Recorded from Kuwait (Dickson et al. 1973) but not seen recently.

2. P. glaucum Boiss. & Hausskn. in Boiss., Fl. orient. 1: 116 (1867). Illustr.: Collenette (1985 p.391); Fl. Iraq 4 (2): 800 (1980).

Glaucous annual or biennial, glabrous throughout, 10–50cm. Leaves mainly towards the base of the stem, oblong, pinnatisect with toothed segments; stem-leaves amplexicaul. Flowers c.6cm diam.; peduncles long, glabrous. Petals crimson with a dark basal spot. Capsule globose, 15–20 × 7–12mm, shortly stipitate, glabrous; stigmatic disc flat to conical. **Map 464, Fig. 65.**

Sandy wadi-beds and rocky ravines; 700m.

Saudi Arabia. SW Asia.

3. P. decaisnei Hochst. & Steud. ex Elkan, Tent. Mon. gen. Papav.: 26 (1839).

Erect glaucous annual, usually glabrous, sometimes sparsely setose, 10–50cm. Leaves oblong, pinnatifid or pinnatisect with entire or toothed segments; stem-leaves amplexicaul. Flowers c.2cm diam.; peduncles long, glabrous. Petals red to purple. Capsule obconical, to 16 × 8mm, ribbed, tapered at the base, glabrous; stigmatic disc flat. **Map 465, Fig. 65.**

Sandy and rocky deserts; 350–1220m.

Saudi Arabia, Oman, UAE. Egypt and SW Asia from Jordan to Pakistan and Afghanistan.

4. P. macrostomum Boiss. & Huet ex Boiss., Fl. orient. 1: 115 (1867). Illustr.: Fl. Iraq 4, 2: 800 (1980).

Erect or ascending annual, 15–40cm, adpressed-setose to glabrous above, hispid below. Leaves oblong, pinnatisect, with entire or toothed segments, hispid on the midrib beneath and sometimes above. Flowers 4–9cm diam.; peduncles adpressed-setose. Petals scarlet with a dark purple base,. Capsule oblong-elliptic, 10–20 × 4–9mm, faintly ribbed, glabrous; stigmatic disc flat, the rays ridged with the sinuses between them extending to over half the radius of the disc. **Map 466, Fig. 65.**

Stony hillsides and wadi-beds in Iraq - no habitat details available for Arabia.

?Kuwait. Syria and Turkey eastwards to Afghanistan.

Recorded from Kuwait (Dickson et al. 1973) but not seen recently.

5. P. rhoeas L., Sp. pl.: 507 (1753). Syn.: *P. polytrichum* Boiss. & Kotschy ex Boiss., Diagn. pl. orient. sér. 2, 5: 14 (1856). Illustr.: Fl. Palaest. 1: pl.334 (1966) as *P. polytrichum*.

Erect, usually branched annual, 20–60cm, usually hispid throughout, sometimes adpressed-setose above. Leaves pinnatisect or pinnate into serrate segments; segments oblong, the terminal the largest, hispid especially on the main veins. Flowers 15–40mm diam.; peduncles hispid with adpressed or spreading hairs. Petals red with a dark blotch at the base. Capsule subglobose, $10-20 \times 7-11$mm, rounded at the base, glabrous; stigmatic disc flat. **Map 467, Fig. 65.**

A weed of cultivation and waste places; (50–)1500–2150m.

Saudi Arabia, ?Kuwait. Europe, N Africa and Asia.

Recorded from Kuwait (Dickson et al. 1973) but not seen recently. Its presence in Kuwait needs confirming.

6. P. umbonatum Boiss., Diagn. pl. orient. sér. 1, 2(8): 11 (1849). Syn.: *P. syriacum* Boiss. & Blanche ex Boiss., op. cit.: 11. Illustr.: Fl. Palaest. 1: pl.332 (1966).

Annual herb; stems erect to ascending, branched from the base, 10–40cm, sparsely hairy below. Leaves obovate, pinnatifid to pinnatisect, with the terminal segment much broader than the laterals. Flowers c.4cm diam.; peduncles hispid with adpressed to spreading hairs. Petals dark purple at the base. Capsule obovoid, $8-20 \times 3-7$mm, glabrous; stigmatic disc clearly umbonate, the stigmatic rays not ridged, the sinuses between the rays shallow. **Map 468, Fig. 65.**

Garden weed; 20m.

Qatar. Turkey, Syria, Lebanon and Palestine.

Recorded as a garden weed from Qatar (Boulos 1978 p.390) but apparently not seen recently.

7. P. dubium L. subsp. **laevigatum** (M. Bieb.) Kadereit, op. cit.: 244 (1988). Illustr.: Collenette (1985 p.390).

Erect annual, 10–75cm, sparsely to moderately setose; basal leaves long-petiolate, up to 20×6cm; stem-leaves sessile, much smaller, obovate, mainly 1–2-pinnatisect. Flowers c.5cm diam.; peduncle adpressed-setose. Petals red with a small to large dark basal spot. Capsule obconical, c.18×6mm, ribbed, glabrous. Stigmatic disc conical, the rays not ridged, the sinuses between the rays shallow. **Map 469, Fig. 65.**

Weed of cultivation, under acacias in a grassy valley; 900–3000m.

Saudi Arabia, Yemen (N), ?Oman. Europe, the Mediterranean region, Ethiopia and SW & C Asia.

PAPAVERACEAE

Blatter (1919–1936 p.6) doubtfully records this species from Oman. However, it has not been recollected and its presence there seems unlikely.

8. P. hybridum L., Sp. pl.: 506 (1753). Illustr.: Fl. Pakistan 61: 11 (1974).

Erect or ascending annual, slightly branched, up to 50cm, glabrous or hispid. Leaves 1–3-pinnatisect, with linear-oblong segments, the segments serrulate, the basal leaves long-petiolate, the stem-leaves sessile, sparsely setose beneath. Flowers solitary, up to 3.5cm diam.; peduncles 5–20cm, adpressed-setose. Petals light to dark red, dark blotched at the base. Capsules ovoid or ellipsoid, 8–18 × c.10mm, densely setose; stigmatic disc cone-shaped, narrower than the capsule, with 4–8 prominent rays. **Map 470, Fig. 65.**

Weed of cultivation and waste places; 50–2050m.

Saudi Arabia, ?Kuwait. Temperate Europe and Asia.

A literature record of this species is noted in the Flora of Kuwait (1985 p.69). However, it has not been seen recently and its presence in Kuwait needs confirming.

5. HYPECOUM L.

Glabrous, glaucous, low-growing annual herbs with watery latex. Leaves 2–4-pinnatisect, with narrow segments. Flowers in dichasial cymes, mainly yellow. Sepals 2. Petals 4, in 2 series, the inner pair 3-lobed, the outer pair lobed or simple. Stamens 4 with winged filaments. Fruit a lomentum, splitting into 1-seeded segments, articulated, transversely ribbed.

H. deuteroparviflorum Fedde has been recorded from Saudi Arabia (Migahid 1978, 1: 44) and Kuwait (Dickson et al., 1973). However, no verified Arabian material has been seen and these records are probably based on misidentifications of either *H. pendulum* or *H. geslinii*. It is distinguished from these species by its 3-lobed, not entire, outer petals.

1. Fruit pendulous; articulations of fruit scarcely thickened; central lobe of the inner petals as broad as or narrower than the entire lower part of the petal; inner petals with dark markings — **1. H. pendulum**
+ Fruit ascending; articulations obviously thickened; central lobe of the inner petals broader than the entire lower part of the petal; inner petals without dark markings — **2. H. geslinii**

1. H. pendulum L., Sp. pl.: 124 (1753). Illustr.: Fl. Kuwait (1985 pls. 80–81).

Ascending to decumbent herb with 1–several stems from a basal rosette, up to 35cm. Basal leaves up to 12cm long, 2–3-pinnatisect, the segments linear. Flowers pale to mid-yellow, 5–6(–10)cm long. Sepals ovate to lanceolate, c.3mm long. Outer petals variable, from lanceolate to obscurely lobed. Inner petals with dark purple markings, the central lobe as broad as or narrower than the base of the petal. Fruit pendulous, linear-cyindrical, 20–60 × 1.5–3mm, straight or slightly curved. **Map 471, Fig. 66.**

Silty and sandy soils, in wadis, deserts and fields; 10–1200m.

Saudi Arabia, Bahrain, Kuwait. S Europe, N Africa and W Asia.

Mandaville (1990 p.51) remarks that specimens of *Hypecoum* with immature fruits can be difficult to distinguish but that the proportional width of the central lobe of the inner petal compared to that of the entire base of the petal is a reasonably reliable character. He also observes that the dark markings on the inner petals of *H. pendulum* are reasonably diagnostic but can be lost in dried specimens. The two species (at least in Eastern Saudi Arabia) have habitat preferences, *H. pendulum* for silty soils and *H. geslinii* for sand.

2. H. geslinii Coss. & Kral. in Bull. Soc. Bot. France 4: 522 (1857). Illustr.: Fl. Kuwait (1985 pl.79); Collenette (1985 p.389); Mandaville (1990 pl.13).

Ascending to decumbent herb with usually many stems from a basal rosette, 10–40cm. Basal leaves up to 15cm, 2–3 pinnatisect, the segments linear. Flowers yellow, 5–7mm long. Sepals lanceolate, c.2–3mm. Outer petals narrowly rhomboid, c. 6×2mm. Inner petals without dark markings, the central lobe broader than the entire lower part of the petal. Fruit ascending, $20-35 \times 1.5-2$mm. **Map 472, Fig. 66.**

Sandy deserts and fields; 250–1000m.

Saudi Arabia, Kuwait. N Africa, Palestine, Jordan and Iraq.

See comments under *H. pendulum*.

Family 55. FUMARIACEAE

D.R. McKEAN

Annual or perennial herbs with clear watery latex. Stems erect, scrambling or scandent. Leaves alternate or opposite, 2–3-pinnatisectly or ternately divided. Stipules absent. Flowers zygomorphic, bisexual, in racemes or spikes. Sepals 2, free, caducous. Petals 4, free or cohering, in two dissimilar pairs, the outer pair often saccate or spurred at the base. Stamens 6, in 2 bundles, the central anther of each bundle 2-locular and the 2 laterals 1-locular; nectary glands often present. Ovary of 2 united carpels, unilocular with parietal placentation; style solitary; stigmas 2–8. Fruit a 2-valved capsule or an indehiscent nutlet.

FUMARIA L.

Glabrous annual herbs, sometimes climbing. Leaves alternate, 2–3-pinnatisect; petioles often twining. Flowers pink, white or purplish, in bracteate racemes in the axils of the upper leaves. Upper petal spurred at the base and the two lateral petals fused at the tips. Fruit an indehiscent 1-seeded nutlet.

FUMARIACEAE

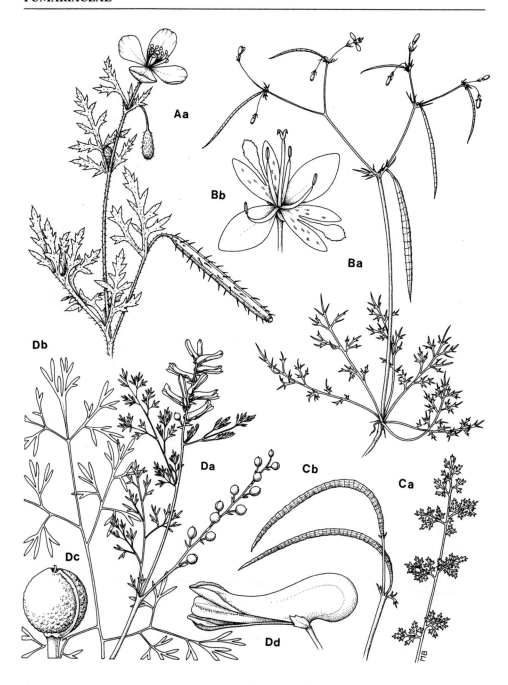

Fig. 66. Papaveraceae. A, *Roemeria hybrida* subsp. *dodecandra*: Aa, flowering and fruiting shoot (×1). B, *Hypecoum pendulum*: Ba, habit (×0.6); Bb, flower (×5). C, *H. geslinii*: Ca, leaf (×1.5); Cb, fruits (×1.5). **Fumariaceae.** D, *Fumaria parviflora*: Da, habit (×2); Db, leaf (×10); Dc, fruit (×10); Dd, flower (×10).

F. densiflora DC., a widespread weed of the E Mediterranean and N Africa, may occur in Arabia. It is recognized by the sepals which are broader than the corolla tube.

1. Leaf segments linear-oblong; peduncles up to 5mm long; lateral petals with dark purple tips **2. F. parviflora**
+ Leaf segments narrowly oblong; peduncles usually more than 5mm long; lateral petals with maroon tips **1. F. abyssinica**

1. F. abyssinica Hamm. in Nova Acta Regiae Soc. Sci. Upsal., ser. 3, 2: 275, t.6 (1857). Illustr.: Fl. Trop. E. Afr.: 4 (1962); Collenette (1985 p.389) as *F. parviflora*.

Erect or straggling herb up to 60cm. Leaf segments narrowly oblong with acute or apiculate apices. Inflorescences densely 10–20-flowered; peduncles usually more than 5mm. Bracts longer than the pedicels. Sepals c. 2 × 1mm, ovate, acute, the margin faintly toothed. Petals pink with maroon tips, c.6mm long. Fruit c. 2.3 × 2mm, glaucous, subglobose with 2 shallow apical pits. **Map 473.**

Rocky slopes in *Juniperus* woodland, and a weed of cultivation; 1800–2900m.

Saudi Arabia, Yemen (N), Oman. Tropical E and NE Africa.

Frequently confused with the following species from which it differs in its more glaucous leaves, broader leaf segments, longer peduncles and the paler (normally white) flowers which have pinkish maroon not dark purple tips.

2. F. parviflora Lam., Encycl. 2: 567 (1788). Illustr.: Fl. Palaest. 1: pl.356 (1966).

Erect or diffuse herb up to 30cm. Leaf segments linear-oblong, sometimes channelled above. Inflorescence, densely 10–20-flowered; peduncles up to 5mm. Bracts equalling the pedicels. Sepals minute, 0.5–1.5 × 0.3 × 0.8mm, laciniate-dentate. Petals white to pink with dark purple tips, c.5mm long. Fruit c. 2 × 2mm, subglobose, rugose, the apex acute or slightly retuse, often apiculate, keeled especially near the apex. **Map 474, Fig. 66.**

Weed of cultivation; (1500–)2000–3000m.

Saudi Arabia, Yemen (N). Europe, N Africa and C Asia.

Family 56. CAPPARACEAE

Herbs, shrubs or trees, sometimes climbing; rarely dioecious or gynomonoecious* (*Dhofaria*). Leaves alternate or rarely opposite, simple or digitately 2–7-foliolate, usually entire. Stipules present. Flowers hermaphrodite or rarely unisexual (*Dhofaria*), actinomorphic or zygomorphic, in axillary or terminal racemes or solitary or clustered. Sepals 4, free or fused. Petals 4 or absent, free, sessile or clawed. Stamens few to many, often borne on an elongated androphore. Ovary sessile or borne on an elongated gynophore, 1–several-celled; placentation parietal; ovules few to many. Fruit a berry,

nutlet or tardily or readily dehiscent capsule. Seeds reniform or angular.
[*gynomonoecious - having female and hermaphrodite flowers on the same plant]

1.	Herbs	2
+	Trees or shrubs	3
2.	Fruit an indehiscent nutlet	**2. Dipterygium**
+	Fruit a dehiscent, linear capsule	**1. Cleome**
3.	Climbing or virgate shrubs, with paired stipular thorns at the base of the petiole	**5. Capparis**
+	Trees or non-climbing shrubs, if climbing then without paired spines	4
4.	Shrub with spinescent branches, soon becoming leafless; fruits tardily dehiscent, globose, bearing large stalked glands	**7. Dhofaria**
+	Leafy trees or shrubs, branches not spinescent; fruits globose or cylindrical, sometimes glandular but not bearing large stalked glands	5
5.	Receptacle (the hypanthium) cylindrical or campanulate; stamens many	**3. Maerua**
+	Receptacle disc-like; stamens 4–9	6
6.	Stamens borne on an androphore; nectary appendage tubular, often petaloid above; fruits cylindrical, torulose, tardily dehiscent	**6. Cadaba**
+	Stamens not borne on an androphore; nectary gland absent; fruits globose, indehiscent	**4. Boscia**

1. CLEOME L.

D.F. CHAMBERLAIN & J. LAMOND

Annual or perennial herbs, or subshrubs. Leaves alternate, simple or digitately compound. Flowers in racemes, often zygomorphic. Sepals 4, free or fused at the base. Petals 4, equal or unequal, sessile or clawed. Stamens 4–numerous, free; androphore usually absent (present in *C. gynandra*). Staminodes present or absent. Gynophore present or absent. Fruit a broadly elliptic to linear capsule, dehiscing by 2 valves.

1.	Most leaves compound, with 3 or more leaflets	2
+	Most leaves simple	12
2.	Stamens 10–20; leaves 3(–5)-foliolate	**19. C. viscosa**
+	Stamens 6–8; leaves 3–7-foliolate	3
3.	Stamens borne on an androphore; gynophore present	**20. C. gynandra**
+	Stamens not borne on an androphore; gynophore present or absent	4
4.	Flowers pink; stamens 7–8; seeds ridged, glabrous	**21. C. hanburyana**
+	Flowers greenish, white or yellow to brownish; stamens 6; seeds smooth or finely reticulate, glabrous or hairy	5

5.	Leaflets filiform	6
+	Leaflets lanceolate, elliptic or oblong	7
6.	Leaves 3-foliolate; capsule not stipitate	**18. C. tenella**
+	Leaves 5–7-foliolate; capsule stipitate	**17. C. angustifolia**
7.	Seeds glabrous; fruit 4–12mm long	**3. C. brachycarpa**
+	Seeds hairy; fruit 10–130mm long	8
8.	Fruit 6–13cm; petals 15–30mm long	**23. C. paradoxa**
+	Fruit 1–10.5cm; petals to 8mm long	9
9.	Fruit erect or erect-spreading	10
+	Fruit pendulous	11
10.	Fruit 2–4mm broad; leaflets elliptic, 2–5 × as long as broad **15. C. albescens** subsp. **omanensis**	
+	Fruit 4–7mm broad; leaflets obovate to broadly elliptic, sometimes only slightly longer than broad **16. C. socotrana**	
11.	Petals 3–4mm long; fruit oblong, 5–8mm broad, sessile or on a gynophore up to 2mm long **5. C. amblyocarpa**	
+	Petals 4–8mm long; fruit linear, 1–4(–8)mm broad, the gynophore 3–5mm long **4. C. ramosissima**	
12.	Stamens 10–14; fruiting pedicels circinnate	**1. C. chrysantha**
+	Stamens 4–6; fruiting pedicels ± straight	13
13.	Lower leaves linear-lanceolate to lanceolate	**22. C. monophylla**
+	Lower leaves ovate to orbicular	14
14.	Stamens 6; petals not appendiculate	15
+	Stamens 4; petals appendiculate	17
15.	Flowers actinomorphic; fruit usually spreading, 0.5–1mm broad **2. C. scaposa**	
+	Flowers zygomorphic; fruit pendulous, 1–8mm broad	16
16.	Leaf indumentum of stipitate glands; fruit (1–)2–5mm broad, 8 × as long as broad **14. C. arabica**	
+	Leaf indumentum of sessile glands; fruit 4–8mm broad, 5–6 × as long as broad **13. C. rupicola**	
17.	Petioles 0–5mm long; seeds pubescent	**6. C. brevipetiolata**
+	Petioles 5–30(–50)mm long; seeds smooth or minutely papillate	18
18.	Leaf margin markedly glandular-crenate	**7. C. macradenia**
+	Leaf margin entire, glandular or not, rarely obscurely crenate	19
19.	Fruit oblong to linear, (8–)15–30mm long; leaves often up to 40mm long	20
+	Fruit elliptic to broadly oblong, rarely linear, 10–18(–22)mm long; leaves 4–20mm long	21

20.	Fruit 5–6mm broad; bracts usually cordate	**12. C. polytricha**
+	Fruit 2.5–4.5mm broad; bracts seldom cordate, sometimes absent **11. C. noeana**	
21.	Inflorescence ± terminal and well-defined, ± bracteate	**9. C. austroarabica**
+	Inflorescence scarcely terminal, flowers scattered in the axils of the leaves	22
22.	Leaves thick-textured, not glaucous, often broader than long, usually with an obtuse apex	**8. C. droserifolia**
+	Leaves thin-textured, glaucous, apex usually acute	**10. C. pruinosa**

1. C. chrysantha Decne in Ann. Sci. Nat. Bot. 3: 274 (1835). Illustr.: Collenette (1985 p.96).

Glandular, woody, aromatic perennial, up to 60cm. Leaves simple, thick, ovate, 5–17 × 3–10mm, densely villous with a mixture of short and long glandular hairs; petioles absent or up to 12mm. Flowers borne in the axils of leaf-like bracts; pedicels 5–15mm, thickening and becoming circinnate in fruit. Sepals oblong-ovate, 2–7 × 1–2mm. Petals shortly appendiculate, obovate, 4–7 × 1.5–3mm, yellow. Stamens 10–14. Style c.1.5mm. Fruit broadly elliptic, compressed, c.8 × 2.5mm, not stipitate. Seeds c.1mm diam., reticulate, glabrous. **Map 475, Figs 67 & 68.**

Moist sand, rocky places and cultivated land; 200–800m.

Saudi Arabia. Libya, Egypt, Sudan, Ethiopia and S Iran.

A distinctive species on account of the circinnate pedicels and the number of stamens.

2. C. scaposa DC., Prodr. 1: 239 (1824). Syn.: *C. papillosa* Steud., Nomencl. bot. ed. 2: 382 (1840), nom. nud. Illustr.: Collenette (1985 p.97); Fl. Qatar pl.35 (1981).

Herb, usually annual; stems simple or branched, 7–30cm, sparsely villous, glandular-hairy above. Leaves simple, broadly ovate to orbicular, up to 20 × 17mm, strigillose, the hairs arising from tuberculate bases, the margin glandular-hairy; petioles 5–25mm. Inflorescence lax, few-flowered, ebracteate above; pedicels elongating to 12mm in fruit. Flowers actinomorphic. Sepals lanceolate to ovate, 1.5–2mm. Petals not appendiculate, obovate to elliptic, 3–4 × c. 1.5mm, pale yellow. Stamens 6. Style 1–3mm. Fruit not stipitate, usually spreading and curved, occasionally erect or recurved, (4–)20–35 × 0.5–1mm. Seeds minutely granulate, c.0.5mm diam., glabrous. **Map 476, Figs 67 & 68.**

Wadi-beds, sand dunes and rocky hillsides; 0–600m.

Saudi Arabia, Yemen (N & S), Socotra, Oman, Qatar. N & tropical Africa, Pakistan.

3. C. brachycarpa Vahl ex DC., Prodr. 1: 240 (1824). Syn.: *C. brevisiliqua* Schultes f. in Schultes & Schultes f., Syst. veg. 7: 40 (1829); *C. diversifolia* Hochst. & Steud. nom. nud. in sched. Illustr.: Collenette (1985 p.96); Fl. Qatar pl.34 (1981); Western (1989 p.57). Type: Yemen (N), *Forsskal* (BM).

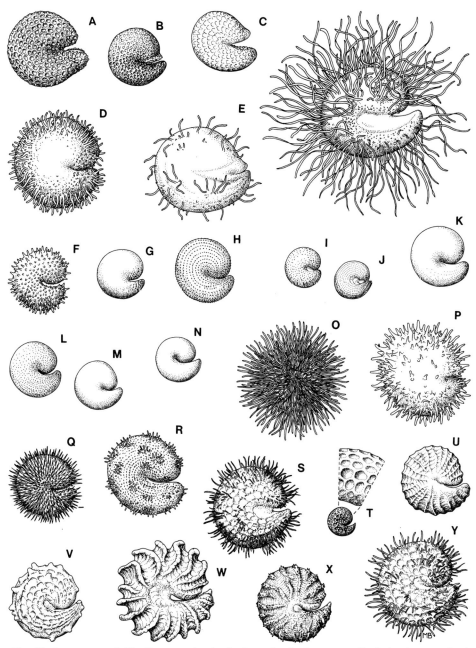

Fig. 67. Capparaceae. A–Y, *Cleome* seeds: A, *C. chysantha*; B, *C. scaposa*; C, *C. brachycarpa*; D, *C. ramosissima*; E, *C. amblyocarpa* (showing variation in indumentum); F, *C. brevipetiolata*; G, *C. macradenia*; H, *C. droserifolia*; I, *C. austroarabica* subsp. *austroarabica*; J, *C. austroarabica* subsp. *muscatensis*; K, *C. pruinosa*; L, *C. noeana* subsp. *noeana*; M, *C. noeana* subsp *brachystyla*; N, *C. polytricha*; O, *C. rupicola*; P, *C. arabica*; Q, *C. albescens* subsp. *omanensis*; R, *C. socotrana*; S, *C. angustifolia*; T, *C. tenella*; U, *C. viscosa*; V, *C. gynandra*; W, *C. hanburyana*; X, *C. monophylla*; Y, *C. paradoxa*. (E, M & Q × 16, others × 24).

Aromatic perennial herb, sometimes flowering in the first year; stems erect or ascending, simple or branched, 10–35cm, shortly stipitate-glandular. Leaves 3(–5)-foliolate; leaflets elliptic, 4–15 × 1–8mm, sparsely stipitate-glandular; petioles 2–30mm. Inflorescence lax, few-flowered, elongating in fruit; bracts leaf-like, minute or up to the same size as the leaves, entire, sometimes orbicular, to trifoliolate; pedicels 5–15mm. Sepals ovate, 1–3 × 0.5–1.5mm, the apex acute or acuminate. Petals not appendiculate, elliptic, 3–8 × 1–2mm, yellow, sometimes with reddish markings, rarely whitish-yellow. Stamens 6. Style 1–6mm. Fruit not stipitate, erect-spreading, oblong, 4–12 × 1–4mm. Seeds 0.5–1mm diam., finely reticulate, glabrous. **Map 477, Figs 67 & 68.**

Gravelly wadi-beds, on limestone and sandstone, lava slopes, etc.; 0–2600m.

Saudi Arabia, Yemen (N & S), Socotra, Oman, UAE, Qatar. Egypt, tropical and NE Africa, S Iran, Afghanistan, Pakistan and NW India.

A distinctive, though variable, species recognized by its short, oblong, long-styled fruits.

4. **C. ramosissima** Webb ex Parl., Fragm. fl. aethiop.-aegypt.: 22 (1854). Syn.: *C. schweinfurthii* Gilg in Notizbl. Königl. Bot. Gart. Berlin 1: 62 (1895). Illustr.: Collenette (1985 p.98) as *C. schweinfurthii*.

Aromatic perennial herb; stems usually woody, copiously branched, 7–100cm, densely stipitate-glandular. Leaves trifoliolate; leaflets narrowly elliptic to elliptic, 6–35(–45) × 2–10(–15)mm, shortly stipitate-glandular; petioles 5–15(–45)mm. Inflorescence lax, many-flowered; bracts simple, obovate to elliptic, 3–10mm; pedicels 5–15mm. Sepals lanceolate, 1.5–3 × 0.5–1.5mm. Petals not appendiculate, elliptic, 4–7(–8) × 1–2mm, yellowish-white to reddish-purple, with purple or brownish veins. Stamens 6. Style 1.5–5(–6)mm. Fruit borne on a 3–5mm gynophore, pendulous, linear, (15–)25–95(–105) × 1–4(–8)mm. Seeds 1–1.5(–2)mm diam., lanate when mature. **Map 478, Fig. 67.**

Rocky hillsides, on volcanics, sandstone and granite, and terrace walls; (550–)1200–2900m.

Saudi Arabia, Yemen (N & S). Ethiopia, Djibouti, Somalia and Sudan.

5. **C. amblyocarpa** Barr. & Murb. in Acta Univ. Lund. n.s. Afl. 2,1,4: 25 (1905). Illustr.: Collenette (1985, p.96).

Annual aromatic herbs; stems much-branched, 10–60(–100) cm, scabrid. Leaves trifoliolate, the basal rarely simple; leaflets elliptic, 5–30(–35) × 2–7(–10)mm, minutely scabrid; petioles 5–25(–30)mm. Inflorescence lax, few-flowered; bracts leaf-like, the uppermost entire and smaller; pedicels to 10mm. Sepals ovate, 2mm. Petals not appendiculate, narrowly oblong to obovate, 3–4 × 1–2mm, pale yellow. Stamens 6. Style to 4mm. Fruit sometimes shortly stipitate, pendulous, oblong, straight or slightly curved, (17–)30–40 × 5–8mm. Seeds c.2mm diam., densely lanate when mature. **Map 479, Fig. 67.**

Damp places in sand dunes, wadis and waste ground; 15–2400m.

Saudi Arabia, Yemen (N & S), Oman, UAE. N Africa, Sudan, Ethiopia, Palestine, Iraq and S Iran.

This species has previously been confused nomenclaturally with *C. arabica*. The application of the name used here follows Botschantzev (in Novosti Sist. Vyssh. Rast. 1968: 236–237).

6. C. brevipetiolata Chamberlain & Lamond in Edinb. J. Bot. 51 (1): 49 (1994). Type: Oman, *Miller* 6021 (E, K, ON, UPS).

Dwarf twiggy shrub; stems 7–50cm, covered with ± sessile glands. Leaves simple, broadly ovate to orbicular, 3–8 × 2–8mm, covered with ± sessile glands; petioles 0–5mm. Inflorescence several-flowered, lax, even in flower; bracts leaf-like but obviously smaller than the leaves; pedicels spreading, 0–8mm. Sepals lanceolate, 1–5 × 0.5–2 mm. Petals appendiculate, dimorphic, the lamina elliptic, 3–6 × 1–3 mm, yellow or greenish, sometimes with a red stripe. Stamens 4. Style (2–)7–10mm. Fruit erect, oblong, slightly curved, 5–20 × 2–5mm. Seeds 0.5–1mm, densely pubescent. **Map 480, Figs 67 & 69.**

Sandy or stony deserts, wadis, etc.; 50–600m.

Oman. Endemic.

A distinctive species among those with four stamens on account of its densely pubescent seeds. A hybrid between this species and *C. noeana* has been reported from Oman.

7. C. macradenia Schweinf. (1896 p.188). Type: Yemen (S), *Schweinfurth* 178 (K).

Foetid dwarf shrub; stems much-branched, 10–90cm, densely glandular-pilose at first. Leaves simple, ± orbicular, 4–10(–13)mm diam., glandular-tuberculate, the margin glandular-crenate; petioles 4–12(–20)mm. Inflorescence lax, few-flowered; bracts leaf-like, not clearly differentiated from the leaves; pedicels 5–15mm. Sepals lanceolate, 3–6 × 1–2mm. Petals appendiculate, dimorphic, the lamina narrowly elliptic or broadly lanceolate, 5–12 × 1–3mm, yellow. Stamens 4. Style 2–7(–10)mm. Fruit erect-spreading, elliptic, 9–15 × 3–4mm, straight. Seeds c.0.5mm diam., smooth. **Map 481, Figs 67 & 68.**

Sand dunes and wadis; 30–800m.

?Saudi Arabia, Yemen (S). Endemic.

Closely allied to *C. droserifolia* but distinguished by its glandular-crenate leaves and somewhat smaller petals. Schwartz (1939) records *C. macradenia* from Saudi Arabia though no specimens have been seen to confirm this.

8. C. droserifolia Del. (1813 p.106). Syn.: ?*C. brachyadenia* Schwartz (1939 p.62).

Aromatic dwarf shrub; stems much-branched, 25–75cm, with a dense '2-storeyed' glandular indumentum. Leaves simple, thick-textured, broadly ovate to orbicular,

CAPPARACEAE

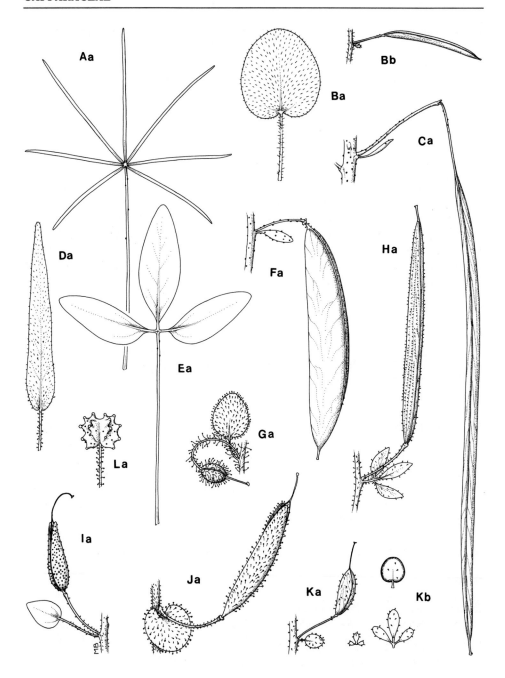

Fig. 68. Capparaceae. A, *Cleome angustifolia*: Aa, leaf. B, *C. scaposa*: Ba, leaf; Bb, fruit. C, *C. paradoxa*: Ca, fruit. D, *C. monophylla*: Da, leaf. E, *C. albescens* subsp. *omanensis*: Ea, leaf. F, *C. rupicola*: Fa, fruit. G, *C. chrysantha*: Ga, fruit. H, *C. viscosa*: Ha, fruit. I, *C. pruinosa*: Ia, fruit. J, *C. polytricha*: Ja, fruit. K, *C. brachycarpa*: Ka, fruit; Kb, bracts. L, *C. macradenia*: La, leaf. (All × 1.5)

often broader than long, 4–14 × 4–14mm, the apex obtuse, green or glaucous, glandular-hairy, the hairs fine and short or longer and coarser; petioles 8–20mm. Inflorescence lax, few-flowered, the flowers apparently axillary; bracts not differentiated from the leaves; pedicels 10–15mm. Sepals lanceolate, dimorphic, 4–7 × 1–2mm. Petals appendiculate, dimorphic, 2 broad and 2 narrow, the lamina lanceolate, 7–10 × 1.5–3mm, greenish yellow to yellow. Stamens 4. Style 4–10mm. Fruit erect, not stipitate, oblong, 10–17 × 3–3.5mm. Seeds 0.5–1mm diam., smooth or minutely papillate, glabrous. **Map 482, Fig. 67.**

Rocky hills, wadis, etc.; 0–1200m.

Saudi Arabia, Yemen (S). Egypt, Palestine, Syria and Jordan.

This species has a disjunct distribution in the Arabian Peninsula, being restricted to NW Saudi Arabia and the Hadramaut in Yemen (S).

Two extreme forms may be recognized. The first, matching the type of the species (originally described from Egypt) and restricted in Arabia to NW Saudi Arabia, is much-branched, with fine stems, short (c.10mm) fruits and comparatively thin leaves. The second, probably referable to *C. brachyadenia*, from Yemen (S), has fewer and thicker branches, longer (15–20mm) fruits and thicker leaves. Plants intermediate between the two extreme forms occur in Saudi Arabia and Egypt. Records of *C. droserifolia* in Schwartz (1939) from Oman refer to *C. austroarabica* subsp. *muscatensis*.

9. C. austroarabica Chamberlain & Lamond in Edinb. J. Bot. 51 (1): 51 (1994). Illustr.: Miller (1988 p.95) as *C. droserifolia*. Type: Oman, *Miller* 6247 (E, K, KTUH, ON, UPS).

Dwarf shrub; stems 15–100cm, densely stipitate-glandular to glandular-villous. Leaves simple, broadly ovate to orbicular, (5–)8–20 × (5–)8–15mm; petioles 7–20mm. Inflorescence lax, few-flowered; bracts leaf-like at least initially, 3–6mm diam.; pedicels 4–15mm. Sepals dimorphic, lanceolate, 3–5 × 1–2mm. Petals appendiculate, lanceolate, dimorphic, the larger broad, the smaller narrow, 5–8 × 1.5–3mm, yellow with a central reddish-brown stripe. Stamens 4. Style 5–7mm (in fruit). Fruit erect to erect-spreading, not stipitate, elliptic, 12–18(–22) × 3–4.5mm, straight or slightly curved. Seeds 0.6–0.8mm, minutely papillate when mature. **Map 483, Fig. 67.**

1.	Indumentum of stems and leaves stipitate-glandular	subsp. **austroarabica**
+	Indumentum of stems and leaves glandular-villous	subsp. **muscatensis**

subsp. **austroarabica**

Rocky slopes, gravelly wadi-beds etc.; 0–600m.

Yemen (S), Socotra, Oman, UAE. Endemic.

In Oman this subspecies is restricted to the south, mainly to Dhofar.

subsp. **muscatensis** Chamberlain & Lamond in Edinb. J. Bot. 51 (1): 51 (1994). Illustr.: Collenette (1985 p.97) as *C.* sp. aff. *droserifolia*. Type: Oman, *Miller & Nyberg* 9569 (E, K, KTUH, ON).

CAPPARACEAE

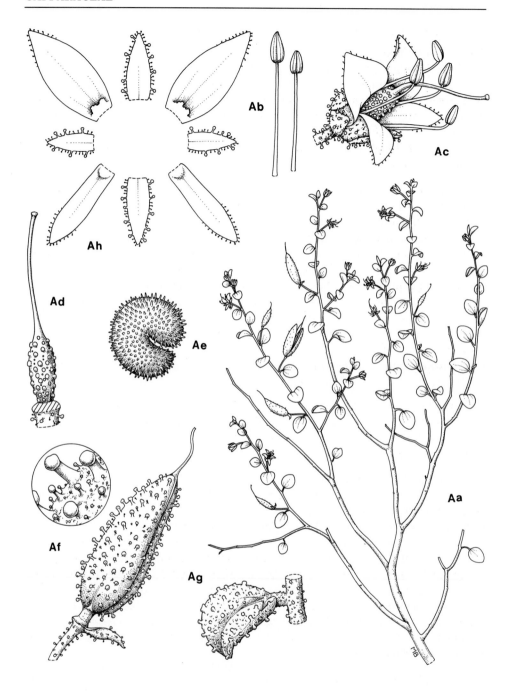

Fig. 69. Capparaceae. A, *Cleome brevipetiolata*: Aa, habit (× 0.6); Ab, stamens (× 5); Ac, flower (× 5); Ad, gynoecium (× 5); Ae, seed (× 32); Af, fruit (× 4) and detail of glands; Ag, leaf (× 4); Ah, dissected perianth parts (× 5).

On limestone and gabbro, cinder cones, rocky wadi-beds etc.; 0–1700m.

Saudi Arabia, Oman, UAE. Endemic.

Subspecies *muscatensis* occurs in N Oman with an outlying station in W Saudi Arabia. The differences between the two subspecies are small but there is no overlap in their geographical distributions. The indumentum of subsp. *muscatensis* may have evolved in response to a drier, non-monsoonal climate; in this character the one specimen from Saudi Arabia is extreme.

C. austroarabica is allied to *C. droserifolia* but exhibits a more open habit, and has a different distribution.

10. C. pruinosa T. Anders. (1860 p.3). Type: Yemen (S), *J.D.Hooker* s.n. (K).

Low aromatic shrub; stems 30–100cm, minutely glandular-scabrid, with a few larger scattered stipitate glands. Leaves simple, thin-textured, broadly ovate to orbicular, 7–15 × 5–14mm, the apex ± acute, apparently glaucous at least when dry, minutely glandular-scabrid with a few stout stipitate glands especially on the margins; petioles 5–15mm. Inflorescence lax, few-flowered, the flowers apparently axillary; bracts leaf-like, not differentiated from the leaves; pedicels 7–15mm. Sepals lanceolate, 3–5 × 1–2mm. Petals appendiculate, dimorphic with 2 broad and 2 narrow, the lamina lanceolate, 4–9 × 1–2mm, yellow. Stamens 4. Style 3–10mm. Fruit erect-spreading, not stipitate, elliptic, straight, 10–16 × 3–4.5mm. Seeds c.0.5mm diam., smooth, glabrous. **Map 484, Figs 67 & 68.**

Rocky basaltic slopes and wadi-beds; 0–15m.

Yemen (?N & S). ?Endemic.

Distinguished from *C. droserifolia* and *C. macradenia* particularly by its thin, glaucous leaves. It is however, only doubtfully distinct from the former and is apparently restricted to the area immediately around Aden, with one doubtful record for Hanish Island in the Red Sea.

11. C. noeana Boiss., Diagn. pl. orient. sér. 2,1: 48 (1853).

Annual or perennial, unpleasantly aromatic herb or dwarf shrub; stems branched, 10–60cm, with dense and spreading glandular hairs. Leaves simple, broadly ovate to orbicular, up to 40 × 40mm, shortly glandular-hairy; petioles 5–20(–50)mm. Inflorescence lax, often elongate even in flower; bracts like the leaves but smaller, to 10 × 10mm, sometimes absent in the upper part of the inflorescence; pedicels erect or spreading, 6–15mm. Sepals ovate, 3–5mm. Petals appendiculate, 4–7 × 0.5–3mm, yellow, sometimes with a central red line, dimorphic, 2 with lamina triangular and 2 with lamina lanceolate. Stamens 4. Style (1–)3–7(–12)mm. Fruit not stipitate, erect to erect-spreading, narrowly oblong to linear, (8–)15–30 × 2.5–4.5mm, straight or curved, glandular. Seeds 0.7–0.8mm diam., minutely granulate to ± smooth, glabrous, shining. **Map 485, Fig. 67.**

1. Inflorescence spicate, overtopping the leaves; bracts usually much-reduced or absent   subsp. **noeana**
+ Inflorescence not an obvious spike; bracts ± leaflike, gradually decreasing upwards   subsp. **brachystyla**

subsp. **noeana**. Syn.: *C. drepanocarpa* O. Schwartz (1939 p.62). Illustr.: Cornes (1989 p.109) as *C.* cf. *quinquenervia*; Western (1989 p.58) as *C.* aff. *dolichostyla*.

Gravel deserts; 0–700m.

Saudi Arabia, Yemen (S), Oman, UAE, Qatar, Bahrain. Iraq, Iran, Afghanistan, Pakistan and C Asia.

While some of the material of this subspecies from the northern part of its range in Arabia has a leaf indumentum composed of long glandular and eglandular hairs (like that of the type), that from Oman consistently differs in having much shorter hairs.

Records of *C. quinquenervia* DC., a species described from Iran which differs in its generally smaller stature and shorter (10–18mm) fruits, are almost certainly referable to this species.

Records of the poorly known *C. fimbriata* Vicary, originally described from Pakistan, are also almost certainly referable to this species.

C. dolichostyla Jafri, a species described from western Iran and recorded from the Arabian Peninsula, is supposed to differ from *C. noeana* in its gland-fringed leaves and styles that are 8–12mm long. There is no evidence that there is any correlation between these two characters in the Arabian material so the older name is used for material that is referable to these two taxa.

subsp. **brachystyla** (Deflers) Chamberlain & Lamond in Edinb. J. Bot. 51 (1): 52 (1994). Syn.: *C. brachystyla* Deflers in Bull. Soc. Bot. France 34: 65 (1887). Illustr.: Collenette (1985 p.96). Type Yemen (S), *Deflers* s.n. (P).

Rocky basaltic slopes and ravines; 0–10m.

Saudi Arabia, Yemen (N & S). ?Somalia.

Subsp. *brachystyla* to some extent intergrades with subsp. *noeana* and is intermediate between it and *C. polytricha*. There is, however, no overlap in the geographical distribution of the two subspecies. It may be distinguished from *C. polytricha* by its narrower fruits borne on shorter erect pedicels.

12. C. polytricha Franchet in J. Bot. (Morot) 1: 41 (1887). Syn.: *C. beckiana* Rech.f. in Österr. Akad. Wiss., Math.-Naturwiss. Kl., Anz. 87: 296 (1950). Syntypes: Yemen (S), *Courbon* (P); *Beaudoin* (P); *Deflers* 43 (K).

Perennial aromatic herb with a woody base; stems divaricately branched, 10–40 (–60)cm, densely glandular-villous. Leaves simple, broadly ovate to orbicular, 10–30 × 10–25mm, densely glandular-strigillose; petioles 8–12mm. Inflorescence lax, several- to many-flowered; bracts like the leaves but smaller, 5–8mm diam.; pedicels 10–17mm, spreading. Sepals narrowly lanceolate, c.4 × 1mm. Petals appendiculate, c.5 × 1.5mm, yellow with red markings, dimorphic, 2 with lamina triangular and 2 with lamina lanceolate. Stamens 4. Style 3–7mm. Fruit not or hardly stipitate, erect, narrowly oblong, 20–27 × 5–6mm, straight or slightly curved, glands with long slender stalks. Seeds c.0.8mm diam., ± smooth, glabrous. **Map 486, Figs 67 & 68.**

Waste ground, basaltic slopes etc.; 30–300m.

Yemen (S). Somalia.

This species can be generally distinguished from its allies by the broad erect fruits borne on spreading pedicels.

13. C. rupicola Vicary in J. Asiat. Soc. Bengal 16: 1158 (1847). Syn.: *C. glaucescens* sensu Mandaville (1990) non DC.; *C. oxypetala* Boiss. var. *micrantha* Boiss., Fl. orient. 1: 415 (1867). Illustr.: Western (1989 p.58); Mandaville (1990 pl.71) as *C. glaucescens*.

Perennial, often glaucous herb with a woody base; stems branched, erect, 15–45cm, shortly stipitate-glandular. Leaves simple (rarely 2–3-foliolate), ovate-elliptic, 15–40 × 8–28mm, thick, glandular; petioles 8–35mm. Inflorescence many-flowered, contracted in flower, elongating in fruit; bracts narrowly elliptic to elliptic, up to 15mm long; pedicels 5–15mm. Flowers zygomorphic. Sepals triangular to ovate, 1–3 × 0.5–1mm, glandular. Petals not appendiculate, broadly ovate to suborbicular, 3–7 × 1–2mm, yellow to orange, with brown to purple veins, the apex rounded to acute. Stamens 6. Style 0.5–1mm. Fruit with gynophore lacking or up to 1mm long, pendulous, linear, straight, 20–45 × 4–8 mm, sparsely glandular, at least when young. Seeds (0.5–)1mm, papillate when young, densely lanate when mature. **Map 487, Figs 67 & 68.**

Rocky places, on sandstone and sandy plains; 30–1750m.

Saudi Arabia, Oman, UAE. SE Iran and Pakistan.

Closely allied to *C. glaucescens* DC. and *C. oxypetala* Boiss. but distinguished from both by the glandular fruit. All the specimens seen from the Arabian Peninsula that have been assigned to these two species are referable to *C. rupicola*, as is the type of *C. oxypetala* Boiss. var. *micrantha* Boiss.

14. C. arabica L., Cent. pl. I: 20 (1755), non L., Syst. Nat. ed. 12: 448 (1767) et auctt. Syn.: *C. trinervia* Fresen., Beitr. Fl. Aegypt. 2: 177, t.11 (1834). Illustr.: Collenette (1985 p.98) as *C. trinervia*.

Strongly aromatic perennial herb with a woody base; stems multiple, simple or with a few branches, 14–50cm, glandular-strigillose. Leaves simple (rarely 3-foliolate), ovate, 10–40 × 5–18mm, stipitate-glandular; petioles 4–25mm. Inflorescence lax, few- to many-flowered, with progressively smaller linear-lanceolate bracts; pedicels 6–14mm. Flowers zygomorphic. Sepals ovate, c.2mm. Petals not appendiculate, oblong-elliptic to elliptic-rhomboid, 5–9mm long, reddish-brown. Stamens 6. Style 0.5–1mm. Fruit not or very shortly stipitate, pendulous, straight or slightly curved, (8–)15–43 × (1–)2–5 mm. Seeds c.1.5mm diam., papillose when young, lanate when mature. **Map 488, Fig. 67.**

Rocky hillsides, often on volcanics; 600–900m.

Saudi Arabia, Yemen (?N & ?S). Egypt, Palestine, S Iran (Islands).

For a discussion of the nomenclature of this species see Kers (*Acta Horti Berg.* 20 (8): 335–342 (1966)). Records from Yemen probably refer to *C. amblyocarpa*. A specimen (*Collenette* 7070) from Saudi Arabia has some of the upper leaves with three leaflets but otherwise resembles *C. arabica*.

15. C. albescens Franchet subsp. **omanensis** Chamberlain & Lamond in Edinb. J. Bot. 51 (1): 52 (1994). Type: Oman, *Miller & Nyberg* 9029 (E).

Glaucous annual or perennial herb; stems 15–30cm, shortly stipitate-glandular. Leaves 3-foliolate, occasionally simple near the base; leaflets elliptic, 5–15(–20) × 1–8mm, ± glabrous; petioles 2–12mm. Inflorescence few-flowered, lax in flower, elongating in fruit; bracts trifoliolate, 2–3mm or (more usually) absent; pedicels 2–10mm, usually glabrous. Sepals ovate, 1–1.5 × 0.5–1mm, glabrous. Petals not appendiculate, narrowly elliptic, 1–4 × 0.5–1mm, orange-yellow. Stamens 6. Style 0.5–2mm. Fruit very shortly stipitate, erect or erect-spreading, narrowly elliptic, 10–40 × 2–4mm, glabrous. Seeds c. 1mm diam., densely hairy when mature. **Map 489, Figs 67 & 68.**

Wadis and rocky slopes; 100–600m.

Yemen (S), Oman. Endemic subspecies.

A single specimen from Yemen (S) (*Guichard* KG/HAD/418) differs only in the stipitate-glandular pedicels and calyces, in which respect it approaches the closely allied *C. socotrana*. However, it agrees with the present taxon in its shorter stature, laxer inflorescences, narrower fruits and narrower leaflets. Specimens in herbaria named '*C. venosa* Hutch.' belong here.

Subsp. *albescens* (from Somalia) differs in its upper leaves which are generally broader and simple or only occasionally 3-foliolate.

16. C. socotrana Balf. f. (1882 p.501). Syntypes: Socotra, *Balfour, Cockburn & Scott* 76 (BM); *Schweinfurth* 659 (K) & 710 (K).

Woody-based, probably perennial, herb; stems branched, 25–70cm, shortly stipitate-glandular, occasionally glabrous. Leaves 3-foliolate; leaflets broadly elliptic to obovate, 5–20 × 4–20mm, very sparsely stipitate-glandular or glabrous; petioles 5–30mm. Inflorescence many-flowered, ± dense in flower, becoming lax in fruit; bracts simple, broadly obovate to orbicular, 2–4mm; pedicels 4–10mm, stipitate-glandular. Sepals ovate, 1-2 × 0.5–1mm, stipitate-glandular. Petals not appendiculate, narrowly elliptic, 2–5 × 0.5–1.5mm, white to yellow, with red or brown markings. Stamens 6. Style 0.5–2 mm. Fruit not stipitate, erect or erect-spreading, narrowly elliptic to narrowly oblong, 10–45 × 4–7mm, sparsely stipitate-glandular. Seeds c.1.5mm diam., densely hairy when mature. **Map 490, Fig. 67.**

Bushland and thicket on granite; 200–1350m.

Socotra. Endemic.

Closely allied to *C. albescens* Franchet.

17. C. angustifolia Forsskal (1775 p.120). Type: Yemen (N), *Forsskal* (BM).

Glaucous annual foetid herb; stems laxly branched, 30–100 cm, glabrous. Leaves 5–7-foliolate; leaflets filiform, 5–33 × 0.5mm, glabrous; petioles 20–50mm. Inflorescence lax, few-flowered; bracts 1–3-foliolate, to 2cm; pedicels 20–25mm. Sepals lanceolate, 3–5 × 1–2mm. Petals not appendiculate, obovate, 3–10 × 1–3mm, yellow, sometimes with red veins. Stamens 6. Style 2–9mm. Fruit borne on a 3–6mm gynophore, spreading or pendent, straight, 20–70 × 1–6mm. Seeds c.1.5mm diam., reticulate, shortly hairy. **Map 491, Fig. 67.**

A weed of fields and roadsides; 1000–2400m.

Yemen (N). Ethiopia and Sudan.

18. C. tenella L.f., Suppl. pl.: 300 (1781).

Annual herb; stems erect, slender, much-branched, 15–30cm, glabrous. Leaves 3-foliolate; leaflets filiform, 7–20 × c.0.5mm, glabrous; petioles 7–20mm. Flowers solitary or up to 5 in lax inflorescences; bracts simple or 3-foliolate, filiform; pedicels 8–15mm. Sepals ovate, c.1mm, glabrous. Petals not appendiculate, obovate, c.2.5mm, colour not known. Stamens 6. Style minute. Fruit without a gynophore, erect or erect-spreading, linear, straight or curved, 20–35 × c.2.5mm. Seeds c.1mm diam., reticulate, glabrous. **Map 492, Fig. 67.**

Weed of date gardens, etc.; c.50m.

Socotra. Tropical Africa from Ethiopia to Senegal, India.

19. C. viscosa L., Sp. pl.: 672 (1753). Illustr.: Collenette (1985 p.98).

Annual foetid herb; stems erect, simple or branched, up to 70cm, densely glandular-villous and stipitate-glandular. Leaves 3(–5)-foliolate; leaflets ovate-elliptic, 10–40 × 3–20mm, shortly glandular-pilose; petioles 7–70mm. Inflorescence few-flowered; bracts leaf-like, trifoliolate; pedicels 10–30mm. Sepals ovate-elliptic, 3–6mm. Petals not appendiculate, broadly elliptic, 4–10 × 1–3mm, yellow drying white. Stamens 10–20. Style 1–4mm. Fruit without a gynophore, erect or erect-spreading, linear, straight or slightly curved, 50–80 × 3–5mm, prominently nerved. Seeds 1–1.5mm diam., transversely ridged, glabrous. **Map 493, Figs 67 & 68.**

Rocky and gravelly slopes, fields and amongst scrub; 50–1500m.

Saudi Arabia, Yemen (N & S), Socotra. A pantropical weed.

20. C. gynandra L., Sp. pl.: 671 (1753). Syn.: *C. pentaphylla* L., Sp. pl. ed.2: 938 (1763); *Pedicellaria gynandra* (L.) Schrank in Bot. Mag. (Römer & Usteri) 3 St. 8: 10 (1790); *Gynandropsis pentaphylla* (L.) DC., Prodr. 1: 238 (1824); *G. gynandra* (L.) Briq., Annuaire Conserv. Jard. Bot. Genève 17: 382 (1914). Illustr.: Collenette (1985 p.99) as *Gynandropsis gynandra*.

Faintly aromatic annual herb; stems branched from the base, 35–100cm, ± villous and stipitate-glandular. Leaves 3–5-foliolate; leaflets obovate, up to 13–70 × 6–33mm, cuneate at base, glandular above or almost glabrous; petioles (19–)30–95(–120)mm. Inflorescence many-flowered; bracts small, the lower 3-foliolate, the upper simple; pedicels 15–25mm, spreading, not elongating in fruit. Sepals ovate, 2–4mm, acute. Petals not appendiculate, the lamina orbicular, 2–6mm diam., white or yellow to mauve-pink. Stamens 6, borne on an 8–20mm anthophore. Style 1–4mm. Fruit borne on a 5–12mm gynophore, linear, straight or slightly curved, erecto-patent, 20–80(–115) × 2–5mm. Seeds c.1mm diam., coarsely reticulate, glabrous. **Map 494, Fig. 67.**

CAPPARACEAE

Cultivated ground; 0–1500m.

Saudi Arabia, Yemen (N & S), Socotra, Oman, UAE. A pantropical weed.

Allied to *C. hanburyana* and sometimes confused with it. Often placed in a separate genus, on account of the long androphore.

21. C. hanburyana Penzig in Penzig (ed.), Atti Congr. Bot. Genova: 330 (1893). Syn.: *C. areysiana* Deflers in Bull. Soc. Bot. France 42: 297 (1895); *C. deflersii* Blatter (1919–36, p.38) nom. nud.

Annual or perhaps perennial herb; stems branched from the base, 20–40cm, densely glandular-villous and shortly stipitate-glandular. Leaves 5-foliolate; leaflets obovate, 5–20 × 2–6mm, densely stipitate- or villous-glandular; petioles (8–)15–60mm. Inflorescence many-flowered; bracts small, the lower 3-foliolate, the upper simple; pedicels 8–15mm. Sepals triangular-ovate, 3–7 × 1–3mm, densely glandular. Petals not appendiculate, the lamina elliptic, 13–25 × 3–6mm, white and crimson to purple with a yellow spot. Stamens 6–8. Style 1–3 mm. Fruit borne on a 17–25mm gynophore, linear, straight or slightly curved, spreading to pendulous, 55–90 × 1–4cm. Seeds 1–2mm diam., ridged, glabrous. **Map 495, Fig. 67.**

Rocky places on hillsides and in wadis; 30–1100m.

Saudi Arabia, Yemen (S). N Kenya and NE tropical Africa.

Allied to *C. gynandra*, differing in the absence of an androphore and in its smaller leaflets. It has a more restricted distribution.

22. C. monophylla L., Sp. pl.: 672 (1753).

Sprawling annual; stems branched, 30–50cm, stipitate-glandular. Leaves simple, linear-lanceolate to lanceolate, 20–35(–60) × 3–5(–20)mm, rounded at the apex, stipitate-glandular; petioles absent or up to 10mm. Inflorescence few-flowered; bracts similar to the leaves but smaller; pedicels c.2mm in flower, elongating to 10mm in fruit. Sepals linear, 2.5–3 × 0.5mm, stipitate-glandular. Petals not appendiculate, linear to obovate, 4–7 × 1–2mm, mauve. Stamens 6. Style 2–5mm. Fruit borne on a 3–6mm gynophore, linear, straight, spreading. Seeds c.1.5mm diam., ridged, the ridges minutely papillate. **Map 496, Figs 67 and 68.**

Roadside; c.2000m.

Yemen (N). Tropical Africa, India and Sri Lanka.

In Arabia only known from a single gathering from Yemen (N).

23. C. paradoxa R.Br. ex DC., Prodr. 1: 241 (1824). Illustr.: Collenette (1985 p.97).

Large upright herb, probably perennial, aromatic or not; stems sometimes woody, 40–250cm, ± glabrous, often with brown punctate glands above. Leaves 5(–6)-foliolate; leaflets linear to narrowly elliptic, 15–65 × 1–7mm, glabrous; petioles 20–110mm. Inflorescence many-flowered; bracts simple, linear, 5–15 × 0.5–1mm; pedicels stout, especially in fruit, 10–30 mm. Sepals dimorphic, lanceolate or triangular-ovate,

5–14 × 2–6mm, punctate-glandular. Petals not appendiculate, dimorphic, the larger 1.5–2 × the smaller, broadly elliptic, 15–30 × 5–15mm, yellow, sometimes with red or brown veins. Stamens 6. Style 1–5(–9)mm. Fruit borne on a 10–15(–25)mm gynophore, linear, 60–120(–130) × 3–7mm, straight or slightly curved, spreading or pendulous. Seeds c.1.5mm diam., reticulate, densely hairy when mature. **Map 497, Figs 67 & 68.**

Rocky slopes, screes and disturbed ground; 30–700(–1650)m.

Saudi Arabia, Yemen (N & S). Ethiopia and Sudan.

The size of the plants and flowers make this a distinctive species.

Species incompletely known

C. digitata Forsskal (1775 p.120). Type: Yemen (N), *Forsskal* (?).

Stems villous-hispid. Lower leaves digitate, upper leaves 3-foliolate; leaflets c.25mm. Stamens 6.

Only known from the type which unfortunately cannot be traced (according to Christensen (1922) it is not at C). The digitate lower leaves limit the number of species with which *C. digitata* might have affinities. Since we have not seen the only known specimen, we hesitate to confirm the status of this species.

2. DIPTERYGIUM Decne

J.A. NYBERG

Glabrous or glandular perennial herbs. Leaves small, simple, shortly petiolate. Flowers small, in lax racemes, actinomorphic. Sepals 4, free. Petals 4, free, shortly clawed. Disc-appendage absent. Stamens 6, equal; androphore absent. Gynophore short. Fruit an indehiscent, 1-seeded, winged nutlet.

1. D. glaucum Decne in Ann. Sci. Nat. Bot. sér. 2, 4: 67 (1835). Syn.: *D. glaucum* Decne var. *macrocarpa* Blatter (1919–36 p.30); *Cleome pallida* Kotschy in Sitzungsber. Kaiserl. Akad. Wiss. Math.-Naturwiss. Cl., Abt. 1, 52: 262 (1866). Illustr.: Collenette (1985 p.98) Type: Saudi Arabia, *Bové* (P).

Woody-based herb or sub-shrub up to 60(–150)cm, with dense slender branches. Leaves linear-oblong to ovate, 2–20 × 1–6mm, acute to obtuse. Flowers on short slender pedicels. Sepals 0.5–2mm long. Petals pale yellow, rarely pink-tinged, 3–4 × 1–2mm. Stigma capitate on a short style. Fruits elliptic, 3.5–5 × 2–4mm, muricate, narrowly winged. **Map 498, Fig. 71.**

Widespread in stony, sandy and gravelly deserts; 0–500(–1350)m.

Saudi Arabia, Yemen (N & S), Oman, UAE. Egypt, Sudan, Ethiopia, Somalia, Djibouti, Iran and Pakistan.

3. MAERUA Forsskal

A.G. MILLER & J.A. NYBERG

Shrubs, trees or climbers. Leaves alternate or clustered, simple or 1–3-foliolate. Flowers in terminal racemes, panicles or clustered in the leaf axils, actinomorphic. Receptacle (the hypanthium) cylindrical or campanulate. Sepals 3–4, free. Petals (3–)4 or absent, subequal. Stamens 6–numerous, free, borne on an androphore. Gynophore present. Stigma capitate. Fruits globose to ellipsoidal or cylindrical, often torulose, indehiscent.

1. Climbing shrubs or, if rarely non-climbing, then a low and stunted shrub; petals present; fruits irregularly globular **4. M. oblongifolia**
+ Tree or non-climbing shrub; petals present or absent; fruits globose to ellipsoidal, or if cylindrical then ± regularly torulose 2

2. Petals present; leaves simple or 1–3-foliolate **3. M. triphylla**
+ Petals absent; leaves simple 3

3. Leaves shortly petiolate (petioles 1–4(–5)mm), usually clustered on short spur-branches; flowers in axillary clusters **1. M. crassifolia**
+ Leaves long-petiolate (petioles (5–)8–20mm), not usually clustered; flowers in terminal or axillary racemes **2. M. angolensis**

1. M. crassifolia Forsskal (1775 p. CXIII & 104). Syn.: ?*M. thomsoni* T. Anders. (1860 p.5); *M. uniflora* Vahl (1790 p.36); *M. arabica* J.F. Gmel., Syst. nat.: 827 (1791); *Wiegmannia arabica* (J.F. Gmel.) Hochst. & Steud. ex Steud., Nomencl. bot. ed. 2, 2: 787 (1841). Illustr.: Collenette (1985 p.99) as *M. oblongifolia*; Miller (1988 p.89); Western (1989 p.59). Type: Yemen (N), *Forsskal* (C).

Tree or shrub, 1–7m (less commonly a large tree up to 20m and with a c.1m diam. trunk), typically with an umbrella-shaped canopy. Leaves solitary on the new shoots and in clusters on short spur-branches on the older wood, simple, obovate to elliptic or oblong-elliptic, 7–20(–32) × (3–)4–10(–18)mm, obtuse or rarely acute to retuse at the tip, mucronate, glabrous or finely pubescent; petioles 1–4(–5)mm. Flowers in 1–4(–6)-flowered clusters arising from the spur-branches. Sepals greenish yellow, (3–)5–7mm. Hypanthium 2–3mm. Petals absent. Gynophore 10–20mm. Fruits cylindrical, torulose, 10–50 × 4–6mm, pale green when mature. **Map 499, Fig. 70.**

Sandy or gravelly plains, stony hills, rocky outcrops, dry volcanic slopes in semi-desert bushland and *Acacia-Commiphora* bushland; 10–1800m.

Saudi Arabia, Yemen (N & S), Oman, UAE. N Africa, W, C, NE & E tropical Africa, Palestine, Iran and Pakistan.

A very variable species, requiring further study in Arabia, and perhaps better treated as several taxa. In Yemen, John Wood (pers. comm.) records two forms with different vernacular names: "MERU" for glabrous plants from the Tihama (coastal plain) and "SIRAH" for pubescent plants from the dry inner plateau. Mats Thulin

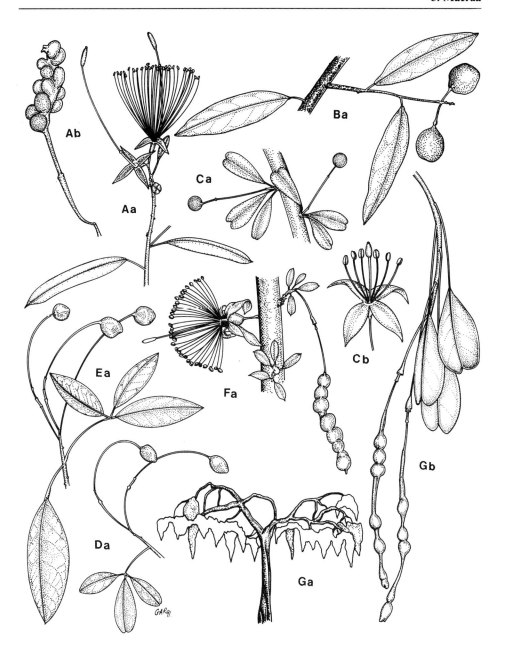

Fig. 70. Capparaceae. A, *Maerua oblongifolia*: Aa, flowering shoot (×1); Ab, fruit. B, *Boscia angustifolia*: Ba, fruiting branch (×1). C, *B. arabica*: Ca, part of fruiting stem (×1); Cb, flower (×4). D, *Maerua triphylla* var. *calophylla* type "A": Da, fruiting stem (×0.3). E, *M. triphylla* var. *calophylla* type "B": Ea, fruiting stem (×0.3). F, *M. crassifolia*: Fa, part of fruiting and flowering branch (×1). G, *M. angolensis* subsp. *socotrana*: Ga, tree; Gb, fruiting shoot (×1).

(pers. comm.) considers the mainly coastal, glabrous plants to be *M. crassifolia* (syn. *M. thomsonii* T. Anders.) and the pubescent plants to be a separate, more widespread taxon, which would have to be called *M. rigida* R. Br. The ranges of both of these taxa overlap in southern and western Arabia.

In the past *M. thomsonii* (described from Aden) has been wrongly misinterpreted as being conspecific with *M. angolensis*. De Wolf (Fl. Trop. E. Afr.: 29 (1964)) states that *M. thomsonii* T. Anders. appears to be an ecological variant of the widespread *M. angolensis*. However, *M. thomsonii* is clearly part of the *M. crassifolia* complex and not at all closely related to *M. angolensis*.

2. M. angolensis DC., Prodr. 1 : 254 (1824).

Shrub or small tree up to 6m. Leaves solitary, simple, obovate to oblong-ovate or subcircular, 1.5–40 × 1–3cm, rounded or retuse at the tip, mucronate, glabrous; petioles (5–)10–20mm. Flowers solitary or in few-flowered axillary and terminal racemes. Sepals cream, greenish or pale yellow, 5–10mm. Hypanthium 2–6 (–15)mm long. Petals absent. Fruit cylindrical, torulose, up to 40–50 × 3–5mm, green. **Fig. 70.**

1. Hypanthium 4–6(–15)mm; leaf-blade ovate to oblong-ovate, 1.5–3.5 × 1–2cm; low shrub   subsp. **angolensis**
+ Hypanthium 2–4mm; leaf-blade narrowly obovate to ovate or subcircular, 2–4 × 1–3cm; small tree with pendulous branches   subsp. **socotrana**

subsp. **angolensis**.

Shrub (in Arabia) up to 1.5m. Leaves ovate to oblong-ovate, 1.5–3.5 × 1–2cm. Sepals cream, 10mm. Hypanthium 4–6(–15)mm. Fruit cylindrical, torulose, up to 22cm long, green. **Map 500.**

Volcanic cones and sandstone hills; 1900–2100m.

Yemen (N). Tropical & southern Africa.

In Arabia known only from two gatherings in the dry inner plateau of Yemen. Fruiting material from Arabia has not been seen and the description has been completed from African material.

subsp. **socotrana** (Schweinf. ex Balf. f.) Kers var. **socotrana** in Novon 3: 54 (1993). Syn.: *M. angolensis* DC. var. *socotrana* Schweinf. ex Balf. f. (1884 p.402); *M. angolensis* sensu Balfour (1888) non DC.; *M. socotrana* (Schweinf. ex Balf. f.) Gilg in Bot. Jahrb. Syst. 33: 228 (1903). Syntypes: Yemen (Socotra) *Balfour, Cockburn & Scott* 193 (BM, K); 588 (B, BM, E, K, P); *Schweinfurth* 251; 457 (B, K, P); 603 (K, P).

Tree up to 6m, with a single trunk and graceful pendulous branches. Leaves narrowly obovate to ovate or subcircular, 2–4 × 1–3cm. Sepals greenish or pale yellow, 5–10mm. Hypanthium 2–4mm long. Fruits cylindrical, torulose, 40–50 × 3–5mm, pale green when mature. **Map 501, Fig. 70.**

Sandy plains, rocky slopes and on cliffs in *Jatropha unicostata-Croton socotranus* bushland; 10–400m.

Socotra. Endemic variety.

Two varieties are recognized within *M. angolensis* subsp. *socotrana*: var. *africana* (Ethiopia and Somalia) differs from var. *socotrana* (Socotra) in its longer sepals (1–2cm) and hypanthia (2–6mm). For a discussion of the *M. angolensis* complex in NE Africa and Arabia see Kers (*Novon* 3: 50–54 (1993)).

3. M. triphylla A. Rich. (1847 p.32, t.6).

Shrub or small tree up to 5m. Leaves simple or trifoliolate, narrowly elliptic to lanceolate or oblong-elliptic, 2–11 × 1–3.5cm, acute or obtuse at the tip, often mucronulate, sometimes retuse, the base cuneate, pubescent or glabrous; petioles 5–50mm. Flowers in axillary or terminal corymbose racemes. Sepals white or cream, 5–7mm. Hypanthium 2—4mm. Petals absent. Fruit globose or oblong-ovoid (in Arabia), 10—15 × 6–12mm, green. **Fig. 70.**

A very variable species with several varieties recognized in Africa. We have found it difficult to assign varieties to the few gatherings made in Arabia. This treatment must remain provisional and awaits a fuller revision of the genus in Africa and Arabia.

1. Twigs and pedicels densely pubescent; ovary usually hairy; fruit cylindrical
       var. **johannis**
+ Twigs and pedicels glabrous or pubescent; ovary usually glabrous; fruit globose or ovoid-ellipsoid   2

2. Twigs and pedicels glabrous   3
+ Twigs and pedicels pubescent   var. **calophylla** type **C**

3. Leaves and leaflets oblong-elliptic, 20–40 × 10–17mm, obscurely reticulate beneath; fruits ellipsoid   var. **calophylla** type **A**
+ Leaves and leaflets elliptic, 50–90 × 25–40mm, distinctly reticulate beneath; fruits globose   var. **calophylla** type **B**

var. **johannis** (Volkens & Gilg) De Wolf in Kew Bull. 16: 82 (1962).

Shrub to 1m, densely pubescent throughout. Leaves mainly trifoliolate; leaflets elliptic to lanceolate, up to 90 × 23mm, strongly reticulate beneath. Ovary hairy. Fruits in Arabia unknown. **Map 501.**

Terrace walls; 1500m.

Yemen (N). Tropical Africa.

Very rare in Arabia. Found only once, on J. Raymah, in the outer escarpment mountains.

var. **calophylla** (Gilg) De Wolf in Kew Bull. 16: 82 (1962).

Within var. *calophylla* three variants, with apparently distinct ecological preferences, are found in Arabia. No attempt has been made to compare these with African material. Outside Arabia var. *calophylla* is found in E and NE tropical Africa.

Type A. including: *M. cylindricarpa* sensu Schwartz (1939 p.68) non Gilg & Benedict; *M. nervosa* sensu Blatter (1919–36 p.40) non Oliv.

Shrub to 2m, glabrous throughout. Leaves simple and trifoliolate; leaves and leaflets, oblong-elliptic, 20–60 × 10–25mm, obscurely reticulate beneath; fruits ellipsoid to club-shaped. **Map 501, Fig. 70.**

Succulent shrubland dominated by *Euphorbia* spp.; 1300–2200m. **Yemen (N).**

Type B. including: *M. variifolia* sensu Schwartz (1939 p.69) non Gilg & Benedict.

Small tree up to 5m, glabrous throughout. Leaves simple and trifoliolate; leaves and leaflets elliptic, 50–90 × 25–40mm, visibly reticulate beneath; fruits globose. **Map 502, Fig. 70.**

Valley forest, cliffs and field margins, 500–1900m. **Yemen (N).**
Restricted to the outer escarpment mountains

Type C. Illustr.: Collenette (1985 p.99)

Shrub up to 1m, pubescent throughout. Leaves simple and trifoliolate; leaves and leaflets elliptic, up to 80 × 40mm, visibly reticulate beneath. Ovary glabrous. Fruits globose. **Map 502.**

On steep road cutting; 1525m. **Saudi Arabia.**

Outer escarpment mountains. Very similar to var. *calophylla* type **B** except that the plants are pubescent.

4. M. oblongifolia (Forsskal) A. Rich., (1847 p.32, t.5). Syn.: *Capparis mithridatica* Forsskal (1775 p.99); *C. oblongifolia* Forsskal (1775 p.99); *M. ovalifolia* sensu Blatter (1919–36 p.41) non Cambess.; *Niebuhria oblongifolia* (Forsskal) DC., Prodr. 1: 244 (1824). Illustr.: Collenette (1985 p.99) as *M. crassifolia*. Type: Yemen (N), *Forsskal* (C, BM).

Climbing shrub up to 3m, rarely non-climbing and then a low stunted shrub. Leaves typically narrowly oblong, rarely broadly elliptic (in non-climbing plants), 25–75 × 5–12(–30)mm, acute or obtuse at the tip, mucronate, the base rounded; petioles 5–10mm. Flowers in axillary or terminal corymbose racemes, rarely solitary and axillary. Sepals green, 8–10mm, with white-puberulous margins. Hypanthium 5–8mm. Petals white or yellowish-green. Gynophore 15–20mm. Fruit unevenly globular, 2–4 × c.1cm, red at maturity. **Map 503, Fig. 70.**

Acacia-Commiphora bushland, semi-desert shrubland, ?salt flats and terrace walls; 0–1750m.

Saudi Arabia, Yemen (N & S), Oman. Tropical Africa.

A single specimen (*Collenette* 6885) growing, together with typical plants, on Farasan Island is a low non-climbing shrub with broadly elliptic leaves up to 30mm across.

Species incompletely known

M. racemosa Vahl (1790 p.36); no specimen has been seen. According to Christensen (1922) there is no specimen at C.

4. BOSCIA Lam.

J.A. NYBERG

Small trees. Leaves alternate, solitary or clustered, simple. Flowers small, in terminal or axillary corymbose racemes, actinomorphic. Sepals 4, free, valvate. Petals absent. Stamens 4–9, free, borne on a very short androphore. Gynophore present. Stigma capitate, borne on a short style. Fruit globose, indehiscent.

1. Leaves narrowly oblong-elliptic to obovate, usually more than 2.5cm long; leaf venation reticulate, prominent on the undersurface   **1. B. angustifolia**
+ Leaves obovate or oblong-obovate, up to 3cm long; leaf venation obscure   **2. B. arabica**

1. B. angustifolia A. Rich. in Guill. & Perr., Fl. Seneg. tent. 1: 26, t.6 (1831). Illustr.: Collenette (1985 p.93).

Tree up to 12m, with a rounded crown; trunk smooth, pale brown or grey; young branches sometimes pubescent. Leaves dark green, narrowly oblong-elliptic to obovate, 2.5–6 × 0.8–2cm, rounded to retuse at the tip, mucronate, glabrous; petioles 1–5mm. Flowers in terminal or axillary clusters; pedicels 3–10mm. Sepals c.4 × 1–1.5mm; disk fimbriate. Stamens (4–)5–9. Gynophore present. Fruit green, ovoid, verrucose, c.1cm diam. **Map 504, Fig. 70.**

Acacia-Commiphora bushland; 200–1950m.

Saudi Arabia, Yemen (N). Tropical Africa.

Very variable in leaf shape.

2. B. arabica Pestalozzi in Bull. Herb. Boissier 6, App. 3: 127 (1898). Illustr.: Miller (1988 p.89).

Tree up to 6m, usually with a flat-topped crown; trunk smooth, greyish or greyish brown. Leaves yellow-green, becoming glaucous with age, obovate to oblong-obovate, 8–30 × 3–12mm, rounded or retuse at the tip, glabrous to finely pubescent, somewhat farinose; petioles 2–4mm. Flowers in axillary racemes; pedicels 10–15mm. Sepals 2.5–4 × 1.5–2.5mm. Stamens 6–8. Gynophore 5–7mm long. Fruit globose, 5–10mm diam. **Map 505, Fig. 70.**

Acacia-Commiphora bushland; 50–1250m.

Yemen (N & S), Oman. Endemic.

The leaves and pedicels of specimens from Oman and the eastern part of Yemen (S) are finely pubescent, whereas those from the western part of the range are glabrous. There are no other consistent differences between the two variants so they have not been given any formal taxonomic status.

Species doubtfully recorded

B. senegalensis (Pers.) Lam. ex Poir. (Migahid 1978); the specimen of this at KSUH is an Asclepiad.

B. minimifolia Chiov. (De Marco & Dinelli p. 222).

5. CAPPARIS L.

J.A. NYBERG

Shrubs or small trees. Leaves simple, entire, often leathery, with a pair of recurved stipular thorns. Flowers solitary in the leaf axils or in terminal or axillary racemes or clusters. Sepals 4, free, ± equal or with the upper sepal larger and strongly hooded. Petals 4, free. Stamens 10–numerous; androphore absent. Gynophore present. Fruit a globose to obovoid berry.

1. Plant soon aphyllous; leaves when present linear-oblong; petals red   **1. C. decidua**
+ Leaves present, broadly ovate to orbicular; petals white or cream, often becoming pink with age   2
2. Flowers in terminal racemes; fruits globose   **4. C. tomentosa**
+ Flowers solitary in the axils of the leaves; fruits obovoid to ellipsoid   3
3. Sepals subequal, ovate to broadly ovate   **2. C. spinosa**
+ Sepals unequal, with 3 oblong-ovate and shallowly boat-shaped and the other ovate and strongly hooded   **C. cartilaginea**

1. C. decidua (Forsskal) Edgew. in J. Linn. Soc. Bot. 6: 184 (1862). Syn.: *Capparis aphylla* Roth, Nov. pl. sp.: 238 (1821) *Sodada decidua* Forsskal (1775 p. 81);. Illustr.: Collenette (1985 p.95). Type: Yemen (N), *Forsskal* (C).

Shrub or small tree up to 3m, usually leafless. Leaves soon deciduous, linear to linear-oblong, 10–25 × 0.5–3mm, sparsely pubescent. Flowers axillary, solitary or in clusters; pedicels 5–15mm. Sepals unequal, 5–10mm long, with 3 shallowly boat-shaped and the upper strongly hooded. Petals pale orange or red; inner pair narrowly ovate, 8–12 × c.2mm; outer pair ovate, 6–12 × 5–8mm. Stamens 8–20, c.15mm. Gynophore 10–15mm. Fruit yellow to scarlet, globose, 5–15mm diam., pubescent. **Map 506.**

Sandy coastal plains, fossil coral; 0–300m.

Saudi Arabia, Yemen (N & S), Socotra, Oman. Tropical NE & N Africa, Palestine, Iran, Pakistan and NW India.

2. C. spinosa L., Sp. pl.: 503 (1753) non *C. spinosa* Forsskal. Illustr.: Mandaville (1990 pls. 69 & 70).

Spreading or sprawling shrub up to 60cm; young shoots velutinous at first, becom-

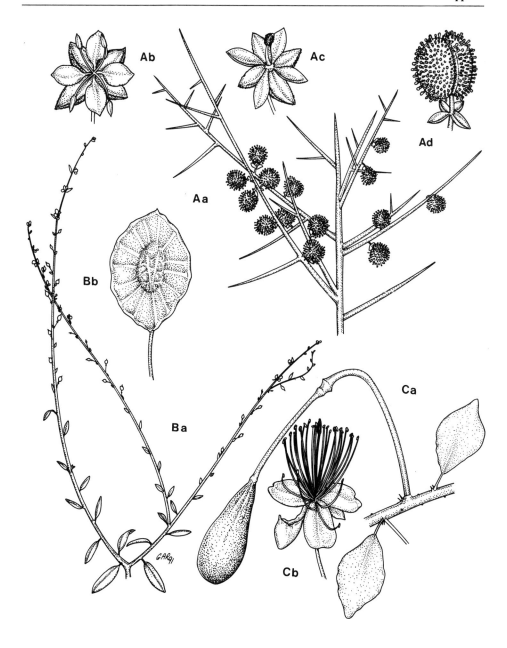

Fig. 71. Capparaceae. A, *Dhofaria macleishii*: Aa, fruiting branch (×1); Ab, male flower (×4); Ac, female flower (×4); Ad, fruit (×2.5). B, *Dipterygium glaucum*: Ba, habit (×1); Bb, fruit (×8). C, *Capparis cartilaginea*: Ca, part of fruiting branch (×1); Cb, flower (×1).

ing glabrous. Leaves narrowly ovate or broadly obovate to subcircular, 10–50 × 5–45mm; petioles 2–10mm. Flowers solitary; pedicels 2–6cm. Sepals subequal, 10–20mm long. Petals white or cream becoming pink, obovate or oblong, 7–25 × 5–15mm. Stamens 1.5–3.5cm. Gynophore 2–3.5cm. Fruit red when ripe, ovoid or obovoid, 2–6 × 1–2cm. **Map 507.**

1. Leaves 2–3.5 × as long as broad, 10–35 × 4–17mm, with acute tips
         var. **mucronifolia**
+ Leaves less than 2 × as long as broad, 8–60 × 6–45mm, with rounded or retuse tips   2
2. Fruits 25–50mm long; petals 15–40mm long; petioles 3–10mm long
         var. **spinosa**
+ Fruits 10–20mm long; petals 8–20mm long; petioles 1.5–5mm long
         var. **parviflora**

var. **spinosa.** Syn.: *C. aegyptia* Lam., Encycl. 1: 605 (1783); *C. spinosa* L. var. *aegyptia* (Lam.) Boiss., Fl. orient. 1: 420 (1867); *C. spinosa* L. var. *canescens* Cosson, Notes pl. crit. 1: 28 (1848); *C. leucophylla* DC., Prodr. 1: 246 (1824). Illustr.: Collenette (1985 p.95).

Rocky slopes, cliffs, waste ground and roadsides; 0–2300m.

Saudi Arabia, Yemen (N & S), UAE, Qatar, Bahrain. Africa and southern Europe eastwards to C Asia and India.

var. **mucronifolia** (Boiss.) Hedge & Lamond, Fl. Iran. 68: 7 (1970). Syn.: *C. elliptica* Hausskn. & Bornm. ex Bornm. in Mitth. Thüring. Bot. Vereins 6: 49 (1894); *C. elliptica* Hausskn. & Bornm. ex Bornm. var. *maskatensis* Hausskn. & Bornm., loc. cit.; *C. mucronifolia* Boiss., Diagn. pl. orient. sér. 1, 1: 5 (1843).

Rocky hillsides and cliffs; 50–1250m.

Saudi Arabia, Oman, UAE, Qatar. Southern Iran and Pakistan.
   This variety usually has slender yellow stems and thorns.

var. **parviflora** (Boiss.) Boiss., Fl. orient. 1: 420 (1867).
   Recorded from Arabia (Fl. Iraq 4(1): 143, 1980) but no specimens have been seen.

3. C. cartilaginea Decne in Ann. Sci. Nat. Bot. sér. 2, 3: 273 (1835). Syn.: *C. galeata* Fresen., Beitr. Fl. Abyssin. 2: 111 (1837); *C. galeata* Fres. var. *montana* Schweinf. (1896 p.191); ?*C. inermis* Forsskal (1775 p.100); *C. spinosa* L. var. *galeata* (Fresen.) Hook. f. & Thoms. in Hook f., Fl. Brit. India 1: 173 (1872); *C. spinosa* sensu Balf.f (1888 p.14) non L.; ?*C. spinosa* sensu Forsskal (1775 p.99) non L. Illustr.: Collenette (1985 p.94); Miller (1988 p.93). **Map 508, Fig. 71.**

Spreading or sprawling shrub up to 2m; young shoots puberulous at first, becoming glabrous. Leaves elliptic to ovate or orbicular, 15–50 × 8–50mm, acute to rounded with a small recurved spine inserted below the tip, leathery; petiole 5–30mm. Flowers solitary, on stout 3–8cm pedicels. Sepals unequal, 1–4cm long, the upper strongly hooded. Petals white becoming pink or purple, unequal, ovate to orbicular, 1–3cm long, the upper pair appearing fused. Stamens many, c.5cm long. Gynophore c.3.5cm. Fruit bright red, obovoid, 2–5 × 2–3cm. **Map 508.**

Cliffs and large boulders; 0–2400m.

Saudi Arabia, Yemen (N & S), Socotra, Oman, UAE. E and SW Africa to India.

4. C. tomentosa Lam., Encycl. 1: 606 (1785). Illustr.: Collenette (1985 p.95).

Shrub or small tree, velutinous throughout. Leaves ovate, 4–7 × 2–3.5cm; petiole 3–10mm. Flowers in short terminal and axillary racemes; pedicels c. 2.5cm. Sepals subequal, 13–20mm. Petals creamy-green, obovate, 2–2.5 × c.1cm. Stamens many, c.5cm. Gynophore c. 4cm. Fruit reddish or yellowish green, globose (pear-shaped in *Wood* 3392), up to 5cm diam. **Map 509.**

Cliffs; 300–1500m.

Saudi Arabia, Yemen (N). Tropical and southern Africa.

6. CADABA Forsskal

J. A. NYBERG

Trees or shrubs. Leaves alternate or clustered, simple. Flowers in terminal corymbs or racemes, rarely solitary, zygomorphic. Sepals 4, free. Petals 4 or absent, clawed. Nectarial appendage tubular, the upper part sometimes petaloid. Stamens 4–5; androphore present. Ovary 1–2-locular; gynophore present; ovules many. Fruits cylindrical, often somewhat torulose, tardily dehiscent.

1. Leaves broadly elliptic to broadly ovate or orbicular, usually more than 1cm long and about as broad as long; stamens attached towards the base of the gynophore   2
+ Leaves linear-elliptic to narrowly oblong or narrowly obovate, usually at least twice as long as broad, if rarely broader then less than 5mm long; stamens attached near the middle of the gynophore   5

2. Petals present   3
+ Petals absent   4

3. Leaves densely and minutely stellate-hairy; fruits densely clothed with sessile glands   **1. C. heterotricha**
+ Leaves glabrous or minutely pubescent, hairs simple; fruits densely clothed with stalked glands   **2. C. mirabilis**

4. Leaves clothed with stalked glandular hairs; fruits glandular   **3. C. glandulosa**
+ Leaves glabrous or sometimes strigose when young, never glandular; fruits non-glandular   **4. C. rotundifolia**

5.	Leaves bright green, glabrous, 30–70mm long	**6. C. longifolia**
+	Leaves grey-green, farinose, 5–25mm long	6
6.	Petals yellowish-green to brownish-green to cream; fruits non-glandular	**5. C. farinosa**
+	Petals clear yellow; fruits glandular	**7. C. baccarinii**

1. C. heterotricha Stocks ex Hook. in Hooker's Icon. pl. ser. 2, 9: t.839 (1852). Illustr.: Miller (1988 p.91).

Small tree up to 5m. Leaves grey-green, broadly obovate to orbicular, 10–35 × 7–20cm, rounded or retuse at the tip, densely clothed with minute stellate hairs and rarely also with simple glandular hairs. Flowers in terminal corymbs. Sepals 4–7 × c.2mm. Petals white, 6–8mm, suborbicular with a long claw ± equalling the blade. Stamens 5, attached at the base of the gynophore. Nectarial appendage c.1cm, tubular, broadened at the tip. Fruits c.25 × 3mm, densely covered with sessile glands and stellate hairs. **Map 510, Fig. 72.**

Acacia-Commiphora bushland on dry rocky slopes and cliff faces; 50–1400m.

Yemen (N & S), Oman. Kenya, Ethiopia, Somalia, Pakistan and India.

The stems and leaves of *Miller* 6368 are unusual in being densely clothed with long-stalked and somewhat strigose glandular hairs which resemble those of *C. glandulosa*. *Miller* 7010 is intermediate between the above specimen and typical *C. heterotricha* in having similar stalked glands but only on the young leaves.

2. C. mirabilis Gilg in Annuario Reale Ist. Bot. Roma 6: 93 (1896).

Shrub up to 3m. Leaves ovate to oblong-elliptic or subcircular, 2.5–4 × 1.5–3cm, rounded or obtuse at the tip, fleshy, glabrous or minutely pubescent. Flowers in terminal corymbs. Sepals 8–10 × 2–3mm. Petals cream, 10–15mm long, subcircular and shortly and abruptly clawed. Stamens 5, attached to the base of the gynophore. Nectarial appendage c.2cm, 'S'-shaped with an oblong, orange, petaloid tip. Fruits 1–3 × c.0.4cm, densely covered with stalked glandular hairs. **Map 511, Fig. 72.**

Acacia-Commiphora bushland; 300m.

Yemen (S). Sudan, Ethiopia, Somalia and Kenya.

3. C. glandulosa Forsskal (1775 p.68). Illustr.: Collenette (1985 p.94). Type: Yemen (N), *Forsskal* (C).

Shrub up to 1.5m. Young shoots densely glandular-hairy. Leaves grey-green, broadly elliptic to orbicular, 5–20 × 5–20mm, mucronate, clothed with long-stalked and somewhat strigose glandular hairs. Flowers in few-flowered corymbs. Sepals 7–9 × 4–6mm. Petals absent. Stamens 5, attached at the base of the gynophore. Nectarial appendage c.15mm, tubular, petaloid at the tip, yellow. Fruits 10–15 × 5mm, densely covered with stalked glands. **Map 512, Fig. 72.**

Acacia-Commiphora bushland on dry rocky slopes and sandy plains; 0–1400m.

6. Cadaba

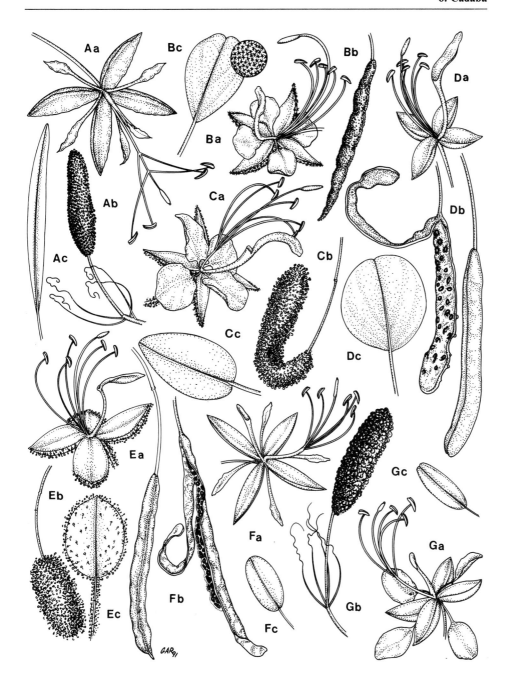

Fig. 72. Capparaceae. *Cadaba* species: A, *C. longifolia*; B, *C. heterotricha*; C, *C. mirabilis*; D, *C. rotundifolia*; E, *C. glandulosa*; F, *C. farinosa*; G, *C. baccarinii*: a, flowers (×2); b, fruits (×2.5); c, leaves (×1.5).

Saudi Arabia, Yemen (N & S). Uganda, Kenya, Tanzania, Somalia, Ethiopia, Sudan and Mali.

4. C. rotundifolia Forsskal (1775 p.68). Illustr.: Collenette (1985 p.94). Type: Yemen (N), *Forsskal* (C, BM).

Shrub up to 2m. Leaves dark green, broadly ovate to orbicular, 17–50 × 12–45mm, retuse or rounded at the tip, sometimes strigose at first but becoming glabrous. Flowers in terminal corymbs. Sepals c. 7 × 4mm. Petals absent. Stamens 4–5, attached at the base of the gynophore. Nectarial appendage c.2cm, tubular, petaloid at the tip, yellow. Fruits 15–50 × 2–7mm, non-glandular. **Map 513, Fig. 72.**

In *Acacia-Commiphora* bushland on sand and rocky slopes, along wadi-margins, coastal plains and escarpment foothills; 0–1000m.

Saudi Arabia, Yemen (N & S), Socotra. Kenya, Sudan, Ethiopia, Djibouti and Somalia.

5. C. farinosa Forsskal (1775 p.68). Illustr.: Fl. Pakistan 34: 14 (1973); Collenette (1985 p.93); Miller (1988 p.91). Type: Yemen (N), *Forsskal* (C).

Shrub up to 2(–3)m. Leaves farinose, grey-green, narrowly elliptic or oblong to obovate or more or less orbicular, 2–20(–25) × 1.5–12mm, rounded at the tip. Flowers solitary or in few-flowered racemes at the tips of branches. Sepals 8–10 × 3–4mm. Petals cream to greenish yellow, 10–15mm, linear-elliptic, clawed. Stamens 4–5. Androgynophore 5–8mm. Nectarial appendage short, tubular, c.5mm long, shortly toothed at the tip. Fruits 25–55 × 3–5mm, somewhat torulose, farinose. **Map 514, Fig. 72.**

Acacia-Commiphora bushland and succulent shrubland on rocky and sandy slopes and plains; 0–1800m.

Saudi Arabia, Yemen (N & S), Socotra (Abd al Kuri), Oman. Widespread in tropical Africa, Egypt, Pakistan and India.

Low and stunted specimens have been placed in subsp. *rariflora* Jafri. However, these seem to be merely grazed forms.

6. C. longifolia DC., Prodr. 1: 244 (1824). Illustr.: Collenette (1985 p.94).

Shrub up to 1.5m. Leaves bright green, linear-elliptic to linear-oblong or narrowly ovate, 30–70 × 5–12mm. Flowers in terminal corymbs. Sepals 8–15mm. Petals bright yellow, ageing red, 10–15mm, narrowly elliptic, clawed. Stamens 4. Androgynophore 10–15mm long. Nectarial appendage c.10mm long, the tip broadened and slightly lobed. Fruits 8–20 × 3–6mm, densely glandular. **Map 515, Fig. 72.**

In *Acacia-Commiphora* bushland on dry rocky slopes, sand and dry coastal foothills; 0–800m.

Saudi Arabia, Yemen (N & S), Socotra. Kenya, Ethiopia, Djibouti, Somalia and Sudan.

7. C. baccarinii Chiov. in Ann. Bot. (Rome) 13: 377 (1915). Illustr.: Miller (1988 p.91).

Shrub up to 1(–2)m. Leaves farinose, grey-green, narrowly oblong to narrowly obovate, 5–25 × 2–8mm, acute to rounded at the tip. Flowers 1–3 at the tips of branches. Sepals 7–10 × 4–6mm. Petals clear yellow, 10–14mm, ovate, clawed. Stamens 4. Androgynophore 7–15mm. Nectarial appendage 8–10mm, tubular, acute at the tip. Fruit 10–17 × 2.5–6mm, conspicuously papillose-glandular. **Map 516, Fig. 72.**

Acacia-Commiphora bushland on dry coastal hills; 5–150m.

Oman. Somalia.

7. DHOFARIA A. Miller

J.A. NYBERG

Dioecious or gynomonoecious shrubs; branches becoming spine-tipped, glabrous. Leaves simple, soon deciduous. Flowers small, in spinose racemes, zygomorphic. Sepals 4, free, unequal, the upper and lower valvate. Petals 4, free, unequal, shortly clawed. Male flowers with 4 stamens, the gynoecium absent. Female flowers with a short gynophore. Fruit a tardily dehiscent capsule, 1–3(–4)-seeded, splitting into (3–)4 valves, densely covered with stalked glands.

A monotypic genus, endemic to Arabia.

1. D. macleishii A.G. Miller in Notes Roy. Bot. Gard. Edinburgh 45(1): 55 (1988). Illustr.: Miller (1988 p.97). Type: Oman, *Miller* 6330 (E, K, ON, UPS, KTUH).

Much-branched spiny shrub up to 1m. Leaves soon deciduous, linear-elliptic to linear-obovate, 4–15 × 1–2mm, acute to obtuse at the tip, glabrous; pedicels 1.5–2.5mm. Sepals 3–4.5 × 1.5–4mm, densely stellate-hairy. Petals cream, 3–4 × 1.3–2.5mm. Gynophore c.5mm. Stigma sessile, capitate. Fruit globose, 6–8mm diam. **Map 517, Fig. 71.**

Dry rocky slopes and wadi-sides in *Acacia-Commiphora* bushland; 450–1600m.

Yemen (S), Oman. Endemic.

An endemic, monotypic genus restricted to the Dhofar region of Oman and adjacent parts of Yemen where it is quite common on the dry, northern dip slopes of the escarpment mountains and is also occasionally found in the wetter, south-draining wadis of the escarpment woodlands. Its spiny, leafless habit and distinctive, pea-sized fruits, which persist on plants throughout the year, make it readily recognizable. It flowers in the dry season (January and February) and has only been collected in flower once. It occupies an isolated position in the Capparaceae and its affinities are obscure.

Family 57. CRUCIFERAE

A.G. MILLER

Annual, biennial or perennial herbs, rarely subshrubs, glabrous or with simple or branched hairs. Leaves alternate, entire or variously divided; stipules absent. Flowers bisexual, actinomorphic or rarely slightly zygomorphic, 4-merous, in racemes, usually ebracteate. Sepals 4, free, in two pairs, the inner pair often saccate at the base. Petals 4, rarely absent, free, usually clawed. Stamens 6, tetradynamous (usually 2 short and 4 longer); filaments free, sometimes winged or dilated at the base. Nectary glands arranged at filament bases. Ovary superior, bicarpellate, divided by a thin membranous septum (the replum) into 2 loculi; placentation parietal; ovules usually many; style solitary; stigma capitate or bilobed, sometimes decurrent. Fruit dry, either a siliqua (elongate) or a silicula (short), dehiscing by two valves, or an indehiscent nut or a lomentum (breaking transversely into parts when ripe). Seeds with radicle incumbent (lying against the face of the cotyledons) or accumbent (lying against the edges of the cotyledons), cotyledons sometimes folded (conduplicate).

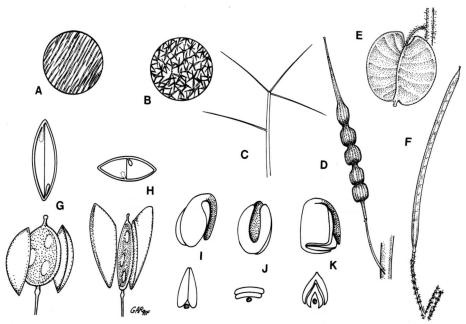

Fig. 73. **Cruciferae.** Indumentum: A, parallel and medifixed hairs; B, stellate hairs; C, branched hair. Fruits: D, lomentum; E, silicula; F, siliqua; G, latiseptate silicula; H, angustiseptate silicula. Radicles: I, accumbent; J, incumbent. Cotyledons: K, longitudinally folded.

See Hedge, I.C. & King, R.A. (1983). The Cruciferae of the Arabian Peninsula: a check-list of species and a key to genera. *Arab. Gulf J. scient. Res.* 1 (1): 41–66.

The Arabian genera can be divided into several tribes:

Tribe BRASSICEAE. Fruit usually a siliqua, often beaked or divided into two segments, either both fertile or one sterile; hairs simple or absent; radicle incumbent, cotyledons conduplicate. Genera 1–20.

Tribe LEPIDEAE. Fruit an angustiseptate silicula; hairs simple or absent; radicle incumbent or accumbent. Genera 21–30.

Tribe EUCLIDEAE. Fruit indehiscent, nut-like; hairs simple, branched or absent; radicle incumbent or accumbent. Genera 31–33.

Tribe ALYSSEAE. Fruit a compressed, latiseptate silicula, hairs branched, stellate, medifixed or simple; radicle accumbent. Genera 34–37.

Tribe ARABIDEAE. Fruit a siliqua; hairs simple or branched; radicle accumbent. Genera 38–41.

Tribe MATTHIOLEAE. Fruit a siliqua; hairs simple or branched; stigma often with decurrent carpidial lobes; sepals erect; radicle accumbent. Genera 42–46.

Tribe HESPERIDEAE. Fruit a siliqua; hairs simple or branched, sometimes glandular; stigma sometimes with decurrent carpidial lobes; sepals erect; radicle incumbent. Genera 47–53.

Tribe SISYMBRIEAE. Fruit a siliqua; hairs simple or branched; stigma capitate or bilobed; sepals spreading; radicle incumbent. Genera 54–57.

Key to genera

1.	Plant with branched hairs present on at least some parts	2
+	Plant with simple hairs only or glabrous	28
2.	Fruit short, less than 3 × as long as broad (silicula)	3
+	Fruit elongated, at least 3 × as long as broad (siliqua)	10
3.	Fruit compressed at right angles to the septum (angustiseptate)	4
+	Fruit compressed parallel to the septum or scarcely compressed (latiseptate)	5
4.	Annual herbs with a thin indumentum including some forked and dendroid hairs; fruit obtriangular (or if oblong-elliptic to obovate see 35a *Erophila*)	**30. Capsella**
+	Low shrub with a very dense indumentum entirely of adpressed stellate-dendroid hairs; fruit ellipsoid	**29. Lachnocapsa**
5.	Indumentum on stem and leaves of adpressed, medifixed hairs	6
+	Indumentum on stem and leaves of adpressed or spreading, branched-dendroid hairs	7
6.	Fruit at least 8mm long, ± ovate; seeds broadly winged; petals c.8mm or more	**34. Farsetia**
+	Fruit up to 3mm long, ± circular; seeds with a narrow wing; petals c.3mm	**36. Lobularia**
7.	Leaves amplexicaul or sagittate at the base; fruit nut-like, indehiscent, ± spherical	**33. Neslia**
+	Leaves attenuate at the base; fruit tardily or readily dehiscent, not spherical	8

8.	Flowers sessile; fruit tardily dehiscent with rounded apical auricles	**31. Anastatica**
+	Flowers clearly pedicellate; fruit readily dehiscent, without appendages	9
9.	Fruiting pedicels erect-spreading	**35. Alyssum**
+	Fruiting pedicels recurved	**37. Clypeola**
10.	Leaves 2–3-pinnatisect with linear segments	**56. Descurainia**
+	Leaves simple or 1-pinnate	11
11.	Upper cauline leaves simply pinnate with linear-filiform segments	**47. Leptaleum**
+	Cauline leaves undivided or at most toothed	12
12.	Indumentum entirely (or almost so) of adpressed medifixed hairs	13
+	Indumentum of forked to stellate-dendroid hairs	15
13.	Fruits apically horned	**42. Notoceras**
+	Fruits without apical horns	14
14.	Delicate annual herb; petals clear yellow; fruits deflexed	**57. Arabidopsis**
+	Annual or perennial herbs; petals white, pink or mauve, rarely yellowish green; fruits ascending	**34. Farsetia**
15.	Stigma with decurrent carpidial lobes	16
+	Stigma without decurrent carpidial lobes	23
16.	Petals pink, mauve, livid or brown	17
+	Petals white, cream or yellow	20
17.	Apex of fruit with two prominent horns	**44. Matthiola**
+	Fruit without apical horns	18
18.	Leaves oblong to obovate, at least the basal dentate or lobed	**49. Malcolmia**
+	Leaves linear or linear-oblong, rarely broader (*Eremobium*), entire	19
19.	Annual with ± glabrous stems; fruit linear-oblong, straight, valves readily separating, never horned	**50. Eremobium**
+	Perennial with dense white indumentum on the stems; fruit narrowly linear, ± torulose, valves not readily separating, sometimes horned	**44. Matthiola**
20.	Apex of fruit with two prominent horns	**43. Diceratella**
+	Apex of fruit without horns	21
21.	Inner sepals distinctly saccate; leaves linear to linear-oblong, not or scarcely attenuate below; valves of fruit readily separating	**50. Eremobium**
+	Inner sepals not saccate; leaves oblong to ovate, attenuate at the base; valves of fruit not readily separating	22
22.	Stigma lobes widely divergent; fruiting pedicels c.10–12mm long, slender (Yemen)	**51. Sterigmostemum**

+	Stigma lobes not divergent; fruiting pedicels c.1–2mm long, thick	**45. Morettia**
23.	Flowers yellow	**57. Arabidopsis**
+	Flowers white, pink or lilac	24
24.	Cauline leaves amplexicaul, sagittate or auriculate; radicles accumbent; petals white or cream	**39. Arabis**
+	Cauline leaves sessile or petiolate, the base cuneate or sometimes minutely rounded but never amplexicaul, sagittate or auriculate; radicle incumbent; petals white or pinkish-lilac	25
25.	Siliqua with a long, slender (3–4mm) style	**48. Eigia**
+	Siliqua with a short (less than 1mm) style	26
26.	Delicate desert annual; stems thinly stellate-hairy or glabrous above; petals pink or lilac	**53. Maresia**
+	Annual or perennial; stems densely tomentose above or if glabrous or thinly hairy then the hairs simple or bifurcate and the flowers white	27
27.	Fruits slender, torulose, often contorted; indumentum of simple and forked hairs only	**55. Neotorularia**
+	Fruits straight, not torulose or contorted; indumentum of branched hairs below and simple and forked hairs above	**57 Arabidopsis**
28.	Plant with spiny branches, lilac flowers and a hard indehiscent globose fruit with conical beak	**14. Zilla**
+	Plant not spiny	29
29.	Fruit compressed at right angles to the septum (angustiseptate), never with a terminal beak	30
+	Fruit compressed parallel to the septum or scarcely compressed (latiseptate), sometimes with a terminal beak	39
30.	Cauline leaves amplexicaul, auriculate or sagittate	31
+	Cauline leaves sessile to petiolate, not amplexicaul	36
31.	Fruit didymous (with two prominent lobes), the lobes flattened	**27. Biscutella**
+	Fruit not didymous	32
32.	Petals yellow; fruit pendent, oblong, 1-seeded	**24. Isatis**
+	Petals white or cream; fruit erect, 2–several-seeded	33
33.	Fruit loculi 2–3-seeded	**28. Thlaspi**
+	Fruit loculi 1-seeded	34
34.	Flowers in radiate corymbs, the two outer petals larger than the two inner	**26. Iberis**
+	Flowers in elongated racemes, never radiate corymbs, all petals of equal length	35

35.	Fruits with rounded turgid valves, cordate or broadly ovate at the base, indehiscent	**23. Cardaria**
+	Fruits with ± flattened valves, rounded at the base, readily dehiscent	**21. Lepidium**
36.	Fruit didymous (with two prominent lobes)	37
+	Fruit not didymous	38
37.	Fruits strongly flattened, the valves readily dehiscent	**27. Biscutella**
+	Fruits with rounded, verrucose, indehiscent valves	**22. Coronopus**
38.	Flowers mauve-purple; fruit indehiscent, winged, keeled	**25. Horwoodia**
+	Flowers white or cream; fruit dehiscent, winged or not, not keeled	**21. Lepidium**
39.	Fruiting pedicels 10–25mm, capillary, spreading or reflexed; fruit shortly stipitate, 10–14 × 5–7mm	**19. Savignya**
+	Fruiting pedicels shorter, not capillary; fruit not as above	40
40.	Cauline leaves clearly amplexicaul, auriculate or sagittate	41
+	Cauline leaves sessile to petiolate, not amplexicaul, sometimes absent	47
41.	Flowers yellow	42
+	Flowers white, pink or mauve	43
42.	Petals c.1mm long; fruit a silicula, 2–5mm long, ovoid, irregularly tuberculate with an oblique flat beak	**32. Schimpera**
+	Petals c.10–15mm long; fruit a siliqua, 50–70mm long	**1. Brassica**
43.	Fruit a siliqua	44
+	Fruit a silicula	45
44.	Fruit 1-membered, dehiscent, many-seeded	**20. Moricandia**
+	Fruit 2-membered; lower segment 2-valved, dehiscent, 2–4-seeded; upper segment elongate, indehiscent, sterile	**16. Dolichorhynchus**
45.	Petals 4.5–5mm long; fruit short, curved, ± 4-angled, irregularly ridged with a short curved beak	**52. Goldbachia**
+	Petals 10–15mm long; fruit 2-membered, either lower part short, stalk-like and inconspicuous and the upper part ± globose with a conical beak or lower part narrowly winged, fertile, many-seeded and upper part with a narrow beak	46
46.	Petals white, c.15mm long; fruit 2-membered; lower part short, stalk-like, inconspicuous; upper part ± globose with a conical beak, 2-seeded	**15. Physorrhynchus**
+	Petals lilac, c.10mm long; fruit 2-membered; lower part narrowly winged, fertile, many-seeded; upper part a narrow beak	**18. Schouwia**
47.	Petals with distinct brown to purple venation	48
+	Petals without distinct darker venation	51

48.	Leaves 2–3-pinnatisect; petals c.8mm; fruit c.8mm long with a spathulate upper member	**17. Carrichtera**
+	Leaves undivided to lyrate-pinnatifid; petals 7–15mm; fruit more than 20mm long, the upper member not spathulate	49
49.	Petals c.7mm; racemes leafy; fruit clearly constricted between the seeds	**9. Enarthrocarpus**
+	Petals more than 12mm; racemes leafless; fruits not clearly constricted between the seeds	50
50.	Upper member of fruit cylindrical with a conical beak; fruiting pedicel c.0.5mm thick; flowers pink, purple or white	**8. Raphanus**
+	Upper member of fruit a sterile compressed beak; fruiting pedicel c.1mm thick, ± adpressed to stem; petals yellow to whitish	**6. Eruca**
51.	Stigma with decurrent carpidial lobes	52
+	Stigma without decurrent carpidial lobes	53
52.	Plant glabrous, glaucous with simple leaves and 2-membered fruits, the upper member conical, c.10mm long	**15. Physorrhynchus**
+	Plant with substrigose hairs and small stipitate glands; leaves pinnatisect or ± lyrate; fruit c.30–60mm long, constricted between the seeds	**46. Chorispora**
53.	Flowers lilac, pink or purple	54
+	Flowers white, cream or yellow	57
54.	Leaves pinnate with a ± linear terminal segment (*Erucaria hispanica* rarely entire)	55
+	Leaves undivided to lyrate	56
55.	Upper segment of fruit a long flattened seedless beak; stigma sessile	**13. Cakile**
+	Upper segment of fruit not beaked, but seeded and constricted between the seeds; style distinct	**12. Erucaria**
56.	Fruit inflated and indehiscent or constricted between the seeds and breaking into 1-seeded segments	**8. Raphanus**
+	Fruit dehiscing by 2 valves, neither inflated nor constricted between the seeds	**5. Diplotaxis**
57.	Plant completely glabrous; leaves pinnately divided; aquatic or plant of marshy places	58
+	Plant with an indumentum (on at least some part of the plant); leaves divided or not; plants of well-drained habitats	59
58.	Petals white; siliqua 10–20mm long	**40. Nasturtium**
+	Petals yellow; siliqua 5–10mm long	**41. Rorippa**

59.	Fruit a silicula, 2-membered with a globose upper segment and a narrower, cylindrical lower segment (or if 1-membered and flattened see 35a *Erophila*)	60
+	Fruit a siliqua, of various shapes but not as above	61
60.	Flowers white; stigma sessile	**10. Crambe**
+	Flowers yellow; stigma borne on a long style	**11. Rapistrum**
61.	Fruits 1-membered, without an obvious beak	62
+	Fruits 2-membered, with a ± prominent beak	65
62.	Fruits slender, torulose, often contorted	**55. Neotorularia**
+	Fruit ± straight, not or scarcely constricted between the seeds	63
63.	Petals white; seeds in one row in each fruit loculus; radicles accumbent	**38. Cardamine**
+	Petals yellow; seeds in one or two rows in each fruit loculus, if petals white then seeds in two rows; radicles incumbent	64
64.	Seeds arranged in two rows in each loculus; cotyledons longitudinally folded	**5. Diplotaxis**
+	Seeds arranged in one row in each loculus; cotyledons not longitudinally folded	**54. Sisymbrium**
65.	Seeds arranged in two rows in each loculus	**5. Diplotaxis**
+	Seeds arranged in one row in each loculus	66
66.	Fruits closely adpressed to stem; fruiting pedicels short and thick (c.1–1.5mm); base of stem with ± stiff retrorse hairs	67
+	Fruits spreading-erect; fruiting pedicels ± elongated and slender; base of stem with not, or scarcely, retrorse hairs or hairs absent	68
67.	Upper segment of fruit ovoid or spherical, clearly broader than the lower segment	**11. Rapistrum**
+	Upper segment of fruit oblong, as wide as the lower segment	**4. Hirschfeldia**
68.	Fruit with convex, not parallel, edges; valves not readily separating at maturity	69
+	Fruit with parallel edges, sometimes somewhat constricted between the seeds; valves readily separating at maturity	70
69.	Upper segment of fruit laterally compressed; leaves undivided; shrub (restricted to cliffs on Socotra)	**7. Hemicrambe**
+	Upper segment of fruit inflated-cylindrical; leaves undivided to lyrate-pinnatisect; annual to biennial (widespread cultivated plant or weed)	**8. Raphanus**
70.	Fruit 4-angled; sepals all equal; stem with retrorse hairs	**2. Erucastrum**
+	Fruit not 4-angled; sepals ± unequal; stem glabrous or with spreading hairs	71

71.	Valves of fruit with one prominent median nerve and a few inconspicuous lateral nerves	**1. Brassica**
+	Valves of fruit with 3–7 nerves (apparent on young fruits)	**3. Sinapis**

1. BRASSICA L.

Annual, biennial or perennial herbs, glabrous or bearing simple hairs. Lower leaves simple or lyrately lobed, upper leaves amplexicaul at base or not. Sepals erect or erect-spreading, the inner slightly saccate. Petals yellow or white. Filaments unappendaged, free. Stigma capitate. Fruit a many-seeded dehiscent siliqua, beaked, usually ± torulose, sometimes a stipe present, glabrous; beak 0–2-seeded; valves 1-nerved; seeds uniseriate, globose; radicles incumbent, the cotyledons longitudinally folded.

Species of *Brassica* have been cultivated for centuries and several are unknown in the wild state. They frequently escape from cultivation and can persist as weeds long after their cultivation has ceased.

1.	Upper stem leaves petiolate or narrowed at the base	2
+	Upper stem leaves amplexicaul, rounded or cordate at the base	4
2.	Siliqua and fruiting pedicels deflexed	**1. B. deflexa**
+	Siliqua and fruiting pedicels erect or spreading	3
3.	Lower leaves hispid, with 5–10 pairs of lateral lobes; petals 5–7mm; beak of siliqua 10–20mm	**3. B. tournefortii**
+	Lower leaves sparsely hairy or glabrous, with 1–3 pairs of lateral lobes; petals 6–10mm; beak of siliqua 6–10mm	**2. B. juncea**
4.	Sepals erect; petals 15–20mm; outer stamens not curved at the base; lower leaves fleshy	**4. B. oleracea**
+	Sepals erect-spreading; petals 6–10mm; outer stamens curved outwards at the base; lower leaves not fleshy	5
5.	Inflorescence with the open flowers overtopping the buds; lower leaves bright green	**5. B. rapa**
+	Inflorescence with the buds overtopping the open flowers; lower leaves glaucous	**6. B. napus**

1. B. deflexa Boiss. in Ann. Sci. Nat. Bot. sér. 2, 17: 87 (1842). Syn.: *B. lasiocalycina* (Boiss. & Hausskn.) Boiss., Fl. orient. suppl.: 66 (1888).

Erect annual herb up to 60(–80)cm, glabrescent above, usually hispid below. Lower leaves oblong-ovate, 5–10 × 1–4cm, coarsely toothed to lyrate-pinnatisect, with 1–3 pairs of lateral lobes, serrate; upper leaves entire, shortly petiolate or narrowed at the base. Petals yellow, (6–)8–13mm. Siliqua deflexed, 30–60(–90) × 1.5–2mm; beak conical, ± swollen, 4–6mm, broader than the stigma, 0–1-seeded. **Map 518, Fig. 74.**

Weed of cereal fields; sea-level.

Saudi Arabia, Kuwait. Syria, Turkey, Iran, Afghanistan and Iraq.

2. B. juncea (L.) Czernj. & Cosson in Czernj., Consp. Pl. Chark.: 8 (1859). Illustr.: Fl. Trop. E. Afr.: 4 (1982).

Erect annual herb up to 100cm, glabrous or sparsely hairy below. Lower leaves petiolate, oblong-ovate, 3–15 × 0.5–4cm, entire to lyrate-pinnatisect with 1–3 pairs of lateral lobes, coarsely and irregularly serrate; upper leaves entire, petiolate. Petals bright yellow, 6–10mm. Siliqua erect-spreading, 20–75 × 2–4mm; beak narrowly conical, 4–10mm, narrower than the stigma, seedless. **Map 519, Fig. 74.**

Weed of gardens and waste places; 0–2300m.

Saudi Arabia, Yemen (N & S), Oman, Kuwait.

Probably a native of C Asia, now widely cultivated in Asia, Africa, N America and S Europe.

3. B. tournefortii Gouan, Ill. observ. bot. 44: t.20A (1773). Illustr.: Fl. Palaest. 1: t.457 (1966); Fl. Kuwait 1: pls. 82–84 (1985); Collenette (1985 p.194).

Erect annual herb up to 75cm, glabrous above and hispid below. Leaves mainly in a basal rosette, narrowly obovate, 5–25(–40) × 2–5(–15)cm, lyrate-pinnatisect with 5–15 pairs of lateral lobes, serrate or dentate; upper leaves narrowly oblong, shortly petiolate or narrowed at the base, often absent. Petals pale yellow, 5–7mm long. Siliqua erect-ascending, (20–)50–60(–70) × 2–3mm; beak slightly tapering, 10–20mm, broader than the stigma, 1–2-seeded. **Map 520, Fig. 74.**

On sand and gravel in deserts; 0–2400m.

Saudi Arabia, Yemen (S), Oman, UAE, Qatar, Bahrain, Kuwait. S & W Europe, N Africa and SW Asia.

4. B. oleracea L., Sp. pl.: 667 (1753).

Erect annual, biennial or perennial herb up to 1m, glabrous throughout. Lower leaves petiolate, fleshy, glaucous, up to 50 × 30cm, lyrate-pinnatipartite, with 1–5 pairs of lateral lobes, dentate to undulate; upper leaves sessile or semi-amplexicaul, entire, rounded at the base. Outer stamens not curved at the base. Sepals erect. Petals yellow or white, 15–20mm. Siliqua erect-ascending, 50–100 × 4–5mm; beak narrowly conical, 5–10(–15)mm, seedless. **Map 521.**

Cultivated.

Saudi Arabia, Yemen (N & S). Cultivated world-wide with many cultivars including cabbage, broccoli, cauliflower and kohlrabi.

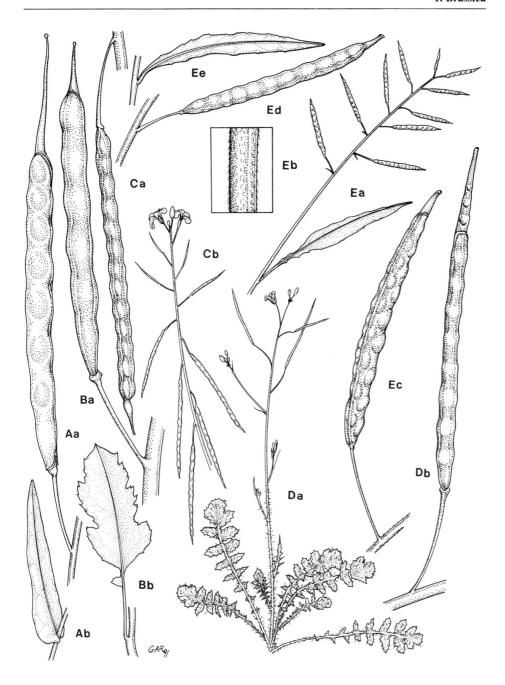

Fig. 74. Cruciferae. A, *Brassica napus*: Aa, fruit (×3); Ab, leaf (×0.6). B, *B. juncea*: Ba, fruit (×3); Bb, leaf (×1). C, *B. deflexa*: Ca, fruit (×3); Cb, flowering and fruiting shoot (×0.6). D, *B. tournefortii*: Da, habit (×0.5); Db, fruit (×3). E, *Erucastrum arabicum*: Ea, fruiting branch (×0.5); Eb, detail of stem (×10); Ec–d, fruits (×3); Ee, leaf.

5. B. rapa L., Sp. pl.: 666 (1753). Syn.: *B. campestris* L., Sp. pl.: 666 (1753). Illustr.: Collenette (1985 p.194).

Erect annual or biennial herb up to 1m, often with a swollen taproot, glabrous or hispidulous below. Lower leaves bright green, petiolate, up to 30 × 10cm, lyrate-pinnatipartite with 1–5 pairs of lateral lobes, undulate to dentate; upper leaves glaucous, entire, amplexicaul, cordate or rounded at the base. Opened flowers overtopping the buds. Sepals spreading. Petals yellow, 6–10mm. Outer stamens curved outwards at the base. Siliqua ± erect, 40–70(–100) × 2.5–4mm; beak slender, 5–20(–30)mm, 0–1-seeded. **Map 522.**

Cultivated and frequently persisting as a weed; 180–2700m.

Saudi Arabia, Yemen (N), Oman, Kuwait. The turnip, cultivated world-wide, often escaping.

6. B. napus L., Sp. pl.: 666 (1753).

Similar to *B. rapa* but lower leaves glaucous; open flowers not overtopping the buds; petals 10–18mm. **Map 523, Fig. 74.**

Cultivated and frequently persisting as a weed; 90–3000m.

Saudi Arabia, Yemen (N & S). The swede and rape, cultivated in Asia, N Africa and Europe.

2. ERUCASTRUM C. Presl

Annual or perennial herbs, glabrous or with simple hairs. Basal leaves lyrately lobed, the upper leaves usually less divided or entire. Sepals erect, the inner slightly saccate. Petals yellow or white. Filaments unappendaged, free. Stigma capitate or bifid; gynophore absent or short. Fruit a dehiscent siliqua, linear, usually 4-angled in section with 1-nerved valves and a seedless conical beak; seeds uniseriate, subglobose to ellipsoid; radicle incumbent; cotyledons longitudinally folded.

Jonsell, B. (1993). Montane taxa of *Erucastrum* in NE tropical Africa and Arabia. *Opera Bot.* 121: 135–143.

1.	Annual; petals 3–6mm long	**1. E. arabicum**
+	Perennial; petals 7–9mm long	2
2.	Stems and leaves hispid or rarely almost glabrous but then with a few scattered hairs on the leaves; stems profusely branched, strongly woody below; flowers in open corymbs; siliquae 30–40 × c.1.5mm, the beak narrowly conical with a small capitate stigma	**2. E. meruense** subsp. **yemenense**
+	Stems and leaves glabrous; stems sparsely branched, somewhat woody at the base; flowers in dense corymbs; siliquae 13–15 × 0.5–0.8mm, the beak narrowly conical with a prominently expanded stigma	**3. E. woodiorum**

1. E. arabicum Fischer & C. Meyer Index sem. hort. petrop. 5: 35 (1838). Syn.: *Brassica arabica* (Fischer & C. Meyer) Fiori in Nuovo Giorn. Bot. Ital., n.s. 19: 445

(1912). Illustr.: Fl. Trop. E. Afr.: 8 (1982). Types: Saudi Arabia, *Schimper* 941 (LE); *Fischer* 189 (LE).

Erect or ascending annual herb up to 50cm; stems glabrescent above and with retrorse hispid hairs below. Lower leaves oblanceolate to lyrate-pinnatifid, up to 15 × 3cm, entire, sinuate or irregularly dentate; upper leaves similar but smaller and less divided, sessile. Racemes densely flowered, ebracteate or bracteate below, lax in fruit; fruiting pedicels slender, ascending, (3–)5–15(–20)mm. Petals yellow, spathulate, clawed, 3–5(–6)mm. Siliquae straight, often somewhat torulose, (15–)20–40(–50) × c.1.5mm; valves glabrous or sparsely hairy; beak 1.5–5mm, narrower than the siliqua, with a capitate stigma. Seeds oblong to elliptic, 0.8–1.2mm, light to dark brown, smooth to finely reticulate. **Map 524, Fig. 74.**

Weed of disturbed and cultivated ground, 650–2700m.

Yemen (N & S), Saudi Arabia, Oman. Tropical and southern Africa.

2. E. meruense Jonsell subsp. **yemenense** Jonsell, op. cit.: 139 (1993). Type: Yemen (N), *J. Wood* 1674 (E, BM).

Erect or ascending perennial herb up to 100cm, glabrous or hispid; stems profusely branched, strongly woody below. Lower leaves elliptic, 6–15 × 18–60cm, lyrate-pinnatifid with 2–3 triangular lateral lobes and the terminal lobe about $\frac{1}{3}$ to $\frac{1}{2}$ as long as the total leaf length, the margins serrate to dentate; upper leaves smaller, lyrate-pinnatifid with a single pair of lateral lobes or undivided, almost linear. Flowers ebracteate, in open corymbs; fruiting pedicels patent, 8–12mm. Petals pure yellow, spathulate, 8–9 × 2.5–4mm. Siliquae mostly curved, rather torulose, 30–40 × c.1.5mm; valves glabrous; beak narrowly conical, 3–4mm. Seeds oblong, c.9–1.5mm, grey to red-brown, reticulate with distinct areoles and coarse ridges which sometimes have a fine reticulum. **Map 525.**

Cliff ledges; 2300–3100m

Yemen (N). Endemic subspecies.

Outside Arabia *Erucastrum meruense* is distributed in N Tanzania (subsp. *meruense*) and Ethiopia (subsp. *balense* Jonsell). Subsp. *yemenense* differs from the African subspecies in having smaller flowers and seeds. Material of subsp. *yemenense* is very variable in its degree of hairiness ranging from almost glabrous to distinctly hispid.

3. E. woodiorum Jonsell, op. cit.: 140 (1993). Type: Yemen (N), *J. Wood* 2646 (E, BM).

Ascending perennial herb up to 40cm, glabrous; stems sparsely branched, somewhat woody at the base. Stem leaves oblong, 30–65 × 10–30mm, pinnate to pinnatifid with 2–3 pairs of oblong lateral lobes with the terminal lobe narrowly ovate and about $\frac{1}{3}$ as long as the total leaf length. Flowers ebracteate, in dense corymbs; fruiting pedicels patent or recurved, 5–9mm. Petals yellow, spathulate, 7–9 × 3–4mm. Siliquae straight or somewhat curved, 13–15 × 0.5–0.8mm, frequently poorly developed; valves glabrous; beak narrowly conical, 2–3mm, with a prominently expanded stigma. Seeds oblong-ellipsoid, 1.4–1.6 × 0.6–0.7mm, light brown, distinctly reticulate with areoles traversed by dense parallel ridges. **Map 526.**

Cliff ledges etc.; 2300–2600m.

Yemen (N). Endemic.

3. SINAPIS L.

Annual herbs, bearing simple hairs. Leaves pinnatifid to pinnatisect. Inner sepals not saccate. Petals white or yellow. Filaments unappendaged. Stigma capitate. Fruit a dehiscent beaked siliqua; valves 3–7-veined; radicles incumbent; cotyledons longitudinally folded.

1. Siliqua with a strongly compressed beak; valves covered with stiff spreading hairs **2. S. alba**
+ Siliqua with a cylindrical or conical beak; valves glabrous or covered with retrorse hairs **1. S. arvensis**

1. S. arvensis L., Sp. pl.: 668 (1753). Illustr.: Collenette (1985 p.206).

Hispid annual herb up to 60cm. Leaves lyrate-pinnatisect, up to 20cm, irregularly toothed; upper leaves usually simple. Petals yellow, 6–12mm. Siliqua 18–45 × 2.5–4mm; beak conical, 10–18mm, 1–2-seeded; lower segment often torulose, 2-valved, 8–16-seeded, glabrous or with stiff retrorse hairs. **Map 527, Fig. 75.**

Weed of cultivation and waste places; 0–1250m.

Saudi Arabia, Oman, UAE, Qatar, Kuwait. W, C and S Europe, N Africa, SW and C Asia.

2. S. alba L., Sp. pl.: 668 (1753).

Hispid annual herb up to 60cm. Leaves lyrate-pinnatisect, up to 20cm, the segments irregularly toothed; upper leaves smaller with fewer lobes or simple. Petals yellow, 8–15mm. Siliqua 20–40 × 3–4mm; beak long-attenuate, compressed, 15–30mm, 0–1-seeded; lower segment 2-valved, 2–8-seeded, covered with stiff spreading hairs. **Map 528, Fig. 75.**

On lava flow by roadside; 1220m.

Saudi Arabia. W, C, & S Europe, SW & C Asia, N Africa, widely introduced elsewhere.
A common weed in SW Asia but only found once in Arabia.

4. HIRSCHFELDIA Moench

Annual or biennial herbs, glabrous or clothed with simple hairs. Leaves lyrate-pinnatifid. Inner sepals slightly saccate. Petals white or pale yellow. Filaments unappendaged. Stigma capitate. Fruit a dehiscent 2-membered linear siliqua; upper segment indehiscent, 0–1-seeded; lower segment dehiscent, 3–9-seeded, the valve 3-nerved; radicles incumbent; cotyledons longitudinally folded.

1. Fruits erect, appressed to the stem, on stout 3–4mm long pedicels, with an abruptly conical beak; petals pale yellow, c.8mm long **1. H. incana**
+ Fruits spreading, on delicate 10–20mm long pedicels, with attenuate beak; petals white, 3–3.5mm long **2. H. rostrata**

1. H. incana (L.) Lagr.-Fossat, Fl. Tarn Garonne: 19 (1847).

Erect annual or biennial herb up to 1m; stems ± hispid below. Leaves lyrate-pinnatifid to pinnatisect, up to 20 × 5cm, the lobes dentate; lower leaves petiolate; upper leaves sessile, ± simple. Petals pale yellow, c.8mm. Fruiting pedicels 3–4mm. Siliqua appressed to the stem, hairy or glabrous; lower segment 6–9 × 1–1.5mm; upper segment 3–7 × 1–1.5mm, conical. **Map 529, Fig. 75.**

Weed of cultivation, roadsides and in upland meadows; 20–2400m.

Saudi Arabia, Yemen (N). Europe & SW Asia, naturalized elsewhere.

An uncommon weed in Arabia and also locally abundant in upland meadows in the Aja' region of Saudi Arabia.

2. H. rostrata (Balf. f.) O. Schulz in Bot. Jahrb. Syst. 54, Beibl. 119: 56 (1916). Syn.: *Brassica rostrata* Balf. f. (1882 p.500). Type: Socotra, *Balfour, Cockburn & Scott* 245 (E).

Annual herb; stems 30–50cm, glabrous or thinly hairy, rarely densely hirsute. Leaves lyrate-pinnatifid to pinnatifid, the uppermost sometimes simple, up to 10 × 4cm, the margins entire to sinuate-dentate or dentate; petioles 2–5cm. Petals white, 3–3.5mm. Fruiting pedicel 10–20mm. Siliqua 10–20 × c.1mm, spreading, glabrous; upper segment 4–8mm, attenuate into an acute tip. **Map 530, Fig. 75.**

Evergreen thickets, *Croton* shrubland, rocky limestone slopes and grassland; 100–800m.

Socotra. Endemic.

Balfour (1888) recognized var. *hirsuta* Balf. f. based on a single specimen (*B.C.S.* 555). This differs from typical plants in its densely hirsute indumentum and strongly dissected, dentate-serrate leaves which serve to give the plant a very different facies. Although many collections of typical *H. rostrata* have now been made on the island var. *hirsuta* has not been recollected and it possibly represents a single anomalous specimen.

5. DIPLOTAXIS DC.

Annual or perennial herbs, glabrous or bearing simple hairs. Leaves simple to lobed, entire to dentate. Sepals erect or erect-spreading; inner sepals not or slightly saccate. Petals white, yellow or pink. Filaments unappendaged, free. Stigma bilobed. Fruit a many-seeded dehiscent, beaked siliqua, linear, compressed; stipe present or not; beak

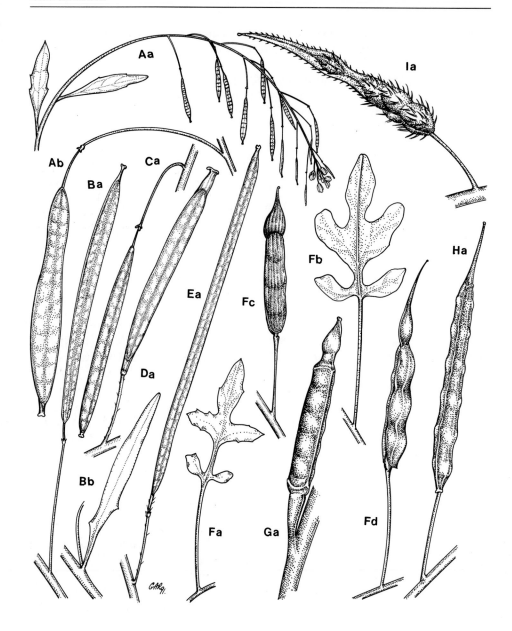

Fig. 75. Cruciferae. A, *Diplotaxis kohlaanensis*: Aa, fruiting and flowering branch (× 0.6); Ab, fruit (× 2.5). B, *D. tenuifolia*: Ba, fruit (× 2.5); Bb, leaf (× 1). C, *D. harra*: Ca, fruit (× 2.5). D, *D. erucoides*: Da, fruit (× 2.5). E, *D. acris*: Ea, fruit (× 2.5). F, *Hirschfeldia rostrata*: Fa–b, leaves (× 5); Fc–d, fruits (× 1). G, *H. incana*: Ga, fruit (× 5). H, *Sinapis alba*: Ha, fruit (× 3). I, *S. arvensis*: Ia, fruit (× 3).

fertile (1-seeded) or sterile; valves 1-nerved; seeds biseriate; radicles incumbent; cotyledons longitudinally folded.

1. Petals white, pink or lilac; annual herbs   2
+ Petals yellow; annual or perennial herbs or subshrubs   3

2. Petals white or pale pink, 6–10.5(–12)mm; stems and leaves clothed with short mainly retrorse hairs; sepals erect-spreading; stipe absent or up to 0.75mm long   **3. D. erucoides**
+ Petals pale purple or lilac, (7–)15–18(–22)mm; stems and leaves subglabrous or with a few spreading hairs; sepals erect; stipe 1–2.5mm long   **4. D. acris**

3. Leaves linear-oblong, entire or pinnately lobed with linear lobes   **5. D. tenuifolia**
+ Leaves ovate to broadly ovate to ovate-oblong   4

4. Erect annual or perennial herb; stems and leaves clothed with simple spreading hairs at least at the base of stem and on young leaves; beak 0.5–1.5mm; mature fruits often delicately reflexed   **1. D. harra**
+ Bushy perennial herb or subshrub; stems and leaves glabrous; beak 1.5–2mm; fruits on flexuous pedicels   **2. D. kohlaanensis**

1. D. harra (Forsskal) Boiss., Fl. Orient. 1: 388 (1867). Illustr.: Fl. Iraq 4(2): 860 (1980); Mandaville (1990 pl.74); Fl. Kuwait 1: pls. 87-8 (1985); Collenette (1985 p.197).

Annual or perennial herb, clothed with long spreading hairs throughout, rarely subglabrous (stems at the base and young leaves usually hairy). Stems erect, (10–)30–60(–75)cm, branched mainly from the base. Leaves ovate to broadly obovate, 2–8 × 1–4cm, obtuse, serrate or sinuate to entire, attenuate into a short petiole below; upper leaves smaller and narrower. Sepals erect-spreading. Petals yellow, 6–10.5mm. Fruits pendent when mature, 25–45 × 1.25–2.5(–3.5)mm; stipe 1–2mm; beak sterile, 0.5–1.5mm. **Map 531, Fig. 75.**

Deserts and stony and sandy plains; weed of cultivation; 10–3000m.

Saudi Arabia, Yemen (N & S), Oman, UAE, Bahrain, Kuwait. Deserts from N Africa and Somalia eastwards to Afghanistan and Pakistan.

A very variable and widespread species in the deserts of N Africa and SW Asia. Plants from the high mountains of Oman (3000m) are glabrous and approach *D. kohlaanensis*.

2. D. kohlaanensis A.G. Miller & J. Nyberg in Edinb. J. Bot. 51 (1): 36 (1994). Type: Yemen (N), *Miller & Long* 3213 (E).

Perennial herb or subshrub, glabrous throughout. Stems erect or hanging, 25–50 (–200)cm, branched throughout. Leaves subsucculent, ovate to ovate-oblong, 1–8 × 0.5–3cm, obtuse, unevenly serrate with 4–6 pairs of teeth or sinuate to entire, attenuate into a petiole below; upper leaves smaller and narrower. Sepals erect-spreading.

Petals yellow, 9.5–13.5mm. Fruits erect or hanging, 15–40 × 2.5–3.5mm; stipe 1.25–3.5mm; beak sterile, 1.5–2mm. **Map 532, Fig. 75.**

On sandstone cliffs; 2300–3000m.

Yemen (N). Endemic.

3. D. erucoides (L.) DC., Syst. Veg. 2: 631 (1821). Illustr.: Collenette (1985 p.196).

Annual herb, sparsely clothed with short mainly retrorse hairs. Stems erect or ascending, 20–60cm, branched mainly from the base. Basal leaves in a loose rosette, narrowly elliptic to narrowly obovate, lyrate or pinnatisect or ± entire, 3–11 × 0.5–4cm, acute or obtuse, the margin serrate or sinuate to entire, attenuate into a short petiole below; upper leaves smaller and sessile. Sepals erect-spreading. Petals white or pale pink, 6–10.5(–12)mm. Fruits erect-spreading, 20–35(–40) × 1.5–2.5mm; stipe absent or up to 0.75mm; beak sterile or 1-seeded, 1.75–4mm. **Map 533, Fig. 75.**

Weed of cultivation; 10–2450m.

Saudi Arabia, Yemen (N & S). SW & C Europe, N Africa, Ethiopia, Syria, Lebanon, Palestine, Jordan, Iraq and Iran.

A specimen from Yemen (*J. Wood* 74/369) is rather problematic. It is an annual weed and matches *D. erucoides* in indumentum, leaf-shape and fruit characters. However, the petals are described as yellow and are smaller (c.4mm) than is usual.

4. D. acris (Forsskal) Boiss., Fl. orient. 1: 389 (1867).

Syn.: *Malcolmia arabica* Velen. in Sitzungsber. Königl. Böhm. Ges. Wiss. Prag., Math.-Naturwiss. Cl. 11: 14 (1912). Illustr.: Collenette (1985 p.196); Mandaville (1990 pls.72 & 73); Fl. Kuwait 1: pls. 85 & 86 (1985).

Annual herb, subglabrous or with a few spreading hairs below. Stems erect or ascending, 5–60cm, branched mainly from the base. Leaves mainly in a basal rosette, simple or rarely lyrately lobed, obovate-oblong, 3–10 × 1–4.5cm, obtuse, the margin unevenly serrate or sinuate to entire, attenuate into a short petiole below; upper leaves few, smaller and narrower. Sepals erect. Petals pale purple, lilac or rarely white, (7–)15–18(–22)mm. Fruits erect-spreading, 20–50 × 2–3mm; stipe (1–)1.5–2.5mm; beak sterile, 1.25–1.75mm. **Map 534, Fig. 75.**

Rocky, sandy and gravelly areas, and in wadis; 10–2400m.

Saudi Arabia, Yemen (N & S), Kuwait. Palestine, Jordan, Egypt, Iraq.

5. D. tenuifolia (L.) DC., Syst. nat. 2: 632 (1821).

Perennial herb, woody-based. glabrous throughout. Stems ascending, up to 50(–80)cm, branched throughout. Leaves linear-oblong, entire or pinnately lobed; lobes linear-oblong, 5–10 × 0.3–1cm, acute or obtuse, the margin entire, narrowly attenuate below; upper leaves linear, entire. Sepals erect-spreading. Petals yellow, (7–)11–12(–14)mm. Fruits erect-spreading, (20–)30(–60) × (1–)1.5(–2)mm; stipe (0.5–)1–1.5(–2)mm; beak sterile, 1–1.5mm. **Map 535, Fig. 75.**

On banks in a village; 2500m.

Yemen (N). W, S & C Europe, NW Africa, Caucasus and Syria; introduced elsewhere. In Arabia only collected once from near Kaukaban in Yemen (N).

6. ERUCA Miller

Annual herbs, glabrous or bearing simple hairs. Leaves lyrate-pinnatisect to subentire. Inner sepals saccate. Petals yellow with brown or violet veins. Filaments unappendaged. Stigma with decurrent carpidial lobes. Fruit a dehiscent siliqua with a flattened sterile beak; valves rounded, with 2 rows of seeds in each loculus; radicles incumbent; cotyledons longitudinally folded.

1. E. sativa Miller, Gard. dict., ed. 8: 1 (1768). Syn.: *E. lativalvis* Boiss., Fl. orient. 1: 396 (1867). Illustr.: Collenette (1985 p.199).

Annual herb; stems erect, up to 50cm, glabrous to somewhat hispid. Leaves up to 15(–30)cm. Petals 15–20mm. Siliqua 15–35 × 3–6mm, beak 5–10mm; valves 1-nerved, glabrous or hispid. **Map 536, Fig. 76.**

Weed of cultivation and waste ground; 0–2500m.

Saudi Arabia, Yemen (N & S), Oman, UAE, Qatar, Bahrain, Kuwait. Native of Europe, N Africa and S Asia, widely introduced elsewhere.

7. HEMICRAMBE Webb

Shrubs, glabrous or with simple hairs on the pedicels and sepals. Leaves simple. Inner sepals somewhat saccate. Petals white or rose. Filaments unappendaged. Stigma capitate. Fruit a siliqua, 2-membered, narrowly elliptic, laterally compressed; lower segment 2-valved, dehiscent, sterile; upper segment beak-like, 1–2-seeded; radicles incumbent; cotyledons longitudinally folded.

1. H. townsendii Gómez Campo in Anales Inst. Bot. Cavanilles 34: 154 (1977). Syn.: *Fabrisinapis fruticosus* C.C. Townsend in Hooker's Icon. Pl. 7(4): t.3673 (1971). Type: *Smith & Lavranos* 533 (K, FI, PRE).

Shrub up to 1m. Leaves ovate, 8–30 × 5–15mm, obtuse, irregularly crenate-dentate, cuneate below; petioles 3–20mm. Petals c.7mm. Siliqua 12–18 × 3–4mm, longitudinally-veined; lower segment much shorter than the upper. **Map 537, Fig. 76.**

On granite pinnacles; 900–1000m.

Socotra. Endemic.

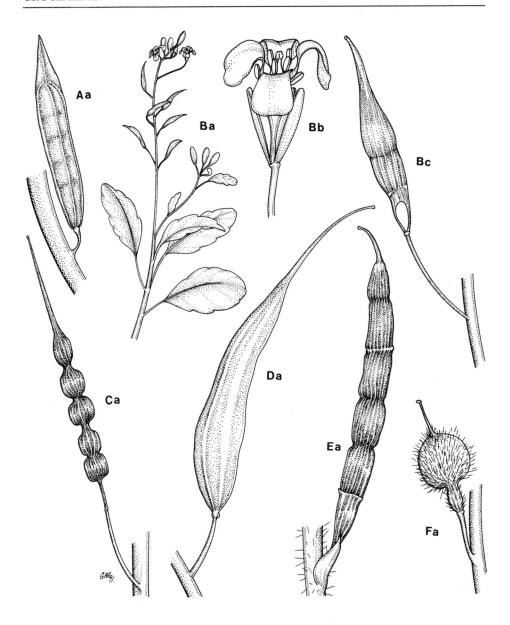

Fig. 76. Cruciferae. A, *Eruca sativa*: Aa, fruit (×3). B, *Hemicrambe townsendii*: Ba, flowering and fruiting branch (×1); Bb, flower (×5); Bc, fruit with lower valve removed (×5). C, *Raphanus raphanistrum*: Ca, fruit (×2). D, *R. sativus*: Da, fruit (×2). E, *Enarthrocarpus lyratus*: Ea, fruit (×4). F, *Rapistrum rugosum*: Fa, fruit (×5).

8. RAPHANUS L.

Annual or biennial herbs, bearing stiff, simple hairs. Leaves entire to lyrate-pinnatisect. Inner sepals slightly saccate. Petals white, yellow, rose or violet. Filaments unappendaged. Fruit an indehiscent siliqua with a narrow beak, 2-membered; upper segment fertile, cylindric or torulose; lower segment stalk-like, sterile; radicle incumbent; cotyledons longitudinally folded.

1.	Siliqua not inflated, constricted between the seeds	**1. R. raphanistrum**
+	Siliqua inflated, not constricted between the seeds	**2. R. sativus**

1. R. raphanistrum L., Sp. pl.: 669 (1753). Illustr.: Cornes (1989 pl.41).

Erect annual up to 50cm. Lower leaves lyrate-pinnatisect, 5–20cm, dentate-serrate, petiolate; upper leaves narrower, simple, sessile. Petals white, yellow or rose, usually with dark veins, 14–17mm. Siliqua 15–55 × 3–7mm, constricted between the seeds and longitudinally ridged, breaking into 1-seeded segments; beak slender, 1–2.5cm. **Map 538, Fig. 76.**

Roadsides and cultivated areas; 0–2900m.

Saudi Arabia, Yemen (N), Oman, Bahrain. Europe, N Africa, SW and C Asia.

2. R. sativus L., Sp. pl.: 669 (1753). Illustr.: Collenette (1985 p.205); Cornes (1989 pl.42).

Annual or biennial herb up to 1m. Lower leaves lyrate-pinnatisect, up to 25 × 7cm, dentate-serrate; upper cauline leaves narrower, simple, sessile. Petals white, pink or violet, 15–20mm. Siliqua inflated, 30–70 × 7–10(–15)mm, not or scarcely constricted between the seeds, the walls spongy. **Map 539, Fig. 76.**

Growing as a weed in cultivated areas and on waste ground; cultivated as a green salad; 10–2400m.

Saudi Arabia, Yemen (N & S), Oman, UAE, Qatar, Bahrain, Kuwait. The radish, cultivated world-wide.

9. ENARTHROCARPUS Labill.

Annual herbs, hispid with simple hairs. Leaves pinnatisect, ± lyrate in outline. Calyx hardly saccate. Petals pale yellow, purple-veined. Filaments unappendaged. Stigma capitate. Fruit an indehiscent, 2-membered siliqua; the lower part persistent, 2-valved, 0–3-seeded; the upper part many-seeded, constricted between the seeds and transversely breaking into 1-seeded segments; radicles incumbent; cotyledons longitudinally folded.

1. E. lyratus (Forsskal) DC., Syst. nat. 2: 661 (1821).

Erect or ascending annual herb up to 60cm. Leaves lyrate-pinnatisect, up to 15 × 5cm, dentate, shortly petiolate, the terminal lobes obovate, the lateral lobes oblong. Petals 6–8mm. Siliqua linear, 20–40 × 2–3mm, longitudinally striate, curved; upper part 2–3 × as long as lower, 3–6-seeded, terminating in a sterile 4–5mm beak; lower part 1–3-seeded. **Map 540, Fig. 76.**

No habitat details available.

Saudi Arabia. Egypt, Palestine, Iran and Pakistan.

A rare weed in Eastern Saudi Arabia. Also collected between Jeddah and Makkah last century.

10. CRAMBE Adans.

Annual, biennial or perennial herbs, bearing simple hairs. Leaves pinnately lobed or divided or simple. Flowers in a panicle. Inner sepals slightly saccate. Petals white (in Arabia) or yellow. Inner filaments usually toothed. Stigma capitate, sessile. Fruit an indehiscent, 2-membered silicula; lower segment short, stalk-like, sterile; upper segment globose, 1-seeded.

1. C. sp. A.

Perennial herb up to 1m; stems much-branched, glabrous; root narrow, smelling of radish. Leaves simple or with 1–3 small lateral lobes; terminal lobes broadly ovate, up to 20 × 20cm, with irregularly sinuate-serrate margins, the base cordate or truncate, hispid; petioles up to 15cm. Sepals oblong, c.2.5 × 0.75mm. Petals white, obovate, 3.5–4mm, shortly clawed. Inner filaments shortly toothed. Fruits erect-spreading on 8–10mm pedicels; upper segment globose, c.3.5mm diam., 4-ribbed, obscurely reticulately-nerved; lower segment cylindrical, c.1mm long. **Map 541.**

On well vegetated hillside with *Olea*; 1980m.

Saudi Arabia. ? Endemic.

This distinctive plant is known only from a single collection (*Collenette* 8226) from near Al-Bahah in SW Saudi Arabia where it was described by the collector as common and possibly an escape. Its fruits resemble those of *C. orientalis* L. (widely distributed in SW and C Asia) in size, shape, their 4 ribs and reticulate patterning, but the cordate-based leaves are quite unlike those of that species and more closely resemble those of *C. kotschyana* Boiss. (distributed from the Caucasus to Central Asia). However, the fruits of *C. kotschyana* are larger those of *Collenette* 8226 and unribbed. Further gatherings are needed so that the status of this species in Arabia can be established.

11. RAPISTRUM Crantz

Annual herbs, bearing stiff simple hairs. Leaves simple to lyrate-pinnatifid. Inner sepals slightly saccate. Petals yellowish-white. Inner filaments unappendaged. Stigma capitate. Fruit a 2-membered silicula; lower segment sterile or 1–3-seeded, dehiscent;

upper segment ± globose, longitudinally ridged, 1-seeded, with a narrow beak; radicle incumbent; cotyledons longitudinally folded.

1. R. rugosum (L.) All., Fl. pedem. 1: 257 (1785). Syn.: *R. orientale* (L.) Crantz, Crucif. emend. 106 (1769).

Erect annual up to 75(–100)cm. Lower leaves lyrate-pinnatifid; upper leaves sessile, entire or dentate. Petals 6–8(–10)mm. Silicula appressed to stem or spreading, 6–10mm, glabrous or hispid; upper segment globose, c.3mm diam., the beak 1–4mm; lower segment oblong, 1–3 × 1.5mm. **Map 542, Fig. 76.**

Weed of cereal fields and disturbed ground; 2100–2500m.

Saudi Arabia, Yemen (N). Europe, N Africa, W & C Asia; widely introduced elsewhere.

12. ERUCARIA Gaertner

Annual herbs, glabrous or bearing simple hairs. Leaves pinnatisect with linear or rarely broader lobes, fleshy. Inner sepals scarcely saccate. Petals white, rose or violet. Filaments without appendages. Stigma capitate. Fruit a 2-membered siliqua; upper part indehiscent, 1–4-seeded, ovoid or rostrate; lower part cylindrical, dehiscing by 2 valves, 1–6-seeded; radicle incumbent.

Ripe fruit is required for certain identification. In flower, species of *Erucaria* can be confused with *Cakile arabica*.

1.	Siliqua with a long attenuate and hooked tip; petals 4–5mm long	**3. E. uncata**
+	Siliqua straight, lacking a long attenuate and hooked tip; petals 3–10mm long	2
2.	Leaves lyrate-pinnatifid with oblong or obovate lobes; petals 3–6mm long	**4. E. sp. A**
+	Leaves 2-pinnatisect with linear or linear-oblong lobes; petals 7–10mm long	3
3.	Siliqua abruptly narrowed at the tip into a filiform style	**1. E. hispanica**
+	Siliqua narrowing gradually at the tip into a conical style	**2. C. crassifolia**

1. E. hispanica (L.) Druce in Bot. Exch. Club Soc. Brit. Isles 3: 418 (1914). Syn.: *Cakile arabica* sensu Mandaville (1985) non Velen. & Bornm.; *E. lineariloba* Boiss. in Ann. Sci. Nat. sér. 2, 17: 290 (1842); *E. aleppica* Gaertn., Fruct. sem. pl. 2: 298 (1791). Illustr.: Collenette (1985 p.198).

Erect annual; stems up to 60cm, branched from the base, glabrous. Leaves 1–2-bipinnatisect, with linear to linear-oblong lobes or rarely entire, petiolate. Petals pale lilac to mauve, 10–13mm. Siliqua (11–)14–15(–17)mm, straight, erect and appressed to the stem; lower segment 2–7 × 1–1.5mm, 1–6-seeded; upper segment (3–)6–8 × 1–

2(–3)mm, 1–3-seeded, torulose, abruptly narrowed at the tip into a (2–)4–5mm filiform style. **Map 543, Fig. 77.**

In sand, gravel and rock crevices, limestone outcrops, wadis, silty depressions and as a weed of cultivation; 0–1530m.

Saudi Arabia, Oman, UAE, Bahrain, Kuwait. E Mediterranean region, N Africa and SW Asia.

2. E. crassifolia (Forsskal) Del. (1813 p.244 t.34,1). Syn.: *Brassica crassifolia* Forsskal (1775 p.118).

Erect annual; stems up to 30(–40)cm, branched from the base, glabrous or pilose below. Leaves 1–2-pinnatisect with linear lobes. Petals rose, violet or white, c.7mm. Siliqua 10–18 × 1.2–2mm, curved, erect or spreading; lower segment 2.5–5 × c.1mm, 2–6-seeded; upper segment 8–13 × c.1.5mm, 3-seeded, narrowing gradually at the tip into a conical style. **Map 544, Fig. 77.**

?Saudi Arabia, ?UAE, ?Qatar. Jordan, Sinai, Egypt and Iraq.

Recorded from Qatar (Batanouny 1981) and the UAE (Western 1989). However, no verified material has been seen and the occurrence of this species in Arabia needs confirmation. The photograph of this species in Collenette (1985 p.198) is of *Cakile arabica*.

3. E. uncata (Boiss.) Asch. & Schweinf., Ill. fl. Égypte: 40 (1887).

Erect annual; stems up to 50cm, glabrous or slightly puberulous. Leaves 1–2-pinnatisect with linear lobes. Petals white to pink, 4–5mm. Siliqua 15–28 × c.1mm, erect or spreading; lower segment 3–4 × c.1mm, c.4-seeded; upper segment 15–20 × c.1mm, 4-seeded, long attenuate and hooked towards the tip. **Fig. 77.**

?Saudi Arabia, ?Kuwait. Sinai.

No verified material of this species has been seen and its occurrence in Arabia is very doubtful; see Fl. Kuwait (1985 p.104) and Hedge & King (op. cit.: 45).

4. E. sp. A.

Annual or ?perennial herb; stems ascending, 25–40cm. Leaves long-petiolate, lyrate-pinnatifid, obovate in outline; lobes obovate to oblong, 5–15 × 4–8mm, rounded, the margin entire or with a few shallow teeth. Petals white, pale blue or violet, 3–6mm. Siliqua (immature) 10–12mm, straight, erect and appressed to stem; lower segment, 5–6mm, c.4-seeded; upper segment 5–6mm, c.1-seeded, narrowing gradually at the tip into a conical style. **Map 545.**

In shade of rocks and date gardens; 650–1800m.

Oman. ? Endemic.

There are several collections of this distinct taxon from northern Oman, but unfortunately none has mature fruits. It is distinguished from the other Arabian species of *Erucaria* by its smaller flowers, lyrate-pinnatifid leaves with broader lobes and apparently distinct fruits.

13. CAKILE Miller

Annual herbs, glabrous or with simple hairs. Leaves 1(–2)-pinnatifid with linear lobes. Inner sepals slightly saccate. Filaments free, without appendages. Stigma capitate. Fruit an indehiscent, longitudinally ridged siliqua, breaking at maturity into 2 unequal parts: the upper part 1-seeded, with a flattened terminal portion; the lower ± tetragonous, 0–1 seeded, persisting on the plant; radicles incumbent or accumbent.

1. C. arabica Velen. & Bornm. in Feddes Repert. Spec. Nov. Regni Veg. 9: 114 (1911). Illustr.: Fl. Iraq 4 (2): 879 (1980); Mandaville (1990 pl.75); Fl. Kuwait 1: pls.90–92 (1985); Collenette (1985 p.195).

Succulent annual herb up to 45cm. Leaves ovate in outline, 4–15 × 2–5cm; lobes 10–40 × 1–4mm, obtuse. Petals violet, 8–10mm. Fruiting pedicels stout, 1–4mm long. Siliqua 10–18 × 2–2.5mm. **Map 546, Fig. 77.**

In rock crevices, sandy ground and on dunes; 10–950m.

Saudi Arabia, Kuwait. S Iran and Iraq.

14. ZILLA Forsskal

Spiny subshrub, glabrous. Leaves fleshy, entire to lyrate-pinnatifid. Inner sepals saccate. Petals lilac. Stigma with decurrent carpidial lobes. Fruit an indehiscent silicula, globose with a conical beak, 2-seeded; radicles incumbent; cotyledons longitudinally folded.

1. Z. spinosa (L.) Prantl in Engl. & Prantl, Nat. Pflanzenfam. ed. 1, 3 (2): 175 (1890). Illustr.: Fl. Kuwait 1: pl. 93 (1985); Collenette (1985 p.208); Mandaville (1990 pl.76).

Glaucous subshrub; stems up to 1m, much-branched, the shoots becoming spiny. Stem leaves few, narrowly oblong, up to 30 × 4mm, the margin entire or with a few small teeth; basal leaves short-lived, lyrate-pinnatifid. Petals 15–19mm. Silicula 5–10mm diam., smooth or verrucose; beak 3–5mm. **Map 547, Fig. 77.**

Sandy and gravelly deserts, stony plains, wadis and silt basins; 100–950m.

Saudi Arabia, Yemen (S), Oman, UAE, Qatar, Kuwait. N Africa, Syria, Palestine, Jordan and Iraq.

Gabali and Al-Gifri (1990) record *Zilla spinosa* from the Mahra Governorate of Yemen (S); however, no verified material has been seen.

15. PHYSORRHYNCHUS Hook. f.

Glaucous perennial herb, sometimes woody-based, glabrous. Leaves fleshy, simple, entire; lower petiolate; upper sessile and amplexicaul. Inner sepals slightly saccate. Petals pink or lilac. Filaments unappendaged. Stigma with decurrent carpidial lobes. Fruit an indehiscent silicula, 2-membered; upper segment ± globose with a conical beak, 2-seeded; lower segment sterile, short, stalk-like, inconspicuous; radicle incumbent; cotyledons longitudinally folded.

1. P. chamaerapistrum (Boiss.) Boiss., Fl. orient. 1: 403 (1867).

Perennial herb up to 2m. Lower leaves elliptic to obovate, 3–15 × 2–9cm, acute or obtuse, shortly petiolate; upper leaves smaller, obovate to ± oblong, acute, amplexicaul. Petals 12–15(–20)mm. Silicula 8–12 × 3–4(–6)mm, the beak 4–6mm, becoming woody at maturity. **Map 548, Fig. 77.**

Rocky hillsides and wadis, and on sand and gravel in open *Acacia* bushland; 30–600m.

Oman, UAE, ?Kuwait. Iran, Pakistan.

There is an old record (*Dickson* 322) of this species from Kuwait. This may represent an isolated introduction, but as there have been no further records, it is probably extinct there now.

16. DOLICHORHYNCHUS Hedge & Kit Tan

Glaucous perennial herb, glabrous. Leaves simple, entire; lower leaves shortly petiolate; upper leaves semi-amplexicaul. Sepals saccate. Petals white. Filaments without appendages. Stigma with decurrent carpidial lobes. Fruit a 2-membered siliqua; lower segment 2-valved, dehiscent, 2–4-seeded; upper segment strongly elongate, indehiscent, sterile; radicles incumbent; cotyledons longitudinally folded.

Endemic monotypic genus.

1. D. arabicus Hedge & Kit Tan in Pl. Syst. Evol. 156: 198 (1987). Illustr.: loc. cit. Type: Saudi Arabia, *Collenette* 5748 (E, K).

Woody-based perennial up to 1m. Stem leaves oblong-elliptic, 5–7 × 1–2cm, entire or remotely toothed, attenuate below into a short petiole; upper stem leaves narrowly elliptic, semi-amplexicaul. Petals c.20mm. Siliqua cylindrical, 30–40 × 3–4.5mm. **Map 549, Fig. 77.**

On sandstone, growing on ledges and amongst rock debris; 610m.

Saudi Arabia. Endemic.

A monotypic genus endemic to a small area in the NW of Saudi Arabia.

17. CARRICHTERA DC.

Annual herbs bearing stiff, simple hairs throughout. Leaves pinnatisect. Inner sepals scarcely saccate. Filaments free, unappendaged. Stigma with decurrent carpidial lobes. Fruit a dehiscent silicula with a flat, spathulate, sterile beak; seed-bearing portion ellipsoid, 2-celled, with c.3 seeds per cell; radicles incumbent; cotyledons longitudinally folded.

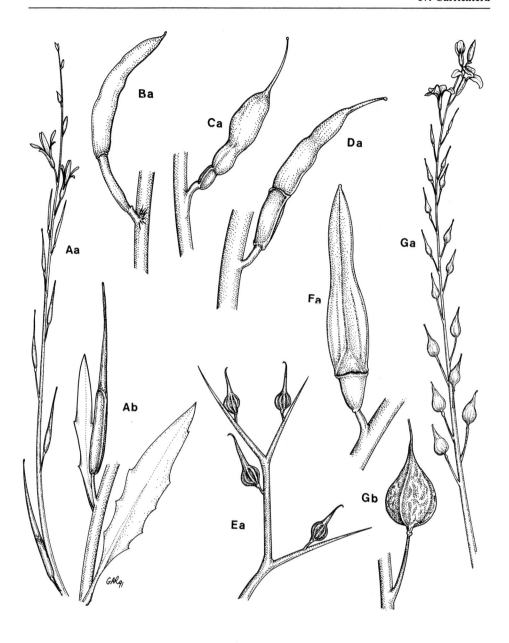

Fig. 77. Cruciferae. A, *Dolichorhynchus arabicus*: Aa, fruiting and flowering shoot (×1); Ab, fruit and leaves (×2). B, *Erucaria crassifolia*: Ba, fruit (×5). C, *E. hispanica*: Ca, fruit (×5). D, *E. uncata*: Da, fruit (×5). E, *Zilla spinosa*: Ea, fruiting branch (×1). F, *Cakile arabica*: Fa, fruit (×6). G, *Physorrhynchus chamaerapistrum*: Ga, fruiting and flowering branch (×1); Gb, fruit (×2.5).

1. C. annua (L.) DC. in Mém. Mus. Hist. Nat. 7: 250 (1821). Syn.: *C. vellae* DC., Syst. nat. 2: 642 (1821) nom. illegit. Illustr.: Collenette (1985 p.195); Mandaville (1990 pls.77–78); Fl. Kuwait 1: pls 94–96 (1985).

Stems erect to ascending, up to 10(–30)cm, branching from the base. Leaves 2–3 pinnatisect, 2–6(–9) × 0.2–2cm; segments linear, obtuse. Petals cream with violet veins, 7–8mm. Silicula 6–8mm, covered with stiff hairs; beak glabrous. **Map 550, Fig. 78.**

Deserts and dry hills, on sand, gravel and silty soils; 10–700m.

Saudi Arabia, Kuwait. Mediterranean area, N Africa, SW Asia to Iran; naturalized in other areas.

18. SCHOUWIA DC.

Glabrous, annual herbs. Leaves fleshy, simple, ± entire, the upper amplexicaul. Inner sepals saccate. Petals white, pink or purple. Filaments unappendaged. Stigma with decurrent carpidial lobes. Fruit a dehiscent silicula, strongly compressed, 2-membered; lower segment narrowly winged, fertile, many-seeded; upper segment a narrow beak; radicle incumbent; cotyledons longitudinally folded.

Moggi, G. (1967). Il genere Schouwia in Africa orientale. *Webbia* 22: 531–538.

1. S. purpurea (Forsskal) Schweinf. (1896 p.183). Syn.: *S. arabica* DC., Syst. nat. 2: 644 (1821); *S. schimperi* Jaub. & Spach, Ill. pl. orient. 3: 145 (1850); *S. thebaica* Webb in Giorn. Bot. Ital. 2, (2): 219 (1847); *Subularia purpurea* Forsskal (1775 p.117). Illustr.: Collenette (1985 p.206). Type: Yemen (N), *Forsskal* (C).

Glaucous bushy annual up to 90cm. Leaves obovate to ovate, 3–10 × 1.5–5cm; lower cuneate at the base; upper amplexicaul with rounded auricles. Petals 7–10mm. Silicula broadly ovate to subcircular in outline, (12–)15–20 × (10–)14–19mm (excl. beak), cordate at the base, the tip emarginate; beak 4–6mm. **Map 551, Fig. 78.**

Coastal sandy plains and a weed of cultivation and irrigated fields; 0–600(–900)m.

Saudi Arabia, Yemen (N & S). Margins of the Red Sea, from Egypt to Ethiopia.

19. SAVIGNYA DC.

Annual herb, glabrous or with simple glandular or eglandular hairs below. Leaves simple, fleshy, sinuate-dentate. Sepals scarcely saccate. Petals white or pale pink. Filaments unappendaged. Stigma capitate. Fruit a dehiscent strongly compressed silicula, shortly stipitate below, shortly beaked; seeds in 2 rows; radicles incumbent; cotyledons folded.

1. S. parviflora (Del.) Webb in Giorn. Bot. Ital. 2: 215 (1847). Syn.: *S. aegyptiaca* DC., Syst. nat. 2: 283 (1821). Illustr.: Fl. Kuwait 1: pls. 97–98 (1985); Cornes (1989 pl.43); Collenette (1985 p.205); Mandaville (1990 pl.79).

Erect annual; stems up to 40cm, branched from the base. Lower leaves oblong-elliptic to obovate, up to 5 × 2cm, sinuate-dentate, petiolate; upper leaves linear, entire. Petals 3–5mm. Fruiting pedicels delicately reflexed, 10–25mm. Silicula oblong-elliptic, 8–15 × 5–8mm; beak 1.5–2(–2.5)mm; stipe c.1mm. **Map 552, Fig. 78.**

Sandy or gravelly deserts; 20–900m.

Saudi Arabia, ?Yemen (S), Oman, UAE, Qatar, Bahrain, Kuwait. N Africa, Palestine, Jordan, Iraq, Iran, Afghanistan and Pakistan.

Hedge & King (op. cit.: 48) record this species from Yemen (S). No material has been seen and its occurrence there seems unlikely.

20. MORICANDIA DC.

Perennial glabrous herbs. Leaves succulent, simple, entire, the upper amplexicaul. Inner sepals saccate. Petals pink. Filaments unappendaged. Stigma bilobed, shortly decurrent. Fruit a linear, dehiscent siliqua; seeds in 2 rows; radicle incumbent, the cotyledons longitudinally folded.

1. M. sinaica (Boiss.) Boiss., Fl. orient. 1: 386 (1867). Syn.: *M. nitens* auctt. non (Viv.) E. Durand & Barratte. Illustr.: Collenette (1985 p.205).

Erect or ascending glaucous, perennial herb; stems 50–75cm, often woody-based. Leaves ovate to oblong-ovate, 2–6(–10) × 2–4.5(–8)cm, obtuse, the upper sessile and amplexicaul with large rounded auricles, the lower cuneate at the base. Petals 12–15mm. Siliqua 50–70 × 1.5–2mm, the valves 1-nerved. **Map 553, Fig. 78.**

Rocky deserts, stony banks and ravines; 600–2600m.

Saudi Arabia, Yemen (N & S), Oman. Egypt, Iran and Pakistan.

21. LEPIDIUM L.

Annual, biennial or perennial herbs, bearing simple hairs. Leaves simple or variously divided. Sepals not saccate. Flowers in cylindrical terminal racemes or panicles. Petals white or lilac, sometimes absent. Stamens 2, 4 or 6; filaments unappendaged. Stigma capitate. Fruit a compressed dehiscent silicula with a narrow septum, 2-seeded, valves often winged; radicle incumbent.

1.	Fruiting racemes dense and cylindrical with the siliculae overlapping	**1. L. aucheri**
+	Fruiting racemes various but never as above	2
2.	Petals equalling or longer than the sepals	3
+	Petals shorter than sepals or absent	6
3.	Perennial; silicula 1.5–2.5mm long; tip of silicula rounded or retuse	4
+	Annual; silicula 3–6mm long; tip of silicula notched	5

CRUCIFERAE

4. Inflorescence a dense terminal panicle; silicula rounded at the tip, 1.5–2mm long **8. L. latifolium**
+ Inflorescence a terminal raceme; silicula retuse at the tip, 2–2.5mm long **3. L. armoracia**

5. Silicula 5–6mm long; cauline leaves mostly pinnately lobed **2. L. sativum**
+ Silicula 3–4mm long; cauline leaves simple, at most coarsely serrate **4. L. virginicum**

6. Leaves all linear-lanceolate to oblanceolate, entire or serrulate with a few acute teeth at the tip **7. L. africanum**
+ Lower leaves 1–3-pinnatisect with narrowly oblong to almost linear lobes, entire or with a few coarse teeth; upper leaves similar to the lower or less divided or entire **7**

7. Silicula 3–4mm long **5. L. bonariense**
+ Silicula 2–2.5mm long **6. L. ruderale**

1. L. aucheri Boiss. in Ann. Sci. Nat. Bot. sér 2, 17: 195 (1842). Illustr.: Fl. Pakistan 55: 57 (1973); Fl. Iraq 4 (2): 889 (1980); Collenette (1985 p.201).

Prostrate or ascending annual herb; stems up to 30cm, much branched from the base, glabrous or with scattered hairs. Leaves pinnately-lobed with oblong ± rounded lobes or subentire with occasional obtuse teeth, up to 10cm. Petals white, 1–1.75mm. Stamens 6. Fruiting racemes dense, cylindrical with fruits overlapping. Silicula erect, appressed to stem, ovate to quadrangular, 2.5–3 × 1.5–2mm, acutely winged or horned at the tip, the stigma included within or equalling the deep apical notch. **Map 554, Fig. 78.**

Clay-pans, date groves and in silty depressions; 0–1200m.

Saudi Arabia, Oman, Qatar, Kuwait. Egypt and SW Asia (excluding Turkey).

2. L. sativum L., Sp. pl.: 644 (1753). Illustr.: Fl. Trop. E. Afr. 18 (1982); Fl. Iraq 4 (2): 889 (1980); Collenette (1985 p.201).

Annual herb; stem erect, up to 50cm, branched above, glabrous or sparsely pilose. Basal leaves short-lived, 1–3-pinnatisect with linear to narrowly ovate or narrowly obovate lobes; cauline leaves similar; uppermost simple, entire. Petals white or lilac, 2–3mm. Stamens 6. Silicula broadly elliptic to broadly elliptic-oblong, 5–6 × 4-5mm, narrowly winged at tip, the stigma included within or level with the top of the narrow apical notch. **Map 555, Fig. 78.**

Weed of cultivation, irrigated ground, clay-pans, roadsides and waste ground; 0–2250m.

Saudi Arabia, Yemen (N & S), Oman, UAE, Kuwait. Northern temperate regions.

3. L. armoracia Fischer & C. Meyer, Index sem. hort. petrop. 9: 77 (1842). Syn.: *L. armoracia* Fischer & C. Meyer subsp. *abyssinicum* (A.Rich.) Thell. in Vierteljahrsschr.

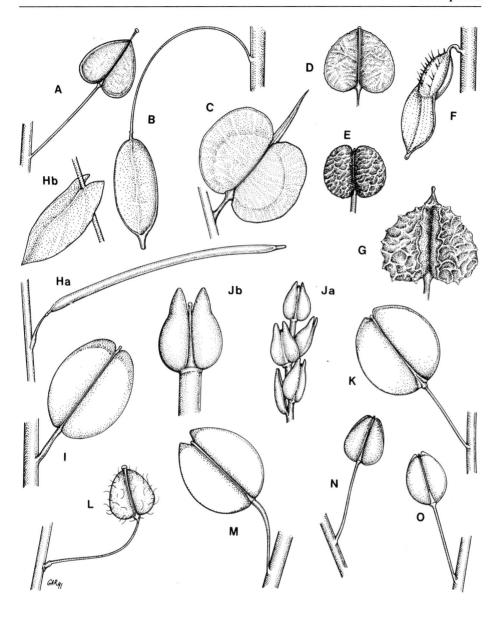

Fig. 78. Cruciferae. A–G, fruits: A, *Cardaria draba* (×4); B, *Savignya parviflora* (×4); C, *Schouwia purpurea* (×2.5); D, *Coronopus niloticus* (×11); E, *C. didymus* (×11); F, *Carrichtera annua* (×7); G, *Coronopus squamatus* (×11); H, *Moricandia sinaica*: Ha, fruit (×2); Hb, leaf (×1); I, *Lepidium sativum* (×8); J, *L. aucheri*: Ja, tip of infructescence (×5); Jb, fruit (×8); K, *L. virginicum* (×8); L, *L. latifolium* (×8); M, *L. bonariense* (×8); N, *L. armoriaca* (×8); O, *L. ruderale* (×8).

Naturf. Ges. Zürich 51: 176 (1906); *L. armoracia* Fischer & C. Meyer subsp. *intermedium* (A.Rich.) Thell. var. *alpigenum* (A.Rich.) Thell. loc. cit.; *L. schweinfurthii* Thell. in Mitt. Bot. Mus. Univ. Zürich 26: 178 (1906). Illustr.: Fl. Trop. E. Afr.: 18 (1982).

Perennial herb, often woody-based, sometimes cushion-forming; stems ascending, up to 30cm, branched from the base, finely puberulous or glabrous. Basal leaves short-lived, 1–2-pinnatisect with ovate or obovate lobes; cauline leaves linear to narrowly ovate or narrowly obovate, acute, entire to serrate, attenuate at the base. Petals white, 1–1.5mm. Stamens 2 or 4. Silicula elliptic to ovate, 2–2.5(–3.8) × 1.5–2(–2.5), the stigma distinctly projecting beyond the retuse tip. **Map 556, Fig. 78.**

Walls, banks and footpaths around habitation; 2100–3650m.

Yemen (N). Tanzania and Ethiopia.

4. **L. virginicum** L., Sp. pl.: 645 (1753).

Erect annual herb; stems up to 50cm, usually unbranched, minutely puberulous. Basal leaves short-lived, lyrate, up to 8cm long; middle and upper leaves simple, narrowly obovate, coarsely serrate to entire. Petals white, 1.5–2mm. Silicula suborbicular, 3–4 × 2.5–3.5mm; stigma sessile, included within the shallow apical notch. **Map 557, Fig. 78.**

Weed of cultivated ground; 50m.

Saudi Arabia. Cosmopolitan weed, originally a native of N America.

5. **L. bonariense** L., Sp. pl.: 645 (1753). Illustr.: Fl. Trop. E. Afr. 18 (1982).

Erect or ascending annual to perennial herb; stems up to 30(–70)cm, branched above, puberulous. Basal and lower leaves up to 7cm, 2–3-pinnatisect with narrowly oblong to almost linear lobes, entire or with a few coarse teeth; upper leaves less divided or entire. Petals shorter than the sepals or absent, up to 0.8mm. Stamens 2. Silicula orbicular, 3–3.5(–4) × 2.5–3mm, the stigma included within the shallow apical notch. **Map 558, Fig. 78.**

Grassy field borders; 2100m.

Yemen (N). Cosmopolitan weed, originally a native of S America.
Known only from a single gathering in Arabia.

6. **L. ruderale** L., Sp. pl.: 645 (1753). Illustr.: Fl. Iraq 4 (2): 889 (1980).

Erect or ascending annual or biennial herb; stems up to 30cm, branched above, glabrous or sparsely hairy. Lower leaves short-lived, 1–2-pinnatisect, with linear to narrowly elliptic lobes, petiolate; upper leaves entire, linear-oblong. Petals absent. Stamens 2. Silicula ovate or broadly elliptic, 2–2.5 × 1.5–2mm, narrowly winged above, the stigma included within the shallow apical notch. **Map 559, Fig. 78.**

Waste ground and rubbish heaps; sea-level.

?Saudi Arabia, Kuwait. A widespread weed of the northern temperate regions.

Recorded from Saudi Arabia by Hedge and King (op. cit.) but no specimens have been seen and the record has not been verified.

7. L. africanum (Burm.f.) DC., Syst. nat. 2: 552 (1821). Illustr.: Fl. Trop. E. Afr.: 22 (1982).

Erect or ascending annual or perennial herb; stems up to 45cm, much-branched. Lower leaves short-lived, in a basal rosette, oblanceolate; upper leaves linear-lanceolate to oblanceolate, 1.5–2.5(–6) × 1.5–3(–6)cm, attenuate below, entire or serrulate with a few acute teeth at the tip. Petals absent or up to 5mm. Stamens 2. Silicula ovate, (2–)2.2–2.8(–3.5) × 1.5–2mm, the stigma included within the shallow apical notch. **Map 559.**

Waste ground; 1980m.

Saudi Arabia. Tropical Africa, also occurs as an alien in Europe.

In Arabia collected only once.

8. L. latifolium L., Sp. pl.: 644 (1753). Illustr.: Fl. Pakistan 55: 61 (1973); Fl. Iraq 4 (2): 889 (1980).

Erect perennial herb; stems up to 1m, much branched above, glabrous or sparsely pubescent. Lower leaves simple, ovate to oblong-ovate, up to 25 × 5cm, entire to dentate, long-petiolate; upper leaves smaller and narrower. Inflorescence many-flowered, forming a dense panicle. Petals white, 1.5–1.75mm. Stamens 6. Silicula orbicular to broadly ovate, 1.5–2 × 1–2mm, without an apical notch, the stigma sessile. **Map 560, Fig. 78.**

No habitat details available.

Yemen (N). Europe, N Africa, SW, C and E Asia.

22. CORONOPUS L.

Annual or biennial herbs, glabrous or clothed with simple hairs. Leaves pinnatifid to bipinnatisect. Sepals not saccate. Petals white, very small or lacking. Stamens 2, 4 or 6; filaments unappendaged. Stigma capitate. Fruit an indehiscent silicula, bilobed with a narrow septum, bilocular; loculi 1-seeded; radicle incumbent.

1.	Silicula 3.5–5mm broad, coarsely reticulately rugose-tuberculate; fruiting pedicels 0.5–2mm	**1. C. squamatus**
+	Silicula 2–2.5mm broad, finely wrinkled (reticulately rugose); fruiting pedicels 3–4mm	2
2.	Silicula emarginate at the tip	**2. C. didymus**
+	Silicula rounded and minutely apiculate at the tip	**3. C. niloticus**

1. C. squamatus (Forsskal) Asch., Fl. Brandenburg 1: 62 (1864). Illustr.: Fl. Iraq 4 (2): 894 (1980).

Prostrate or ascending annual or biennial herb; stems up to 20cm, much-branched, glabrous. Leaves 1–2-pinnatifid, up to 10cm; segments narrowly ovate to narrowly obovate, entire or toothed. Petals 1–2mm, longer than sepals. Fertile stamens 6. Fruiting pedicels 0.5–2mm. Silicula 2.5–4 × 3.5–5mm, with a cordate base and acute tip, reticulately rugose-tuberculate. **Map 561, Fig. 78.**

Weed of lawns; 20m.

Saudi Arabia. Europe, N Africa and SW Asia; introduced elsewhere.

2. C. didymus (L.) Smith, Fl. brit. 2: 691 (1804). Illustr.: Fl. Trop. E. Afr.: 25 (1982).

Prostrate to ascending annual or biennial herb; stems up to 40cm, puberulous, branched mainly below. Leaves 1–2-pinnatisect, up to 3cm; segments narrowly ovate to elliptic, entire or serrate. Petals c.0.5mm, shorter than sepals. Fertile stamens 2 or 4. Fruiting pedicels 3–4mm. Silicula 1.3–1.7 × 2–2.5mm, with a cordate base and emarginate tip, finely wrinkled (reticulately rugose). **Map 562, Fig. 78.**

Weed of irrigated cultivation and waste ground; 20m.

Saudi Arabia, Oman, Kuwait. Cosmopolitan weed.

3. C. niloticus (Del.) Sprengel, Syst. veg. 2: 853 (1825).

Similar to *C. didymus* but the leaves less finely divided, from serrate to bipinnatifid; stems glabrescent to puberulous; silicula rounded and minutely apiculate at the tip. **Fig. 78.**

No habitat details available.

?Saudi Arabia. Egypt.

Recorded from Saudi Arabia by Hedge and King (op. cit.) but no specimens have been seen and the record has not been verified.

23. CARDARIA Desv.

Perennial herbs bearing simple hairs. Leaves simple, ± entire to dentate. Sepals not saccate. Petals white. Filaments free, unappendaged. Stigma capitate. Fruit an indehiscent silicula with a narrow septum, ovoid with a cordate base and inflated valves, (1–)2-seeded; radicles incumbent.

1. C. draba (L.) Desv. in J. Bot. Agric. 3: 163 (1815). Syn.: *Lepidium draba* L., Sp. pl.: 645 (1753). Illustr.: Collenette (1985 p.195).

Stems erect, up to 60cm, much branched above, clothed with short appressed hairs. Basal leaves obovate, 8–12 × 1.5–4cm, dentate, petiolate; stem leaves sessile, oblong-ovate, decreasing in size upwards, dentate to ± entire, amplexicaul or auriculate. Petals white, 3–3.5mm. Silicula 3–5 × 3.5–4.5mm. **Map 563, Fig. 78.**

Weed of cultivation; 10–2600m.

Saudi Arabia, Yemen (N & S), Oman, Kuwait. Widespread as a weed of cultivation throughout the world.

24. ISATIS L.

Annual or perennial herbs, glabrous or bearing simple hairs. Basal leaves entire to pinnately lobed; stem leaves sessile, ± entire, auriculate. Sepals not saccate. Petals yellow. Filaments without appendages. Stigma 2-lobed, sessile. Fruit indehiscent, strongly compressed, winged, unilocular, 1-seeded; radicle incumbent or accumbent.

1. I. lusitanica L., Sp. pl.: 670 (1753). Illustr.: Collenette (1985 p.201).

Erect annual; stems up to 20–30cm, glabrous or hispid below. Basal leaves up to 15 × 5cm, sinuately lobed to subentire, shortly petiolate; stem leaves oblong to oblong-ovate, entire to dentate, auriculate at the base. Petals 2–4mm. Fruit pendent, linear-oblong, 15–30 × 2–5mm, truncate or retuse, the base cuneate, glabrous or covered with short retrorse hairs. **Map 564, Fig. 79.**

On dry rocky slopes, granite and sand, wadi-bottoms and date groves; 850–2350m.

Saudi Arabia. SW Europe, N Africa and SW Asia.

25. HORWOODIA Turrill

Annual herbs, glabrous or bearing simple hairs. Leaves pinnately lobed or dentate. Inner sepals saccate. Petals purplish mauve, rarely white. Filaments without appendages. Stigma capitate. Fruit an indehiscent 1-seeded silicula, ± orbicular, the central loculus sharply keeled and surrounded by a broad wing; radicles incumbent; cotyledons longitudinally folded.

1. H. dicksoniae Turrill in J. Bot. 77: 117 (1939). Syn.: *Malcolmia musilii* Velen. pro parte in Sitzungsber. Königl. Böhm. Ges. Wiss. Prag 11: 13 (1913). Illustr.: Fl. Iraq 4 (2): 899 (1980); Mandaville (1990 pls. 81-2); Collenette (1985 p.200); Fl. Kuwait 1: pls. 99-100 (1985).

Ascending or decumbent annual; stems up to 45cm, branched from the base. Leaves oblong, up to 10 × 3cm, dentate to pinnately lobed, petiolate. Petals 12–15mm. Silicula orbicular, 14–18 × 15–17mm, emarginate to truncate, cordate at the base, prominently nerved and somewhat shining. **Map 565, Fig. 79.**

Sandy and stony deserts; 150–750m.

Saudi Arabia, Kuwait. Sinai, Jordan, Iraq.

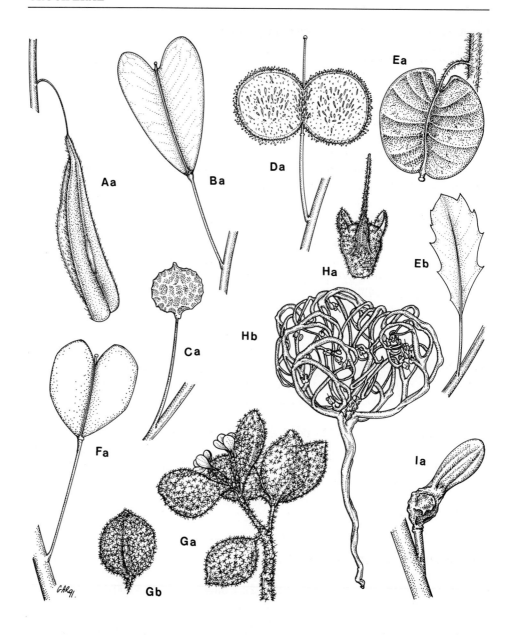

Fig. 79. Cruciferae. A, *Isatis lusitanica*: Aa, fruit (×3). B, *Capsella bursa-pastoris*: Ba, fruit (×4.5). C, *Neslia apiculata*: Ca, fruit (×8). D, *Biscutella didyma*: Da, fruit (×4.5). E, *Horwoodia dicksoniae*: Ea, fruit (×2); Eb, leaf (×1.5). F, *Thlaspi perfoliatum*: Fa, fruit (×6). G, *Lachnocapsa spathulata*: Ga, flowering branch (×3); Gb, fruit (×3). H, *Anastatica hierochuntica*: Ha, fruit (×4); Hb, fruiting habit (×0.6). I, *Schimpera arabica*: Ia, fruit (×6).

26. IBERIS L.

Annual or perennial herbs, bearing simple hairs or glabrous. Leaves simple or pinnatifid. Inflorescence corymbose. Sepals not saccate. Petals white, pink or magenta, the outer 2 larger than the inner 2. Filaments unappendaged. Stigma capitate. Fruit a compressed dehiscent silicula with a narrow septum, the valves winged at the tip, 2-seeded; radicle accumbent.

1. I. umbellata L., Sp. pl.: 649 (1753).

Glabrous annual; stems erect, up to 40cm. Leaves linear. Inflorescence dense, corymbose even in fruit. Petals bright pink or purple, the outer pair longer. Silicula ovate, c.10mm long; valves winged from the base with acute triangular lobes at the tip. **Map 566.**

Cultivated garden ornamental, rarely escaped and naturalized; 2860m.

Saudi Arabia. A native of the Mediterranean region, widely cultivated elsewhere.

Only known from a single gathering in a natural habitat under *Juniperus* but near old abandoned gardens.

27. BISCUTELLA L.

Annual herbs bearing simple hairs. Leaves entire or dentate. Sepals not saccate. Fruit a strongly compressed 2-valved silicula with a narrow septum; valves flat and ± orbicular, 1-seeded, indehiscent; radicles accumbent.

1. B. didyma L., Sp. pl.: 653 (1753). Illustr.: Collenette (1985 p.194).

Stems erect, up to 15(–50)cm, simple or branched, hirsute with stiff hairs. Basal leaves narrowly obovate, 4–5(–10) × 0.5–1(–2)cm, entire or dentate, attenuate at the base, hirsute with stiff hairs. Stem leaves smaller, narrowly oblong-ovate, entire to dentate, half-clasping at the base. Petals yellow (?white), 2–3mm. Fruit valves (4–)5(–8)mm diam., glabrous or pubescent with short clavate hairs. **Map 567, Fig. 79.**

On rocky slope in *Acacia* bushland and on ungrazed terraces; 550–2150m.

Saudi Arabia. S Europe, Egypt and SW Asia.

28. THLASPI L.

Glabrous annual herb. Leaves simple, the cauline amplexicaul. Inner sepals not saccate. Petals white. Filaments unappendaged. Stigma capitate. Fruit a dehiscent silicula with a narrow septum, compressed, winged, with 2–3 seeds in each loculus; radicle accumbent.

1. T. perfoliatum L., Sp. pl.: 646 (1753).

Annual herb; stems up to 25cm. Basal leaves petiolate; stem leaves ovate to narrowly

CRUCIFERAE

ovate, up to 2(–4) × 0.5(–2)cm, entire or slightly toothed, amplexicaul. Petals c.2mm. Silicula obcordate, 4–7 × 4–5mm, with broad wings narrowing towards the base. **Map 568, Fig. 79.**

No habitat details available; 2130m.

Saudi Arabia. Europe, N Africa, SW and C Asia.
Found only once in Arabia.

29. LACHNOCAPSA Balf.f.

Low shrub or woody-based herb, the whole plant white-tomentose with ± appressed stellate and branched hairs. Leaves fleshy, simple, entire. Inner sepals saccate. Petals pale yellow. Filaments unappendaged. Stigma sessile, bilobed. Fruit a silicula, compressed, with a narrow septum, bilocular, with 1 seed per loculus; radicle incumbent.

A monotypic genus endemic to Socotra.

1. L. spathulata Balf. f. (1882 p.500). Illustr.: Balf.f (1888 t.3). Type: Socotra, *Balfour, Cockburn & Scott* 587 (E).

Stems up to 30cm. Leaves broadly ovate to orbicular or broadly obovate, 5–25 × 3–15mm, obtuse or rounded, truncate or cuneate below. Petals 12–13mm. Silicula 10–12 × 6–7mm, compressed, densely white-tomentose, closely resembling the leaves. **Map 569, Fig. 79.**

Sandy coastal plains and rocky wadi-sides; 0–20m.

Socotra. Endemic.
Known only from coastal areas at the east end of Socotra and from the island of Semha.

30. CAPSELLA Medikus

Annual or biennial herbs with simple, forked or stellate hairs. Leaves entire to pinnatipartite. Sepals not saccate. Filaments free, without appendages. Fruit a many-seeded silicula with a narrow septum, obtriangular to obcordate, compressed, dehiscent; radicles incumbent.

1. C. bursa-pastoris (L.) Medikus, Pfl.-Gatt.: 85 (1792). Illustr.: Fl. Iraq 4 (2): 928 (1980); Mandaville (1990 pl.83); Collenette (1985 p.195).

Erect herb up to 50cm. Basal leaves in a rosette, narrowly obovate in outline, 2.5–8(–15) × 0.4–2cm, pinnatipartite to dentate or ± entire, attenuate at the base into a short petiole; stem leaves smaller, sessile, narrowly oblong to ovate, dentate to entire, amplexicaul at the base with acute auricles. Petals white, 2–2.5mm. Silicula 5–10 × 4–6mm, up to 30-seeded. **Map 570, Fig. 79.**

Weed of cultivated and disturbed ground; 1100–2600m.

Saudi Arabia, Yemen (N & S), Socotra, Oman, UAE. Cosmopolitan.

31. ANASTATICA L.

Annual herb, stellate-hairy throughout. Leaves simple. Flowers minute, in short axillary racemes. Sepals not saccate. Petals white. Filaments unappendaged. Fruits persistent, appressed to the stem, ± glabrous with a long persistent style and an 'ear-like' appendage on each valve, bilocular with 2 seeds per loculus; radicle accumbent.

1. A. hierochuntica L., Sp. pl.: 641 (1753). Illustr.: Fl. Iraq 4 (2): 934 (1980); Mandaville (1990 pls. 84 & 85); Cornes (1989 pl.39); Collenette (1985 p.192).

Stems 5–12cm, branching from the base, with age the branches hardening and becoming incurved to form a ball. Leaves obovate, up to 40 × 12mm including the petiole, the margin entire or toothed, attenuate into a petiole below. Fruit (excluding appendages) 3–4mm diam.; style persistent, 2–4mm. **Map 571, Fig. 79.**

Rock and sand deserts; 0–2300m.

Saudi Arabia, Yemen (N & S), Oman, UAE, Qatar, Bahrain, Kuwait. In deserts from N Africa east to Pakistan.

With age the branches of *A. hierochuntica* harden and become incurved forming a ball enclosing the fruit. On moistening the ball opens, exposing the fruits which dehisce with wetting.

32. SCHIMPERA Hochst. & Steud.

Annual herbs, bearing simple hairs. Leaves dentate or 1–2-pinnatifid. Sepals not saccate. Petals yellow. Filaments unappendaged. Fruit an indehiscent silicula, ovoid with a flat, oblique beak, irregularly tuberculate, 1-seeded; radicle incumbent.

1. S. arabica Hochst. & Steud., Pl. Arab. exsicc. no. 144 (1836). Syn.: *S. persica* Boiss., Diagn. pl. orient. sér. 1, 1 (6): 18 (1845). Illustr.: Fl. Kuwait 1: pls. 101–103 (1985); Collenette (1985 p.206); Mandaville (1990 pls. 86–7).

Stems 3–33(–40)cm, much-branched, sparsely pubescent. Basal leaves oblong to narrowly ovate, 5–25 × 0.5–5cm, runcinate-dentate to 1–2-pinnatifid or rarely ± simple, subsessile; stem-leaves linear-oblong, entire to dentate, amplexicaul. Petals c.1mm. Fruiting pedicel erect, c.2mm. Silicula 5–7(–10)mm (including beak); beak 3–5(–7)mm. **Map 572, Fig. 79.**

On sand and gravel in deserts; ?salt marshes; 10–850m.

Saudi Arabia, UAE, Qatar, Kuwait. Egypt, Palestine, Jordan, Syria, Iraq and S Iran.

33. NESLIA Desv.

Annual herb, bearing simple and branched hairs. Leaves simple, amplexicaul. Inner sepals not saccate. Petals yellow. Filaments unappendaged. Fruit an indehiscent silicula, obovoid-spherical, tipped by the apiculate style, wrinkled, 1-seeded; radicle incumbent.

1. N. apiculata Fischer, C. Meyer & Avé-Lall., Index sem. hort. petrop. 8: 68 (1842). Illustr.: Collenette (1985 p.204).

Annual herb; stems up to 30(–70)cm, thinly tomentose. Leaves oblong to narrowly ovate, 5–20(–60) × 1–4(–20)mm, acute, entire to toothed, the base amplexicaul and sagittate. Petals 2–3.5mm. Silicula 2–2.5mm diam., the persistent style c.1mm. **Map 573, Fig. 79.**

Weed of cultivation; 0–2150m.

Saudi Arabia, Kuwait. S Europe, N Africa and SW & C Asia.

In Saudi Arabia only known from a single gathering but reported to be a common weed in the Asir mountains.

34. FARSETIA L.

Annual or perennial herbs or subshrubs, densely covered with parallel medifixed hairs. Leaves entire, linear to broadly obovate. Sepals not saccate. Petals clawed or not. Stamens 6, filaments free, unwinged. Style distinct; stigma decurrent to globose. Ovary 5–many ovulate. Fruit a dehiscent siliqua, linear-oblong to broadly oblong, the septum with characteristic patterning of fibres which are either dense and more or less opaque or thinner and transparent. Seeds uniseriate, broadly winged, not or faintly mucilaginous; radicle accumbent.

Jonsell, B. (1986). A monograph of Farsetia. *Symb. Bot. Upsal.* 25 (3): 1–107.

1. Fruit broadly-oblong, (5–)6–9mm broad; sepals 10–12mm long
   1. F. aegyptia
+ Fruit narrowly oblong to narrowly ovate, up to 5mm broad; sepals 2–10.5mm long
   2
2. Leaves ovate or oblong-elliptic to broadly obovate, more than 4mm broad; plants generally very leafy
   3
+ Leaves linear to narrowly obovate, up to 3mm broad; plants sparsely leafy 5
3. Fruits up to 17mm long, ovate-oblong with an acute tip and rounded base
   8. F. burtoniae
+ Fruits usually more than 15mm long, narrowly oblong with a rounded tip and base
   4
4. Subshrub; leaves subsucculent; fruit more than 4mm broad **2. F. socotrana**

| + | Annual or perennial herb sometimes woody-based; leaves not subsucculent; fruit up to 3mm broad | **3. F. latifolia** |

| 5. | Fruit septum transparent | 6 |
| + | Fruit septum opaque | 8 |

| 6. | Sepals 3–4.5(–5)mm long; delicate annual or woody-based perennial | **4. F. stylosa** |
| + | Sepals 5–10mm long; woody-based herb or subshrub, if sepals 5mm or shorter then a subshrub | 7 |

| 7. | Grey-green woody-based herb; fruiting-style usually more than 1mm | **6. F. longisiliqua** |
| + | Silvery grey subshrub; fruiting-style usually less than 1mm | **7. F. heliophila** |

| 8. | Fruit less than 17mm long, ovate-oblong, valves very firm | 9 |
| + | Fruit more than 20mm long, narrowly oblong, valves thin | 10 |

| 9. | Leaves elliptic, never succulent; inflorescence elongating in fruit; sepals more than 4mm long | **8. F. burtoniae** |
| + | Leaves linear, subsucculent; inflorescence not or little elongating in fruit; sepals less than 4mm long | **9. F. dhofarica** |

| 10. | Fruiting style usually more than 1mm long; stigma globose | **5. F. linearis** |
| + | Fruiting style usually less than 1mm long; stigma decurrent | **7. F. heliophila** |

1. F. aegyptia Turra, Farsetia: 5 (1765). Syn.: *F. ovalis* Boiss., Diagn. pl. orient. sér. 1, 1 (8): 32 (1849). Syn.: *Lunaria scabra* Forsskal (1775 p.117). Illustr.: Mandaville (1990 pl.88); Fl. Kuwait 1: pls. 104–5 (1985); Collenette (1985 p.199).

Subshrub or woody-based perennial, 30–100cm, much-branched throughout. Leaves linear to linear-elliptic, 10–60 × 1–3mm, acute to obtuse, attenuate below. Racemes 5–10-flowered; pedicels erect, 3–9mm. Sepals 10–12mm. Petals white to violet-grey or greenish brown, clawed, 17–20mm; blade narrowly rectangular, truncate or emarginate, entire. Stamens 9–13mm. Siliqua broadly oblong, 8–25 × (5–)6–9mm; style 1–3mm, stigma decurrent; valves firm; septum translucent. Seeds 1.5–3mm (excluding wings) across, the wing 1.5–2.5mm broad. **Map 574, Fig. 80.**

Shallow sand on rocky slopes and gullies; 15–1950m.

Saudi Arabia, Oman, UAE, Kuwait. N Africa, Palestine, Jordan, Syria, Iraq, Afghanistan, Pakistan.

All Arabian material is referable to subsp. *aegyptia*.

2. F. socotrana B.L. Burtt in Kew Bull. 3: 162 (1948). Syn.: *F. prostrata* Balf.f. (1882) non *F. prostrata* (Steud.) Hochst. Type: Socotra, *Balfour, Cockburn & Scott* 205 (K).

Subshrub; stems up to 30cm, branched mainly from the base. Leaves subsucculent, broadly obovate, 7.5–35 × 3.5–20mm, obtuse or truncate, attenuate below into a winged petiole. Racemes 3–8-flowered, subumbellate in fruit. Pedicels (in fruit) 2–4mm. Sepals linear, 5.5–6.5mm, acute. Petals light violet, 1.5 × as long as sepals.

Fig. 80. **Cruciferae**. A, *Farsetia aegyptia*: Aa, fruit (× 5); Ab, stigma. B, *F. longisiliqua*: Ba, fruit (× 5); Bb, fruit septum (× 10); Bc, stigma. C, *F. latifolia*: Ca, fruit (× 5); Cb, stigma; Cc, fruit septum (× 10); Cd, fruiting shoot (× 1). D, *F. burtonae*: Da, fruit (× 5); Db, stigma; Dc, fruiting branch (× 3). E, *F. stylosa*: Ea, fruit (× 5); Eb, stigma. F, *F. socotrana*: Fa, fruit (× 5); Fb, leaf (× 2). G, *F. heliophila*: Ga, fruiting branch (× 1). H, *F. dhofarica*: Ha, part of fruiting branch (× 2); Hb, fruit (× 5); Hc, stigma. I, *F. linearis*: Ia, fruit (× 5); Ib, stigma.

Siliqua narrowly oblong, (10–)15–28 × 4–6mm; style c.0.5mm; stigma decurrent; valves firm; septum opaque. Seeds (excluding wings) c.1.5mm across, the wing 0.8–1.1mm broad. **Map 575, Fig. 80.**

Sandy and rocky places by the sea and dry hills; 0–500m.

Socotra. Endemic.

3. F. latifolia Jonsell & A.G. Miller in Symb. Bot. Upsal. 25 (3): 57 (1986). Type: Oman, *Miller* 2404 (E, UPS).

Annual to perennial herb, sometimes woody-based; stems up to 50cm, branched mainly from the base. Leaves ovate to oblong-elliptic or broadly obovate, 10–45 × 4–20mm, obtuse, the base cuneate into a winged petiole. Racemes 5–15-flowered, elongating up to 20cm in fruit. Pedicels in flower 1–2mm elongating to 3–7mm in fruit. Sepals 4.5–6.5mm. Petals pale pink to pinkish purple, narrowly to broadly spathulate, clawed, 8.5–11mm. Stamens 5–6mm. Siliqua narrowly oblong, 20–48 × 1.5–2.7mm; style 0.5–2.1mm; stigma decurrent; valves thin; septum opaque. Seeds (excluding wings) 0.8–1mm across, the wing 0.2–0.5mm broad. **Map 576, Fig. 80.**

Open rocky or sandy places, *Euphorbia balsamifera* shrubland and *Boswellia sacra* open bushland; 0–1200m.

?Saudi Arabia, Yemen (S), Oman. Endemic.

Lavranos & Collenette 18308 from Saudi Arabia and *Smith & Lavranos* 842 from Yemen may also be referable to this species. They differ in having non-clawed petals, longer stamens and ovaries with large subglobose stigmas.

4. F. stylosa R. Br. in Denham & Clapp., Narr. travels Africa, app.: 216 (1826). Syn.: *F. depressa* Kotschy in Sitzungsber. Kaiserl. Akad. Wiss., Math.-Naturwiss. Cl., Abt. 1, 52: 261 (1866); *F. hamiltonii* Royle, Ill. bot. Himal. Mts., 71 (1834); *F. prostrata* (Steud.) Hochst. in Flora 31: 176 (1848); *F. ramosissima* Fourn. in Bull. Soc. Bot. France 11: 57 (1864); *Matthiola prostrata* Hochst. & Steud. ex Steud., Nomencl. bot. 2: 106 (1841). Illustr.: Collenette (1985 p.200) as *F. ramosissima*.

Annual or perennial herb, sometimes woody-based, 20–60(–90) cm, branched from the base with ascending somewhat twiggy shoots. Leaves linear to linear-elliptic, 10–85 × 0.5–7mm, obtuse, attenuate below, ± sessile. Racemes 10–15-flowered, elongating to 40cm in fruit. Pedicels ascending, in flower 1.2–2.5mm elongating to 2.5–7mm in fruit. Sepals 3.5–4.5(–5.5)mm. Petals white to mauve or yellowish to brownish orange, narrowly oblong or subspathulate, 3.8–6.7mm. Stamens 3–6mm. Siliqua narrowly oblong, 7–32 × 2.5–4mm; style 1.2–3.5mm; stigma capitate; valves thin; septum translucent. Seeds (excluding wings) 0.8–1.8mm across, the wing 0.5–1.7mm broad. **Map 577, Fig. 80.**

Gravelly and sandy deserts, dry rocky slopes and wadis; 30–2350m.

Saudi Arabia, Yemen (N & S), Socotra, Oman, UAE. Tropical N & NE Africa; Pakistan and India.

A widespread and variable species as regards petal colour and siliqua shape.

5. F. linearis Decne ex Boiss. in Ann. Sci. Nat. Bot. sér. 2, 17: 150 (1842); Syn.: *F. hamiltonii* sensu Schwartz (1939) non Royle; *Cheiranthus linearis* Forsskal (1775 p.120). Type: Oman, *Aucher* 4069 (P, BM, FI, K).

Perennial herb or delicate subshrub; stems ascending, 15–75cm, branched from the base. Leaves linear to linear-spathulate, 5–20 × 0.3–1.5(–2.5)mm, acute, attenuate below. Racemes 10–15-flowered, elongating up to 20cm in fruit. Pedicels ascending, in flower 1.5–5mm. Sepals 4–7mm. Petals white, yellow-green or pink, fading brownish, narrowly spathulate with a rounded tip, 6.4–10mm. Stamens 2.6–7mm. Siliqua narrowly oblong, (15–)20–40 × 1.5–2.5mm; style 1–3.5mm; stigma semiglobose; valves thin; septum opaque. Seeds (excluding wings) 1–1.7mm across, the wing 0.2–1mm broad. **Map 578, Fig. 80.**

Deserts, sandy and stony places, rocky slopes, *Acacia-Commiphora* bushland; 0–2200m.

Saudi Arabia, Yemen (N & S), Oman, UAE. Endemic.

6. F. longisiliqua Decne in Ann. Sci. Nat. Bot., sér. 2, 4: 69 (1835). Syn.: *F. stylosa* (Steud.) T. Anderson (1860) non R.Br. (1826). Illustr.: Collenette (1985 p.200). Type: Yemen (N), *Bové* s.n. (P, K).

Woody-based perennial; stems 30–100cm, branched throughout. Leaves linear, 4–40(–70) × 0.5–5mm, acute, attenuate below. Racemes 10–15-flowered, elongating to 30cm in fruit. Pedicels ascending or spreading, 1.5–3mm elongating to 3–10mm in fruit. Sepals 6–10mm. Petals pink or pale yellow fading to purple, slightly clawed and with a rounded tip, 13–19mm. Stamens 5.4–10mm. Siliqua narrowly oblong, 20–50 × 3–5mm; style 0.5–3mm; stigma decurrent; valves rather firm; septum translucent. Seeds (excluding wings) 1–2mm across, the wing 0.5–1.5mm broad. **Map 579, Fig. 80.**

Dry rocky slopes, gravelly and sandy deserts; 0–2700m.

Saudi Arabia, Yemen (N & S), Socotra, Oman, UAE. Borders of the Red Sea from Egypt to Somalia.

7. F. heliophila Bunge ex Cosson, Ill. fl. atlant. 2: 227 (1884). Syn.: *F. arabica* Boulos in Webbia 32: 379 (1978). Illustr.: Cornes (1989 pl.40).

Subshrub up to 50cm, stiffly branched throughout. Leaves linear to narrowly elliptic or narrowly obovate, (7–)15–20 × 1–3mm, acute to rounded, attenuate below. Racemes 5–10-flowered, elongating to 10cm in fruit. Pedicels erect, 2–3.5mm, elongating to 6mm in fruit. Sepals 5–9.5mm. Petals pale or dirty yellow to mauve or maroon, fading to dark brownish, slightly clawed and with an obtuse tip, 9–14mm. Stamens 6–12mm. Siliqua narrowly oblong, 15–60 × 2–3mm; style 0.1–1mm, stigma decurrent; valves firm; septum opaque to translucent. Seeds (excluding wings) 1.2–2mm across, the wing 0.3–1mm broad. **Map 580, Fig. 80.**

Rocky slopes in deserts; 0–50m.

Saudi Arabia, Oman, UAE, Qatar, Bahrain. Iran and Pakistan.

8. F. burtoniae Oliver in Hooker's Icon. Pl. 14: t.1310 (1880). Illustr.: Fl. Iraq 4 (2): 947 (1980). Illustr.: Mandaville (1990 pls. 89–90); Collenette (1985 p.200). Type: Saudi Arabia, *Burton* s.n. (K).

Ascending annual to perennial herb or subshrub up to 25cm. Leaves narrowly elliptic to narrowly obovate, 10–40 × 1.5–8mm, acute, attenuate below. Racemes 10–20-flowered, condensed, 1–2cm long, sometimes elongating to 20cm in fruit. Pedicels stout, patent or ascending, c.1mm elongating to 1.5–3mm in fruit. Sepals 4–7mm. Petals white to pink fading to mauve, obtriangular with a truncate tip, 7.5–8mm. Stamens 3.7–4.5mm. Siliqua narrowly oblong to narrowly ovate with an acute tip and rounded base, 7–17 × 2–4mm; style 1.8–4.5mm; stigma decurrent; valves firm; septum opaque. Seeds (excluding wings) 1–2mm across, wing 0.2–0.5mm broad. **Map 581, Fig. 80.**

Gravelly and sandy deserts; 50–1700m.

Saudi Arabia, Kuwait. Iraq.

Two specimens (*Zeller* 23604 and *Thesiger* s.n. 2/1/1949) from southern Saudi Arabia have rather narrow leaves and are somewhat intermediate between this species and *F. dhofarica*.

9. F. dhofarica Jonsell & A.G. Miller in Symb. Bot. Upsal. 25(3): 98 (1986). Type: Oman, *Popov* 68/53 (BM).

Herb or subshrub up to 30cm, with many spreading shoots from the base. Leaves somewhat fleshy, linear to linear-elliptic, 10–25 × 1–3mm, acuminate, attenuate below. Racemes 10–15-flowered, very condensed, not or little elongating in fruit, subumbellate. Pedicels stout, patent, 4–5.5mm in fruit. Sepals 2.5–3.8mm. Petals white, strap-shaped, rounded at the tip, 4–4.5mm. Stamens 2.7–3.5mm. Siliqua narrowly ovate to oblong with an acute tip and rounded base, 3.5–14 × 2–3.4mm; style 1–2.1mm; stigma decurrent; valves firm; septum opaque. Seeds (excluding wings) 1–1.7mm across, the wing 0.2–0.3mm broad. **Map 582, Fig. 80.**

Sandy and rocky deserts and semi-deserts; 0–1050m.

Yemen (S), Oman. Endemic.

Several specimens (*Guichard* 20a, *Miller* 12331, *Thulin et al.* 8778) from Yemen (S) have broader leaves and somewhat elongating inflorescences and are intermediate between this species and *F. burtoniae*.

35. ALYSSUM L.

Annual or perennial herbs, stellate-hairy throughout (except sometimes on the fruit). Leaves entire. Flowers in short racemes which elongate in fruit. Sepals not saccate. Petals yellow or creamy white, entire or retuse, oblong to spathulate. Filaments simple or winged or toothed below. Stigma capitate. Fruit a latiseptate silicula, dehiscent (in Arabia), the valves compressed or inflated, bilocular with 1–8 seeds per loculus. Seeds often mucilaginous; radicle accumbent.

CRUCIFERAE

1.	Fruit stellate-hairy	2
+	Fruit glabrous	5
2.	Annual; fruits orbicular	3
+	Perennial; fruits elliptic	4
3.	Filaments toothed; fruiting style less than 0.5mm	**4. A. marginatum**
+	Filaments entire; fruiting style c.1mm	**5. A. damascenum**
4.	Shrublet, the branches becoming spinescent; fruits 2–2.25 × 1.25–2mm	**6. A. subspinosum**
+.	Woody-based herb, not spinescent; fruits 3–6 × 1.5–3mm	**7. A. singarense**
5.	Silicula broadly elliptic to obovate; seeds 4–6(–8) per loculus	**1. A. linifolium**
+	Silicula orbicular to broadly obovate; seeds (1–)2 per loculus	6
6.	Silicula broadest at the middle, 3–4mm in diam., entire on the upper margins; short filaments toothed	**2. A. desertorum**
+	Silicula broadest above the middle, 3.5–6mm in diam., minutely denticulate on the upper margins; all filaments without teeth	**3. A. homalocarpum**

1. A. linifolium Stephan ex Willd., Sp. pl. 3: 467 (1800). Illustr.: Fl. Palaest. 1: t.419 (1966); Collenette (1985 p.192).

Annual herb. Stems erect or ascending, up to 10(–20)cm, branched from the base and becoming bushy. Leaves linear to linear-oblong, 5–15 × 1–3mm. Petals creamy white or pale yellow, 0.8–1.5(–2)mm, scarcely exceeding sepals. Filaments all distinctly toothed. Silicula broadly elliptic to obovate, rounded, (3–)4–5(–7) × 2.5–3mm, glabrous; seeds 4–6(–8) per loculus. **Map 583, Fig. 81.**

Rocky outcrops and sandy soil on limestone plains; 20–900m.

Saudi Arabia, Kuwait. Mediterranean region, SW & C Asia.

2. A. desertorum Stapf in Denkschr. Kaiserl. Akad. Wiss., Math.-Naturwiss. Kl. 51: 302 (1886). Illustr.: Collenette (1985 p.192).

Annual herb. Stems erect or ascending, up to 10(–15)cm, branched at the base. Leaves linear to linear-obovate, 5–12(–25) × 0.5–3mm. Petals yellow, 2–2.5mm, clearly exceeding the sepals. Short filaments with 2 teeth; long filaments expanded but not toothed. Silicula orbicular, entire, 3–4mm diam., the valves equally inflated at the centre, glabrous; seeds (1–)2 per loculus. **Map 584, Fig. 81.**

Rocky slopes and wadis; 1950–2150m.

Saudi Arabia. SE Europe, SW & C Asia.

3. A. homalocarpum (Fischer & C. Meyer) Boiss., Fl. orient. 1: 285 (1867). Illustr.: Fl. Palaest. 1: t.422 (1966); Mandaville (1990 pl.91); Collenette (1985 p.192).

Similar to *A. desertorum* but filaments all untoothed; siliculae orbicular to broadly

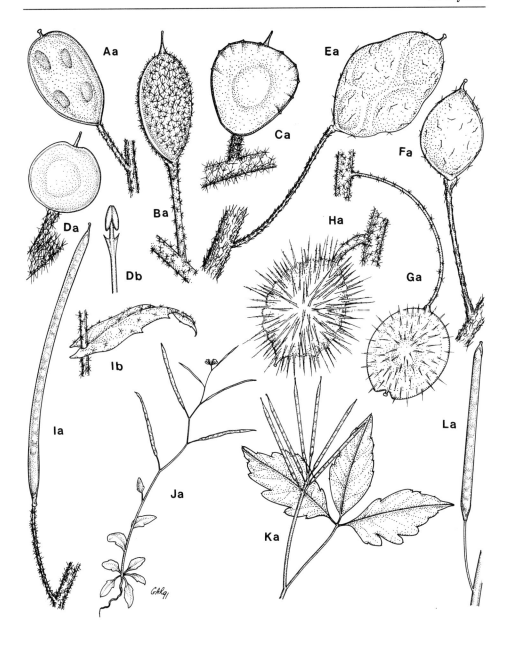

Fig. 81. Cruciferae. A, *Alyssum linifolium*: Aa, fruit (×6). B, *A. singarense*: Ba, fruit (×6). C, *A. homalocarpum*: Ca, fruit (×6). D, *A. desertorum*: Da, fruit (×6); Db, stamen (×30). E, *Lobularia libyca*: Ea, fruit (×7). F, *L. maritima*: Fa, fruit (×7). G, *Clypeola jonthlaspi*: Ga, fruit (×10). H, *C. aspera*: Ha, fruit (×10). I, *Arabis alpina*: Ia, fruit (×2); Ib, leaf. J, *A. nova*: Ja, habit (×1). K, *Cardamine africana*: Ka, fruits and leaf (×0.6). L, *C. hirsuta*: La, fruit (×4).

obovate, usually broadest above the middle, 3.5–6mm diam., minutely denticulate on the upper margin. **Map 585, Fig. 81.**

Rocky and stony slopes and limestone plains; a weed of wheat fields; *Juniperus* woodland; 100–1850m.

Saudi Arabia, Kuwait. Egypt, Palestine, Syria, Jordan, Iraq, Iran and Pakistan.

4. A. marginatum Steud. ex Boiss. in Ann. Sci. Nat. Bot. sér. 2, 17: 157 (1842). Illustr.: Fl. Iraq 4 (2): 968 (1980).

Annual herb. Stems ascending, up to 6(–10)cm, branched from the base. Leaves oblanceolate, 5–25 × 2–5mm. Petals yellow, retuse, subequalling sepals. Short filaments with 2 teeth; long filaments slightly dilated at the base. Fruiting inflorescence short, cylindrical. Silicula orbicular, 3–4mm diam., biconvex, stellate-hairy; style up to 0.5mm; seeds 2 per loculus. **Map 586.**

Steep rocky slopes and clay-pans; 800–1250m.

Saudi Arabia. Egypt, SW & C Asia.

5. A. damascenum Boiss. & Gaill. in Boiss., Diagn. pl. orient. sér. 2, 3(6): 18 (1859). Illustr.: Fl. Palaest. 1: t.420 (1966).

Annual herb. Stems erect or ascending, up to 12cm, branched from the base. Leaves lanceolate, 10–20 × 2–4mm. Petals pale yellow. Filaments toothless. Silicula orbicular, 3–5mm diam., biconvex, stellate-hairy; style c.1mm; seeds 1–2 per loculus.

Saudi Arabia. Palestine and Syria

Recorded from the Al-Harrah protected area in the NW of Saudi Arabia. However, all the material examined from that area has proved to be *A. marginatum* and it is very doubtful whether this species occurs in Arabia.

6. A. subspinosum T. Dudley in Notes Roy. Bot. Gard. Edinburgh 24: 160 (1962).

Shrublet up to 35cm, the branchlets becoming spiny. Leaves linear-oblanceolate, 5–15 × 1–3.5mm. Petals yellow, 1.5–2mm. Long-filaments winged. Silicula elliptic, 2–2.25 × 1.25–2mm, compressed, stellate-hairy; seeds 1 per loculus. **Map 587.**

Among shrubs under boulders; 1770m.

Saudi Arabia. Palestine and Jordan.

7. A. singarense Boiss. & Hausskn. in Boiss., Fl. orient. Suppl.: 49 (1888). Syn.: *A. anamense* Velen. in Sitzungsber. Königl. Böhm. Ges. Wiss. Prag, Math.-Naturwiss. Cl. 11: 12 (1911). Type: ?Saudi Arabia, *Haussknecht* s.n. (BM).

Perennial herb, often woody at the base. Stems erect or ascending, up to 15(–30)cm. Leaves of the basal sterile branches obovate to elliptic, those of the flowering stem linear to linear-obovate, 5–20 × 1.5–4mm. Petals yellow, 2.5–4mm. Long filaments with a broad 2–4-toothed wing; short filaments with an entire tooth free to the base.

Silicula elliptic, 3–6 × 1.5–3mm, compressed, stellate-hairy; seeds 1 per loculus. **Fig. 81.**

?Saudi Arabia. Iraq.

Doubtfully occurring in Arabia, known from a single gathering from 'Abar Ikuk Anama', a locality which may be in Iraq.

35a. EROPHILA DC.

Ephemeral herbs bearing simple, branched or stellate hairs. Leaves confined to a basal rosette, simple, entire or toothed. Petals white, deeply bifid. Sepals not saccate. Filaments free, without appendages. Fruit a latiseptate silicula, the valves compressed or inflated. Seeds in 2 rows; radicle accumbent.

1. E. verna (L.) Bess., Enum. Pl.: 71 (1822).

Stems solitary or several, to 5cm. Leaves ovate to oblanceolate. Petals c.2mm. Fruit oblong-elliptic to obovate, 3–5 × 1.5–3, glabrous. **Map 587.**

In shady places on terrace walls; 500–1000m.

Oman. Europe, Asia and N Africa, introduced into Australia and temperate America.

Only recently recorded for the first time in Arabia from the Musandam Peninsula. A very variable species from a taxonomically difficult genus.

36. LOBULARIA Desv.

Annual or perennial herbs, bearing appressed, medifixed hairs. Leaves simple, entire. Sepals not saccate. Petals white. Filaments unappendaged but dilated below. Stigma capitate. Fruit a dehiscent, compressed silicula with a broad septum. Radicle accumbent.

1.	Silicula 2–3.5mm long with 1 seed per loculus	2
+	Silicula 4–6mm long, with 4–6 seeds per loculus	**2. L. libyca**
2.	Fruiting style up to 1mm long	**1. L. maritima**
+	Fruiting style 2–2.5mm	**3. L. sp. A**

1. L. maritima (L.) Desv. in J. Bot. Agric. 3: 162 (1814).

Canescent annual or perennial herb; stems ascending, up to 15(–55)mm, much branched from the base. Leaves linear to linear-obovate, 5–20(–35) × 1–1.5(–4)mm. Petals 2–3.5mm. Silicula broadly ovate to broadly elliptic, c.2 × 1.5mm, thinly hairy, with 1 seed per loculus, the fruiting style up to 1mm. **Map 588, Fig. 81.**

No habitat notes recorded but probably cultivated as an ornamental.

Saudi Arabia. Mediterranean coasts, widely cultivated as an ornamental.

2. L. libyca (Viv.) Webb & Berth., Hist. nat. Iles Canaries 1: 90 (1837). Illustr.: Collenette (1985 p.201).

Canescent annual or perennial herb; stems prostrate or ascending, up to 20(–40)cm, branched from the base. Leaves linear-obovate, 10–30 × 1–4mm. Petals 2–2.5mm. Silicula broadly elliptic, 4(–6) × c.3mm, thinly hairy, with 4–6 seeds per loculus. **Map 589, Fig. 81.**

In damp sand; 1980m.

Saudi Arabia. N Africa, Mediterranean region and S Iran.

3. L. sp. A.

Similar to *L. maritima* but the leaves linear and the fruiting style 2–2.5mm. **Map 588.**

Fossiliferous rocky area; c.900m.

Saudi Arabia.

Known only from the region of Al Arid in the western Rub' al-Khali. The single specimen (*Chaudhary & Al-Juwayed* RIY 1390) is insufficient for a formal description and further gatherings of this interesting plant are desirable.

37. CLYPEOLA L.

Erect or ascending annual herbs, bearing appressed stellate hairs. Leaves simple, entire. Petals yellow to whitish. Inner sepals not saccate. Inner filaments toothed. Stigma capitate. Fruit an indehiscent silicula, pendulous, suborbicular, strongly compressed; radicle accumbent.

1. Fruit with a pale margin, glabrous or with an indumentum of short, simple hairs   **1. C. jonthlaspi**
+ Fruit without a pale margin, with a conspicuous indumentum of long minutely barbed, setae-like hairs   **2. C. aspera**

1. C. jonthlaspi L., Sp. pl.: 652 (1753). Illustr.: Collenette (1985 p.196).

Stems simple or branched from the base, up to 8(–15)cm. Leaves narrowly obovate, 3–10(–15) × 1–2(–3)mm, attenuate below. Silicula pendulous, (1.75–)3.5–4(–5) × (1.5–)3–3.5(–5)mm, with a pale margin, emarginate with the style included in the notch, glabrous or with an indumentum of short simple hairs. **Map 590, Fig. 81.**

Shallow gullies, wadi-banks, and stony and sandy ground amongst rocks; 1920m.

Saudi Arabia, Oman. S Europe and N Africa to SW and C Asia.

2. C. aspera (Grauer) Turrill in J. Bot. 60: 269 (1922).

Similar to *C. jonthlaspi* but the silicula conspicuously covered with minutely barbed setae-like hairs, without a pale margin. **Map 591, Fig. 81.**

Rocky and sandy wadi-banks; 1100–1150m.

Saudi Arabia, Oman. SW & C Asia.

Less common than the preceding species in Arabia.

38. CARDAMINE L.

Annual, biennial or perennial herbs, glabrous or with simple hairs. Leaves pinnate or tripartite. Inner sepals slightly saccate. Petals white to pinkish. Filaments free, without appendages. Fruit a linear siliqua, dehiscing by the valves explosively coiling from the base upwards; stigma bilobed. Seeds uniseriate; radicle accumbent.

1.	Leaves tripartite	**1. C. africana**
+	Leaves with 2 or more pairs of lateral leaflets	**2. C. hirsuta**

1. C. africana L., Sp. pl.: 655 (1753).

Perennial herb; stems erect or ascending, up to 80cm, finely hairy below, glabrescent above. Leaves tripartite; petiole 10–15cm; leaflets ± equal, ovate, (1–)6–7 × (0.5–)2.5–4cm, acute to acuminate, unevenly crenate, truncate to cuneate below, shortly petiolulate, sparsely hairy. Petals white or pink. Stamens 6. Siliqua (20–)30(–50) × (1.5–)2(–2.5)mm, glabrous. **Map 592, Fig. 81.**

In shade by water; 1400m.

Yemen (N). Widespread on tropical mountains throughout the world.

Very rare in Arabia, known only from a single gathering from J. Raymah in the SW escarpment mountains.

2. C. hirsuta L., Sp. pl.: 655 (1753). Illustr.: Fl. Iraq 4(2): 998 (1980).

Annual herb; stems erect, up to 10–25cm, branched from the base, glabrous or sparsely hairy. Basal leaves forming a rosette, pinnate with 2–6 pairs of lateral leaflets and a larger terminal leaflet, 1.5–5(–10)cm; leaflets obovate to orbicular, entire, 2–10(–15) × 3–10(–20)mm; stem-leaves few, similar to the basal leaves but smaller and with narrower leaflets, all leaves glabrous or sparsely hairy. Petals white, 2–3mm. Stamens 4–6. Siliqua 12–20(–25) × 0.5–1mm. **Map 593, Fig. 81.**

Alpine meadows and shady cliffs; 1500—2400m.

Saudi Arabia. Widespread as a weed of cultivation throughout the world.

39. ARABIS L.

Annual or perennial herbs, bearing branched hairs. Leaves entire or toothed; cauline leaves amplexicaul or auriculate. Inner sepals saccate. Petals white or cream. Filaments unappendaged. Stigma capitate. Fruit a linear, dehiscent siliqua. Seeds uniseriate, winged or not; radicle accumbent.

CRUCIFERAE

1.	Mat-forming perennial herb; petals 5–15mm long	**1. A. alpina**
+	Erect annual herb; petals up to 2(–4.5)mm long	**2. A. nova**

1. A. alpina L., Sp. pl.: 664 (1753). Syn.: *A. caucasica* Willd., Enum. pl., Suppl.: 45 (1813). Illustr.: Fl. Trop. E. Afr.: 46 (1982); Collenette (1985 p.193).

Loosely mat-forming perennial; stems up to 30cm, much-branched below, hairy throughout although more thinly so above. Leaves oblong to obovate, 10–80 × 4–25mm, acute to obtuse, the margins entire to toothed; basal leaves attenuate into a short petiole; cauline leaves sessile and auriculate. Flowering shoots erect. Petals 5–15mm. Fruits erect-spreading, linear, compressed and somewhat torulose, 22–50 × 1–2mm; seeds wingless or with a narrow wing, 1–1.5mm. **Map 594, Fig. 81.**

On cliffs and rock faces; 2000–3000m.

Saudi Arabia, Yemen (N). Mountains of E Africa, Europe & N America.

A very variable species across its entire range. In Arabia, there seem to be fairly consistent differences in flower size between plants from North Yemen and those from Saudi Arabia, the former having petals 5–9mm long and those of the latter having petals 8–15mm in length.

2. A. nova Villars, Prosp. Hist. pl. Dauphiné: 39 (1779).

Erect annual herb; stems up to 20cm, simple or rarely branched below, thinly hairy throughout or glabrescent above. Leaves ovate to obovate, 5–20 × 3–10mm, rounded, the margins entire or obscurely toothed; basal leaves attenuate into a short petiole; cauline leaves sessile, amplexicaul. Petals 2(–4.5)mm. Siliqua erect-spreading, linear, ± round or flattened in section, 10–20(–35) × c.1mm; seeds wingless or with a narrow wing. **Map 595, Fig. 81.**

Shady rock crevices and sand pans; *Juniperus* woodland; 1350—2300m.

Saudi Arabia. Europe and SW Asia.

A very variable species which can flower and fruit when only a few centimetres tall. Small plants have been mistaken for *Arabidopsis pumila* which, however, can be distinguished by its yellow flowers. Most specimens from Arabia are typically small for the species when examined across its entire range.

40. NASTURTIUM R.Br.

Perennial herbs, glabrous or bearing simple hairs. Leaves pinnate. Inner sepals slightly saccate. Petals white. Filaments unappendaged. Stigma capitate. Fruit a dehiscent siliqua; seeds in 1 or 2 rows, distinctly reticulate; radicle accumbent.

1.	Siliqua 2–2.5mm broad with the seeds in 2 rows; seeds with 25–50 areoles on each face	**1. N. officinale**
+	Siliqua 1.5–2mm broad with the seeds in 1 row; seeds with c.100 areoles on each face	**2. N. microphyllum**

1. N. officinale R.Br. in Aiton f., Hort. Kew. ed. 2, 4: 110 (1812). Syn.: *Rorippa nasturtium-aquaticum* (L.) Hayek, Sched. fl. stiriac. exs. no. 170 (1905). Illustr.: Collenette (1985 p.204).

Prostrate to ascending, glabrous perennial herb; stems up to 1m, juicy and hollow. Leaves pinnate with 2–9 pairs of leaflets, 5–10cm long, petiolate; upper leaves minutely auriculate; leaflets elliptic to ± orbicular, entire to sinuate, the terminal lobe larger. Petals white, 4–6mm. Siliqua 10–18 × 2–2.5mm; seeds in 2 rows, reticulate with 25–50 areoles on each face. **Map 596, Fig. 82.**

Damp and sandy wadis; 1500–2150m.

Saudi Arabia. Originally Eurasian, but now cultivated throughout the world.

2. N. microphyllum Boenn. in Reichenb., Fl. germ. excurs.: 683 (1832). Syn.: *Rorippa microphylla* (Boenn.) Hylander in Rit Landbúnaoard. Atvinnud. Háskólans, B 3:109 (1948).

Similar to *N. officinale* but siliqua 1.5–2mm broad; seeds in 2 rows, reticulate with c.100 areoles on each face. **Map 597, Fig. 82.**

Swampy areas in wadis; 2500m.

Yemen (N). Originally Eurasian, but now cultivated throughout the world.

41. RORIPPA Scop.

Annual or perennial herbs, glabrous or bearing simple hairs. Leaves lyrate-pinnatisect or the upper leaves entire and toothed. Inner sepals saccate. Petals yellow. Filaments unappendaged. Stigma 2-lobed. Fruit a dehiscent silicula or siliqua; seeds in 2 rows, minutely reticulate; radicle accumbent.

1. R. palustris (L.) Besser, Enum. Pl.: 27 (1822).

Erect annual herb up to 50(–100)cm. Lower leaves with 2–6 pairs of ± oblong segments, petiolate; upper leaves minutely auriculate at the base. Petals yellow, 1.25–2.5mm, ± equalling sepals. Fruit ellipsoid-oblong, 5–10 × 1.5–2mm. **Map 598, Fig. 82.**

Gravel and mud by ponds and streams; (0–)2500–2700m.

Yemen (N), UAE. Cosmopolitan.

The record from the UAE refers to a single population growing in a very specialized niche (in damp sand under a dripping air-conditioner) at sea-level on Das Island in the Arabian Gulf.

42. NOTOCERAS R. Br.

Annual herb, bearing appressed medifixed hairs. Leaves simple, entire. Inner sepals not saccate. Petals yellow. Filaments expanded near the base. Stigma capitate. Fruit

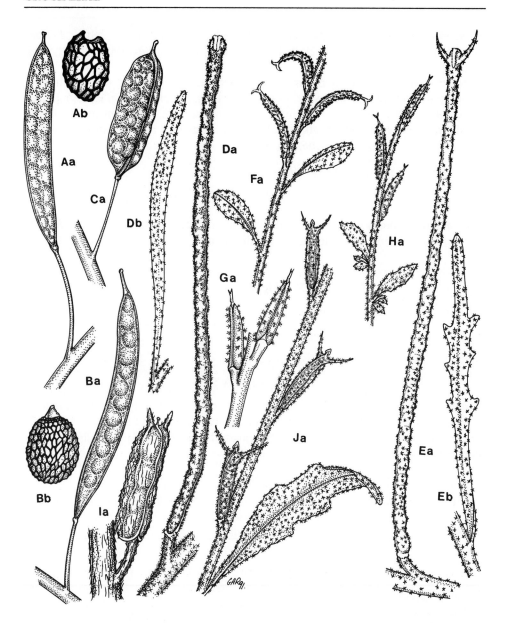

Fig. 82. Cruciferae. A, *Nasturtium officinale*: Aa, fruit (×6); Ab, seed (×30). B, *N. microphyllum*: Ba, fruit (×6); Bb, seed (×30). C, *Rorippa palustris*: Ca, fruit (×6). D, *Matthiola arabica*: Da, fruit (×5); Db, leaf (×1.5). E, *M. longipetala*: Ea, fruit (×5); Eb, leaf (×1.5). F, *Morettia parviflora*: Fa, part of infructescence. G, *M. philaeana*: Ga, fruits (×1.5). H, *M. canescens*: Ha, part of infructescence (×2). I, *Notoceras bicorne*: Ia, fruit (×7). J, *Diceratella incana*: Ja, part of infructescence (×3.5).

a linear, dehiscent siliqua, the valves produced terminally into 2 horns; radicles accumbent.

1. N. bicorne (Aiton f) Amo, Fl. fan. Penins. Iberica 6: 536 (1853). Syn.: *N. canariensis* R.Br. in Aiton f., Hort. Kew. ed. 2, 4: 117 (1812). Illustr.: Mandaville (1990 pl. 80); Collenette (1985 p.205).

Prostrate or decumbent canescent herb; stems up to 20cm, much-branched from the base. Leaves narrowly obovate, 5–50 × 2–6mm, attenuate into a short petiole below. Petals c.1.5mm. Siliqua erect, appressed to the stem, 5–8 × 1.5mm, appressed hairy, the horns 0.5–1mm. **Map 599, Fig. 82.**

Rocky slopes, wadis, abandoned fields and clay pans; 100–2100m.

Saudi Arabia, Yemen (N), Oman, UAE, Kuwait. SW Europe, N Africa and Syria to W Pakistan and Afghanistan.

43. DICERATELLA Boiss.

Subshrubs with a dense covering of branched and stellate hairs. Leaves entire. Inner sepals not saccate. Petals lilac. Filaments unappendaged. Stigma bifid, cone-shaped. Fruit a tardily dehiscent siliqua, with the valves prolonged into 2 horns at the tip; seeds in 1 row with transverse septa separating them; radicles accumbent.

1. D. incana Balf. f. (1882 p.500). Illustr.: Fl. Trop. E. Afr.: 58 (1982). Type: Socotra, *Balfour, Cockburn & Scott* 136 (E).

Densely tomentose subshrub; stems to 60cm. Leaves ovate, 1–3 × 0.5–1.25cm, entire to sinuate-dentate, attenuate below into a short petiole. Petals c.15(–25)mm. Siliqua linear, 10–17(–26) × 1.5–2.5mm, the horns c.1mm. **Map 600, Fig. 82.**

In sand on coastal plains; 10m.

Socotra. Somalia, Ethiopia and Kenya.

44. MATTHIOLA R. Br.

Annual or perennial herbs bearing branched hairs and sometimes stipitate glands. Leaves simple to pinnatifid, entire to dentate. Inner sepals saccate. Petals white, yellow or purple. Inner filaments winged below. Stigma with decurrent carpidial lobes. Fruit a linear siliqua often with 2 horn-like appendages present on either side of the stigma in ripe fruit; seeds uniseriate; radicle accumbent.

Matthiola incana (L.) R.Br. – the Garden Stock – is widely cultivated as an ornamental in gardens. It is an annual or perennial herb with erect, hornless fruits.

1. Leaves entire; fruit without apical horns **1. M. arabica**
+ Leaves dentate to pinnatifid at least below; fruit with apical horns **2. M. longipetala**

1. M. arabica Boiss. in Ann. Sci. Nat. Bot. sér. 2, 17: 49 (1842). Syn.: *M. arabica* Velen. in Sitzungsber. Königl. Böhm. Ges. Wiss. Prag 11: 12 (1911).

Canescent annual; stems erect or ascending, up to 75(–100)cm, densely tomentose throughout. Leaves narrowly oblong, 30–80 × 2–5mm, entire. Petals rosy, lilac or yellowish, linear, c.20mm. Siliqua 4–7cm, spreading, curved or twisted, horns absent, tomentose. **Map 601, Fig. 82.**

Stable sand; 10–1400m.

Saudi Arabia. Sinai, Jordan.

Mandaville (1990) notes that *M. arabica* is often found on sandy soils whilst the following species, *M. longipetala*, prefers silty, gritty soils.

2. M. longipetala (Vent.) DC., Syst. nat. 2: 174 (1821). Syn.: *M. bicornis* (Smith) DC., Syst. veg. 2: 177 (1821); *M. oxyceras* DC., Syst. nat. 2: 173 (1821). Illustr.: Fl. Kuwait 1: pls. 108-111; Collenette (1985 p.203); Mandaville (1990 pls. 92 & 93).

Canescent annual; stems erect or ascending, up to 60cm, densely tomentose, often with scattered glandular hairs present. Leaves linear-oblong, 3–8(–10) × 0.3–2cm, entire to dentate or pinnatifid, attenuate below. Petals variable in colour, yellow, greenish-yellow, pink or purple, linear to narrowly obovate, 15–20mm. Siliqua (2–)4–6(–10)cm, ascending to spreading, straight, rarely curved, the horns 1–5(–10)mm, tomentose with scattered glandular hairs. **Map 602, Fig. 82.**

Deserts and plains, on limestone, sandy and gravelly soils, amongst rocks and in wadi-beds; 15–1700m.

Saudi Arabia, Kuwait. SE Europe, Libya, Syria, Lebanon, Jordan and Palestine to Iran.

Species doubtfully recorded

The following species are recorded by Hedge and King (op. cit.) but no specimens have been examined and the records remain doubtful. *M. livida* (Del.) DC. (recorded from Kuwait) is similar to *M. longipetala* but differs in the contorted siliquas which bear short (up to 2mm) stigmatic horns and *M. humilis* DC. (recorded from Saudi Arabia) which is often considered to be a variety of *M. longipetala*.

45. MORETTIA DC.

Annual or perennial herbs, densely covered with stellate hairs. Leaves simple, entire or dentate. Sepals not saccate. Petals white, pink or yellow. Filaments unappendaged. Stigma bilobed, the lobes decurrent. Fruit a siliqua, tipped by the persistent style, valves transversely septate within; radicle accumbent.

Stork, A.L. & Wüest, J. (1981). *Bol. Soc. Brot.* 53: 241–273.

1. Coarsely scabrid, yellow-green herb with large stellate hairs which adhere to the finger if touched; siliqua c.2–3.5mm across at the base; sepals c.6mm
   3. M. philaeana

+   Softly tomentose to scabrid, canescent herb with hairs not adhering to the finger if touched; siliqua 1–2mm across at the base; sepals 3–4.5mm   2

2. Stems ascending; leaves distinctly petiolate, ± entire   **1. M. parviflora**
+   Stems prostrate to decumbent; leaves sessile, with 2–4 pairs of teeth
    2. M. canescens

1. M. parviflora Boiss. in Ann. Sci. Nat. Bot. sér. 2, 17:60 (1842). Illustr.: Collenette (1985 p.204).

Canescent herb; stems ascending, up to 30(–40)cm; densely tomentose throughout. Leaves ovate to elliptic, 1–4 × 0.5–1.5cm, entire or with 1–2 small teeth, attenuate below into a short petiole. Sepals 2.5–3mm. Petals white, 4–5mm. Siliqua 10–22 × 1–1.5mm, usually curved away from the stem. **Map 603, Fig. 82.**

On wadi-sides, rocky hills and sand dunes; amongst boulders, scree and in watercourses and irrigated fields; sandy and gravelly soils; 20–2700m.

Saudi Arabia, Yemen (N & S), Oman, UAE, Bahrain. Egypt, Sinai, Palestine and Jordan.

Recorded from Bahrain by Hedge & King (op. cit.: 47) but no specimens have been seen.

2. M. canescens Boiss., Diagn. pl. orient. sér. 1, 2(8): 17 (1849). Illustr.: Collenette (1985 p. 203).

Similar to *M. parviflora* but stems prostrate to decumbent; leaves sessile, with 2–4 pairs of teeth; sepals 4–6mm; petals white or pink; siliqua (5–)8–15(–20) × 1–2mm, erect and appressed to the stem or slightly curved. **Map 604, Fig. 82.**

Gravel and sand; 900–1700m.

Saudi Arabia. N Africa, Palestine and Sinai.

3. M. philaeana (Del.) DC., Syst. nat. 2: 427 (1821). Syn.: *M. asperrima* Boiss. in Ann. Sci. Nat. Bot. sér. 2, 17: 60 (1842).

Similar to *M. parviflora* but very coarsely scabrid, the hairs adhering to the finger if touched; yellowish-green; sepals c.7mm; siliqua (7–)10–15(–20) × 2–4mm.

Stony plains and gravelly wadi-beds in open *Acacia* bushland; 10–700m. **Map 605, Fig. 82.**

Saudi Arabia, Oman. N & NE Africa south to Somalia.

Recorded by Schwartz (1939) from Saudi Arabia; however, no recent specimens have been seen and the record is very doubtful.

46. CHORISPORA R.Br. ex DC.

Annual or perennial herbs bearing simple and glandular hairs. Leaves pinnatisect. Inner sepals saccate. Petals purple. Filaments not winged. Stigma with decurrent carpidial lobes. Fruit a cylindrical indehiscent siliqua, torulose; seeds biseriate. Radicle accumbent.

1. C. purpurascens (Banks & Sol.) Eig in J. Bot. 75: 189 (1937). Syn.: *C. syriaca* Boiss. in Ann. Sci. Nat. sér. 2, 17: 384 (1842). Illustr.: Fl. Iraq 4 (2): 1026 (1980).

Erect or ascending annual herb up to 30cm. Leaves oblong to lanceolate-elliptic, up to 7cm long, pinnatisect with entire or toothed lobes, long-attenuate below. Calyx 7–10mm. Petals 15–17mm. Fruit ascending, tapering to a seedless beak, 35–60 × 2–5mm. **Map 606.**

Clay pans; 900m.

Saudi Arabia, ?Kuwait. Lebanon, Turkey, Iraq, Jordan and Iran.

Also recorded by Blatter (1912—1939) from the Zor Hills in Kuwait, but no recent material has been seen.

47. LEPTALEUM DC.

Annual herbs, glabrous or bearing simple and branched hairs. Leaves entire, filiform or pinnately divided into filiform segments. Inner sepals not saccate. Petals white or pink. Stamens 6 or reduced to 2, the longer filaments fused in pairs. Lobes of stigma fused into a minute cone. Fruit a siliqua, the valves opening only at the apex; seeds biseriate; radicle incumbent.

1. L. filifolium (Willd.) DC., Syst. nat. 2: 511 (1821). Illustr.: Fl. Pakistan 55: 223 (1973); Fl. Kuwait 1: pl.111 (1985).

Prostrate or ascending annual herb; stems up to 10(–18)cm, much branched from the base, glabrous or thinly hairy. Leaves up to 6cm, the segments filiform. Flowers axillary, solitary or in 2–4-flowered racemes. Petals linear, 6–10mm. Siliquae linear, compressed, 12–22(–30) × 2–3mm, often crowded at the base. **Map 607, Fig. 83.**

Clay pans, sandy and silty soils; 150–950m.

Saudi Arabia, Kuwait. Egypt to Pakistan and Central Asia.

48. EIGIA Soják
Stigmatella Eig.

Dwarf annual herb, bearing branched hairs throughout. Leaves mainly in a basal rosette, simple, ± entire. Inner sepals saccate. Petals lilac. Filaments dilated at the

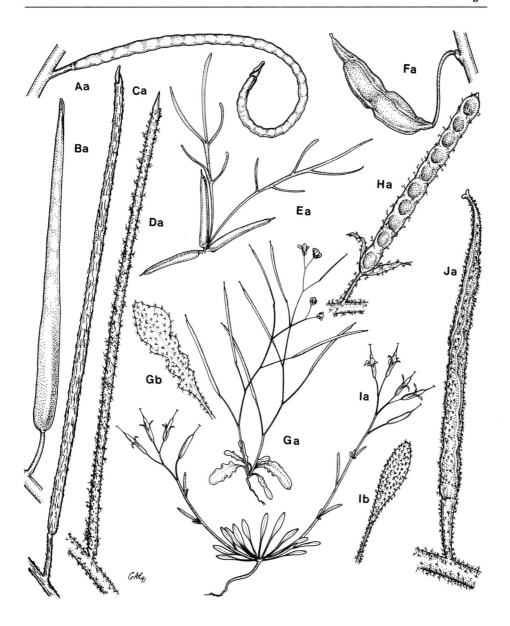

Fig. 83. Cruciferae. A, *Malcolmia grandiflora*: Aa, fruit (×3.5). B, *M. crenulata*: Ba, fruit (×3.5). C, *M. chia*: Ca, fruit (×3.5). D, *M. africana*: Da, fruit (×3.5). E, *Leptaleum filifolium*: Ea, fruits and leaves (×1.5). F, *Goldbachia laevigata*: Fa, fruit (×4). G, *Maresia pygmaea*: Ga, habit (×1.5); Gb, leaf (×4.5). H, *Eremobium aegyptiacum*: Ha, fruit (×4). I, *Eigia longistyla*: Ia, habit (×1); Ib, leaf (×5.5). J, *Sterigmostemum sulphureum*: Ja, fruit (×4).

base. Stigma 2-lobed. Fruit a dehiscent siliqua, terminating in a long persistent style; radicles accumbent.

1. E. longistyla (Eig) Soják in Čas. Nár. Mus., Odd. Přír., 148 (3–4): 193 (1980). Syn.: *Stigmatella longistyla* Eig in Palestine J. Bot., Jerusalem Ser. 1: 80 (1938). Illustr.: Fl. Palaest. 1: t.385 (1966).

Annual, ± canescent herb; stems up to 7(–10)cm, branching from the base. Basal leaves narrowly obovate, 5–10(–20) × 1–2mm, entire or rarely denticulate, attenuate into a short petiole below; stem-leaves few, smaller, linear-oblong, sessile. Petals c.10mm. Siliqua linear, 10–15(–18)mm; style 3–4mm. **Map 608, Fig. 83.**

In gravel and sand, amongst sandstone buttes; 950m.

Saudi Arabia. Palestine.

49. MALCOLMIA R. Br.

Annual herbs, bearing simple, bipartite or branched hairs. Leaves simple, entire, dentate or pinnatifid. Inner sepals saccate or not. Petals white, pink or lilac to pale purple. Filaments unappendaged, connate in pairs or free. Stigma with decurrent carpidial lobes or conical. Fruit a dehiscent, linear siliqua; seeds uniseriate; radicle incumbent.

1.	Sepals 10–12(–18)mm; petals 15–20(–25)mm; stem-leaves amplexicaul	**4. M. crenulata**
+	Sepals 3–6mm; petals 6–13(–16)mm; stem-leaves sessile	2
2.	Plant canescent with appressed, medifixed bipartite and 3–4-armed hairs	**3. M. chia**
+	Plants green and glabrous or tomentose with simple and branched hairs	3
3.	Fruiting pedicel 2–3mm; plant glabrous or with scattered hairs on the calyx and leaf margins	**2. M. grandiflora**
+	Fruiting pedicel 0.5–1.5(–2)mm; plant tomentose	**1. M. africana**

1. M. africana (L.) R.Br. in Aiton f., Hort. Kew. ed. 2, 4: 121 (1812). Syn.: *Strigosella africana* (L.) Botsch. in Bot. Zhurn. (Moscow & Leningrad) 57: 1038 (1972).

Annual herb up to 40cm, thinly to densely tomentose with branched hairs. Leaves elliptic to oblong-elliptic or obovate, 2–8(–13) × 0.5–2.5(–4)cm, entire to repand-dentate, attenuate below. Sepals 3.5–6mm. Petals white, pale lilac or pale purple, (5–)6–10(–12)mm. Siliqua straight or curved, 40–60 × c.1mm, ± square in section, densely hairy, the fruiting pedicels 0.5–1.5(–2)mm. **Map 609, Fig. 83.**

Sandy and fallow ground, drifted sand, wadi-banks and abandoned gardens; 0–2000m.

Saudi Arabia, Oman, UAE, Kuwait. Mediterranean region to India and N China.

A specimen (*Chaudhary* s.n.) from Al-Kharj in Saudi Arabia may be referable to this species; however, it is completely glabrous and without mature fruits, and if it does belong here, is somewhat anomalous.

2. M. grandiflora (Bunge) Kuntze in Acta Hort. Petrop. 10: 167 (1887). Syn.: *Strigosella grandiflora* (Bunge) Botsch. in Bot. Zhurn. (Moscow & Leningrad) 57: 1044 (1972). *M. behboudiana* Rech.f. & Esfand. in Phyton 3: 64 (1951). Illustr.: Mandaville (1990 pl.94); Fl. Kuwait 1: pls. 112–114 (1985); Collenette (1985 p.202).

Annual herb up to 45cm, ± glabrous or with scattered simple and forked hairs. Leaves oblong-elliptic, 2–10 × 0.5–2.5cm, entire to repand-dentate, shortly attenuate below. Sepals 3–4.5mm. Petals pink to mauve, 10–13(–16)mm. Siliqua straight to circinnately coiled, (25–)30–60 × c.1mm, glabrous, the fruiting pedicels 2–3mm. **Map 610, Fig. 83.**

Rock crevices in limestone and in hard and drifted sand; 0–800m.

Saudi Arabia, Qatar, Kuwait. Syria to C Asia and Pakistan.

All Arabian plants are referable to var. *glabrescens* (Boiss.) B.L. Burtt & Lewis. The type variety, which has densely hairy leaves, comes from Iran and C Asia. Recorded from Qatar by Hedge and King (op. cit.: 47) but no specimens seen.

3. M. chia (L.) DC., Syst. nat. 2: 440 (1821). Illustr.: Collenette (1985 p.202).

Erect annual herb, stems to 25cm, canescent with appressed medifixed bipartite and 3–4-armed hairs. Leaves elliptic to narrowly obovate, 1.5–5 × 0.5–1.5cm, entire to denticulate, attenuate below. Sepals 3.5–5.5mm. Petals white or lilac, 9–10mm. Siliquae straight or curved, 35–60 × c.1mm, appressed hairy, the fruiting pedicels 5–8mm. **Map 611, Fig. 83.**

Shade among rocks, *Juniperus* woodland; 1800–2150m.

Saudi Arabia. E Mediterranean.

4. M. crenulata (DC.) Boiss., Fl. orient. 1:229 (1867). Illustr.: Collenette (1985 p.202).

Annual herb, 5–15cm, glabrous or scabrid with simple hairs. Leaves oblong elliptic to obovate, 2–5 × 0.5–1.5cm, entire to repand-dentate; basal leaves attenuate; cauline leaves amplexicaul with small auricles. Sepals 10–12(–18)mm. Petals lilac with a yellow centre, 15–20(–25)mm. Siliqua straight or slightly curved, 40–80(–160) × 2–3mm, glabrous or scabrid, the fruiting pedicels 3–10mm. **Map 612, Fig. 83.**

Clay pans; 880m.

Saudi Arabia. Levant to Iran.

50. EREMOBIUM Boiss.

Annual herbs, bearing appressed stellate hairs. Leaves simple, entire. Inner sepals distinctly saccate. Petals white to lilac. Stamens unappendaged. Stigma 2-lobed. Fruit a linear, dehiscent siliqua, ± constricted between the seeds. Seeds uniseriate; radicles obliquely incumbent or accumbent.

1. E. aegyptiacum (Sprengel) Asch. & Schweinf. ex Boiss., Fl. orient. Suppl.: 30 (1888). Syn.: *E. diffusum* (Decne) Botsch. in Novosti Sist. Vyssh. Rast 1: 359 (1964); *E. lineare* (Del.) Boiss., Fl. orient. 1: 157 (1867); *E. nefudicum* (Velen.) B.L. Burtt & Rech.f. in Bot. Not. 115: 38 (1962); *Malcolmia nefudica* Velen. in Sitzungsber. Königl. Böhm. Ges. Wiss. Prag 11: 13 (1911); *M. aegyptiaca* Sprengel, Syst. nat. 2: 898 (1825). Illustr.: Mandaville (1990 pl.95); Collenette (1985 p.197).

Annual herb; stems prostrate to ascending, up to 25cm, canescent, sometimes viscid. Leaves linear to narrowly oblong, 5–35 × 0.5–6mm, obtuse, attenuate at the base. Petals 5–10mm. Siliqua linear, 10–35 × 1–2mm, terete or compressed, abruptly narrowed at the tip into a short style. **Map 613, Fig. 83.**

Gravelly and sandy deserts; 15–1100m.

Saudi Arabia, Yemen (N & S), Oman, UAE, ?Qatar, ?Kuwait. N Africa, Palestine, Jordan, Iraq, Iran and Pakistan.

Two varieties (var. *aegyptiacum* and var. *lineare* (Del.) Zoh.) are recognized by Townsend (Fl. Iraq 4(2): 1038 (1980)); however, the differences he gives do not work in Arabian plants (see also Mandaville 1990 p.153). Sheila Collenette (pers. comm.) recognizes two forms: the first, from stony and hard sands, is bushy in habit and has pinkish or white flowers; the second, from the "Red Sands", has a spreading habit and pink to magenta flowers. Recorded from Kuwait by Townsend (loc. cit.) and Qatar (Hedge & King op. cit.: 45) but no material has been seen from either country.

51. STERIGMOSTEMUM M.Bieb.

Annual or biennial herbs, bearing short branched hairs. Leaves simple, entire to pinnatifid. Sepals not saccate. Petals yellow. Longer filaments fused in pairs. Stigma with decurrent carpidial lobes. Fruit a siliqua, breaking transversely into 2-seeded segments at maturity; radicle incumbent.

1. S. sulphureum (Banks & Sol.) Bornm. in Beih. Bot. Centralbl. 28(2): 110 (1911). Illustr.: Fl. Iraq 4 (2): p.1047 (1980).

Erect annual or biennial herb; stems up to 1m, ± canescent (sometimes glandular-stipitate). Leaves narrowly ovate to narrowly oblong-ovate, 5–10(–30) × 1–4(–5.5)cm, entire to repand-dentate, attenuate below. Petals 7–10mm. Siliqua linear, 30–70 × 1–3(–5)mm, ± torulose and often somewhat contorted. **Map 614, Fig. 83.**

Dry volcanic slopes; 2700m.

Yemen (N). SW Asia.

Known in Arabia from a single flowering specimen which is referable to subsp. *sulphureum*.

52. GOLDBACHIA DC.

Glabrous annual herbs. Leaves simple. Sepals somewhat saccate. Petals white to pale purple. Filaments unappendaged. Stigma capitate. Fruit indehiscent, ± 4-angled in section, constricted in the middle, irregularly ridged, with a short flat beak; radicle incumbent.

1. G. laevigata (M.Bieb.) DC., Syst. nat. 2: 577 (1821).

Erect annual herb; stems 15–20cm. Basal leaves 3–10 × 1–3cm, denticulate to pinnatisect, attenuate into a short petiole; stem-leaves smaller and narrower, entire to dentate, amplexicaul. Petals 4.5–5mm. Fruit curved upwards from the reflexed pedicel, 10–12 × 2.5–4mm. **Map 615, Fig. 83.**

Depressions in gravel deserts.

Saudi Arabia. SW and C Asia.

Rare in Arabia, only known from a single gathering.

53. MARESIA Pomel

Low annual herbs, bearing branched and stellate hairs. Leaves entire to pinnatifid. Inner sepals saccate. Petals pink or white. Longer filaments narrowly winged below. Stigma capitate or 2-lobed. Fruit a dehiscent siliqua, linear, torulose; seeds uniseriate; radicle incumbent.

1. M. pygmaea (Del.) O. Schulz in Engl. & Prantl, Pflanzenr. 86 (IV. 105): 210 (1924). Syn.: *Malcolmia pygmaea* (Del.) Boiss., Fl. orient. 1: 222 (1867). Illustr.: Mandaville (1990 pl.96); Fl. Kuwait 1: pls. 115-116 (1985); Collenette (1985 p.203).

Erect or ascending annual herb; stems up to 10(–15)cm, thinly tomentose or glabrous above. Leaves mainly basal, narrowly oblong to narrowly obovate, up to 10(–20) × c.0.5mm, entire to pinnatifid with ± sinuate lobes. Sepals 3–3.5mm, often lilac tinged. Petals lilac, 7–8(–12)mm. Siliqua 20–30(–35) × c.0.75mm. **Map 616, Fig. 83.**

Stony and sandy deserts; 10–950m.

Saudi Arabia, Kuwait. Egypt, Syria, Jordan, Palestine and Iran.

Species doubtfully recorded

M. pulchella (Del.) O. Schulz and *M. nana* (DC.) Battand. have been recorded from Saudi Arabia and Kuwait but no verified material has been seen. Both are similar to *M. pygmaea* but *M. pulchella* has larger flowers (sepals 4–5mm; petals 10–15mm) and mainly cauline leaves, and *M. nana* is grey-tomentose with smaller flowers (sepals 2–2.5mm; petals 3–5mm). A single specimen from Saudi Arabia (*Chaudhary* 8379, from near Riyadh), with petals c.10mm long, is rather intermediate between *M. pygmaea* and *M. pulchella* but has mainly cauline leaves and on balance is closer to *M. pygmaea*.

54. SISYMBRIUM L.

Annual or rarely biennial herbs, glabrous or bearing simple hairs. Leaves pinnately lobed or divided, often lyrate, rarely entire or 3-lobed; stem-leaves not amplexicaul. Sepals erect or erect-spreading, not or scarcely saccate. Petals yellow. Filaments unappendaged, free. Stigma bilobed or capitate. Fruit a many-seeded dehiscent siliqua, linear to linear-conical; valves 1–3-nerved; seeds uniseriate; radicle incumbent.

1.	Fruits erect, appressed to the stem	**6. S. officinale**
+	Fruits erect-spreading, never appressed to the stem	2
2.	Petals up to 5mm long (see also *S. loeselii*)	3
+	Petals more than 5mm long	5
3.	Siliqua 8–10cm long, with a clavate tip; petals 5mm long; upper leaves with a linear or lanceolate terminal lobe	**3. S. orientale**
+	Siliqua 1.5–5.5cm long with an attenuate tip; petals 1.5–4mm long; upper leaves with an ovate or triangular terminal lobe	4
4.	Fruit 15–40mm, tapering from the base to the tip; petals 1.25–2.5mm; fruiting pedicel 1.5–3.5(–5.5)mm, $\pm$ as thick as the fruit	**5. S. erysimoides**
+	Fruit (25–)35–55mm, parallel sided; petals (2.5–)3–4mm; fruiting pedicel 3–10mm, distinctly thinner than the fruit	**1. S. irio**
5.	Young fruits overtopping the buds in the inflorescence; fruit 8–10cm with a clavate tip	**3. S. orientale**
+	Young fruits not overtopping the buds in the inflorescence; fruit 1–8cm, with an attenuate or bi-lobed tip, rarely $\pm$ clavate in *S. septulatum*	6
6.	Pedicel thinner than the siliqua, c.10mm long; upper leaves entire or with an ovate-triangular terminal lobe	**2. S. loeselii**
+	Pedicel thicker than the siliqua, 3–5mm long; upper leaves pinnate with linear lobes	**4. S. septulatum**

1. S. irio L., Sp. pl.: 659 (1753). Syn.: *S. pinnatifidum* Forsskal (1775 p.118). Illustr.: Fl. Palaest. 1: t.366 (1966); Fl. Pakistan p.254 (1973); Fl. Kuwait 1: pl. 117 (1985); Collenette (1985 p.207).

Stems erect or prostrate, 15–50cm, sparsely pubescent to glabrescent. Lower leaves 3–15 × 1–4cm, pinnately to lyrately pinnatipartite to pinnatisect with 2–4(–5) pairs of lateral lobes and a large often hastate or narrowly triangular terminal lobe; upper leaves with a narrowly triangular terminal lobe and 1–3 pairs of lateral lobes. Young fruits overtopping the flowers. Petals yellow, narrowly obovate, (2.5–3)–4mm.

Fruiting pedicels 3–10mm, thin. Siliqua erect-spreading, often curved, (25–) 35–55 × c.1mm, glabrous; style 0.5–1.25mm. **Map 617, Fig. 84.**

Weed of cultivated, disturbed and waste ground; 0–3650m.

Saudi Arabia, Yemen (N & S), Oman, UAE, Qatar, Kuwait. Europe, N Africa, SW and C Asia; widely introduced elsewhere.

S. irio is sometimes difficult to distinguish from *S. erysimoides*. However, it is usually a weed, with fruits borne on thin pedicels and leaves with a distinctive hastate or narrowly triangular terminal lobe. See also comments under *S. erysimoides*.

2. S. loeselii L., Cent. pl. I: 18, n. 49 (1755). Illustr.: Collenette (1985 p.207).

Stems erect, up to 50cm, densely villous or hispid below, glabrescent above. Lower leaves lyrate-pinnatisect, up to 8 × 4cm; upper leaves with a large ovate-triangular terminal lobe. Flowers overtopping the young fruits. Petals yellow, obovate, 10–15mm. Fruiting pedicels 10-12mm, thin. Fruits erect-spreading, straight, 50-60 × c.1mm, glabrous; style short, bilobed. **Map 618, Fig. 84.**

Sandy deserts; 850–950m.

Saudi Arabia. C and SE Europe, SW and C Asia; introduced elsewhere.

3. S. orientale L., Cent. pl. II: 24, n. 173 (1756). Illustr.: Fl. Palaest. 1: t.369 (1966); Collenette (1985 p.207).

Stems erect, up to 75cm, pubescent to villous. Basal leaves lyrate-pinnatipartite, soon withering; lower stem-leaves up to 12 × 6cm, pinnately pinnatisect with a large linear to narrowly lanceolate terminal lobe and 1–4 pairs of narrow lateral lobes; upper stem-leaves entire or trilobed with narrowly elliptic lobes. Young fruits equalling or overtopping buds. Petals yellow, narrowly obovate, 5–10mm. Fruiting pedicel 3–5mm, as thick as the siliqua. Fruits erect-spreading, 80–100 × c.1mm, glabrescent; style short, clavate. **Map 619, Fig. 84.**

Weed of cultivation and waste ground; 0–2500m.

Saudi Arabia, Yemen (N), Qatar, Bahrain, Kuwait. Europe, N Africa, SW & C Asia; introduced elsewhere.

4. S. septulatum DC., Syst. nat. 2: 471 (1821). Illustr.: Fl. Iraq 4 (2): 1071 (1980); Collenette (1985 p.207).

Stems erect, up to 50cm, sparsely villous or glabrous. Basal leaves 5–25 × 1–5cm, pinnatisect sometimes runcinate with up to 8 pairs of lateral lobes; upper leaves pinnatisect with linear lobes. Petals yellow, broadly obovate, 10–15mm. Flowers overtopping young fruits. Fruiting pedicel 3–5mm, as thick as the siliqua. Fruits erect-spreading, 20–80 × 1–2.5mm, glabrous; style short, stigma bi-lobed. **Map 620, Fig. 84.**

Silty desert soils, waste and disturbed ground; 0–1250m.

Saudi Arabia, Kuwait. SW and C Asia.

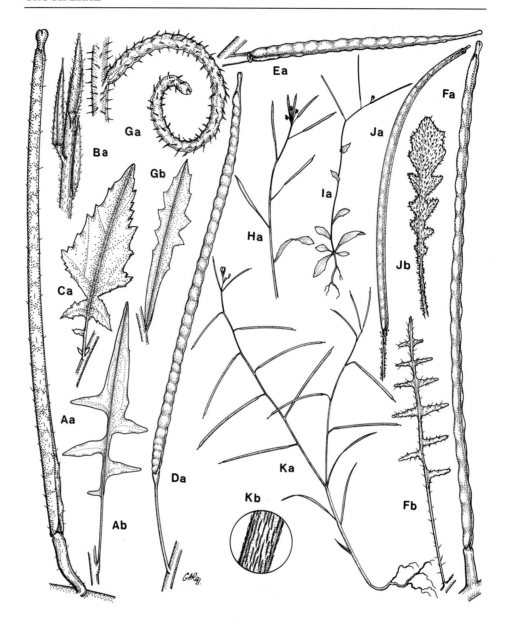

Fig. 84. Cruciferae. A, *Sisymbrium orientale*: Aa, fruit (×4); Ab, leaf (×1). B, *S. officinale*: Ba, fruits (×4). C, *S. loeselii*: Ca, leaf (×1). D, *S. irio*: Da, fruit (×4). E, *S. erysimoides*: Ea, fruit (×4). F, *S. septulatum*: Fa, fruit (×4); Fb, leaf (×1.5). G, *Neotorularia torulosa*: Ga, fruit (×4); Gb, leaf (×1.5). H, *Arabidopsis thaliana*: Ha, part of fruiting shoot (×1). I, *A. pumila*: Ia, habit (×1). J, *A. kneuckeri*: Ja, fruit (×3); Jb, leaf (×1.5). K, *A. erysimoides*: Ka, habit (×0.6); Kb, detail of branch showing indumentum (×15).

5. S. erysimoides Desf., Fl. atlant. 2: 84 (1798). Illustr.: Fl. Palaest. 1: t.370 (1966); Collenette (1985 p.206).

Stems erect, 10–45(–70)cm, sparsely pubescent to glabrescent. Lower leaves 2–15 × 1–5cm, lyrate-pinnatipartite to lyrate-pinnatisect with 1–4 pairs of lateral lobes and a larger ovate terminal lobe; upper leaves similar to lower. Young fruits equalling or overtopping flowers. Petals yellow, narrowly obovate, 1.25–2.5mm. Fruiting pedicel 1.5–3.5(–5.5)mm, ± as thick as the siliqua. Fruits erect-ascending, straight, narrowly tapering from the base to the tip, 15–40 × c.1mm, glabrous or sparsely hairy; style 0.5–1.25mm. **Map 621, Fig. 84.**

In gravel, amongst rocks and in silty soils, cultivated ground and on wadi-sides; 20–3650m.

Saudi Arabia, Yemen (N & S), Socotra, Oman, UAE, Qatar, Kuwait. Mediterranean region, Africa and SW Asia.

S. erysimoides can sometimes be difficult to separate from *S. irio*; the difference in fruit shape is often difficult to see. However, *S. erysimoides* usually has shorter, straight fruits borne on short, thick (± equalling the fruit) pedicels, smaller petals and the leaves usually coarsely serrate (not undulate to serrate) with terminal lobes ovate not narrowly triangular. See also comments under *S. irio*.

6. S. officinale (L.) Scop., Fl. carniol. ed. 2, 2: 26 (1772). Illustr.: Fl. Palaest. 1: t.369 (1966).

Annual herb; stems erect, up to 80cm, retrorsely hairy. Lower leaves lyrate-pinnatisect with up to 5 pairs of lateral lobes; upper leaves with a hastate terminal lobe. Flowers overtopping the young fruits. Petals pale yellow, narrowly obovate, 3–6mm. Fruiting pedicels 2–3mm, as thick as the siliqua. Fruits erect, appressed to the stem, 10–22mm, narrowly conical, attenuate towards the tip, glabrous or hairy; style short with a capitate stigma. **Map 622, Fig. 84.**

Wadi-banks; 2500m.

Yemen (N), UAE, ?Bahrain. Europe, N Africa and SW Asia; introduced elsewhere.

Recorded from Bahrain by Hedge and King (op. cit.: 48) but no specimens seen.

55. NEOTORULARIA Hedge & Léonard
Torularia O. Schulz

Annual or perennial herbs, bearing simple and branched hairs. Leaves simple to pinnatifid. Inner sepals not saccate. Petals white, lilac or violet. Filaments unappendaged. Stigma bilobed. Fruit a linear dehiscent siliqua, straight or contorted; radicle incumbent.

1. N. torulosa (Desf.) Hedge & Léonard in Bull. Jard. Bot. Nat. Belg. 56: 395 (1986). Syn.: *Malcolmia torulosa* (Desf.) Boiss., Fl. orient. 1: 225 (1867); *Torularia torulosa*

(Desf.) O. Schulz in Engl. & Prantl, Pflanzenr. 86 (IV. 105): 214 (1924). Illustr.: Fl. Kuwait 1: pl.118 (1985); Collenette (1985 p.208); Mandaville (1990 pl.97).

Erect or ascending annual herb; stems up to 30cm, with long simple hairs and short branched hairs. Leaves narrowly elliptic, 20–50(–90) × 2–6mm, sinuate-dentate to pinnatifid, attenuate into a short petiole below. Petals white, 2–4mm. Siliqua contorted or coiled, 10–20 × c.0.75mm, torulose, hispid or glabrous. **Map 623, Fig. 84.**

Loose sand and gravel on plains and in cultivated areas; 0–900m.

Saudi Arabia, Kuwait. SE Europe, N Africa and C & SW Asia.

56. DESCURAINIA Webb & Berth.

Annual or biennial herbs bearing branched and simple hairs. Leaves 2–3-pinnatisect with linear segments; cauline leaves sessile. Sepals not saccate. Petals yellow. Filaments unappendaged, free. Fruit a many-seeded dehiscent siliqua. Seeds uniseriate, mucilaginous on wetting; radicles incumbent.

1. D. sophia (L.) Webb & Berth. in Engl. & Prantl, Nat. Pflanzenfam. 3 (2): 192 (1891).

Stems erect, up to 80cm, thinly hairy below, glabrous above. Petals 2.5–4 × 0.5–1mm. Siliqua linear, 10–15mm, straight or curved, ± torulose, with a very short style. **Map 624.**

Weed of cultivation; 50m.

Kuwait. Europe, N Africa, SW & C Asia. Introduced elsewhere.

Only recorded once from Kuwait.

57. ARABIDOPSIS Heynh.

Annual or perennial herbs, bearing branched or mixed simple and branched hairs or appressed medifixed hairs. Basal leaves often forming a rosette, simple to pinnatifid; cauline leaves sessile, narrowed below or sagittate. Sepals not saccate. Petals pale yellow, white or pinkish lilac. Filaments unappendaged. Stigma capitate. Fruit a linear, dehiscent siliqua, glabrous; seeds 1-seriate; radicles incumbent.

1.	Leaves linear; indumentum of appressed, medifixed hairs	**4. A. erysimoides**
+	Leaves broader, obovate to narrowly oblong; indumentum of simple or branched hairs, never medifixed	2
2.	Flowers yellow; cauline leaves sagittate	**3. A. pumila**
+	Flowers white or lilac-pink; cauline leaves not sagittate	3
3.	Fruit 1–1.5(–1.8)cm long; basal leaves entire or toothed; indumentum of branched hairs below and becoming glabrous or with simple and forked hairs above	**1. A. thaliana**

\+   Fruit (1.5–)2.5–5.5cm long; basal leaves pinnatifid; indumentum of branched hairs throughout   **2. A. kneuckeri**

1. A. thaliana (L.) Heynh. in Holl & Heynh., Fl. Sachsen 1: 538 (1842). Illustr.: Fl. Trop. E. Afr.: 44 (1982); Collenette (1985 p.193).

Annual herb, with an indumentum of branched hairs below and glabrous or with simple and forked hairs above. Stems simple or branched from the base, up to 15(–30)cm. Leaves of the basal rosette obovate, up to 20(–50) × 5(–10)mm, entire or toothed; cauline leaves narrowly obovate, decreasing in size upwards. Petals white or pinkish, 2–4mm. Siliqua linear, 1–1.5(–1.8)cm, glabrous. **Map 625, Fig. 84.**

Usually in weedy habitats; 1050–2150m.

Saudi Arabia. Widespread in N Africa, Europe & Asia.

2. A. kneuckeri (Bornm.) O. Schulz in Engl., Nat. Pflanzenfam. ed. 2, 17b: 641 (1936). Illustr.: Collenette (1985 p.193).

Annual or perennial herb, with an indumentum of branched hairs throughout; stems erect to ascending, up to 20cm, branched from the base. Leaves of the basal rosette narrowly obovate, 2–5 × 0.5–1.2cm, pinnatifid; cauline leaves narrowly oblong, pinnatifid or toothed, decreasing in size upwards. Petals lilac-pink, c.4.5mm. Siliqua linear, (1.5–)2.5–5.5cm, glabrous. **Map 626, Fig. 84.**

Steep rocks and moist places; 2750–2800m.

Saudi Arabia. Sinai.

3. A. pumila (Stephan ex Willd.) N. Busch, Fl. Cauc. Crit. 3 (4): 457 (1909). Illustr.: Fl. Iraq 4 (2): 1009 (1980).

Annual herb, with an indumentum of branched hairs; stems erect, up to 8(–50)cm, simple or branched from the base. Leaves of the basal rosette obovate, 1–3(–10) × 0.3–0.6(–1.5)cm, entire or pinnatifid; cauline leaves ovate, entire or toothed, sagittate at the base. Petals pale yellow. Siliqua linear, 10–15 × c.1mm. **Map 627, Fig. 84.**

Amongst rocks on exposed mountain slopes; 2040m.

?Saudi Arabia, Oman. E Europe, SW & C Asia.

The occurrence of this species in Arabia has been confirmed only from a single collection from the Musandam Peninsula in Oman. Plants from Arabia previously named *A. pumila* have been confused with *Arabis nova* which differs in having white flowers. *Arabidopsis pumila* is also likely to occur in the western mountains of Saudi Arabia.

4. A. erysimoides Hedge & Kit Tan in Pl. Syst. Evol. 156: 202 (1987). Illustr.: loc. cit. Type: Saudi Arabia, *Collenette* 5713 (E, K).

Delicate annual herb, with an indumentum of appressed, medifixed hairs; stems erect, little-branched, up to 20cm. Leaves linear, 35–40 × 0.5–1mm. Petals pale yellow, c.4.2mm. Siliqua linear, c.4.5 × 1cm, deflexed, appressed-hairy. **Map 628, Fig. 84.**

On sand dunes; 900m.

Saudi Arabia. Endemic.

This remarkable plant has only been found once, in the great sand desert of the Nafud, between Qulban and Qana, in Saudi Arabia. It is unusual in the genus because of its indumentum of appressed, medifixed hairs. Al-Shehbaz (*Novon* 4: 1–2, 1994) has proposed that it should be transferred to the genus *Erysimum*, proposing the name *E. hedgeanum*.

Family 58. RESEDACEAE

A.G. MILLER

Herbs or shrubs, sometimes dioecious. Leaves alternate, sometimes fasciculate, entire or variously dissected. Stipules present. Inflorescence a terminal raceme or spike. Flowers zygomorphic, subtended by a bract. Sepals 2–8, free or fused at the base. Petals 0–8, free, unequal, usually laciniate with a basal appendage. Disc present or not. Stamens 10–numerous. Carpels 2–7, superior, ± free or fused and forming a 1-locular ovary which is often gaping above; ovules numerous; placentation parietal or basal; stigma sessile. Fruit a berry or capsule. Seeds reniform, numerous.

Abdallah, M.S., The Resedaceae, a taxonomical revision of the family. *Meded. Landbouwhoogeschool* 67 (8): 1–98 (1967); Abdallah, M.S & de Wit, H.C.D. op. cit. 78 (14): 99–416 (1978).

1.	Usually leafless shrubs, often spiny; leaves when present linear or very reduced	**2. Ochradenus**
+	Leafy herbs, if subshrubs then leaves broader	2
2.	Carpels more or less free, fused only at the base; fruits distinctly stipitate	**1. Caylusea**
+	Carpels fused together; fruits not stipitate	3
3.	Leaves simple, linear; disc absent; petals without a basal appendage	**3. Oligomeris**
+	Leaves usually variously divided, if simple then broad not linear; disc present; petals with a basal appendage	**4. Reseda**

1. CAYLUSEA A. St. Hil.

Annual or perennial herbs. Leaves simple, entire. Flowers bisexual, in spike-like racemes. Sepals 5, fused at the base. Petals 5, free, unequal, laciniate, with a basal appendage. Disc present. Stamens 10–15. Ovary stipitate; carpels 5–6, fused only at the base and gaping above; placentation basal. Fruiting carpels spreading and exposing the seeds; seeds 1(–2) per carpel.

Taylor, P. (1959). The genus Caylusea in Tropical Africa. *Kew Bull.* 13: 283–286.

1.	Seeds minutely tuberculate; upper petal 5–9-lobed	**1. C. hexagyna**
+	Seeds rugulose; upper petal 4–5-lobed	**2. C. abyssinica**

1. C. hexagyna (Forsskal) M.L. Green Nom. Prop. Brit. Bot.: 102 (1929). Syn.: *C. canescens* A. St. Hil., Deux. Mém. Réséd.: 38 (1837); *C. jaberi* Abedin in Willdenowia 15: 433 (1986); *Reseda hexagyna* Forsskal (1775 p.92). Illustr.: Abdallah, op. cit.: f.3 (1967); Fl. Iraq 4 (2): 1088 (1980); Collenette (1985 p.418); Mandaville (1990 pl.102).

Erect or decumbent herb, 20–60(–80)cm, whitish- or greyish-hairy or glabrous. Leaves narrowly obovate to linear-obovate, $10–50 \times 2–10$mm, acute or obtuse, the margin flat or wavy, attenuate below, glabrous or with a ciliate margin, subsessile. Racemes up to 20cm; pedicels 1–3mm. Sepals ovate, $1.2–2.0 \times 0.5–0.6$mm. Petals white; the upper 5–9-lobed, 2–3mm long. Carpels strigose-ciliate, 2–3mm long, on a 2–3(–5)mm stipe. Seeds $1 \times 0.75–1$mm, minutely tuberculate, black or brown. **Map 629, Fig. 85.**

Weed of cultivation, by wadis, on sand and rocky slopes; 0–3000m.

Saudi Arabia, Yemen (N & S), Socotra, Kuwait. N Africa to Ethiopia, Crete, Palestine, Jordan, Iraq and Iran.

Sheila Collenette (pers. comm.) comments that there are two forms of *C. hexagyna* in Saudi Arabia. The first, an erect, bushy, annual herb with hairy stems, is found at low altitudes from Jeddah north to Sawawin and the second, with glabrous leaves and ascending stems, is found mainly at higher altitudes in the mountains south from Madinah. The ranges of the two forms overlap at lower altitudes. The photographs in Collenette (1985 p.418) are of the first form on the right and the second on the left.

2. C. abyssinica (Fresen.) Fischer & C. Meyer, Index sem. hort. petrop. 7: 43 (1840). Illustr.: Abdallah, op. cit.: f. 2, 4 (1967).

Similar to *C. hexagyna* but the petals less divided with the upper 4–5-lobed; seeds rugulose, transversely wrinkled. **Map 630.**

In the escarpment mountains; 2000–2200m.

Yemen (N). Tropical E Africa from Sudan to Malawi.

Much rarer than the preceding species; most literature records in Arabia have proven to be *C. hexagyna*.

2. OCHRADENUS Del.

Dioecious or hermaphrodite shrubs, often spinescent, sometimes scandent, soon aphyllous. Leaves linear to narrowly ovate or circular, entire, sessile, sometimes fascicled. Flowers unisexual or bisexual, in spikes or racemes. Sepals 5–8. Petals

RESEDACEAE

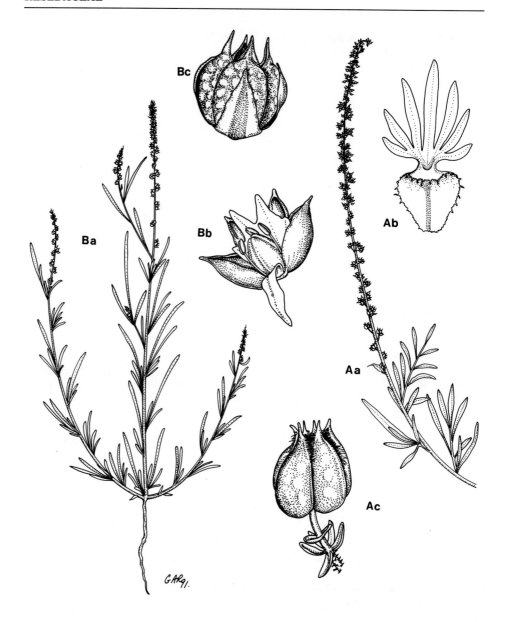

Fig. 85. Resedaceae. A, *Caylusea hexagyna*: Aa, habit (×1); Ab, upper petal (×30); Ac, fruit (×15). B, *Oligomeris linifolia*: Ba, habit (×1); Bb, flower (×15); Bc, fruit (×15).

absent or linear with an auriculate base. Disc excentrically surrounding the ovary, fleshy, single or double. Stamens 10–numerous. Ovary sessile, of 3 fused carpels; placentation parietal. Fruit a capsule or berry, sometimes gaping at the apex; seeds numerous.

Miller, A.G. (1984). A revision of Ochradenus. *Notes Roy. Bot. Gard. Edinburgh* 41: 491–504.

1. Plant glandular-pubescent in the region of the inflorescence; racemes dense (flowers touching) **5. O. spartioides**
+ Plant glabrous throughout; racemes dense or lax 2
2. Stamens 20–80; disc double, the inner part cup-shaped; fruits oblong-ovoid to ellipsoid, at least 1.5 × as long as broad, ascending; pedicels at least 2mm long 3
+ Stamens 10–18; disc single; fruits globose, ± as long as broad, subsessile or on pedicels up to 1.5mm long 4
3. Stamens 28–80; anthers yellow; seeds smooth, glossy **4. O. aucheri**
+ Stamens c.25; anthers orange; seeds finely tuberculate, dull **6. O. gifrii**
4. Fruit papery, inflated, ± 3(–6)-lobed in transverse section; leaves spathulate, 1.5–3mm broad, or much reduced and ± round; sepals 1.5–2mm long
 3. O. harsusiticus
+ Fruit either baccate or papery and inflated, round in transverse section; leaves linear, up to 1.75mm broad; sepals 1–1.5mm long 5
5. Fruit baccate, white (often drying red); seeds minutely tuberculate; male flowers usually on short (c.1mm) pedicels **1. O. baccatus**
+ Fruit papery, ± inflated, yellow or straw-coloured; seeds smooth, glossy (as if varnished); male flowers subsessile **2. O. arabicus**

1. O. baccatus Del. (1813 p.63). Syn.: *O. baccatus* var. *monstruosa* Muell. Arg., Monogr. Résédac.: 95 (1857); *O. baccatus* var. *scandens* Hochst. & Steud. ex Muell. Arg., loc. cit. Illustr.: Abdallah op. cit.: 7, 8 (1967); Fl. Kuwait 1: pls. 119-121; Cornes (1989 pl.46); Mandaville (1990 pls. 98 & 99); Phillips (1988 p.130).

Dioecious shrub with slender branches, sometimes scandent, 1–3m tall, glabrous throughout. Leaves linear, 1–4cm. Racemes with distant or touching flowers, 8–20cm long; pedicels 1–2mm. Flowers yellow. Sepals oblong-ovate, 1–1.5mm. Petals absent or rarely present. Disc ± reflexed. Stamens 10–18. Fruit baccate, globose, 3–6mm across, subsessile to sessile, white when mature (often red when dried). Seeds 1.2–1.8 × 1–1.5mm, minutely tuberculate, brown to black. **Map 631, Fig. 86.**

Rocky and sandy deserts, semi-deserts and mountains; 10–2750mm.

Saudi Arabia, Yemen (N & S), Socotra, Oman, Qatar, Bahrain, Kuwait. In deserts from Libya east to Pakistan, Iran and Jordan south to Somalia.

Plants from Socotra are somewhat intermediate with *O. arabicus*; they have the

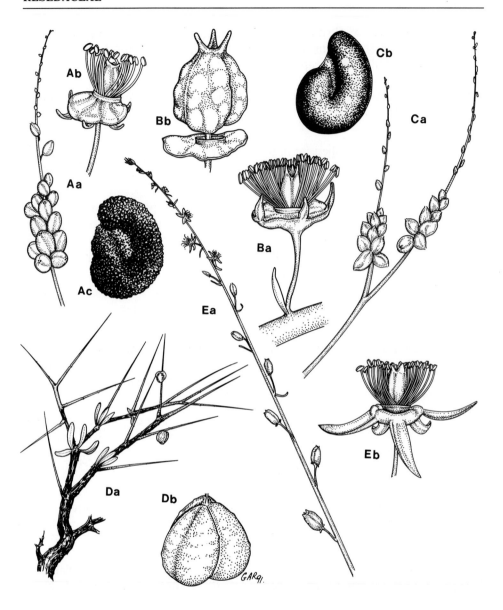

Fig. 86. Resedaceae. A, *Ochradenus baccatus*: Aa, fruiting branch (×1); Ab, flower (×10); Ac, seed (×25). B, *O. gifrii*: Ba, flower (×10); Bb, fruit (×10). C, *O. arabicus*: Ca, fruiting branch (×1); Cb, seed (×25). D, *O. harsusiticus*: Da, habit (×1); Db, fruit (×7); E, *O. aucheri*: Ea, fruiting and flowering branch (×1); Eb, flower (×5).

smooth, shiny seeds of *O. arabicus* but the fleshy fruits of *O. baccatus*. Further investigation will perhaps show them to merit formal taxonomic recognition, perhaps as a subspecies of *O. baccatus*.

2. O. arabicus Chaudhary, Hillc. & A.G. Miller in Notes Roy. Bot. Gard. Edinburgh 41: 494 (1984). Syn.: ?*Reseda spinescens* Schwartz (1939 p.76). Illustr.: Western (1989 p.66) Type: UAE, *Edmondson* 3129 (E).

Dioecious shrub, 20–50(–100)cm tall, intricately branched, the branches often becoming spinose, glabrous throughout. Leaves linear-oblong, (2–)15–25 × 0.5–1.75mm, sometimes fasciculate. Racemes with touching flowers, the tips often becoming spinose, 2–6 × c.0.5cm. Flowers yellow, unisexual, subsessile. Sepals (5–)6–7, oblong-ovate, 0.5–1 × 0.3–0.4mm. Petals absent or rarely 1–2 present. Disc entire to undulate, somewhat reflexed. Stamens 12–16, sterile in female flowers. Capsule leathery or papery, globose to ovoid, 4.5–5.5 × 4.5–6mm, green becoming yellow. Seeds 1.2–1.8 × 1–1.5mm, smooth, glossy, black to reddish black. **Map 632, Fig. 86.**

Open rocky and sandy places, roadsides; 0–2000m.

Saudi Arabia, Yemen (S), Oman, UAE. Endemic.

Schwartz (1939 p.76) described *Reseda spinescens* from a single specimen collected in the Hadramaut Governorate of Yemen. The specimen cannot be traced but, from the description, it seems more likely to have been a species of *Ochradenus* very similar to, if not the same as, *O. arabicus*.

3. O. harsusiticus A.G. Miller in Notes Roy. Bot. Gard. Edinburgh 41: 497, fig. 3A (1984). Type: Oman, *Maconochie* 3430 (E).

Spiny shrub, up to 1m, intricately branched and glabrous throughout. Leaves narrowly elliptic to spathulate, sometimes very reduced and ± round, 3–10 × 1–3.5mm. Spikes with 5–8 distant flowers, c.4cm long. Flowers bisexual, sessile. Sepals 6, ovate to narrowly elliptic, 1.5–2 × 0.5–1.4mm. Petals 4, white, linear with 1–2 basal teeth, c.2mm long. Disc entire, pentagonal, flat. Stamens c.12. Capsule papery, brittle, ovoid, 4–6mm across, 3- or 6-lobed in transverse section, yellowish brown. Seeds 1.4 × 1mm, smooth but not glossy, blackish brown. **Map 633, Fig. 86.**

Open *Acacia-Prosopis* bushland on limestone; 100–150m.

Oman. Endemic.

Found only on the low plateau of the Jiddat al Harasis in central Oman.

4. O. aucheri Boiss., Diagn. pl. orient. sér. 2, 3 (1): 50 (1854). Syn.: *Homalodiscus aucheri* Boiss., Fl. orient. 1: 422 (1867); *Ochradenus dewittii* Abdallah, op. cit. p.61 (1967). Illustr.: Abdallah, op. cit. fig. 9 (1967) as *O. dewittii*.

Slender shrub, up to 1m, with straight ascending branches, glabrous throughout. Leaves linear, 1–6cm. Racemes with distant flowers, 10–30cm; pedicels 1–2mm, lengthening in fruit. Flowers yellow, bisexual. Sepals 6, obovate, 1.5–2mm. Petals with 1–2 basal teeth. Disc double, the outer part flat, the inner cup-shaped. Stamens 28–80. Capsule leathery, erect, oblong to ovoid, c.10 × 7–8mm, yellowish brown.

Seeds 1.5–1.8 × 1–1.4mm, shiny, smooth, pale brown to black. **Map 635, Fig. 86.**

Open rocky slopes; 10–1900m.

Oman, UAE, Qatar. Iran, Pakistan.

Plants from Arabia belong to subsp. *aucheri* which is distinguished by its double disc and is endemic to Arabia.

5. O. spartioides (Schwartz) Abdallah, op. cit., p. 67, fig. 12 (1967). Syn.: *Randonia spartioides* Schwartz (1939 p.75); ?*Reseda micrantha* Schwartz (1939 p.75). Type: Yemen (S), *Wissmann* 1474 (WU).

Slender shrub, up to 2m, with straight ascending branches and glandular-pubescent in the region of the inflorescence. Leaves linear, 1.5–5cm. Racemes dense, (2–)5–12cm; pedicels 2–5mm. Flowers bisexual. Sepals 6, oblong, 2–2.5mm. Petals with ovate appendage. Disc double, cup-shaped, the inner part membranous, the outer fleshy. Stamens c.25. Capsule leathery, ovoid or oblong, 5–8 × 5–6mm, erect on a 4–6mm pedicel. Seeds c.1.5 × 1mm, tuberculate, dark brown to black, dull. **Map 635.**

Open rocky slopes and wadi-beds; 50–1050m.

Yemen (S). Endemic.

Schwartz (1939 p.76) described *Reseda spinescens* from a single specimen collected in the Hadramaut Governorate of Yemen. The specimen cannot be traced but, from the description, it seems more likely to have been a species of *Ochradenus* very similar to, if not the same as, *O. spartioides*.

6. O. gifrii Thulin, Nord. J. Bot. 14: 383 (1994). Type: Yemen (S), *Thulin et al.* 8417 (UPS, K).

Slender shrub, up to 2m, with straight ascending branches, glabrous throughout. Leaves linear, 0.7–1.7cm. Racemes lax, 15–40cm; pedicels 2–5mm. Flowers bisexual. Sepals 6, oblong, 2–2.5mm. Petals with ovate appendage. Disc double, cup-shaped, the inner part membranous, the outer fleshy. Stamens c.25. Capsule leathery, obovoid or subglobose, erect or ascending on a 4–5mm pedicel. Seeds c.1.5 × 1mm, tuberculate, dark brown to black, dull. **Map 635, Fig. 86.**

Open rocky slopes and wadi-beds; 50–1300m.

Yemen (S), Oman. Endemic.

3. OLIGOMERIS Cambess.

Annual or perennial herbs. Leaves linear, simple, entire, sometimes fasciculate. Flowers unisexual, in spike-like racemes. Sepals 2–5. Petals 2, free, entire or shallowly lobed, without a basal appendage. Disc absent. Stamens 3. Ovary sessile, of (3–)4(–5) fused carpels, gaping at the apex; placentation parietal. Capsule gaping; seeds numerous.

1. O. linifolia (Vahl) J. F. Macbr. in Contr. Gray Herb. 53: 13 (1918). Syn.: *O. subulata* Webb, Fragm. fl. aethiop.-aegypt.: 26 (1854). Illustr.: Abdallah, op. cit.: figs. 14 & 16 (1967); Fl. Iraq 4 (2): 1090 (1980); Fl. Kuwait 1: pl.122 as *O. subulata*; Collenette (1985 p. 419); Cornes (1989 pl.47); Phillips (1988 p.131).

Erect or decumbent, annual or perennial herb, glabrous throughout. Leaves linear-elliptic to narrowly oblong, 10–50 × 1–3mm, sessile, sometimes fasciculate. Sepals 1.5–2mm. Petals white, 1–2(–3), entire or bilobed, 1.25–1.5mm. Stamens inserted to one side of the flower opposite the petals. Fruits sessile, depressed-globose, 3–4mm across, 4-lobed in section and four toothed at the apex. Seeds 0.3–0.5mm, smooth, glossy, black or greenish-black. **Map 636, Fig. 85.**

Rocky slopes, sandy ground, salt flats and a as weed of cultivation; 0–2300m.

Saudi Arabia, Yemen (N), Oman, UAE, Qatar, Bahrain, Kuwait. SW Asia, Africa and America.

4. RESEDA L.

Annual or perennial herbs or subshrubs. Leaves variously dissected or entire. Flowers bisexual in spike-like racemes, 4–6(–9)-merous. Petals free, unequal, laciniate, with a basal appendage. Disc present. Stamens 10–numerous. Ovary shortly stipitate, of 3–6 carpels, usually gaping at the apex; placentation parietal. Capsules gaping, with numerous seeds.

1.		Leaves pinnatisectly divided into numerous pairs of lobes; limb of the upper petal divided to less than halfway; capsule 4-toothed at the apex	**1. R. alba**
	+	Leaves entire or pedately or ternately divided; limb of upper petal divided to the middle; capsule 3- or 5–6-toothed at the apex	2
2.		Central lobe of the upper petal shorter than the lateral lobes	3
	+	Central lobe of the upper petal equalling or longer than the lateral lobes	4
3.		Lateral lobes of the upper petal entire or shallowly toothed, semi-lunate; fruit oblong, oblong-ovoid or ellipsoid, erect or nodding; filaments deciduous	**5. R. lutea**
	+	Lateral lobes of the upper petal deeply incised into 5–9 smaller lobes; fruit subglobose, nodding; filaments persistent	**2. R. arabica**
4.		Leaves mainly ternately or pedately divided; lobes linear to linear-obovate (*R. aucheri* may rarely key out here)	5
	+	Leaves entire or sometimes a few with 2–3 lobes; lobes broader	6
5.		Stems and leaves papillose; capsule subglobose to ovoid	**6. R. muricata**
	+	Stems and leaves glabrous to thinly papillose; capsule cylindrical to obconical	**7. R. stenostachya**

6.	Capsule 5–6-toothed at the apex	**4. R. pentagyna**
+	Capsule 3-toothed at the apex	7
7.	Stems glabrous	8
+	Stems papillose	9
8.	Capsule rounded in section; stamens more than 20	**8. R. sphenocleoides**
+	Capsule 3-angled in section; stamens 12–15	**10. R. viridis**
9.	Shrub or robust perennial herb, densely leafy, with short (1–3mm) internodes; flowers 6-merous; seeds minutely tuberculate	**9. R. amblyocarpa**
+	Annual or perennial herb, not densely leafy, with longer (usually more than 1cm) internodes; flowers 6–7(–8)-merous; seeds smooth	**3. R. aucheri**

1. R. alba L. subsp. **decursiva** (Forsskal) Maire in Cat. pl. Maroc. 2: 315 (1932). Syn.: *R. decursiva* Forsskal (1775 p.LXVI). Illustr.: Abdallah, op. cit.: figs. 18b & 21a–k (1978); Fl. Kuwait 1: pls. 123–125; Collenette (1985 p.419) as *R. decursiva*.

Glabrous or thinly papillose-scabridulous herb; stems erect or ascending, 10–30cm, simple or branched mainly from the base. Leaves pinnatisect, up to 15cm long, the lobes linear-oblong with undulate margins; basal leaves larger, often forming a rosette. Spike up to 15cm. Flowers 5-merous, sweetly scented. Petals conspicuous, white, equalling or just exceeding the sepals; upper petals 3–5mm, 3-lobed, the central lobe equalling or just exceeding the lateral lobes. Stamens c.10. Capsules erect, obovoid, $5–6 \times c.3$mm, 4-toothed at the apex. Seeds $c.0.8–1 \times 0.8$ mm, black, dull, minutely tuberculate. **Map 637, Fig. 87.**

Sandy and gravelly deserts; 50–900m.

Saudi Arabia, Kuwait. N Africa and SW Europe to Iran.

Subsp. *alba* may also be expected in Arabia. It is a taller, more robust, plant with its petals greatly exceeding the sepals.

2. R. arabica Boiss., Diagn. pl. orient. sér 1, 1 (1): 6 (1843). Illustr.: Abdallah & de Wit, op. cit.: figs. 26-7 (1978); Collenette (1985 p.419); Fl. Kuwait 1: pls. 126–8; Mandaville (1990 pl.100).

Annual herb; stems erect, ascending or decumbent, 10–20(–30)cm, simple or branched mainly from the base, glabrous or thinly papillose-scabridulous. Basal leaves entire, narrowly obovate, $20–50 \times 3–7$mm, obtuse, with entire or undulate margins and long-attenuate base; upper leaves ternately or pedately divided into linear-obovate or linear-oblong lobes. Flowers 6-merous, scentless. Petals white; upper petals 2–3mm, 3-lobed, the central lobe simple and shorter than the laciniate lateral lobes. Stamens c.20, orange. Capsule nodding, $\pm$ globose, $4–10 \times 4–7$mm, 3(–4)-toothed at the apex. Seeds $c.1.5 \times 1$mm, whitish brown, wrinkled-rugose. **Map 638, Fig. 87.**

Sandy and stony deserts and in clay pans; 0–700m.

Saudi Arabia, Qatar, Kuwait. N Africa, Palestine, Syria, Jordan, Iraq and Iran.

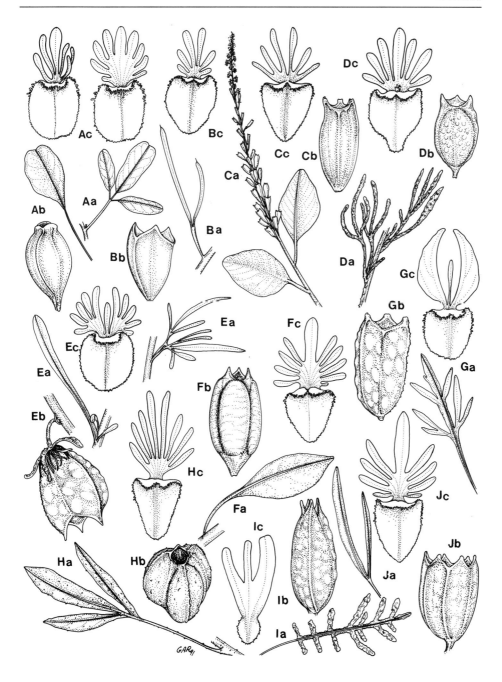

Fig. 87. Resedaceae. A–J, *Reseda* species: A, *R. viridis*; B, *R. stenostachya*; C, *R. sphenocleoides*; D, *R. muricata*; E, *R. arabica*; F, *R. amblyocarpa*; G, *R. lutea*; H, *R. aucheri*; I, *R. alba*; J. *R. pentagyna*. a, leaves (× 1); b, fruits (× 4); c, upper petals (× 20).

3. R. aucheri Boiss. var. **bracteata** (Boiss.) Abdallah & De Wit, op. cit.: 167 (1978). Syn.: *R. bracteata* Boiss., Diagn. pl. orient. sér. 1, 1 (6): 22 (1845). Illustr.: Abdallah & de Wit, op. cit.: fig. 29 (1978); Western (1990 p.67).

Annual or perennial herb, papillose throughout; stems erect, 5–60cm, simple or branched. Leaves narrowly obovate or ovate to broadly ovate, 1–10 × 0.3–4cm, entire or with 2–3 lobes, the lobes narrowly obovate to linear-obovate often with undulate margins. Flowers (6–)7(–8)-merous. Petals cream or yellow; upper petals 3–5mm, deeply divided with 5–11 lobes, the central lobe equalling the lateral lobes. Stamens 12–17. Capsule erect, globose to cylindrical, 3–6 × c.3mm, 3-toothed at the apex. Seeds 0.6–0.8 × c.0.5mm, dark brown, shiny, smooth. **Map 639, Fig. 87.**

Dry rocky slopes, wadi-beds and sandy irrigated areas; 10–1050m.

Saudi Arabia, Oman, UAE, ?Kuwait. Syria, Palestine and Iran to C Asia.

There is also a literature record of this species in the Flora of Kuwait (1985: 109) but no specimens have been seen recently.

4. R. pentagyna Abdallah & A.G. Miller in Edinb. J. Bot. 51 (1): 42 (1994). Type: Saudi Arabia, *Collenette* 4812 (E, K).

Annual herb, papillose or glabrous throughout; stems erect or ascending, up to 30cm, sparsely branched. Leaves narrowly oblong or oblanceolate, 30–80 × 3–12mm, entire or with 2–3 lobes. Flowers c.7-merous. Petals white; upper petals c.4mm, deeply divided into 7–9 lobes, the central lobe longer than the lateral lobes. Stamens c.17. Capsules erect, oblong or triangular, c.5 × 4mm, the apex truncate and 5–6-toothed. Seeds c.0.6 × 0.5mm, black, tuberculate. **Map 640, Fig. 87.**

On hard sand and low rocky hills; 600m.

Saudi Arabia. Endemic.

Readily distinguished from all other species in Arabia by its 5–6-toothed capsule. Known from three separate gatherings in NW Saudi Arabia.

5. R. lutea L., Sp. pl.: 449 (1753). Illustr.: Abdallah & de Wit op. cit.: figs. 57–59 (1978).

Annual or perennial herb, glabrous or papillose-scabridulous throughout; stems erect or ascending, up to 60cm, numerous. Leaves pedately or ternately divided, up to 10cm long; lobes linear-oblong to spathulate. Flowers yellow, sweet-scented, 6–8-merous. Upper petals 3–5mm, 3-lobed; central lobe linear-oblong; lateral lobes entire or shallowly toothed, semi-lunate, longer than the central lobe. Stamens 15–20. Capsule erect or nodding, oblong, oblong-obovoid or ellipsoid, 8–15 × 3–5mm, 3-toothed at the apex. Seeds c.2mm, dark brown, glossy, smooth. **Map 641, Fig. 87.**

Weed of wheat field; 2100m.

Saudi Arabia. W & S Europe, N Africa, and SW Asia.

Rare in our area, known only from a single gathering.

6. R. muricata C. Presl in Abh. Königl. Böhm. Ges. Wiss., Ser. 5, 3: 438 (1845). Syn.: *R. patzakiana* Rech.f. in Anz. Osterr. Akad. Wiss. Math.-Naturwiss. Kl. 8: 246 (1961).

Illustr.: in Abdallah & de Wit, op. cit.: figs. 65 & 66 (1978); Fl. Kuwait 1: pls. 129–131; Collenette (1985 p. 420); Mandaville (1990 pl.101).

Erect or ascending perennial herb, often woody-based, simple or branched mainly from below, papillose throughout. Leaves ternately or pedately divided, up to 3cm long; lobes linear to linear-obovate, 1–2.5mm broad. Flowers 6(–8)-merous. Petals white; the upper petals c.2.5mm, divided to the middle into 7–9 lobes; central lobe broader and longer than the lateral lobes. Stamens 10–16(–18), the filaments persistent. Capsule erect, subglobose to ovoid, 2–5 × 2–4mm, 3-toothed at the apex. Seeds 0.6–0.8 × c.0.6mm, dark brown, shiny, tuberculate. **Map 642, Fig. 87.**

Dry stony hills, sandy and gypseous soils in deserts; 0–1250m.

Saudi Arabia, Oman, Qatar, Bahrain, Kuwait. Egypt, Jordan, Palestine and Iraq.

7. R. stenostachya Boiss., Diagn. pl. orient. sér. 1, 1 (1): 5 (1843). Illustr.: Abdallah & de Wit, op. cit.: fig. 78 (1978).

Similar to *R. muricata* but the stems and leaves glabrous or thinly papillose; filaments deciduous; capsules cylindrical or obconical with truncate apices. **Fig. 87.**

Deserts.

?Saudi Arabia, ?Bahrain. Palestine, Sinai.

Doubtfully recorded from our area; no verified material has been seen.

8. R. sphenocleoides Deflers in Bull. Soc. Bot. France 42: 298 (1895). Illustr.: Abdallah & de Wit, op. cit.: fig. 76 (1978); Collenette (1985 p. 420). Type: Yemen (S), *Deflers* 530 (P).

Robust perennial herb or shrub, glabrous throughout; stems erect, 50–150cm. Leaves entire, fleshy, ovate to oblong-elliptic, 3–10 × 1.5–5cm, acute or obtuse, the base obtuse or attenuate into a 1–2cm petiole. Flowers pale yellow, 6(–9)-merous. Upper petals 3–4mm, divided to the base into 7–11 lobes; central lobe longer than the lateral lobes. Stamens 20–numerous. Capsule erect, obovoid to obconical, 7–12 × 3–4mm, 3-toothed at the apex. Seeds 0.6–0.7 × c.0.5mm, dark brown, glossy, minutely tuberculate. **Map 643, Fig. 87.**

Dry rocky slopes; 50–2450m.

Saudi Arabia, Yemen (N & S), Oman. Endemic.

9. R. amblyocarpa Fresen. in Mus. Senckenberg. 2: 108 (1837). Illustr.: Abdallah & de Wit, op. cit.: figs. 24a & 25 (1978).

Robust perennial herb or shrub, minutely papillose and branched throughout; stems erect, 50–100cm, densely leafy. Leaves entire or rarely the upper shallowly 3-lobed, narrowly obovate, 1.5–5 × 0.3–1cm, acute or obtuse, the base attenuate into a 1–3cm petiole. Flowers yellow-green, 6-merous. Upper petals c.2mm, divided from $\frac{1}{2}$ to $\frac{3}{4}$ the way to the base into 5–9 lobes; central lobe longer than the lateral lobes. Stamens c.20. Capsule erect, obovoid to ellipsoid, 3-grooved in section, 5–7 × 2–4mm,

3-toothed at the apex. Seeds c.0.6 × 0.5mm, black, glossy, minutely tuberculate. **Map 644, Fig. 87.**

Dry rocky slopes; 0–600m.

Yemen (S). Djibouti.

All Arabian plants belong to var. *adenensis* Perk. which is characterized by its smaller, densely crowded leaves.

10. R. viridis Balf. f. (1882 p.501). Illustr.: Abdallah & de Wit, op. cit.: fig. 89 (1978). Type: Socotra, *Balfour, Cockburn & Scott* 230 (E).

Shrub, glabrous and branched throughout; stems erect or ascending, up to 150cm. Leaves entire or 2–3-lobed, ovate or broadly ovate to ± circular, 1.5–6 × 1–3cm, rounded at the tip, the base acute or truncate; petiole 1–3cm. Flowers greenish white, 6-merous. Upper petals 2–3mm, divided from the middle into 6–9 lobes; central lobe longer and broader than the lateral lobes. Stamens 12–15. Capsule erect or nodding, ellipsoid to obovoid, 8–15 × 3–4mm, strongly 3-angled in section, 3-toothed at the apex. Seeds c.7 × 7mm, dark brown, minutely tuberculate. **Map 645, Fig. 87.**

On cliffs and amongst large boulders in dense deciduous bushland; on limestone; 30–750m.

Socotra. Endemic.

Family 59. MORINGACEAE

A.G. MILLER

Deciduous trees. Leaves large, alternate, 2–3-pinnate; leaflets mainly opposite, entire, the base of the petiolules and pinnae articulated and with stipitate glands. Stipules minute or absent. Flowers bisexual, zygomorphic, in axillary panicles. Sepals 5, free above the cup-shaped receptacle. Petals 5, free, unequal. Stamens 5, epipetalous or inserted on the margin of the disc, alternating with 3–5 staminodes. Ovary superior, 1-locular, with numerous ovules on 3 parietal placentas; style terminal, filiform with a minute stigma. Fruit a pendant 3-valved capsule. Seeds wingless or 3-winged.

Verdcourt, B. (1985). A synopsis of Moringaceae. *Kew Bull.* 40: 1–23.

MORINGA Adanson

1. Seeds wingless; leaflets soon deciduous, narrowly oblong-obovate, 1–6mm broad; leaf axes persistent **1. M. peregrina**
+ Seeds 3-winged; leaflets usually persistent, elliptic to obovate, 3–15mm broad; leaf axes falling with the leaflets **2. M. oleifera**

1. M. peregrina (Forsskal) Fiori in Agric. Colon. 5: 59 (1911). Syn.: *Hyperanthera peregrina* Forsskal (1775 pp.CVII & 67; *Moringa aptera* Gaertn., Fruct. sem. pl. 2: 315 (1791); *M. arabica* Pers., Syn. pl. 1: 461 (1805). Illustr.: Miller (1988 p.211);

Western (1989 p.84). Type: Yemen (N), *Forsskal* (C).

Tree up to 4(–10)m. Leaves up to 30cm long, the leaf axes persistent but the leaflets soon deciduous; leaflets narrowly oblong-obovate, 10–20 × 1–6mm, obtuse. Sepals narrowly oblong, 8–9 × 1.5–2mm, acuminate. Petals white or streaked red or pink, oblong-obovate, 8–15 × 2–4mm, obtuse. Capsule cylindrical, 10–30 × 1–1.5cm, deeply longitudinally ribbed, somewhat torulose; seeds ovoid, slightly 3-angled, 13–15 × 8–10mm, whitish brown. **Map 646, Fig. 88.**

Cliffs and rocky slopes, sandy and rocky wadi-beds; 50–1400m.

Saudi Arabia, Yemen (N & S), Oman, UAE. NE tropical Africa, SW Asia to ?Pakistan.

2. M. oleifera Lam., Encycl. 1: 398 (1785). Syn.: *M. pterygosperma* Gaertn., Fruct. sem. pl. 2: 314 (1791). Illustr.: Fl. Trop. E. Afr.: 10 (1986).

Similar to *M. peregrina* but the leaflets persistent, the leaf axes deciduous together with the leaflets; leaflets elliptic to obovate, 5–25(–30) × 3–15(–20)mm; petals white, cream or tinged yellow; capsule 3-angled in section; seeds 3-winged. **Map 647, Fig. 88.**

Cultivated in gardens; 30–50m.

Yemen (N & S), Oman. A native of India and Pakistan now widely cultivated in the tropics.

Cultivated as an ornamental, and for the oil extracted from its seeds.

Family 60. CRASSULACEAE

A.G. MILLER

Annual, biennial or perennial herbs or small shrubs, often succulent. Leaves opposite or alternate, usually simple and entire; stipules absent. Flowers bisexual, regular, axillary or in terminal cymes or spikes, 4–10-merous. Sepals and petals free or united. Stamens equal to or twice as many as the petals, free or adnate to the petals; scale-like nectaries usually present. Ovary superior; carpels free or slightly fused at the base, equal in number to the sepals, tapering above into the style; ovules few to many. Fruit of free follicles.

1.	Leaves circular, peltate with long petioles	**6. Umbilicus**
+	Leaves various, narrowed at the base into a short petiole or sessile, never peltate or circular	2
2.	Petals free or united only at the base	3
+	Petals united at least to the middle	5
3.	Shrubs; flowers 7–10-merous	**3. Aeonium**
+	Annual or biennial herbs; flowers 3–5-merous	4

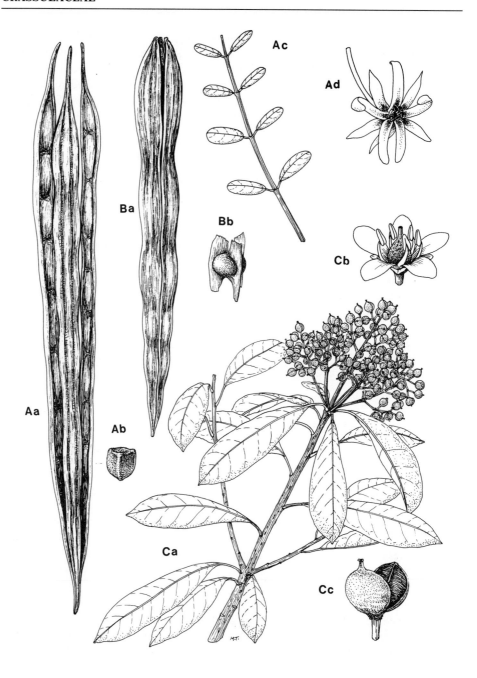

Fig. 88. Moringaceae. A, *Moringa peregrina*: Aa, fruit (×0.6); Ab, seed (×0.6); Ac, leaf (×0.6); Ad, flower (×1.5). B, *M. oleifera*: Ba, fruit (×0.6); Bb, seed (×0.6). **Pittosporaceae.** C, *Pittosporum viridiflorum*: Ca, fruiting branch (×0.6); Cb, flower (×5); Cc, fruit (×8).

4.	Leaves opposite or in a basal rosette, united at the base to form a sheath around the stem	**1. Crassula**
+	Leaves alternate, not united at the base	**2. Sedum**
5.	Flowers 4-merous	**4. Kalanchoe**
+	Flowers 5-merous	**5. Cotyledon**

1. CRASSULA L.

Annual or perennial, often fleshy, herbs. Leaves opposite or in a basal rosette, sessile or shortly petiolate, united at the base to form a sheath around the stem. Flowers 3–5(–6)-merous, in 1–4-flowered axillary clusters or a terminal panicle. Sepals free or shortly united at the base. Petals very shortly united at the base. Stamens equal in number to the petals, attached to the base of the corolla. Nectary scales small. Carpels 3–5(–6), free or slightly united at the base; ovules numerous.

1.	Leaves in a basal rosette; flowers in a terminal corymbose panicle	**6. C. alba**
+	Leaves opposite, not in a basal rosette; flowers axillary	2
2.	Leaves shortly petiolate, the blade ovate to elliptic or rhombic	**4. C. alsinoides**
+	Leaves sessile, the blade linear to lanceolate	3
3.	Sepals rounded; herb of moist habitats, either submerged or in mud at the edge of pools	**5. C. granvikii**
+	Sepals acute or acuminate; herbs of well-drained habitats	4
4.	Perennial herb; leaves acute or obtuse at the tip, never aristate; seeds smooth	**3. C. schimperi**
+	Ephemeral herb; leaves with a short, white arista at the tip; seeds longitudinally ridged	5
5.	Petals about half as long as the sepals	**1. C. alata**
+	Petals equalling or barely exceeding the sepals	**2. C. tillaea**

1. C. alata (Viv.) Berger in Engl. & Prantl, Nat. Pflanzenfam., ed. 2, 18a: 389 (1930). Syn.: *Tillaea alata* Viv., Pl. aegypt. dec. 4: 16 (1830).

Glabrous ephemeral herb; stems erect, up to 5(–12)cm. Leaves sessile, lanceolate, 2.5–4 × 0.5–1.5mm, shortly aristate. Flowers 3–5-merous, 2 per leaf axil; pedicels 0.5–2mm. Calyx lobes lanceolate, 1–1.5 × 0.2–0.4mm, acuminate, aristate, shortly united at the base. Petals white or greenish, narrowly ovate, 0.6–0.8 × 0.2–0.3mm, acuminate. Follicles 2-seeded; seeds longitudinally ridged. **Map 648, Fig. 89.**

1.	Flowers 3–4-merous	subsp. **alata**
+	Flowers 5-merous	subsp. **pharnaceoides**

subsp. **alata**

In gravel and hard sand; on limestone hillsides; 280–500m.

Saudi Arabia, Oman, UAE, Kuwait. N Africa eastwards to Iran and NW India.

subsp. **pharnaceoides** (Fischer & C. Meyer) Wickens & Bywater in Kew Bull. 34: 633 (1980). Syn.: *C. pharnaceoides* Fischer & C. Meyer, Index sem. hort. petrop. 8: 56 (1842); *Tillaea pharnaceoides* Steud., Nomencl. bot. ed. 2, 2: 687 (1841) nom. nud.

Rocky slopes; 1800–2600m.

Saudi Arabia, Yemen (N), Socotra. Cameroon, tropical NE & E Africa.

2. C. tillaea Lester-Garl., Fl. Jersey: 87 (1903).

Similar to *C. alata* but a minute, moss-like ephemeral herb; stems 1–5cm, often reddish tinged; flowers 3–4-merous; petals equalling or barely exceeding the sepals. **Map 649.**

In shallow bogs over granite and in damp sand by pools; 1980–2050m.

Saudi Arabia. S & W Europe, Turkey, N Africa, Canary Is. & Madeira.

3. C. schimperi Fischer & C. Meyer in Index sem. hort. petrop. 8: 56 (1841).

Glabrous perennial herb; stems erect or sprawling, often rooting at the nodes, up to 15cm, succulent, sometimes woody-based, with peeling brown bark at the base. Leaves sessile, linear to lanceolate, 3–8(–12) × 0.5–2(–3)mm, acute to obtuse. Flowers (4–)5-merous, 1–4 in axils of leaves, subsessile or on short (up to 1mm) pedicels. Calyx lobes triangular-ovate, (0.7–)1.4–1.5(–1.8)mm, acute to acuminate, shortly united at the base. Petals whitish, greenish or pinkish, ovate, 1–1.5 × 0.3–0.6mm, acute to acuminate. Follicles 2-seeded; seeds smooth. **Map 650, Fig. 89.**

1. Leaves thin when dry, concolorous, not rigid; sheath spurred
   　　　　　　　　　　　　　　　　　　　　　　　　　　subsp. **schimperi**
+ Leaves thick when dry, discolorous, rigid; sheath not spurred
   　　　　　　　　　　　　　　　　　　　　　　　　　　subsp. **phyturus**

subsp. **schimperi**. Syn.: *C. pentandra* (Edgew.) Schönl. in Engl. & Prantl, Nat. Pflanzenfam. ed. 1, 3 (2a): 37 (1891); *Tillaea pentandra* Edgew. in Trans. Linn. Soc. London 20: 50 (1846). **Fig. 89.**

Rocky slopes and in gravel; 1300–3200m.

Saudi Arabia, Yemen (N & S). Tropical Africa and India.

subsp. **phyturus** (Mildbr.) Ros. Fernandes in Bol. Soc. Brot. 52: 172 (1978). Illustr.: Fl. Trop. E. Afr.: 9 (1987).

Differs from the type subspecies in the characters given in the key; a 'Lycopodium-like' plant.

In gravel amongst rocks, in dwarf shrubland; 600–1370m.

1. Crassula

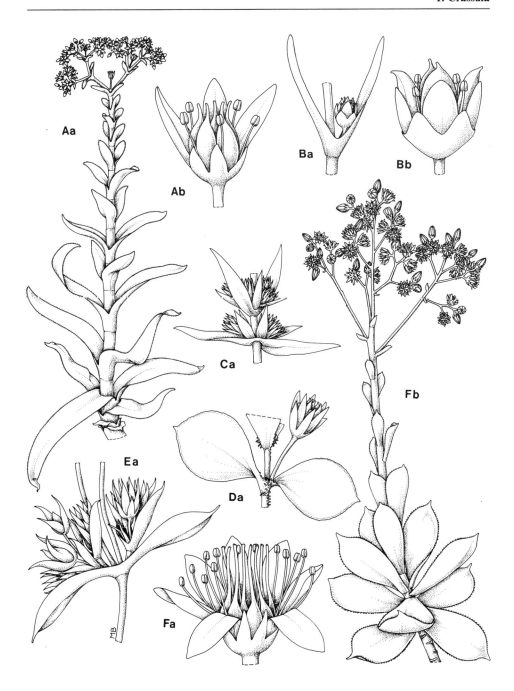

Fig. 89. Crassulaceae. A, *Crassula alba*: Aa, habit (×0.6); Ab, flower with 2 petals removed (×8). B, *C. granvikii*: Ba, leaves and flower (×6); Bb, flower (×16). C, *C. schimperi* subsp. *schimperi*: Ca, part of inflorescence (×6). D, *C. alsinoides*: Da, part of inflorescence (×2.5). E, *C. alata*: Ea, part of inflorescence (×6). F, *Aeonium leucoblepharum*: Fa, flower with a petal and 3 stamens removed (×5); Fb, habit (×0.6).

Socotra. Uganda, Kenya, Tanzania, Sudan and Ethiopia.

4. C. alsinoides (Hook. f.) Engl., Hochgebirgsfl. Afrika: 231 (1892). Syn.: *C. pellucida* L. subsp. *alsinoides* (Hook. f.) Tölken in J. S. African Bot. 41: 114 (1975); *Tillaea alsinoides* Hook. f. in J. Linn. Soc., Bot. 7: 192 (1864).

Creeping perennial herb, glabrous except for longitudinal lines of papillose hairs along the stems; stems up to 50cm, rooting at the lower nodes. Leaves shortly petiolate, ovate to elliptic or rhombic, 7–20 × 2–12mm, acute or rarely obtuse. Flowers 5(–6)-merous, usually solitary in the leaf axils; pedicels 3–17mm. Calyx lobes lanceolate, 3–5 × 0.75–1mm, acute or acuminate, shortly united at the base. Petals white or pink, oblong-lanceolate, 5–6 × 1.2–1.8mm, acute. Follicles many-seeded. **Map 651, Fig. 89.**

Moist ground by water, terrace-walls; 2000m.

Yemen (N). Tropical Africa and Madagascar.

This species has only been recorded once from Yemen (J. Wood, pers. comm.). No Arabian material has been seen and the above description is based on African material.

5. C. granvikii Mildbr. in Notizbl. Bot. Gart. Berlin-Dahlem 8: 227 (1922). Syn.: *Bulliarda abyssinica* A. Rich. (1847–8 p.306), non *Crassula abyssinica* A. Rich. (1847–8); *C. hedbergii* Wickens & Bywater in Kew Bull. 34: 631 (1980).

Aquatic annual herb, glabrous throughout. Stems erect or ascending, rooting at the nodes. Leaves sessile, linear-oblong to oblong-spathulate, 2–10(–20) × 0.3–3(–5)mm, acute to obtuse. Flowers 4-merous, solitary in the leaf axils; pedicels stout, c.1mm. Calyx lobes triangular, c. 0.7 × 0.5mm, rounded, connate at the base. Petals pink (or white), ovate, c.1 × 0.5mm, rounded. Follicles (4–)6–7-seeded. **Map 652, Fig. 89.**

Rooted in mud in pools; 3000m.

Yemen (N). E tropical Africa from Ethiopia to Malawi.

6. C. alba Forsskal (1775 p.60). Illustr.: Fl. Trop. E. Afr. 14 (1987). Type: Yemen (N), *Forsskal* (C).

Perennial rosette-leaved herb; stems erect, up to 100cm, glabrous or retrorsely papillose-hairy. Basal leaves fleshy, ovate to linear-lanceolate or oblong-lanceolate, 1.5–10(–20) × 0.5–1.5(–2.5)cm, acute, with papillose-hairy margins; stem-leaves similar but becoming smaller above. Flowers 5-merous, in a terminal corymbose panicle; pedicels 2–10(–15)mm. Calyx lobes triangular-ovate to ovate, 1.5–4 × 1–1.25mm, acute, fringed with papillae or glabrous. Petals white, pink or yellowish, oblong, 3–4 × 1.25–1.5mm, mucronate and slightly hooded at the tip. Follicles 8–30-seeded. **Map 653, Fig. 89.**

Cliffs and open rocky slopes; 1500–2850m.

Saudi Arabia, Yemen (N). E Africa, from Ethiopia south to South Africa.

A very variable species particularly as regards the ratio of calyx to petal length and the presence or absence of papillose hairs on the margins of the sepals and leaves.

2. SEDUM L.

Annual or rarely biennial herbs. Leaves succulent, alternate, entire, often spurred at the base. Flowers 4–5(–9)-merous, in few-flowered terminal cymes. Sepals free or connate at the base. Petals free. Stamens equal to or twice the number of the petals. Nectary scales small. Carpels 4–5, free; ovules numerous.

1. Plant glandular-pubescent above; leaves ± oblong, 5–15mm long; stamens 10; follicles stellately spreading **1. S. hispanicum**
+ Plant glabrous; leaves broadly ovoid to ellipsoid, 3–6mm long; stamens 4–5; follicles erect-spreading **2. S. caespitosum**

1. S. hispanicum L., Cent. pl. I: 13 (1755) & Amoen. acad. 4: 273 (1759). Illustr.: Collenette (1985 p.191).

Annual or rarely biennial herb, thinly glandular-hairy above. Stems erect or ascending, up to 10(–15)cm. Leaves terete or somewhat flattened, ± oblong, 5–15 × 1–4mm, green or glaucous. Flowers 5-merous, subsessile, in simple or 2–5-forked lax, spicate cymes. Calyx lobes ovate-triangular, 1–2(–2.5) × 0.8–1.5mm. Petals white, with or without a pink midrib, lanceolate, 4.5–5(–6) × 1–1.5(–2)mm, acuminate. Stamens 10. Nectary scale lobed. Follicles stellately spreading. **Map 654, Fig. 92.**

Rocky slopes, granite sand pans and stony wadi-banks; 1130–3000m.

Saudi Arabia, Yemen (N), Oman. S & C Europe, Caucasus, Turkey, Lebanon, Palestine and N Iran.

Arabian plants all seem referable to var. *semiglabrum* Fröder. which has 5-merous (not 6–9-merous) flowers and simple stems below the inflorescences.

2. S. caespitosum (Cav.) DC., Prodr. 3: 405 (1828). Syn.: *S. rubrum* (L.) Thell. in Repert. Spec. Nov. Regni Veg. 10: 290 (1912).

Glabrous annual herb. Stems erect, up to 3cm. Leaves globular or ± oblong, 3(–6) × 1.5–2mm, green or reddish-tinged. Flowers 4–5-merous, sessile, in simple (or forked), lax, spicate cymes. Calyx lobes ovate-triangular, 1 × 0.8–1mm. Petals white or reddish white, lanceolate, c.3 × 0.8mm, mucronate. Stamens 4–5. Nectary-scale entire. Follicles erect-spreading. **Map 655, Fig. 92.**

Clay pans and margins of alpine meadows; 2370m.

Saudi Arabia. Europe and SW Asia.

3. AEONIUM Webb & Berth.

Succulent shrub. Leaves alternate or spirally arranged. Flowers 8–10-merous, in a terminal corymb. Sepals and petals fused at the base. Stamens twice as many as the petals. Nectary scales small. Carpels 8–10, united at the base.

Liu, Ho-Yih (1989). Systematics of Aeonium. Special Publ. Natl. Mus. Nat. Sci. Taiwan 3: 75–77.

1. A. leucoblepharum Webb ex A. Rich. (1848 p.314). Syn.: *Sempervivum chrysanthum* Britten, Fl. Trop. Afr. 2: 400 (1871); *Aeonium chrysanthum* (Britten) Berger in Engl. & Prantl, Nat. Pflanzenfam. ed. 2, 18a: 432 (1930). Illustr.: Fl. Ethiopia 3: 17 (1989).

Stems up to 50cm, $\pm$ glabrous or glandular-pubescent. Leaves arranged in rosettes at the tips of sterile shoots and spirally arranged along the flowering shoots, spathulate, 1.2–12 × 1–3.5(–5)cm, rounded and apiculate or shortly acuminate at the tip, the margins ciliate. Calyx lobes triangular, 1.25–2 × 0.8–1.25mm, glabrous or glandular-pubescent. Petals yellow, oblong to oblanceolate, 5–6 × 1.25–3mm. **Map 656, Fig. 89.**

On cliffs; 2500–2800m.

Yemen (N). Uganda, Kenya, Somalia & Ethiopia.

4. KALANCHOE Adans.

Annual or perennial succulent herbs or sub-shrubs. Leaves opposite, simple, deeply incised or 3-foliolate, entire or crenate to serrate. Flowers 4-merous, in a terminal, many-flowered panicle. Calyx lobes $\pm$ free or fused at the base. Corolla tubular with spreading lobes. Stamens 8 in 2 whorls of 4, epipetalous. Nectary scales present. Carpels 4, free; ovules numerous.

A difficult genus where species limits are often not clear cut. The difficulties are increased by the generally inadequate herbarium material and the problem of matching Arabian plants with African species. In Africa species with local polyploids have been found (Raadts 1983) and an understanding of the situation in Arabia will probably require extensive field observations and cytotaxonomic studies.

Cufodontis, G. (1965). The species of Kalanchoe occurring in Ethiopia and Somalia Republic. *Webbia* 19: 711–744; Raadts, E. (1977). The genus Kalanchoe in tropical East Africa. *Willdenowia* 8: 101–157. Raadts, E, (1983). Cytotaxonomische Untersuchungen an Kalanchoe 1. *Willdenowia* 13: 373–385.

1.	Leaves $\pm$ linear, terete or sub-triangular in section	**11. K. bentii**
+	Leaves broader, $\pm$ flat in section (although usually fleshy)	2
2.	Corolla tube 20–30mm; corolla lobes rounded; all anthers exserted; flowers slightly zygomorphic and held horizontally (Socotra)	**12. K. robusta**
+	Corolla tube 6–19mm; corolla lobes acute, apiculate or mucronate; anthers usually included or at most 4 slightly exserted; flowers actinomorphic, held $\pm$ erect	3

3.	Young leaves and stem white or whitish green, scaly and peeling when dry; corolla tube uniformly cylindrical, not abruptly narrowed above the ovary (Socotra)	**13. K. farinacea**
+	Leaves and stem green or glaucous, not scaly or peeling when dry; corolla tube abruptly narrowed above ovary	4
4.	Plant hairy, at least in the region of the inflorescence	5
+	Plant totally glabrous	10
5.	Hairs non-glandular	**4. K. citrina**
+	Hairs glandular	6
6.	Leaves trifoliolate or deeply incised	**3. K. laciniata**
+	Leaves entire or at most deeply toothed	7
7.	Leaves sessile, clasping the stem at the base	8
+	Leaves shortly petiolate, not clasping the stem	9
8.	Stem square at the base, rounded above; stamens included within the corolla tube; corolla lobes 1.25–2.5mm broad; indumentum of long-stalked glandular hairs (more than 0.4mm long); annual	**1. K. lanceolata**
+	Stem round at the base; upper stamens held at the mouth of the corolla tube; corolla lobes 2.5–4.5mm broad; indumentum of short-stalked glandular hairs (up to 0.2mm long); perennial	**2. K. yemensis**
9.	Corolla orange-red or pink (rarely yellow); corolla lobes 3.5–4mm wide; upper anthers held at the mouth of the corolla tube	**6. K. deficiens**
+	Corolla yellow or orange; corolla lobes 1.25–2.5mm wide; all anthers included within the corolla tube	**9. K. crenata**
10.	Calyx lobes 1–2.5mm long; corolla lobes linear-ovate, 0.5–0.8mm broad (Socotra)	**10. K. rotundifolia**
+	Calyx 3–10mm long; corolla lobes broader, 1.25–4mm broad	11
11.	Corolla tube 17–19mm; styles c.6mm	**7. K. sp. A**
+	Corolla tube 8–16mm; styles 2–4mm	12
12.	Corolla yellowish, pinkish or reddish green; corolla (with lobes included) more than 25mm long; corolla lobes 6–18mm long, acute; leaves crowded on swollen base of stem	**5. K. alternans**
+	Corolla red, orange or yellow; corolla (with lobes included) less than 20mm long; corolla lobes 4–7mm long; leaves not usually crowded	13
13.	Corolla scarlet, orange-red, orange or pink; corolla lobes 2.25–4mm wide; upper anthers held at the mouth of the corolla tube	**6. K. deficiens**
+	Corolla yellow, orange or pinkish orange; corolla lobes 1.5–2mm wide; all anthers included within the corolla tube	**8. K. glaucescens**

1. K. lanceolata (Forsskal) Pers., Syn. pl. 1: 446 (1805). Syn.: *Cotyledon lanceolata* Forsskal (1775 p.89); *K. laciniata* sensu Schwartz (1939) non DC. (1802); *Verea* *lanceolata* (Forsskal) Sprengel, Syst. veg. 2: 260 (1825). Illustr.: Collenette (1985 p. 191). Type: Yemen (N), *Forsskal* 68 (C).

Annual herb, up to 1m, glabrous below, glandular-pubescent with long (0.4–1mm) hairs above. Stem square below, rounded in the region of the inflorescence. Leaves lanceolate to narrowly oblong or obovate, 6–17 × 1–7cm, usually obtuse, entire or sinuate to bluntly crenate or serrate, sessile with a clasping base. Calyx lobes connate at the base, ovate or ovate-triangular to lanceolate, 4–8(–13) × 1–3(–4.5)mm. Corolla yellow, orange, pink or flame red; tube 9–15mm; lobes ovate, 3.5–4 × 1.25–2.5mm, mucronate. All anthers included within the corolla tube. Nectary scales linear. Styles c.1mm. **Map 657, Fig. 90.**

Rocky slopes, ravines and below bushes in open *Acacia-Commiphora* bushland; 550–1700m.

Saudi Arabia, Yemen (N & S). Tropical and southern Africa, Madagascar and India.

Readily distinguished from all other kalanchoes in Arabia by its stems which are square below and rounded above. There is apparently a geographically based variation in flower colour throughout the range of the species: in Yemen the flowers are predominantly orange, pink or red; in Saudi Arabia the only flower colour recorded is yellow; in Africa the flowers are typically yellow or orange, and only rarely pink. *K. lanceolata* is restricted to the dry foothills of the outer escarpment mountains and records from the high inner plateau are probably referable to the following species.

2. K. yemensis (Deflers) Schweinf. (1896 p. 203). Syn.: *K. brachycalyx* A. Rich. var. *yemensis* Deflers (1889 p.138). Syntypes: Yemen (N), *Deflers* 632, 694 (P).

Perennial herb up to 1m, glabrous below, glandular-pubescent with short (up to 0.2mm) hairs above. Stems round. Leaves narrowly oblong to lanceolate or ovate, 4–8(–16) × 1.5–3(–6)cm, obtuse or acute, entire to sinuate or rarely bluntly serrate, sessile with a clasping base. Calyx lobes connate at the base, oblong-ovate to ovate-triangular, 4–10 × 1.5–4mm. Corolla bright yellow, pale yellow or orange-yellow; tube 8–12mm; lobes ovate, 5–8 × 2.5–4.5mm, mucronate. Upper anthers held at the mouth of the corolla tube. Nectary scales linear. Styles 2–4mm. **Map 658, Fig. 90.**

Rocky slopes, volcanic cones, wadi-banks and field margins; 1900–2600m.

Yemen (N). Endemic.

Closely related to and sometimes confused with *K. lanceolata* but occurring further east, mainly on the high inner plateau and inner escarpment mountains.

3. K. laciniata (L.) DC., Pl. hist. succ. 2: 100 (1802) non sensu Schwartz (1939). Syn.: *K. schweinfurthii* Penzig in Atti Congr. Bot. Genova 1892: 341 (1893). Illustr.: Collenette (1985 p.190) as *K.* sp. aff. *laciniata*.

Perennial herb up to 1m, glandular-pubescent above, glabrous below. Leaves divided, usually trifoliolate or 3-lobed, bluntly serrate, petiolate with the petiole broadened or semi-clasping at the base; leaflets lanceolate, up to 14 × 6cm, acute, dentate or lobed. Calyx lobes free ± to the base, narrowly triangular, acute,

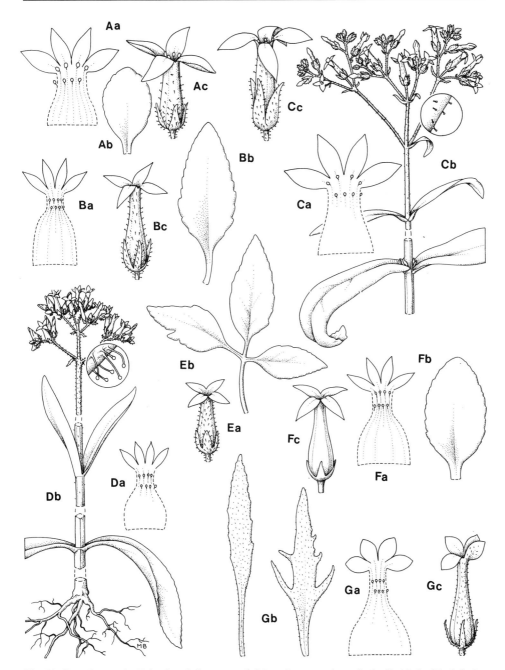

Fig. 90. Crassulaceae. A, *Kalanchoe deficiens* var. *deficiens*: Aa, opened corolla (×2); Ab, leaf (×0.6); Ac, flower (×2). B, *K. crenata*: Ba, opened corolla (×2); Bb, leaf (×0.6); Bc, flower (×2). C, *K. yemensis*: Ca, opened corolla (×2); Cb, flowering shoot (×0.6); Cc, flower (×2). D, *K. lanceolata*: Da, opened corolla (×2); Db, habit (×0.6). E, *K. laciniata*: Ea, flower (×2); Eb, leaf (×2). F, *K. glaucescens*: Fa, opened corolla (×2); Fb, leaf (×0.6); Fc, flower (×2). G, *K. citrina*: Ga, opened corolla (×2); Gb, leaves (×0.6); Gc, flower (×2).

(2.5–)3(–5) × 0.5(–1.8)mm. Corolla pale yellow, deep yellow or pink; tube 8.5–11mm; lobes oblong to lanceolate, 3.5–5 × 1.5–3mm, apiculate. Upper anthers included within the corolla tube. Nectary scales linear. Styles 0.75–2mm. **Map 659, Fig. 90.**

Wooded areas and rocky hillsides; 950–1600m.

Saudi Arabia. NE, E and southern tropical Africa; India.

Only recorded twice from Saudi Arabia. Readily recognized by its deeply divided leaves and glandular-pubescent inflorescence. See also note under *K. glaucescens*.

4. K. citrina Schweinf. (1896 p.199). Type: Yemen (N), *Schweinfurth* 1831 (n.v.).

Tomentose perennial herb up to 60(–90)cm; hairs non-glandular, ± totally covering the plant. Leaves lanceolate to rhombic, simple or deeply incised to lobed, up to 10 × 3cm, acute, entire or dentate, sessile or shortly petiolate, the petiole not broadened at the base. Calyx lobes free ± to the base, narrowly triangular to subulate, 4–8 × 0.5–2mm. Corolla pale yellow; tube 9–13mm; lobes obovate, 3–5(–7) × 1.25–3.5mm, mucronate. All anthers included within the corolla tube. Nectary scales linear. Styles 1.5–2mm. **Map 660, Fig. 90.**

Rocky hillsides and wadis; 950–1600mm.

Saudi Arabia, Yemen (N). E and NE tropical Africa.

5. K. alternans (Vahl) Pers., Syn. pl. 1: 446 (1805). Syn.: *Cotyledon alternans* Vahl (1791 p.51) non Haw. (1819); *C. orbiculata* Forsskal (1775 p.CXII) non L. (1753); *K. rosulata* Raadts in Bot. Jahrb. Syst. 91: 480 (1972); *Verea alternans* (Vahl) Sprengel, Syst. veg. 2: 260 (1825). Illustr.: Collenette (1985 p.189). Type: Yemen (N), *Forsskal* (C).

Glabrous perennial herb, up to 50cm. Leaves mostly crowded at the base of the stem, broadly ovate or lanceolate, 2.5–12 × 1–6.5cm, acute or obtuse, entire or crenate, subsessile with the base not broadened. Calyx lobes free ± to the base, narrowly triangular to lanceolate, 4–10 × 1–2.5mm. Corolla yellowish to pinkish or reddish green, cream or ?white; tube 12–16mm; lobes lanceolate to oblong-lanceolate, 6–18 × 1–4.5mm, acute. Upper anthers held at the mouth of the corolla tube. Nectary scales linear. Style 3–5mm. **Map 661, Fig. 91.**

Dry and exposed places, stony ground, rocky slopes and grassy hillsides; 1800–2900m.

Saudi Arabia, Yemen (N & S). Endemic.

There are apparently two distinct forms of this species: the first (resembling the type) with broadly ovate leaves; and the second with lanceolate leaves.

6. K. deficiens (Forsskal) Asch. & Schweinf. in Mém. Inst. Égypt. 2: 79 (1887). Syn.: *Cotyledon deficiens* Forsskal (1775 p.89); *K. aegyptiaca* (Lam.) DC., Pl. hist. succ.: tab. 64 (1801); *K. crenata* sensu J. Wood non (Andrews) Haw.; *K. glaucescens* Britten var. *deficiens* (Forsskal) Senni in Boll. Reale Orto Bot. Giardino Colon. Palermo 4: 12 (1905); *K. laciniata* sensu Schwartz (1939) pro parte non DC.; *Cotyledon nudicaulis* Vahl (1791 p.51). Neotype: Yemen (S), *Deflers* 572 (P).

Perennial, clump-forming herb up to 1.5m, glabrous throughout or glandular-pubescent above. Leaves ovate to broadly obovate, 4–10 × 2.5–4.5cm, obtuse, entire or crenate, shortly petiolate, the base of the petiole not broadened, glabrous. Calyx lobes free ± to the base, narrowly ovate to lanceolate, 3–6.5 × 1–2mm. Corolla scarlet, orange-red or pink; tube 8–12mm; lobes ovate to obovate, 5–7 × 2.25–4mm, mucronate. Upper anthers held at the mouth of the corolla tube. Nectary scales linear. Styles 2–4mm. **Fig. 90.**

	1.	Plants glandular-pubescent above, glabrous below	var. **deficiens**
	+	Plants glabrous throughout	var. **glabra**

var. **deficiens**

Plant glandular-pubescent above, glabrous below; corolla orange-red or pink; corolla lobes 5–7 × 3.5–4mm. **Map 662, Fig. 90.**

Open scrub and steep rocky slopes; 1400–1800 m.

Yemen (N & S). Endemic.

var. **glabra** Raadts in Willdenowia 18: 423 (1989). Syn.: *K. glaucescens* Britten subsp. *arabica* Cuf. in Webbia 19: 719 (1965); *K. glaucescens* sensu Schweinf. (1896) pro parte non Britten (1871); *K. laciniata* sensu Schwartz (1939) pro parte non DC. (1802). Type: Yemen (N), described from cultivated material grown at the Berlin Botanical Garden, *Müller-Hohenstein & Deil* 635, exs. 20. 1. 1987 (B).

Similar to var. *sabaea* but glabrous throughout; corolla orange, scarlet, or salmon-red; corolla lobes 5–7 × 2.25–4mm broad. **Map 663.**

Rocky slopes, terrace-walls and grassy banks around fields; 600–2200mm.

Yemen (N). Endemic.

7. K. sp. A.

Similar to *K. sabaea* but glabrous throughout; flowers yellow, larger; calyx lobes 6–8 × 1.5–2.5mm; corolla tube 17–19mm; corolla lobes c.8 × 3mm; styles c.6mm.

Known only from a single cultivated gathering of Socotran origin.

8. K. glaucescens Britten in Oliv., Fl. Trop. Afr. 2: 393 (1871), non Blatter (1919–36). Illustr.: Collenette (1985 p.190).

Glaucous, clump-forming perennial, up to 1m, glabrous throughout. Leaves succulent, sometimes red-tinged, ovate to broadly ovate, 7–15 × 3–8cm, obtuse, subentire to crenate; petioles 1–3cm. Calyx lobes free ± to the base, narrowly lanceolate, 3.25–5.5 × 1–1.5mm, acute. Corolla deep yellow, orange-yellow, orange or pinkish orange; tube 9–10mm; lobes 4.5–5.5 × 1.5–2mm, mucronate. All anthers included within the corolla tube. Nectary scales linear. Styles 1–2mm. **Map 664, Fig. 90.**

Rocky slopes and open shrubland; 400–1300m.

Saudi Arabia, Oman. NE & E tropical Africa.

Two gatherings (*Collenette* 3293 and *Baierle et al.* 82-1110) from J. Fayfa in the extreme SW of Saudi Arabia differ from typical material in being possibly annual, having more open inflorescences and smaller, orange flowers. They are possibly referable to *K. densiflora*, an African species. In Arabia the flower colour of *K. glaucescens* is recorded as orange, orange-yellow, yellow or rarely (*Collenette* 7438) bright pinkish orange. However, in E Africa the flower colour is more variable with yellow, orange, pink, red or crimson flowers being found. Further gatherings and field observations of *K. glaucescens*, *K. crenata* and *K. laciniata* from the mountains of SW Saudi Arabia are desirable. This treatment of these species must be considered provisional.

9. K. crenata (Andrews) Haw., Syn. pl. succ.: 109 (1812).

Perennial herb up to 1.5m, glandular-pubescent in the region of the inflorescence, glabrous below. Leaves ovate, up to 15 × 6cm, obtuse, crenate; petiole up to 4cm. Calyx lobes free ± to the base, narrowly lanceolate, 3–7 × 1–1.25. Corolla bright yellow or orange; tube 10–12mm; lobes 4.5–6 × 1.25–2.5mm, mucronate. All anthers included within the corolla tube. Nectary scales linear. Styles 1–2mm. **Map 665, Fig. 90.**

Rocky areas, in drought-deciduous bushland on hillsides and in wadis; 750–1600m.

Saudi Arabia. Throughout tropical Africa; naturalized elsewhere.

Flowers of Arabian plants are generally smaller than those of specimens examined from E Africa. Two specimens (*Collenette* 3289 and 3290) from Saudi Arabia which were collected at the same time and place are almost identical except that the former has simple leaves and had been placed under *K. crenata*, whereas the latter has trifoliolate leaves and has been conditionally placed under *K. laciniata* (see also note under *K. glaucescens*).

10. K. rotundifolia (Haw.) Haw. in Philos. Mag. J. 66: 31 (1825).

Glabrous annual or perennial herb up to 40cm, glabrous throughout. Leaves ovate or rhombic to oblanceolate or broadly obovate, 2–5 × 0.5–2.5cm, acute to obtuse or rounded, entire, shortly petiolate, the petiole not broadened at the base. Calyx lobes free ± to the base, subulate or narrowly triangular, 1–2.5 × 0.2–0.5mm. Corolla scarlet or red; tube 8–10mm; lobes linear, 5–8 × 0.5–0.8mm, acute. All anthers included within the corolla tube. Nectary scales linear. Style c. 1mm. **Map 666, Fig. 91.**

Low shrubland on steep rocky slopes and sand dunes; 0–750m.

Socotra. Tanzania, Mozambique, Zimbabwe, Botswana, Namibia and South Africa.

The Socotran populations of *K. rotundifolia* are a distant outlier of the species. The nearest station outside Arabia is in Tanzania.

11. K. bentii C.H. Wright in Bot. Mag. 127: t. 7765 (1901) subsp. **bentii**. Syn.: *K. deflersii* Gagnepain in Notul. Syst. (Paris) 3: 221 (1916); *K. teretifolia* Deflers in Bull. Soc. Bot. France 40: 299 (1893) non Haw. ex Wallich (1831). Type: Yemen (S), described from cultivated material grown at Kew Gardens, *Bent* s.n. (K).

Glabrous, perennial herb up to 1m. Leaves ± cylindrical, terete to ± triangular in section, 7–40 × 0.5–1.5cm, acute, entire, semi-clasping at the base. Sepals free ± to the base, lanceolate, 5–12 × 1.5–2.5mm. Corolla whitish or pinkish tinged; tube 20–30mm; lobes ovate, 10–15 × 2–6mm, mucronate. Upper anthers held at the mouth of the corolla tube. Nectary scales linear. Styles c.5mm. **Map 667, Fig. 91.**

Rocky slopes and on limestone in *Acacia-Commiphora* bushland; 700–1550m.

Yemen (S). Endemic subspecies.

Two subspecies are recognized within *K. bentii*: subsp. *bentii* which is completely glabrous and endemic to the Arabian Peninsula; and subsp. *somalica* Cuf. which has glandular-pubescent calyces and corollas and is endemic to Somalia.

12. K. robusta Balf. f. (1882 p.512). Syn.: *K. abrupta* Balf. f. (1882 p.512). Illustr.: Bot. Mag. 179 (4): t.657 (1973). Type: Socotra, *Balfour, Cockburn & Scott* 151 (K).

Glabrous perennial herb or subshrub up to 50cm. Leaves obovate to oblong-oblanceolate, 3.5–6.5 × 2.5–3.5cm, acute, entire, subsessile or shortly petiolate, the petiole slightly broadened at the base. Calyx lobes free ± to the base, triangular, 3–4 × 2–2.5mm. Flowers held horizontally. Corolla orange-scarlet, slightly zygomorphic; tube cylindrical, 20–30mm; lobes ovate-oblong, rounded, 6–8 × 3–5mm. Anthers all exserted. Nectary scales suborbicular. Styles c.25mm. **Map 668, Fig. 91.**

Rocky slopes, in steep gullies and amongst limestone boulders; 300–550m.

Socotra. Endemic.

A very distinctive species readily distinguished from all other species in the region by its horizontally-held, slightly zygomorphic flowers with the anthers exserted from the corolla.

13. K. farinacea Balf. f. (1882 p.512). Illustr.: Bot. Mag. 127: t.7769 (1901). Syntypes: Socotra, *Schweinfurth* 753 (K); *Balfour, Cockburn & Scott* 521 (K), *Hunter* 15 (E).

Glabrous perennial herb or subshrub up to 30cm; young stems and leaves scaly and peeling when dry. Leaves held erect, imbricate, whitish green, obovate to orbicular, 2–5.5 × 1.5–3.5cm, rounded, entire, subsessile, the petiole not broadened at the base. Calyx lobes free ± to the base, triangular, 1–2 × 1–1.5mm. Corolla bright red; tube cylindrical, 10–15mm; lobes ovate-oblong, apiculate. Upper anthers slightly exserted, the lower anthers held at the mouth of the corolla tube. Nectary scales narrowly triangular, rounded. Styles c.4mm. **Map 669, Fig. 91.**

Amongst rocks in drought-deciduous shrubland; 100–400m.

Socotra. Endemic.

Readily distinguished from all other species of *Kalanchoe* in the region by the stems and leaves which are white or whitish green in fresh material and scaly and peeling when dry.

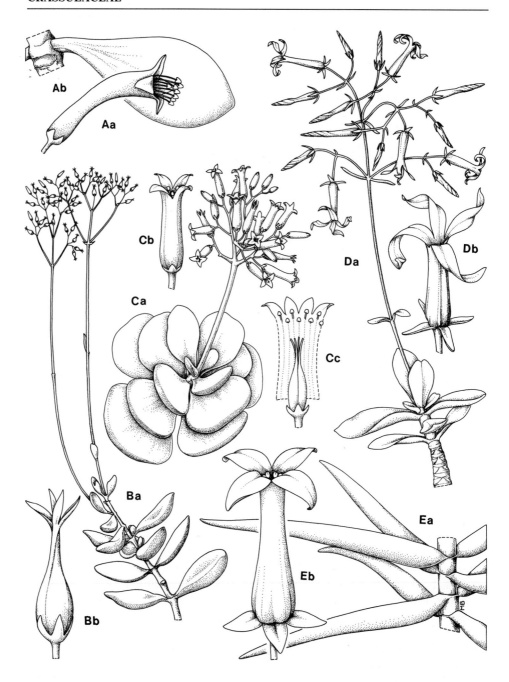

Fig. 91. Crassulacae. A, *Kalanchoe robusta*: Aa, flower (×1.5); Ab, leaf (×0.6); B, *K. rotundifolia*: Ba, habit (×0.5); Bb, flower (×4). C, *K. farinacea*: Ca, habit (×0.6); Cb, flower (×1.5); Cc, dissected corolla and ovary (×1.5). D, *K. alternans*: Da, habit (×1.5); Db, flower (×1.5). E, *K. bentii*: Ea, leaves (×0.6); Eb, flower (×1.5).

5. COTYLEDON L.

Succulent shrub. Leaves opposite, entire, succulent. Flowers 5-merous, pendulous, in an erect terminal, pedunculate panicle. Sepals connate at the base. Petals united into a tube below, gibbous at the base. Stamens 10, attached at the base of the corolla tube, exserted or included. Nectary scales clavate, depressed at the tip. Carpels 5, free or united at the base; ovules numerous.

1. C. barbeyi Baker in Gard. Chron., ser. 3, 13: 624 (1893). Illustr.: Fl. Ethiopia 3: 18 (1989). Type: Yemen (N), *Schweinfurth* 1493 (K, BM, G).

Grey-green shrub up to 1m, glabrous. Leaves flattened, obovate, 5–12 × 2–6cm, obtuse, shortly petiolate. Calyx lobes ovate, 7–10 × 3–4mm, acute, glandular-pubescent. Corolla red; tube tubular-urceolate, c.15mm; lobes oblong-lanceolate. 9–15 × 3–5mm, acute or apiculate, glandular-pubescent. Nectary scales 4 × 3mm. **Map 670, Fig. 92.**

Cliffs, rock-faces and terrace-walls; 2100–2600m.

Yemen (N & S). NE, E & S tropical Africa and South Africa.

6. UMBILICUS A. DC.

Glabrous, fleshy, perennial herbs with tuberous or rhizomatous rootstocks. Basal leaves long-petiolate, circular, peltate (in Arabia). Inflorescence a terminal, many-flowered raceme. Flowers 5-merous. Sepals small. Corolla tubular or urceolate with short lobes. Stamens (5–)10, inserted on corolla. Nectary scales small. Carpels 5; ovules many. Follicles many-seeded.

1.	Corolla 5.5–7mm; pedicels 0–3mm	**1. U. horizontalis**
+	Corolla 8–10mm; pedicels 2.5–5mm	**2. U. rupestris**

1. U. horizontalis (Guss.) DC., Prodr. 3: 400 (1828). Syn.: *Cotyledon umbilicus* auct. non L. (1753). Illustr.: Collenette (1985 p.191) as *U. erectus* DC.

Stems erect, up to 30cm. Basal leaves circular, 3–9cm diam., the margin repand-crenate; petioles up to 10cm. Flowers horizontal or pendulous, pedicels 0–3mm. Sepals 1–1.5mm. Corolla pale green sometimes reddish tinged, tubular or urceolate, 5.5–7 × 3–4.5mm; corolla lobes ovate, 1.5–2.5, shortly acuminate. **Map 671, Fig. 92.**

Damp and shady rock crevices, under boulders and on terrace-walls; 1000–2950m.

Saudi Arabia, Yemen (N), Socotra, Oman, UAE. Mediterranean region and SW Asia.

Two varieties have been recognized within *U. horizontalis*: var. *horizontalis* with horizontally spreading flowers and tubular corollas; and var. *intermedius* (Boiss.) Chamberlain (syn. *U. intermedius* Boiss.) with pendulous flowers and urceolate corollas. In Arabia these differences apparently break down and it has proven impossible to assign any specimens to variety.

CRASSULACEAE

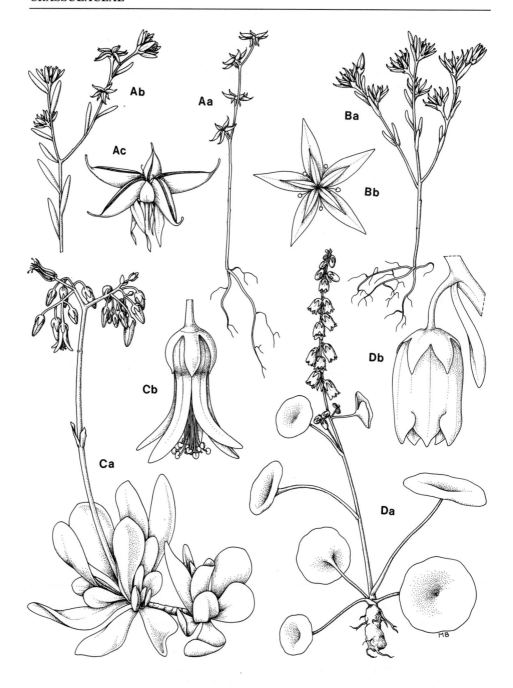

Fig. 92. Crassulaceae. A, *Sedum hispanicum*: Aa, habit (×1.5); Ab, inflorescence (×1.5); Ac, fruit (×6). B, *S. caespitosum*: Ba, habit (×2); Bb, flower (×8). C, *Cotyledon barbeyi*: Ca, habit, (×0.3); Cb, flower (×1.5). D, *Umbilicus horizontalis*: Da, habit (×1); Db, flower (×6).

2. U. rupestris (Salisb.) Dandy in Riddelsdell, Hedley & Price, Fl. Gloucestershire: 611 (1948).

Similar to *U. horizontalis* but the flowers pendulous; corollas tubular, 8–10mm; pedicels 2.5–5mm. **Map 672.**

Damp, shady crevices amongst boulders and on terrace-walls; 1800–2150m.

Saudi Arabia. S and W Europe, Egypt.

Very similar to and easily confused with the preceding species. Another species, *Umbilicus botryoides* A. Rich., from Egypt and tropical Africa, has also been recorded from Yemen. It closely resembles *U. rupestris* but has a narrower corolla. However, all the Yemen material that has been examined is here regarded as *U. horizontalis.*

Family 61. PITTOSPORACEAE

A.G. MILLER

Evergreen trees or shrubs. Leaves alternate, simple, entire. Stipules absent. Flowers actinomorphic, functionally unisexual (in Arabia), in terminal or axillary panicles. Sepals 5, free or fused at the base. Petals 5, free. Stamens 5, alternating with the petals. Ovary superior, unilocular, with 2 parietal placentas (in Arabia); style simple; stigma truncate. Fruit a loculicidally dehiscing capsule (in Arabia); valves coriaceous or woody, containing 4–8-seeds; seeds covered with a sticky resin.

Friis, I. (1987). A reconsideration of the genus *Pittosporum* in Africa and Arabia. *Kew Bull.* 42: 319–335.

PITTOSPORUM Soland.

Description as for the family.

1. P. viridiflorum Sims in Bot. Mag. 41: t.1684 (1815). Syn.: *P. viridiflorum* subsp. *arabicum* Cufod. in Feddes Repert. 55: 72 (1952); *P. abyssinicum* sensu Schwartz (1939) non Hochst. Type (of *P. viridiflorum* subsp. *arabicum*): Yemen, *Deflers 408* (MPU).

Shrub or tree up to 10m; young branches and petioles pubescent to glabrescent. Leaves crowded towards the ends of the branches, narrowly oblanceolate, 8–15 × 1.5–4.5cm, the tips acuminate, attenuate below, prominently veined beneath; petioles up to 2cm. Panicles many-flowered. Flowers fragrant. Sepals slightly fused at the base, narrowly ovate, up to 1.5 × 0.8mm. Petals white, cream or yellowish green, narrowly oblong, 4–7mm, rounded. Capsule ± globose, 5–10mm diam.; capsule-valves spreading or reflexed, up to 2mm thick, 4-seeded. **Map 673, Fig. 88.**

Evergreen bushland; 1500–2600m.

Yemen (N), Socotra. Southern and tropical Africa; southern India.

PITTOSPORACEAE

A polymorphic and widespread species in tropical Africa. The Arabian plants differ somewhat from the African and were placed in subspecies *arabicum* by Cufodontis (1952). Friis (1987) considered the Arabian plants were better treated as an informal 'arabicum' grouping within the species. The description given above is for the Arabian plants.

Family 62. ROSACEAE

D.F. CHAMBERLAIN

Herbs, shrubs or trees. Leaves generally alternate, simple or compound, stipulate. Flowers bisexual, actinomorphic, solitary or in simple or compound corymbs, cymes, fascicles or umbels. Receptacle flat, concave, convex or hollow, sometimes enlarged and fleshy in fruit. Sepals 4 or 5, sometimes with alternating epicalyx lobes. Petals 5 or rarely absent, free, inserted on the margin of the disc. Stamens 4 or numerous. Carpels one to many, free or united, sometimes fused with the receptacle (hypanthium). Styles as many as the carpels, free or united. Ovules 1–2 per carpel, anatropous. Fruit an achene, follicle, drupe or pome (with the carpels adnate to and enclosed by the fleshy receptacle), rarely a capsule.

1.	Leaves simple, sometimes deeply lobed or dissected	2
+	Leaves compound, palmate or imparipinnate	9
2.	Herbs	**8. Alchemilla**
+	Trees or shrubs	3
3.	Ovary superior; fruit a drupe	**10. Prunus**
+	Ovary inferior; fruit a pome	4
4.	Leaves entire	5
+	Leaves toothed, sometimes also lobed	7
5.	Leaves sparsely hairy or soon glabrescent beneath	**3. Pyrus**
	Leaves densely tomentose or villous beneath	6
6.	Flowers 4–6cm diam.; fruit 3–12 cm long	**2. Cydonia**
+	Flowers to 1cm diam.; fruit 0.6–0.8cm long	**1. Cotoneaster**
7.	At least some leaves lobed; fruit walls woody	**5. Crataegus**
+	Leaves not lobed; fruit fleshy	8
8.	Fruit globose to pyriform, with numerous grit cells; styles free (the pear)	**3. Pyrus**
+	Fruit globose, grit cells few or absent (the apple)	**4. Malus**
9.	Unarmed herb; flowers yellow	**7. Potentilla**
+	Spiny shrub; flowers white or pink to yellow	10

10.	Carpels and fleshy fruit exposed on the receptacle	**6. Rubus**
+	Carpels enclosed in a flask-shaped hypanthium	**9. Rosa**

1. COTONEASTER Medicus

Unarmed shrubs. Leaves deciduous, simple, entire. Stipules soon falling. Flowers small, solitary or in cymes or corymbs. Petals white or pink. Stamens c.20. Ovary semi-inferior, styles and carpels 2–5, ± free. Fruit globose or turbinate, with mealy flesh, containing 2–5 pyrenes (seeds), crowned by the persistent calyx.

1. C. nummularia Fischer & C. Meyer, Index sem. hort. petrop. 2: 34 (1836). Syn.: *C. racemiflorus* auct. non (Desf.) Bosse. Illustr.: Collenette (1985 p.425) as *C. racemiflorus*.

Shrub, semi-prostrate or up to 2m. Leaves broadly ovate to orbicular, 0.5–2 (–4) × 0.5–1.2(–3.5)cm, the apex retuse or acute, thinly pilose or glabrescent above, densely pilose-tomentose beneath. Flowers in 3–7-flowered cymes. Pedicels, receptacle and sepals white-tomentose. Petals white. Fruit 6–8mm long when ripe, red at first, becoming purple or purplish black at maturity; pyrenes 2. **Map 674, Fig. 93.**

Rocky hillsides and ravines in *Juniperus* woodland and evergreen bushland; 1700–3000m.

Saudi Arabia, Yemen (N), Oman. Crete, Cyprus, Lebanon, Syria, Iraq, Iran, Turkey, Caucasia, Ethiopia.

Records of *C. racemiflorus* (Desf.) Bosse from the Arabian Peninsula are assumed to refer to *C. nummularia*. The former was described from cultivated material and differs from *C. nummularia* in its 6–12-flowered inflorescences. None of the Arabian material seen shows this feature.

2. CYDONIA Miller

Unarmed shrubs or small trees. Leaves deciduous, simple, entire. Stipules soon falling. Flowers large, solitary, terminal. Petals white or pink. Stamens 15–25. Styles 5, free. Ovary inferior, 5-locular. Fruits pyriform or subglobose, with a leathery carpel wall and many seeds.

1. C. oblonga Miller, Gard. dict. ed. 8, no. 1 (1768). Syn.: *Pyrus cydonia* L., Sp. pl.: 480 (1753); *Cydonia vulgaris* Pers., Syn. pl. 2: corrigenda (1807).

Large shrub or small tree, up to 8m. Leaves ovate to oblong, up to 10 × 7cm, entire, becoming glabrous above but with a dense white villous indumentum beneath; petioles 1–2cm. Flowers 4–6cm diam. Sepals glandular, toothed, reflexed. Fruit (3–)5–12cm, yellowish, fragrant. **Map 675.**

Fig. 93. Rosaceae. A, *Prunus arabica*: Aa, fruiting and flowering branch (×0.6); Ab, flower (×2); Ac, section through flower (×3). B, *P. korshinskyii*: Ba, fruit and leaves (×0.6). C, *Cotoneaster nummularia*: Ca, fruiting branch (×0.6); Cb, flowers (×0.6); Cc, section through flower (×8). D, *Rubus sanctus*: Da, flowering branch (×0.6); Db, fruit (×1). E, *R. arabicus*: Ea, flowering branch (×0.6); Eb, leaf from sterile branch (×0.6).

Cultivated in the highlands for its edible fruits.

Yemen (N). Native in Caucasia, N Iran and possibly N Iraq; cultivated elsewhere for its edible fruits (the Quince).

3. PYRUS L.

Spiny or unarmed trees. Leaves deciduous, simple, entire or toothed. Flowers in umbel-like clusters. Sepals deciduous or persistent; epicalyx absent. Petals white. Stamens 15-30. Carpels united with each other and with the receptacle; styles 2–5, free. Fruits fleshy, with numerous grit cells

1. P. communis L., Sp. pl.: 479 (1753).

Unarmed tree. Leaves ovate-elliptic to ovate-orbicular, $3-5(-7) \times 1.5-4$cm, entire or toothed, sparsely hairy when young, becoming glabrous; petioles up to 5cm. Flowers 2–3cm diam. Fruit pyriform, up to 15cm long, yellowish-green; calyx persistent. **Map 676**.

Cultivated in the highlands for its edible fruits.

Yemen (N). *P. communis* is widely cultivated for its edible fruit (the pear) and may have arisen as a naturally occurring hybrid between *P. nivalis* Jacq. and *P. cordata* Desv. The cultivated pear belongs to subsp. *sativa* (DC.) Hegi. Subsp. *communis* occurs naturally in an area extending from Europe through Turkey, the Caucasus and Iran to C Asia and differs in its smaller fruits (2–4cm long) and spinescent habit.

4. MALUS Miller

Shrubs or small trees. Leaves deciduous, simple, toothed. Flowers in umbel-like clusters. Petals clawed, white or pink. Stamens 15–50. Carpels 3–5, walls cartilaginous, united with each other and with the receptacle; styles 5, united below. Fruit a globose fleshy pome, with a depression at the base, without grit-cells.

1. M. sylvestris Miller, Gard. Dict. ed. 8: no.1 (1768). Syn.: *Pyrus malus* L., Sp. pl.: 479 (1753); *M. communis* Poir. in Lam., Encycl. 5: 560 (1804).

Tree up to 12m, unarmed or spinescent when young. Leaves elliptic to sub-orbicular, $3-8 \times 2-4$cm, crenate or serrate, pilose when young, soon glabrescent. Flowers in 4–6-flowered clusters, 3–4cm diam. Calyx densely tomentose-pilose, persistent. Fruit subglobose, up to 15cm diam., green, greenish yellow or reddish. **Map 677**.

Cultivated in the highlands for its edible fruit.

Saudi Arabia, Yemen (N). Native in Europe and NW Asia, now extensively cultivated for its edible fruit (the apple) and naturalized elsewhere.

5. CRATAEGUS L.

Spiny shrubs or small trees. Leaves alternate, simple or lobed, entire or serrate. Flowers in corymbs, borne on spur shoots. Sepals persistent in fruit; epicalyx absent. Petals white. Stamens 20–25. Carpels 1–5, free at the apex, united on the inner margin, fused to the hypanthium; styles 1–5, free. Fruit fleshy, containing 1–2 bony pyrenes; flesh mealy.

1. C. sinaica Boiss., Diagn. pl. orient. sér. 2, 3(2): 48 (1856). Illustr.: Collenette (1985 p.426).

Leaves obovate, up to 3 cm long, narrowly cuneate at the base, 3(–5)-lobed, occasionally with the lobes replaced by teeth, glabrous or with a few hairs on the lower surface of the midrib, upper surface shining. Flowers c.15mm diam. Sepals triangular-lanceolate, reflexed, glabrous. Petals white. Styles 1–2. Fruit ovate-ellipsoid, 6–8mm diam., glabrous. **Map 678.**

Field margins; 2150 m.

Saudi Arabia. Syria and Sinai.

In Arabia only known from a single record.

6. RUBUS L.

Shrubs, frequently armed with sharp prickles and acicles (needle-like prickles); vegetative stems (turions) arching or procumbent, with the usually ± erect flowering stems branching off them. Leaves imparipinnate or ternate, the leaflets toothed. Flowers few to many, arranged in panicles. Sepals 5, ovate-lanceolate; epicalyx absent. Petals 5, white to pink, sometimes absent. Stamens many. Receptacle domed, bearing few to many 1-seeded carpels each developing into a fleshy druplet.

A taxonomically difficult group owing to facultative apomixis. As a result there is a large number of microspecies that do not regularly reproduce sexually. The sexual species, *R. caesius* and *R. sanctus* are, however, clear-cut.

1.	Leaflets green on both surfaces (concolorous)	2
+	Leaflets green or grey-green above, whitish-grey to white beneath (discolorous)	4
2.	Low shrub with procumbent turions; drupelets pruinose	**1. R. caesius**
+	Scrambling shrub up to 4m, with arching turions; drupelets pruinose or not	3
3.	Inflorescences with up to 60 flowers; carpels and drupelets glabrous	**3. R. arabicus**
+	Inflorescences with up to 10 flowers; carpels and drupelets pilose	**2. R. asirensis**
4.	Sepals acute to cuspidate; anthers glabrous	**4. R. apetalus**
+	Sepals acute, not cuspidate; anthers pilose	**5. R. sanctus**

1. R. caesius L., Sp. pl.: 493 (1753).

Low shrub; turions usually procumbent, terete, pruinose, usually glabrous; prickles slender, short, straight or curved; flowering stems erect, 15–40cm, pubescent. Leaves ternate (the upper sometimes simple); leaflets broadly ovate, 3–6 × 2.5–5cm, sometimes lobed below, bicrenate-serrate, concolorous, upper surface glabrescent, lower surface laxly pilose; stipules linear to narrowly lanceolate. Flowers 6–10, in lax simple or compound corymbs. Pedicels stipitate-glandular and densely pubescent, with a few acicles. Sepals ovate, 8–10mm, cuspidate, densely pubescent, sometimes also glandular. Petals white, 8–10mm. Fruit with few (2–20) large black pruinose glabrous drupelets. **Map 679.**

Damp shaded rocks; 1800–3000m.

Yemen (N). Most of Europe, extending through the Caucasus and SW Asia to the Altai.

2. R. asirensis D.F. Chamb. in Edinb. J. Bot. 51 (1): 57 (1994). Illustr.: Collenette (1985 p.427) as *R.* cf. *canescens*. Type: Saudi Arabia, *Collenette* 3668 (E).

Large scrambling shrub, 3.5–4m, with long tangled turions; flowering stems ridged, pilose-tomentose, with scattered minute red glands; prickles flattened, broad-based, curved. Leaves ternate; leaflets ovate to elliptic, the terminal often broader, the laterals 3–6 × 2.7–5.5cm, the apex shortly cuspidate, sharply biserrate, sometimes lobed towards the petiole, concolorous, upper surface sparsely pilose or glabrescent, lower surface densely pilose, especially on the veins; stipules linear. Flowers 8–10, in lax compound corymbs. Pedicels and calyx densely tomentose with slender prickles and a few stipitate glands. Sepals 7–10mm, including the long (up to 3mm) cuspidate point. Petals white, 7–10mm. Stamens with glabrous anthers. Carpels several, pilose. **Map 680.**

River banks and on walls; 1800–2500m.

Saudi Arabia. Endemic.

The illustration cited above may be of the type specimen. It was originally referred to *R.* cf. *canescens* DC., a species that differs in several important aspects.

A specimen from Saudi Arabia (J. Raidah, *Collenette* 5161), which differs in its glabrous carpels, and leaves with more rounded teeth and acute apex, may belong to this species, but, without more material, the significance of the differences mentioned above cannot be ascertained.

3. R. arabicus (Deflers) Schweinf. (1896 p.204). Syn.: *R. glandulosus* Bell. var. *arabicus* Deflers (1889 p.136). Type: Yemen (N), *Deflers* 375 (P).

Scrambling shrub up to 2m; turions long and trailing, not ridged, villous; prickles curved. Leaves ternate or imparipinnate and then with 5 leaflets; leaflets ovate-lanceolate, 5–10 × 3.5–8cm, the apex acute to cuspidate, sometimes lobed below, serrate to biserrate, concolorous, laxly pilose on both surfaces; stipules linear. Flowers 10–60, in lax compound corymbs. Pedicels densely tomentose, with varying numbers of stipitate glands, acicles and prickles. Sepals ovate-lanceolate, 6–8mm, cuspidate, densely tomentose, with at least some stipitate glands, sometimes also with a few acicles, especially at the base. Petals white to pale pink, 6–8mm, emarginate. Fruit with 5–10 large black and shining glabrous drupelets. **Map 681, Fig. 93.**

Banks by streams, coffee plantations, etc.; 900-2750m.

Yemen (N). Endemic.

A variable species closely allied to *R. apetalus* and possibly hybridizing with it (q.v.).

4. **R. apetalus** Poir., Encycl. 6: 242 (1804). Syn.: *R. petitianus* A. Rich. (1847 p.256). Illustr.: in Lam., Fl. Ethiopia 3: 32 (1989).

Scrambling shrub; turions long and arching, ridged or sulcate, thinly pubescent; prickles stout, from a wide base, curved. Leaves ternate or imparipinnate and then with 5 leaflets; leaflets broadly ovate, 2–5 × 1.7–4cm, the apex acute, biserrate, the upper surface laxly pilose, the lower surface with a grey to whitish-pubescent indumentum interspersed with longer villous hairs and also with some red punctate glands especially on the veins; stipules linear. Flowers up to 20, in cylindrical compound corymbs. Pedicels densely tomentose, with a few slender prickles and minute red punctate glands. Sepals ovate-lanceolate, 4–5mm long, acute to shortly cuspidate, with an indumentum like that of the pedicels. Petals absent or up to 5mm, white to pale pink. Fruit with c.50 small black glabrous drupelets. **Map 682.**

Terrace-walls, swampy undergrowth etc.; 1600–2800 m.

Yemen (N). C, N and NE tropical Africa.

Allied to *R. arabicus*. Two specimens from Yemen (*J. Wood* 2812 and 2117) are intermediate between *R. apetalus* and *R. arabicus*, having the inflorescence indumentum of the former and the leaf indumentum of the latter. It is not clear whether these are referable to a distinct species or whether they have a hybrid origin.

5. **R. sanctus** Schreber, Icon. descr. pl.: 15, t.18 (1766). Illustr.: Fl. Iraq 2: 121 (1966); Collenette (1985 p.427).

Shrub, 1–2 m; turions arching, rooting at the tips, angled, grooved, densely covered with a white compacted tomentum and prickles, eglandular; prickles stout, hooked, flattened and broad-based. Leaves ternate or palmate and then with 5 leaflets; leaflets broadly obovate, 2–4(–10) × 1.7–3.5cm, the apex obtuse to truncate, biserrate, the upper surface greyish-green, arachnoid-tomentose, the lower surface greyish-white with a dense tomentose indumentum interspersed with pilose hairs on the veins; stipules linear. Flowers 10–50, in lax compound racemes. Pedicels and calyx densely tomentose, eglandular. Sepals ovate, 4–5mm, acute but not cuspidate. Petals 8–13mm, normally pink, often intensely so. Stamens with pilose anthers. Fruit with few black, scarcely juicy, glabrous drupelets. **Map 683, Fig. 93.**

Terrace-walls etc.; 2000–2150m.

Saudi Arabia. W & C Europe, the Mediterranean region, extending through SW Asia to the western Himalayas.

A variable species and, like the closely allied European *R. ulmifolius*, a sexual species. In Arabia it is restricted to the Western Escarpment mountains of Saudi Arabia.

7. POTENTILLA L.

Erect or creeping perennial herbs (in Arabia). Leaves digitate or imparipinnate. Flowers solitary or in dense terminal cymes, 5-merous. Epicalyx present. Petals yellow. Stamens c.20. Carpels numerous, inserted on a convex or conical receptacle. Fruit a head of achenes; styles terminal, long-filiform, deciduous as the achenes ripen.

1.	Plants erect; flowers in 8–20-flowered cymes	**1. P. dentata**
+	Plants creeping; flowers solitary in the leaf axils	**2. P. reptans**

1. P. dentata Forsskal (1775 p.98). Syn.: *P. pennsylvanica* auct non L.; *P. hispanica* auctt. non Zimmet. Illustr.: Collenette (1985 p.426) as *P. hispanica*; Fl. Ethiopia 3: 36 (1989). Type: Yemen (N), *Forsskal* 1602 (C).

Perennial herb; flowering stems 30–50cm, arising from a thick rhizome. Leaves imparipinnate; leaflets 7–10(–15), 10–40 × 5–15mm, coarsely dentate, sparsely pilose and glandular. Flowers in 8–20-flowered lax or dense cymes. Sepals oblong-lanceolate, densely pilose and glandular; epicalyx segments narrowly lanceolate. Petals yellow, 8–12mm. **Map 684, Fig. 94.**

Rocky slopes and terraces, often by water; 2100–3500m.

Saudi Arabia, Yemen (N). Ethiopia, Kenya.

This species has been confused with *P. hispanica* and *P. pennsylvanica*, to both of which it is closely allied. References to Arabian records of *P. viscosa* Donn., a species endemic to the Urals, date back to Lehmann (Rev. Potentilla, 1856) who incorrectly treated *P. dentata* as a synonym of *P. viscosa*.

2. P. reptans L., Sp. pl.: 499 (1753).

Perennial herb, with persistent rosettes from which creeping stems up to 1m long arise. Leaves digitate; leaflets 5–7, 5–20(–50) × 3–10(–23)mm, dentate, glabrous above, thinly pilose beneath especially on the midrib and margins. Flowers solitary in the leaf axils; pedicels 1–10cm. Sepals ovate-lanceolate, 4–5mm, pilose; epicalyx segments lanceolate. Petals yellow, 8–12mm. **Map 685, Fig. 94.**

By streams and springs, in grassland and in woodland margins; 1500–3000m.

Yemen (N). A widespread species, throughout Europe and N Africa, extending to C Asia.

8. ALCHEMILLA L.

Perennial ascending herbs (in Arabia). Leaves palmately lobed; stipules of cauline leaves foliaceous. Inflorescence a compound cyme. Flowers small, greenish. Sepals 4, inserted on the rim of a flask-shaped receptacle, alternating with 4 epicalyx segments. Petals absent. Stamens 4, alternating with the sepals. Carpels (1–)5–8(–20), enclosed by the receptacle. Fruit a group of achenes, included within the receptacle.

1. A. cryptantha Steud. ex A.Rich. (1847 p.259).

Stems herbaceous, up to 50cm, rooting at the nodes. Leaves reniform, 15–20 × c.25mm, sparsely villous; lobes 5–7, truncate, each with 8–15 crenate-serrate teeth. Inflorescence few-flowered, spreading-hairy. Calyx lobes c.1mm. Achenes 5–8. **Map 686, Fig. 94.**

Damp ground by streams, etc.; 2400–3500m.

Saudi Arabia, Yemen (N). Mountains of tropical Africa and Madagascar.

9. ROSA L.

Shrubs, generally deciduous; stems armed with stout prickles, sometimes also with stalked glands. Leaves imparipinnate; leaflets elliptic to broadly ovate-lanceolate, serrate. Flowers solitary or in corymbs. Sepals 5, cuspidate, reflexed in fruit. Epicalyx absent. Petals 5, large, white to pale yellow or reddish-purple. Stamens numerous. Styles numerous, free or fused into a stylar column. Receptacle flask-shaped, enclosing the numerous carpels, becoming fleshy and coloured when ripe.

Widespread in temperate and subtropical parts of the Northern Hemisphere. Cultivated and much hybridized as ornamentals. *Rosa × damascena* Miller, in particular, is cultivated as a source for Attar of Roses and Rose-water.

Boulenger, G.A. (1933). Les Roses du Yemen. *Verh. Naturf. Ges. Basel* 44: 275–284.

1.	Leaflets 10–45mm long; sepals to 18mm	**1. R. abyssinica**
+	Leaflets 48–75mm long; sepals to 13mm	**2. R. barbeyi**

1. R. abyssinica Lindley, Ros. monogr. 116, t.13 (1820). Syn.: *R. moschata* Herrm. var. *abyssinica* (Lindley) Crépin in Bull. Roy. Soc. Bot. Belgique 28(2): 47 (1889); *R. bottaiana* Boulenger, op. cit.: 280 (1933); *R. schweinfurthii* Boulenger, op. cit.: 283 (1933). Illustr.: Collenette (1985 p.427).

Shrub up to 3m; stems with curved prickles, occasionally also with stalked glands. Leaflets 5–7, elliptic to broadly ovate-lanceolate, 10–45 × 7–16mm, crenate-serrate, the teeth gland-tipped and sometimes also with 1–2 lateral glands, glabrous or with hairs restricted to the main veins beneath. Flowers (1–)3–20, in dense corymbs; pedicels 1.5–3.5cm, glabrous to densely villous. Sepals ovate with cuspidate tips, usually simple, 10–18mm, glabrous or villous (rarely glandular) on the back; petals white to pale yellow, 1.2–2.6cm; stylar column hairy. Fruit red or orange-red when

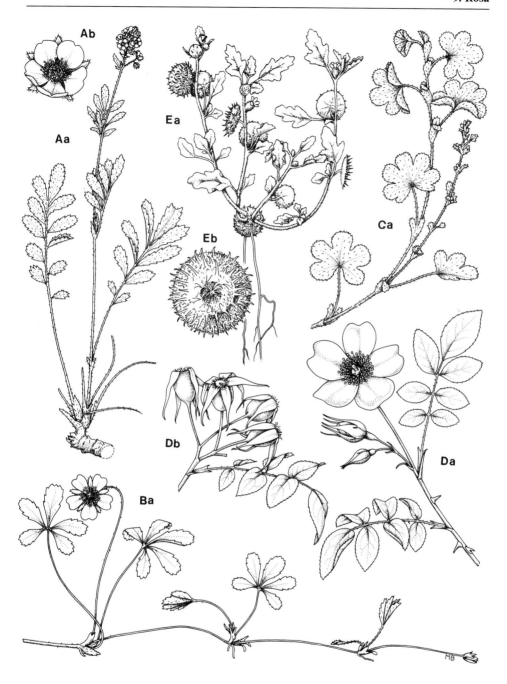

Fig. 94. Rosaceae. A, *Potentilla dentata*: Aa, habit (× 0.3); Ab, flower (× 1.5). B, *P. reptans*: Ba, habit (× 1). C, *Alchemilla cryptantha*: Ca, flowering shoot (× 0.6). D, *Rosa abyssinica*: Da, flowering shoot (× 1); Db, fruiting shoot (× 1). **Neuradaceae.** E, *Neurada procumbens*: Ea, habit (× 0.6); Eb, fruit (× 2).

ripe, 0.7–1.8cm long. **Map 687, Fig. 94.**

Thickets, hillsides, terrace-walls, etc.; (800–)1800–2800m.

Saudi Arabia, Yemen (N). Ethiopia, Somalia, Sudan.

A variable species that shows some geographical variation in the Arabian Peninsula. All the Saudi Arabian material seen has broad leaflets that are usually less than 2cm long and hairy (often densely so) pedicels. The material seen from Yemen often has narrower leaflets, 2.5–3.5cm long, with glabrous pedicels, though one specimen from near Manakhah has leaflets up to 4.5cm long and sparsely hairy pedicels.

Boulenger recognized three species in Yemen: *R. abyssinica, R. bottaiana* and *R. schweinfurthii*. These are reduced to a single species in this account as there is complete overlap between them. See also Browicz, K. & Zieliński, J. (1991). On the geographical distribution of Rosa abyssinica. *Fragm. Flor. Geobot.* 36, 1: 51–55.

2. R. barbeyi Boulenger, op. cit.: 281 (1933). Type: Yemen (N), *Schweinfurth* 584 (K).

Differs from *R. abyssinica* in its larger leaves with leaflets 48–75mm long; flowers 30–40mm diam.; pedicels glabrous; sepals up to 13mm. **Map 688.**

c.1000m.

Yemen (N). Endemic.

Closely allied to *R. abyssinica* and possibly only a well-grown form of that species.

10. PRUNUS L.

Evergreen or deciduous trees or shrubs. Leaves simple, serrate or rarely entire, petiolate; stipules free, usually deciduous. Flowers 5-merous, solitary or in clusters. Receptacle concave or cup-shaped. Sepals deciduous. Petals pink or white. Stamens numerous. Carpel 1, style simple. Fruit a drupe containing a hard 1-seeded stone.

The genus *Prunus* has been split up into a series of segregate genera by several recent authors.

1. Flowers pedicellate, usually appearing with the leaves; fruits and ovaries glabrous **1. P. × domestica**
+ Flowers sessile or subsessile, usually appearing before the leaves; fruits and ovaries hairy 2

2. Broom-like shrub or small tree; twigs ribbed; leaves linear-lanceolate, up to 5mm broad **3. P. arabica**
+ Shrub or tree, not broom-like; twigs not ribbed; leaves more than 10mm broad 3

3. Leaves broadly ovate **2. P. armeniaca**
+ Leaves linear or lanceolate to narrowly oblong, at least twice as long as broad 4

4. Petals bright pink or red; fruits fleshy; stone deeply furrowed and pitted **4. P. persica**

+   Petals pink or white; fruits leathery; stones finely pitted   5
5. Unarmed shrub or small tree up to 8m; leaves 2.5–9(–12)cm long **5. P. dulcis**
+   Subspiny shrub or tree up to 5m; leaves up to 2.5(–3)cm long
                                                                     6. P. korshinskyi

1. P. × domestica L., Sp. pl.: 475 (1753).

Deciduous shrub or tree up to 10m. Leaves obovate to elliptic, 3–10 × 2–6cm, acute or obtuse, crenate-serrate, pubescent beneath. Flowers in clusters of 2–3, appearing with the leaves; pedicels 5–20mm. Petals white, 7–12mm. Fruit globose to oblong, 3–8cm long, purple, red, yellow or green, glabrous, fleshy; stone somewhat rugose or pitted. **Map 689.**

Saudi Arabia, Yemen (N), Oman. Cultivated in the highlands (no specimens have been seen).

Grown for its edible fruit (the plum) throughout the temperate regions of the world.

2. P. armeniaca L., Sp. pl.: 474 (1753). Syn.: *Armeniaca vulgaris* Lam., Encycl. 1: 2 (1783).

Deciduous shrub or small tree up to 8m. Leaves broadly ovate, 5–10 × 5–8cm, acute or shortly acuminate, crenate-serrate, rounded or subcordate at the base, glabrescent. Flowers subsessile, solitary or occasionally in pairs, appearing before the leaves. Petals white or pale pink, 10–15mm. Fruit subglobose, 3–5cm long, yellowish orange, shortly velutinous, fleshy; stone smooth. **Map 690.**

Saudi Arabia, Yemen (N), Oman. Cultivated in the highlands.

Originally a native of W and C Asia, now commonly cultivated for its edible fruit (the apricot) which is usually used dried.

3. P. arabica (Oliv.) Meikle in Kew Bull. 19: 229 (1965). Syn.: *Amygdalus arabica* Oliv., Voy. emp. Othoman 3: 460 (1804). Illustr.: Fl. Iraq 2: 159 (1966); Western (1989, p.88) as *Amygdalus arabicus*.

Broom-like shrub or small tree up to 4m; twigs green, ribbed. Leaves often absent, linear-lanceolate, 10–30(–40) × 0.5–3(–5)mm, acute, entire to crenulate-serrate, glabrous or thinly pubescent. Flowers solitary, sessile, appearing before the leaves. Petals white or pale pink, 5–10mm. Fruit ovoid, 1.5–2.5cm long, brownish, pubescent, leathery; stone smooth. **Map 691, Fig. 93.**

Dry rocky slopes and wadi-sides, sometimes cultivated as a windbreak; (300–)500–2000m.

Saudi Arabia, Oman, UAE. Syria, Lebanon, Turkey, Iraq, Iran, Jordan.

4. P. persica (L.) Batsch, Beytr. Entw. Gewächsreiche 1: 30 (1801). Syn.: *Amygdalus persica* L., Sp. pl.: 472 (1753).

Shrub or small tree up to 8m. Leaves oblanceolate to narrowly oblong, 5–15 × 1–

3cm, acuminate, glandular-serrate, glabrescent. Flowers solitary, rarely paired, subsessile, usually appearing before the leaves. Petals deep pink or red, 10–15mm. Fruit globose, 4–8cm, yellow or pale green, velutinous, fleshy; stone deeply furrowed and pitted. **Map 692.**

Cultivated in the Highlands; 1000–2300m.

Saudi Arabia, Yemen (N & S), Oman.

Probably a native of China now widely cultivated for its edible fruit (the peach and nectarine) throughout southern Asia and Europe.

5. P. dulcis (Mill.) D.A. Webb in Feddes Repert. 74: 24 (1966). Syn.: *Amygdalus communis* L., Sp. pl.: 473 (1753); *Amygdalus dulcis* Miller, Gard. dict. ed. 8, no. 2 (1768); *Prunus amygdalus* Batsch, Beytr. Entw. Pragm. Gesch. Natur-Reiche, 1: 30 (1801); *Prunus communis* (L.) Arcang., Comp. fl. ital. ed. 1: 209 (1882) non Huds. (1778).

Shrub or tree up to 8m. Leaves narrowly ovate to oblong, 3.5–6(–10) × 1–2.5cm, acute, crenate-serrate, glabrescent. Flowers solitary but often congested along the twigs, subsessile, usually appearing before the leaves. Petals pink or white, 15–20mm. Fruit ovoid, 3–6cm long, grey-green, tomentose, leathery; stone finely pitted. **Map 693.**

Cultivated in the Highlands.

Saudi Arabia, Yemen (N), Oman.

A native of SW and C Asia, now widely cultivated elsewhere for its edible fruits (the almond) and as an ornamental.

6. P. korshinskyi Hand.-Mazz. in Ann. K.K. Naturhist. Hofmus. 27: 71 t.2 f.1 (1913). Syn.: *Amygdalus korshinskyi* (Hand.-Mazz.) Bornm. in Beih. Bot. Centralbl. 31(2): 212 (1914). Illustr.: Mout., Nouv. Fl. Lib. Syr. Atlas 2: t.74 f.6 (1970) as *Amygdalus korshinskyi*.

Shrub or small tree up to 5m; shoots subspiny. Leaves ovate-lanceolate to elliptic, up to 2.5(–3) × 1(–1.5)cm, glandular-crenate, sparsely pubescent at first, soon glabrous. Flowers solitary but congested along the twigs and so appearing paired, subsessile, usually appearing before the leaves. Petals pale pink, c.15mm. Fruit ovoid, up to 3cm long, grey-green, tomentose, leathery; stone pitted in the lower part, sometimes with the pits running into short grooves. **Map 694, Fig. 93.**

Granite gullies and steep rocky hillsides; 1650–2300m.

Saudi Arabia. Turkey, Syria, Lebanon.

Closely allied to *P. dulcis* but distinguished by its smaller size and spinescent habit. The records of this species in Saudi Arabia, close to the Gulf of Aqaba, represents an interesting extension of its range; the specimens seen are a good match with material from Turkey.

Family 63. NEURADACEAE

A.G. MILLER

Annual herbs. Leaves alternate, pinnately-lobed. Stipules minute. Flowers small and inconspicuous, axillary, solitary, bisexual, actinomorphic, pedicellate. Receptacle discoid, echinate, enlarging and becoming woody in fruit. Calyx lobes 5, alternating with 5 epicalyx lobes. Petals 5, free. Stamens 10. Ovary immersed in the receptacle, 5–10-celled with 1 ovule per cell; styles 5–10, becoming spinescent. Fruit dry, woody, indehiscent, few-seeded, flattened, orbicular or 5-angled, smooth beneath, spiny above.

Sometimes treated as a subfamily, the Neuradoideae, of the family Rosaceae.

NEURADA L.

Description as for the family.

1. N. procumbens L., Sp. pl.: 441 (1753). Illustr.: Fl. Iraq 2: 152 (1966); Fl. Ethiopia 3: 45 (1989).

Densely grey-tomentose herb; stems prostrate, 6–20cm. Leaves oblong-ovate, 5–20 × 3–13mm, sinuate-pinnatifid. Sepals triangular-ovate, c.2.5 × 1.25mm, acute. Petals white, cream or pink, obovate, c.1 × 0.8mm. Fruit 10–20mm in diameter with 2–10mm spines. **Map 695, Fig. 94.**

Deserts and semi-deserts, mainly on sand; 0–800m.

Saudi Arabia, Yemen (S), Oman, UAE, Qatar, Bahrain, Kuwait. N Africa, Cyprus, Sudan, Ethiopia, and throughout the deserts of SW Asia to NW India.

The characteristic spiny, disc-shaped fruits are commonly found scattered about the desert. They stick to the feet of animals (and car tyres) and act as dispersal units.

DISTRIBUTION MAPS

Distribution Maps

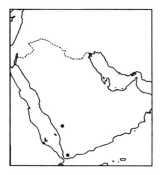

Map 7. Psilotum nudum

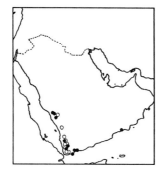

Map 8. Selaginella imbricata

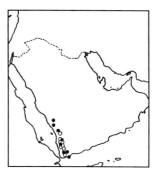

Map 9. S. yemensis

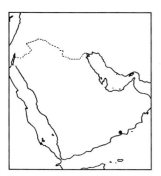

Map 10. S. perpusilla

Map 11. S. goudatana

Map 12. Equisetum ramosissimum

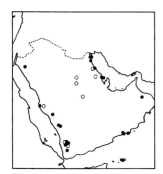

Map 13. Ophioglossum polyphyllum

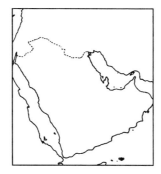

Map 14. Marsilea coromandeliana

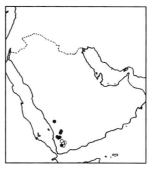

Map 15. M. aegyptiaca

Distribution Maps

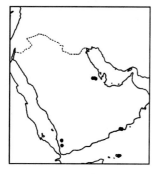
Map 16. Ceratopteris cornuta

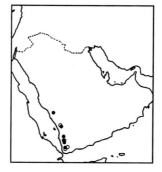
Map 17. Actiniopteris radiata

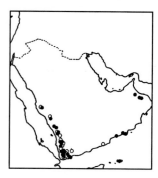
Map 18. A. semiflabellata

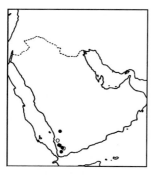
Map 19. Cheilanthes farinosa

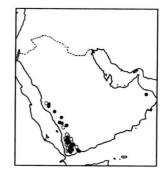
Map 20. C. coriacea

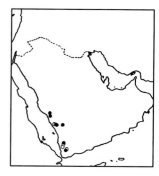
Map 21. C. marantae

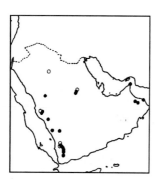
Map 22. C. vellea

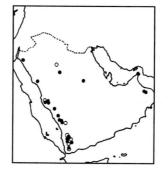
Map 23. C. pteridioides

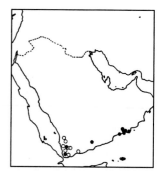
Map 24. Negripteris sciona

Distribution Maps

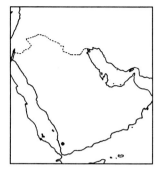

Map 25. Pellaea quadripinnata

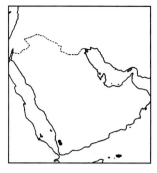

Map 26. P. viridis

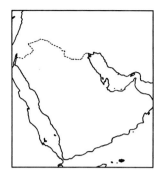

Map 27. P. involuta

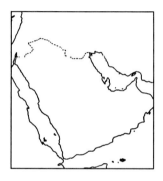

Map 28. Doryopteris concolor

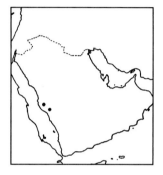

Map 29. Anogramma leptophylla

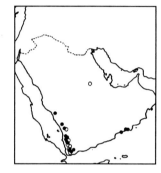

Map 30. Adiantum incisum

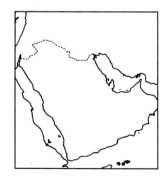

Map 31. A. balfourii

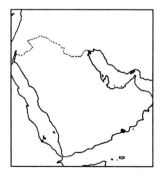

Map 32. A. philippense

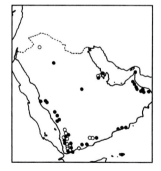

Map 33. A. capillus-veneris

499

Distribution Maps

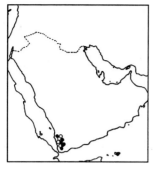 **Map 34.** A. poiretii

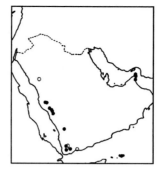 **Map 35.** Onychium divaricatum

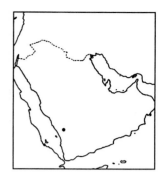 **Map 36.** Acrostichum aureum

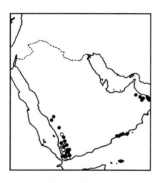 **Map 37.** Pteris vittata

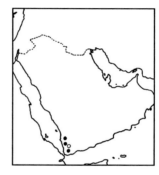 **Map 38.** P. cretica

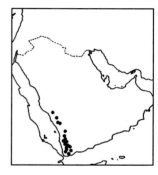 **Map 39.** P. dentata

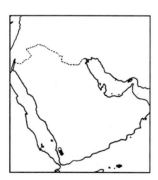 **Map 40.** P. quadriaurita

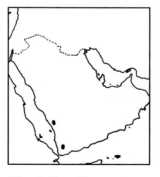 **Map 41.** Pleopeltis macrocarpa

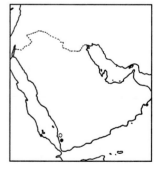 **Map 42.** Loxogramme lanceolata

Distribution Maps

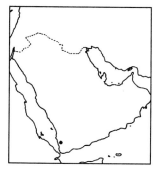

Map 43. Pteridium aquilinum

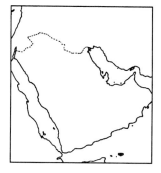

Map 44. Nephrolepis undulata

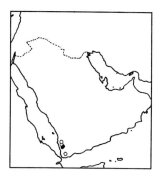

Map 45. Arthropteris orientalis

Map 46. Asplenium trichomanes

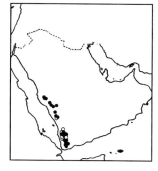

Map 47. A. aethiopicum

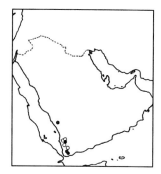

Map 48. A. adiantum-nigrum

Map 49. A. varians

Map 50. A. protensum

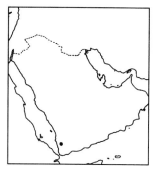

Map 51. A. rutifolium

Distribution Maps

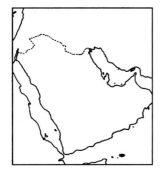
Map 52. A. schweinfurthii

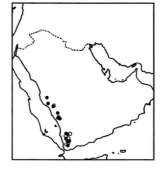
Map 53. Ceterach officinarum

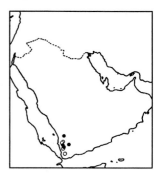
Map 54. C. dalhousiae

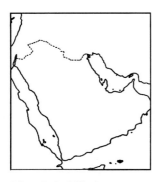
Map 55. C. phillipsianum

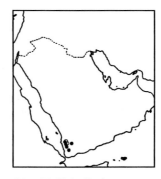
Map 56. Christella dentata

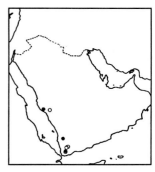
Map 57. Cystopteris fragilis

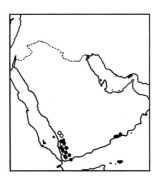
Map 58. Hypodematium crenatum

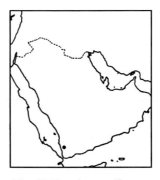
Map 59. Tectaria gemmifera

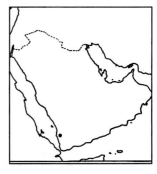
Map 60. Polystichium fuscopaleacum

Distribution Maps

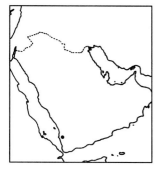

Map 61. P. sp. A

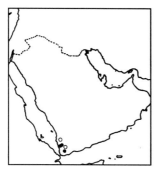

Map 62. Dryopteris schimperiana

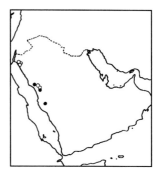

Map 63. Juniperus phoenicea

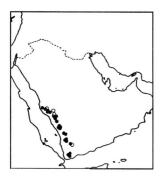

Map 64. J. procera

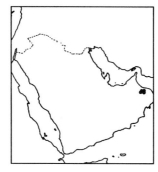

Map 65. J. excelsa subsp. polycarpos

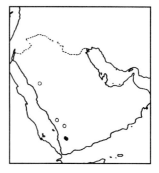

Map 66. Cupressus sempervirens

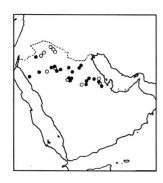

Map 67. Ephedra alata

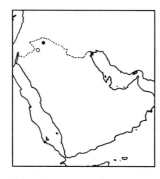

Map 68. E. transitoria

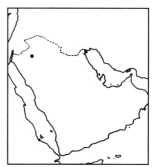

Map 69. E. aphylla

Distribution Maps

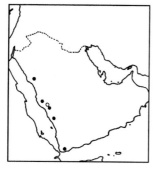
Map 70. E. foeminea

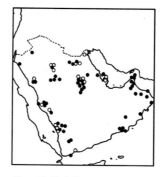
Map 71. E. foliata

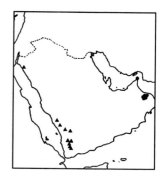
Map 72. E. pachyclada subsp. pachyclada ▲ subsp. sinaica ●

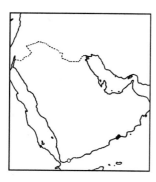
Map 73. E. milleri

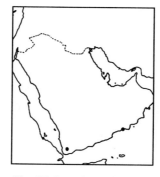
Map 74. Casuarina equisetifolia

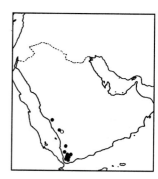
Map 75. Myrica humilis

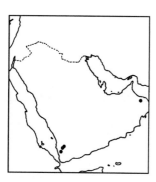
Map 76. Juglans regia

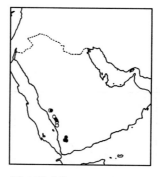
Map 77. Salix mucronata

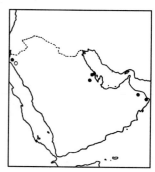
Map 78. S. acmophylla

Distribution Maps

Map 79. S. excelsa

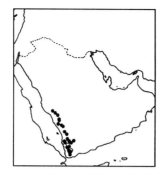
Map 80. Celtis africana

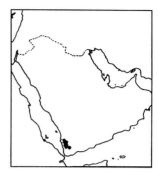
Map 81. C. toka

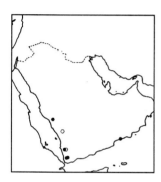
Map 82. Trema orientalis

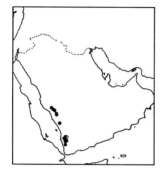
Map 83. Barbeya oleoides

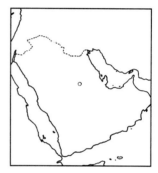
Map 84. Morus nigra

Map 85. M alba

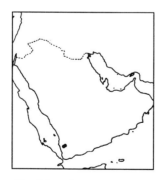
Map 86. Antiaris toxicaria

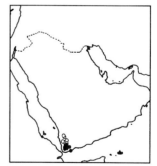
Map 87. Dorstenia barnimiana ●
D. socotrana ▲

Distribution Maps

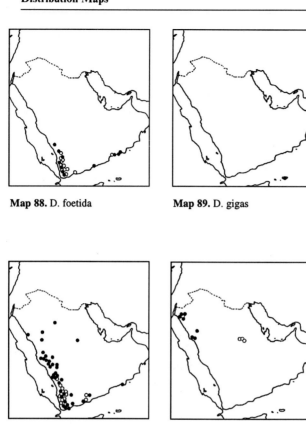

Map 88. D. foetida

Map 89. D. gigas

Map 90. Ficus carica

Map 91. F. palmata subsp. palmata

Map 92. F. palmata subsp. virgata

Map 93. F. johannis

Map 94. F. exasperata

Map 95. F. sycomorus

Map 96. F. sur

Distribution Maps

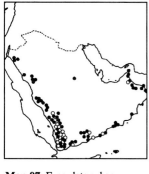

Map 97. F. cordata subsp. salicifolia

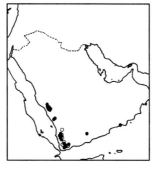

Map 98. F. ingens

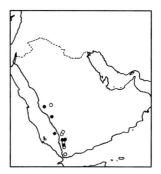

Map 99. F. glumosa

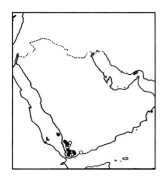

Map 100. F. populifolia

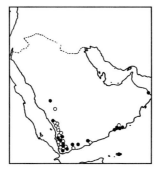

Map 101. F. vasta

Map 102. Cannabis sativa

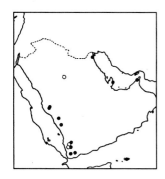

Map 103. Urtica urens

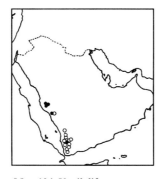

Map 104. U. pilulifera

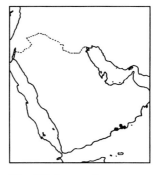

Map 105. Laportea interrupta

Distribution Maps

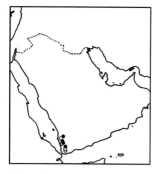
Map 106. L. aestuans

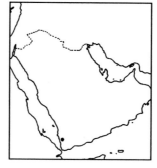
Map 107. Girardinia diversifolia

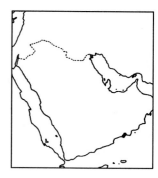
Map 108. Pilea tetraphylla

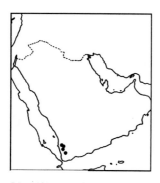
Map 109. Pouzolzia mixta

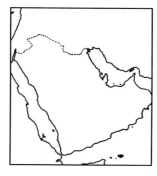
Map 110. P. auriculata

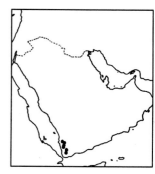
Map 111. P. parasitica

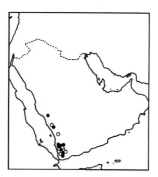
Map 112. Debregeasia saeneb

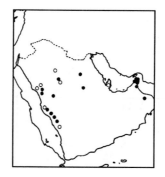
Map 113. Parietaria alsinifolia

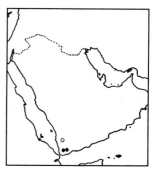
Map 114. P. debilis

Distribution Maps

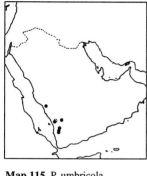
Map 115. P. umbricola

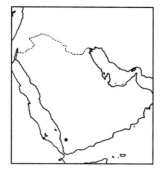
Map 116. P. judaica

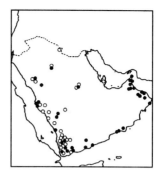
Map 117. Forsskaolea tenacissima

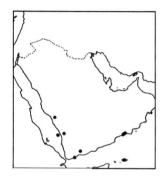
Map 118. F. viridis

Map 119. F. griersonii

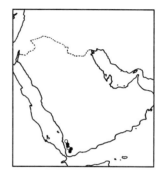
Map 120. Droguetia iners

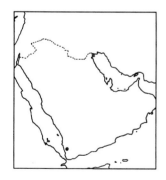
Map 121. Thesium stuhlmannii

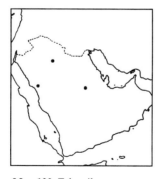
Map 122. T. humile

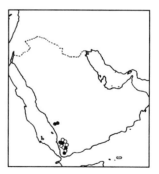
Map 123. T. radicans

Distribution Maps

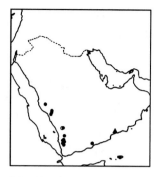 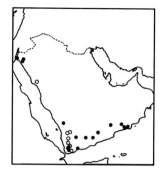 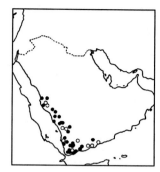

Map 124. Osyris quadripartita ●
O. sp. A. ▲

Map 125. Plicosepalus acaciae

Map 126. P. curviflorus

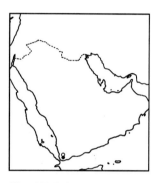 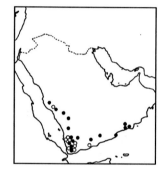 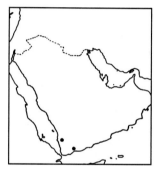

Map 127. Helixanthera thomsonii

Map 128. Oncocalyx schimperi

Map 129. O. doberae

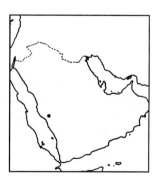 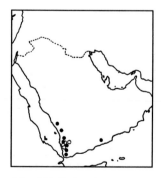 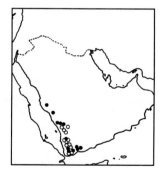

Map 130. O. glabratus

Map 131. Tapinanthus globiferus

Map 132. Phragmanthera austroarabicus

Distribution Maps

Map 133. Viscum triflorum ●
V. cruciatum ▲

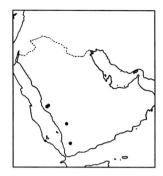
Map 134. V. schimperi

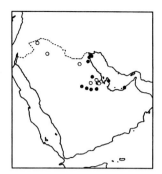
Map 135. Polygonum argyrocoleum

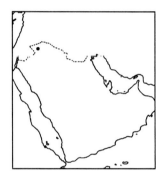
Map 136. P. palaestinum

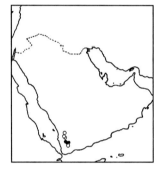
Map 137. P. corrigioloides

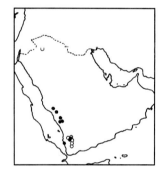
Map 138. P. aviculare

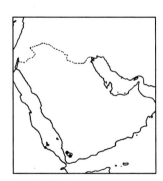
Map 139. Persicaria nepalensis

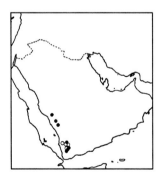
Map 140. P. amphibia

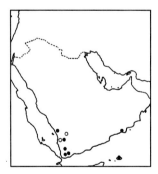
Map 141. P. glabra

Distribution Maps

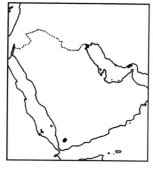
Map 142. P. lapathifolia

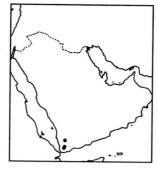
Map 143. P. senegalensis

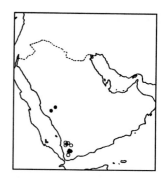
Map 144. P. maculosa

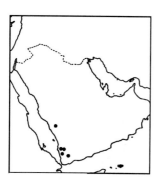
Map 145. P. barbata

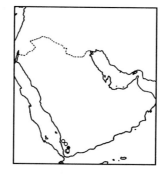
Map 146. P. decipiens

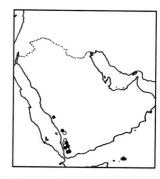
Map 147. Oxygonum sinuatum

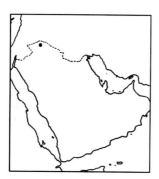
Map 148. Rheum palaestinum

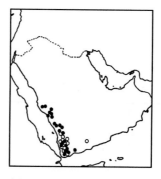
Map 149. Rumex nervosus

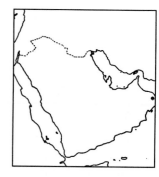
Map 150. R. limoniastrum

Distribution Maps

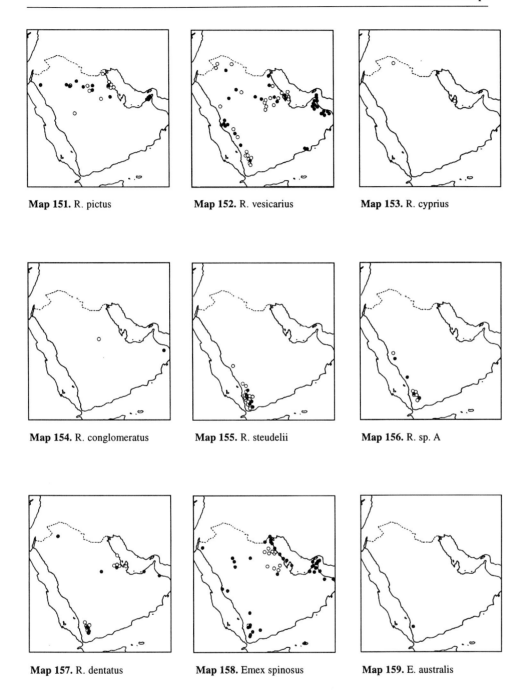

Map 151. R. pictus **Map 152.** R. vesicarius **Map 153.** R. cyprius

Map 154. R. conglomeratus **Map 155.** R. steudelii **Map 156.** R. sp. A

Map 157. R. dentatus **Map 158.** Emex spinosus **Map 159.** E. australis

Distribution Maps

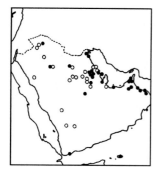

Map 160. Calligonum comosum

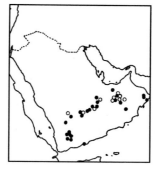

Map 161. C. crinitum subsp. arabicum

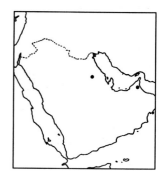

Map 162. C. tetrapterum

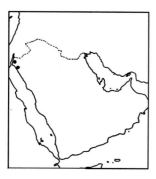

Map 163. Atraphaxis spinosa

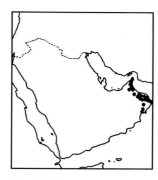

Map 164. Pteropyrum scoparium

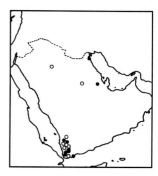

Map 165. Mirabilis jalapa

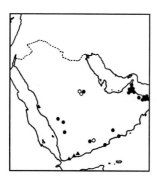

Map 166. Boerhavia elegans subsp. stenophylla ● subsp. elegans ▲

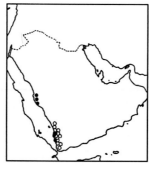

Map 167. B. repens

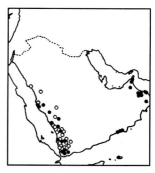

Map 168. B. diffusa

Distribution Maps

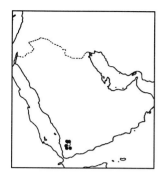

Map 169. Commicarpus pedunculosus

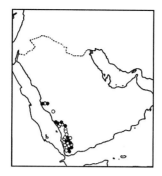

Map 170. C. grandiflorus

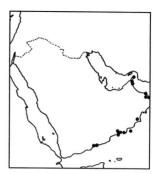

Map 171. C. stenocarpus

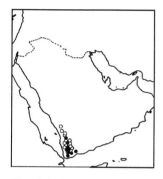

Map 172. C. arabicus

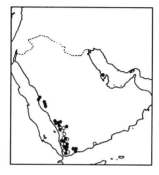

Map 173. C. plumbagineus

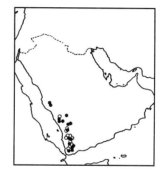

Map 174. C. sinuatus

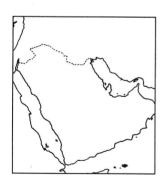

Map 175. C. simonyi

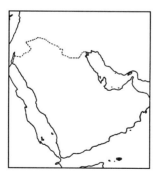

Map 176. C. heimerlii

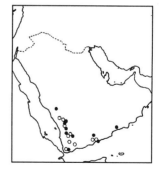

Map 177. C. mistus

Distribution Maps

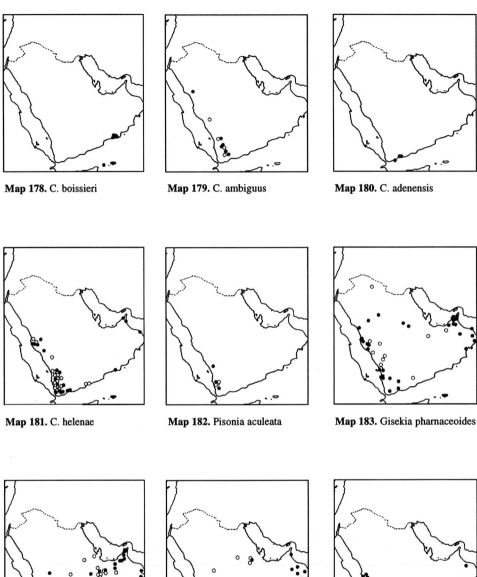

Map 178. C. boissieri

Map 179. C. ambiguus

Map 180. C. adenensis

Map 181. C. helenae

Map 182. Pisonia aculeata

Map 183. Gisekia pharnaceoides

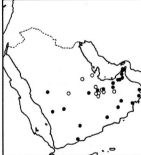

Map 184. Limeum arabicum

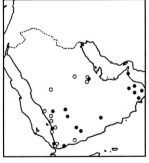

Map 185. L. obovatum

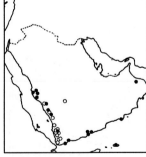

Map 186. Corbichonia decumbens

Distribution Maps

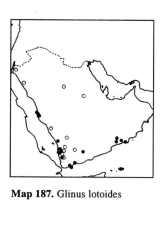

Map 187. Glinus lotoides

Map 188. G. setiflorus

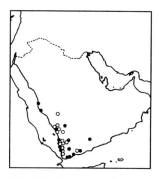

Map 189. Mollugo cerviana

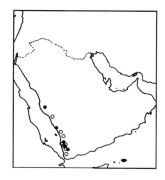

Map 190. M. nudicaulis

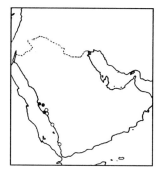

Map 191. Sesuvium sesuvioides

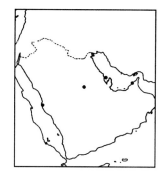

Map 192. S. verrucosum

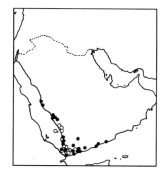

Map 193. Trianthema crystallina

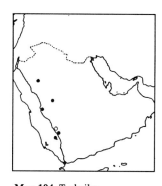

Map 194. T. sheilae

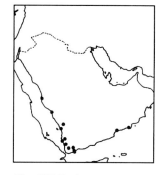

Map 195. T. triquetra

Distribution Maps

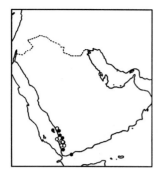
Map 196. T. portulacastrum

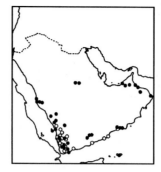
Map 197. Zaleya pentandra

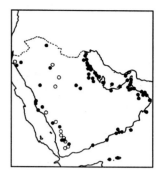
Map 198. Aizoon canariense

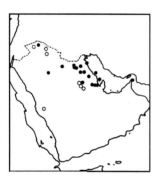
Map 199. A. hispanicum

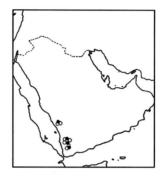
Map 200. Delosperma harazianum

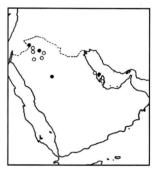
Map 201. Mesembryanthemum forsskalei

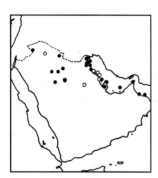
Map 202. M. nodiflorum

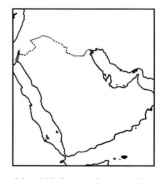
Map 203. Tetragonia pentandra

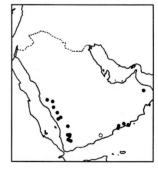
Map 204. Telephium sphaerospermum

Distribution Maps

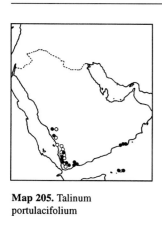

Map 205. Talinum portulacifolium

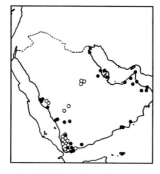

Map 206. Portulaca oleracea

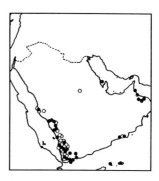

Map 207. P. quadrifida

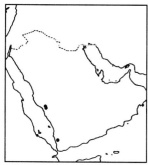

Map 208. P. foliosa

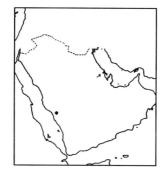

Map 209. P. kermesina ●
P. pilosa ▲

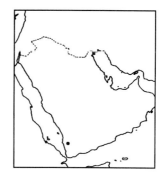

Map 210. Anredera cordifolia

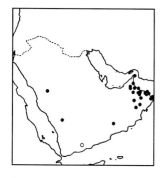

Map 211. Cometes surattensis

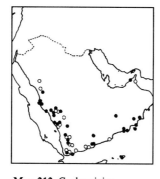

Map 212. C. abyssinica

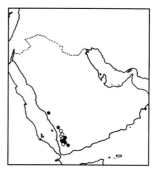

Map 213. Pollichia campestris

Distribution Maps

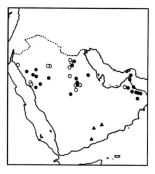

Map 214. Gymnocarpos decandrus ● G. argenteus ▲

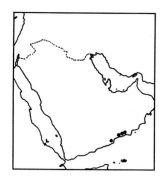

Map 215. G. bracteata ●
G. dhofarensis ▲

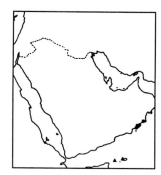

Map 216. G. rotundifolius ●
G. kuriensis ▲

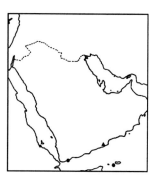

Map 217. Sphaerocoma hookeri ●
Gymnocarpos mahrana ▲

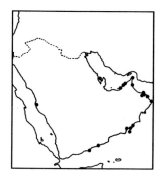

Map 218. S. aucheri

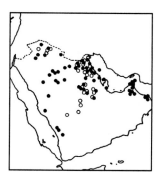

Map 219. Paronychia arabica

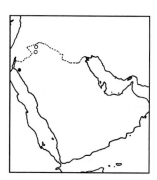

Map 220. P. sinaica

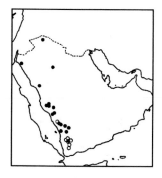

Map 221. P. chlorothyrsa

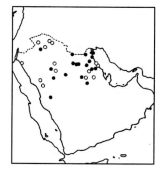

Map 222. Pteranthus dichotomus

Distribution Maps

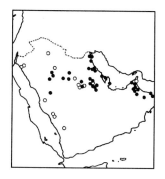

Map 223. Sclerocephalus arabicus

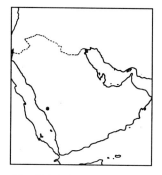

Map 224. Scleranthus orientalis

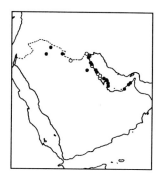

Map 225. Herniaria hemistemon

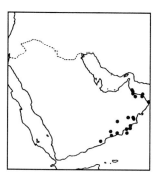

Map 226. H. maskatensis

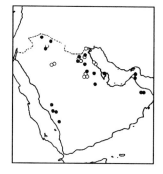

Map 227. H. hirsuta

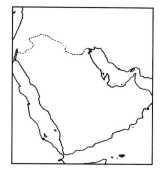

Map 228. Haya obovata

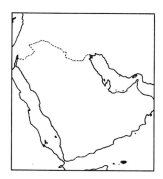

Map 229. Polycarpaea corymbosa

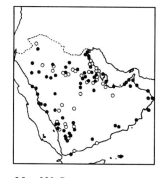

Map 230. P. repens

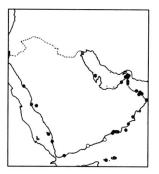

Map 231. P. spicata var. spicata

Distribution Maps

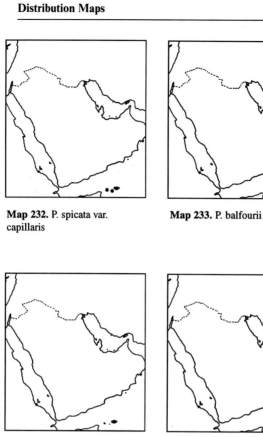

Map 232. P. spicata var. capillaris

Map 233. P. balfourii

Map 234. P. paulayana

Map 235. P. hayoides

Map 236. P. caespitosa ●
P. pulvinata ▲

Map 237. P. kuriensis ●

Map 238. P. hassalensis ●
P. haufensis ▲

Map 239. P. jazirensis

Map 240. P. robbairea

Distribution Maps

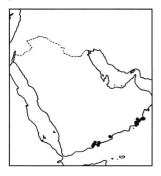

Map 241. Xerotia arabica

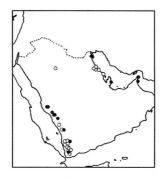

Map 242. Polycarpon tetraphyllum

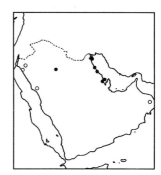

Map 243. P. succulentum

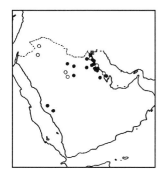

Map 244. Loeflingia hispanica

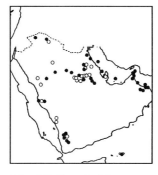

Map 245. Spergula fallax

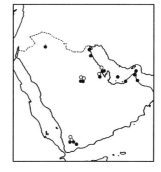

Map 246. Spergularia salina

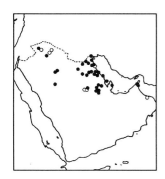

Map 247. S. diandra

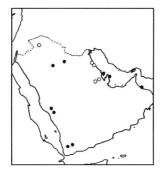

Map 248. S. bocconei

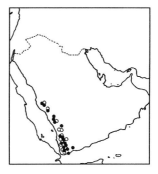

Map 249. Minuartia filifolia

523

Distribution Maps

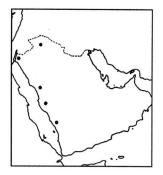
Map 250. M. picta

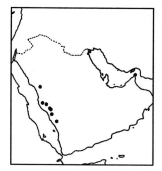
Map 251. M. hybrida

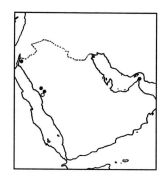
Map 252. M. meyeri

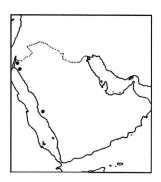
Map 253. Holosteum glutinosum

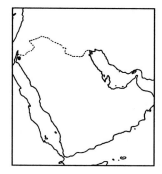
Map 254. Arenaria deflexa

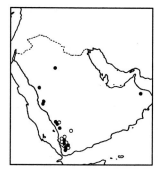
Map 255. A. serpyllifolia agg.

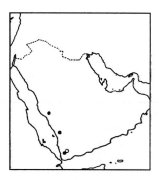
Map 256. Cerastium glomeratum

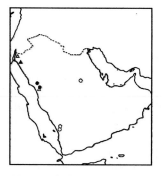
Map 257. C. dichotomum subsp. dichotomum ● subsp. inflatum ▲

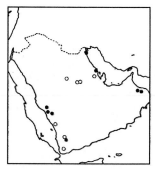
Map 258. Stellaria media

Distribution Maps

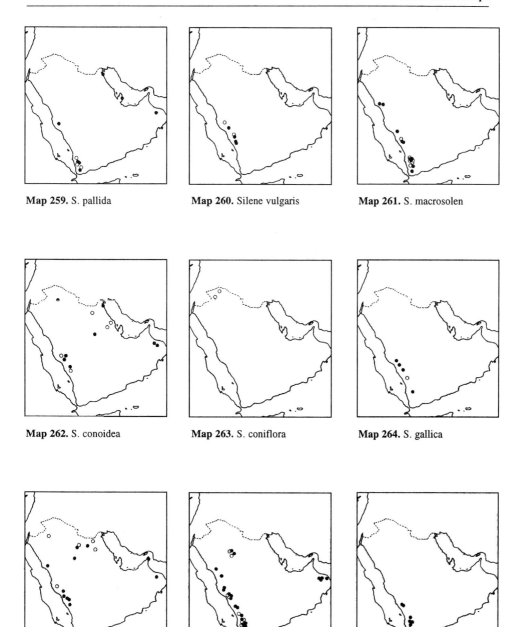

Map 259. S. pallida

Map 260. Silene vulgaris

Map 261. S. macrosolen

Map 262. S. conoidea

Map 263. S. coniflora

Map 264. S. gallica

Map 265. S. apetala

Map 266. S. burchellii agg.

Map 267. S. yemensis

Distribution Maps

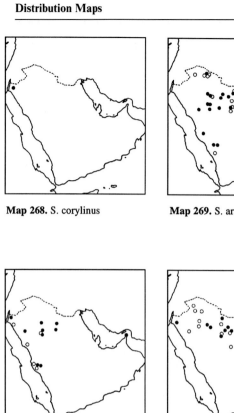

Map 268. S. corylinus

Map 269. S. arabica

Map 270. S. colorata subsp. olivierana

Map 271. S. villosa Form 'A'

Map 272. S. villosa Form 'B'

Map 273. S. asirensis

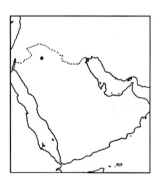

Map 274. S. crassipes

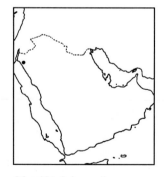

Map 275. S. hussonii

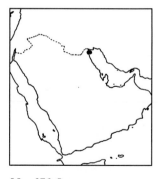

Map 276. S. arenosa

Distribution Maps

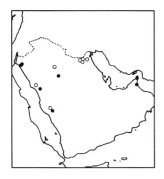

Map 277. S. linearis

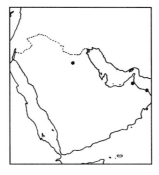

Map 278. S. austro-iranica

Map 279. S. aff. peduncularis

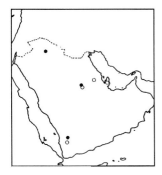

Map 280. Vaccaria hispanica

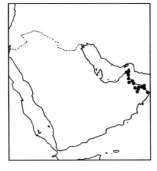

Map 281. Gypsophila bellidifolia

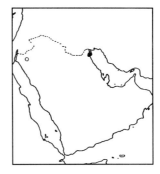

Map 282. G. pilosa

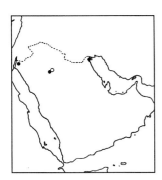

Map 283. G. viscosa

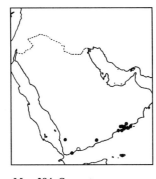

Map 284. G. montana

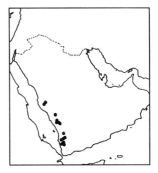

Map 285. G. umbricola

Distribution Maps

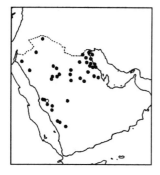
Map 286. G. capillaris

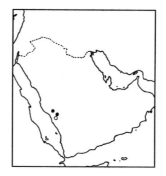
Map 287. Velezia rigida

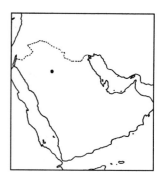
Map 288. Petrorhagia cretica

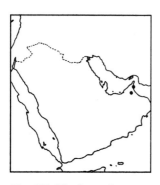
Map 289. Dianthus cyri

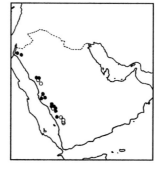
Map 290. D. strictus subsp. sublaevis

Map 291. D. judaicus

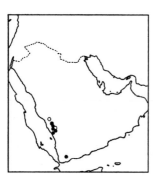
Map 292. D. deserti

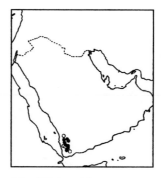
Map 293. D. uniflorus

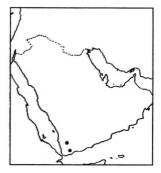
Map 294. D. longiglumis

Distribution Maps

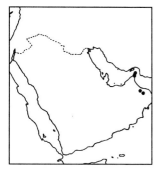
Map 295. D. crinitus

Map 296. D. sinaicus

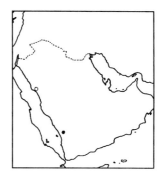
Map 297. Chenopodium carinatum

Map 298. C. ambrosioides

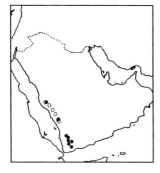
Map 299. C. schraderianum

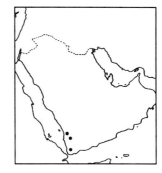
Map 300. C. procerum

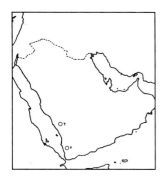
Map 301. C. botrys

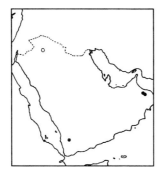
Map 302. C. vulvaria

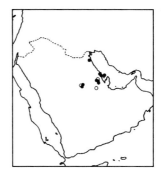
Map 303. C. glaucum

Distribution Maps

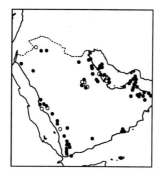

Map 304. C. murale

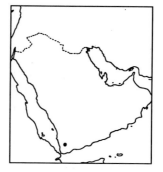

Map 305. C. fasciculosum var. muraliforme

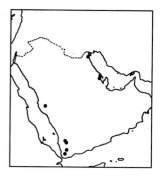

Map 306. C. opulifolium

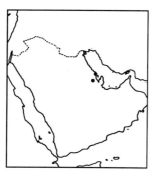

Map 307. C. ficifolium

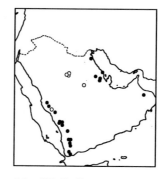

Map 308. C. album

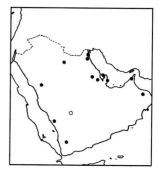

Map 309. Beta vulgaris subsp. maritima

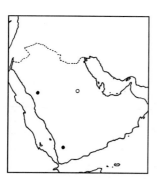

Map 310. Spinacia oleracea

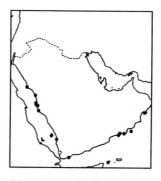

Map 311. Atriplex farinosa

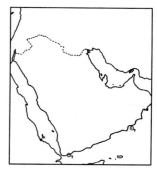

Map 312. A. coriacea

Distribution Maps

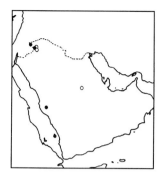

Map 313. A. halimus

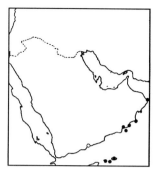

Map 314. A. griffithii subsp. stocksii

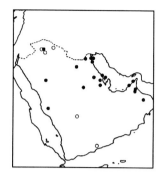

Map 315. A. leucoclada var. inamoena

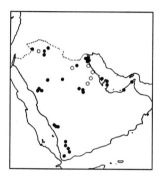

Map 316. A. leucoclada var. turcomanica

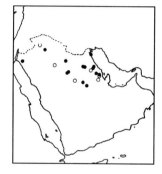

Map 317. A. dimorphostegia

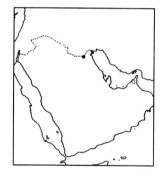

Map 318. A. tatarica

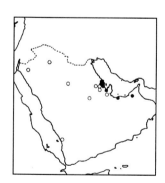

Map 319. Agriophyllum minus

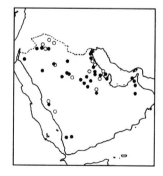

Map 320. Bassia muricata

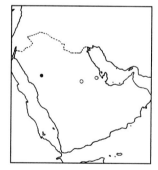

Map 321. B. hyssopifolia

Distribution Maps

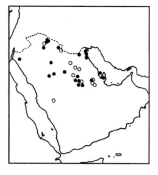 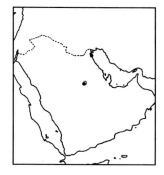 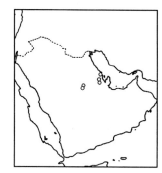

Map 322. B. eriophora   **Map 323.** B. scoparia   **Map 324.** B. indica

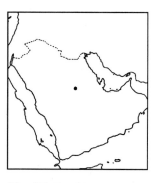 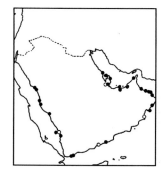 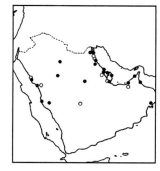

Map 325. B. arabica   **Map 326.** Halopeplis perfoliata   **Map 327.** Halocnemum strobilaceum

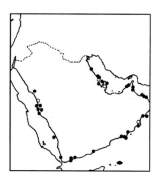 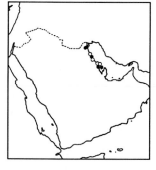 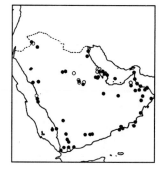

Map 328. Arthrocnemum macrostachyum   **Map 329.** Salicornia europaea   **Map 330.** Suaeda aegyptiaca

Distribution Maps

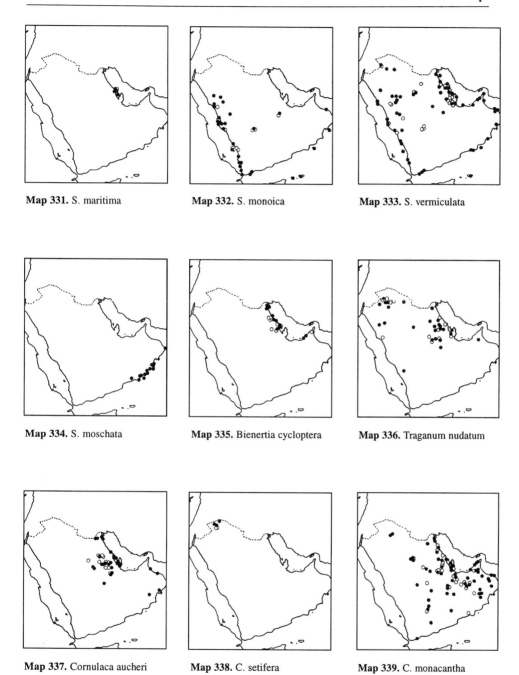

Map 331. S. maritima

Map 332. S. monoica

Map 333. S. vermiculata

Map 334. S. moschata

Map 335. Bienertia cycloptera

Map 336. Traganum nudatum

Map 337. Cornulaca aucheri

Map 338. C. setifera

Map 339. C. monacantha

Distribution Maps

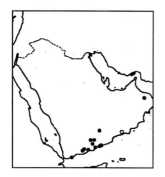
Map 340. C. amblyacantha

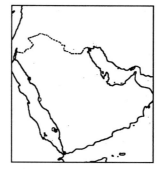
Map 341. C. ehrenbergii

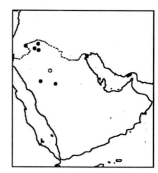
Map 342. Agathophora alopecuroides var. alopecuroides

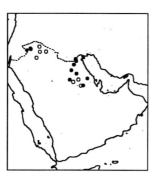
Map 343. A. alopecuroides var. papillosa

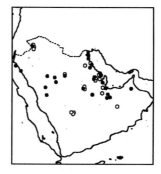
Map 344. Seidlitzia rosmarinus

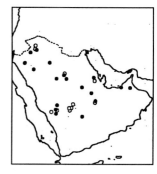
Map 345. Haloxylon persicum

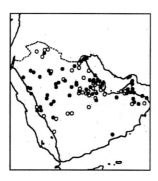
Map 346. H. salicornicum

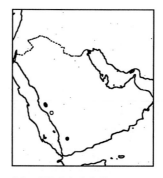
Map 347. Salsola kali

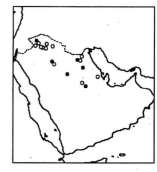
Map 348. S. volkensii

Distribution Maps

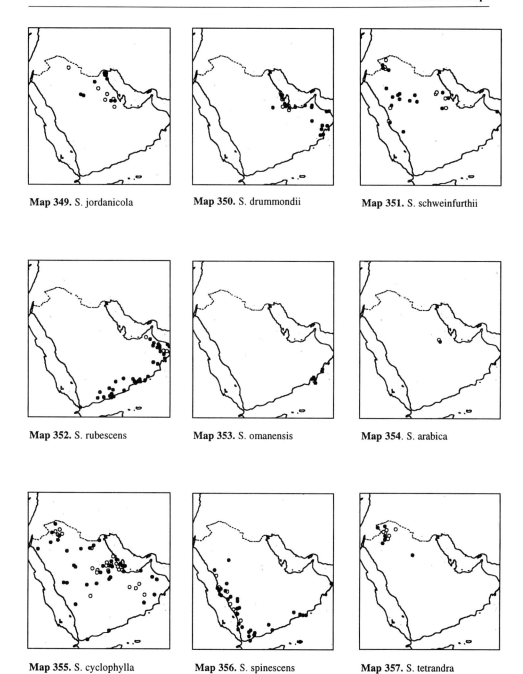

Map 349. S. jordanicola

Map 350. S. drummondii

Map 351. S. schweinfurthii

Map 352. S. rubescens

Map 353. S. omanensis

Map 354. S. arabica

Map 355. S. cyclophylla

Map 356. S. spinescens

Map 357. S. tetrandra

Distribution Maps

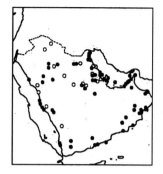
Map 358. S. imbricata

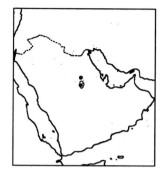
Map 359. S. lachnantha

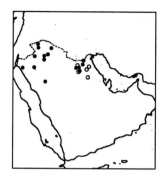
Map 360. S. villosa

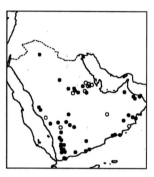
Map 361. Halothamnus bottae

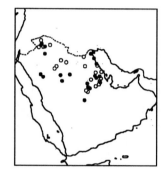
Map 362. H. iraqensis

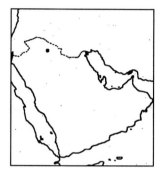
Map 363. H. lancifolius

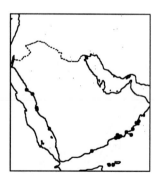
Map 364. Sevada schimperi

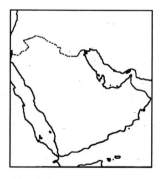
Map 365. Lagenantha cycloptera

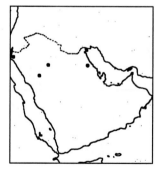
Map 366. Noaea mucronata

Distribution Maps

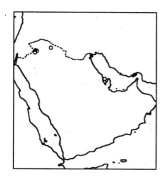
Map 367. Anabasis articulata

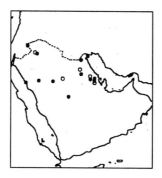
Map 368. A. lachnantha

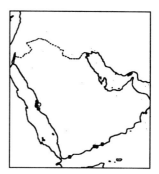
Map 369. A. ehrenbergii

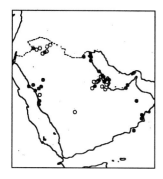
Map 370. A. setifera

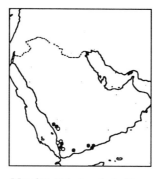
Map 371. Celosia polystachia

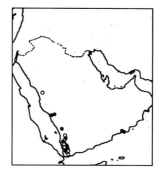
Map 372. C. trigyna

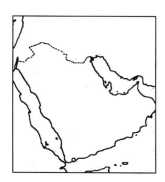
Map 373. C. anthelminthica

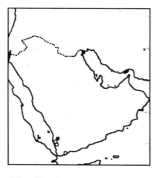
Map 374. C. argentea

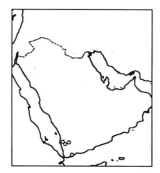
Map 375. Amaranthus caudatus

Distribution Maps

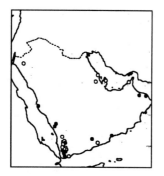

Map 376. A. hybridus subsp. hybridus

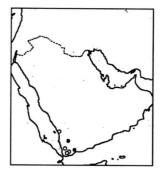

Map 377. A. hybridus subsp. cruentus

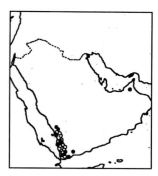

Map 378. A. spinosus

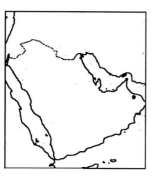

Map 379. A. dubius

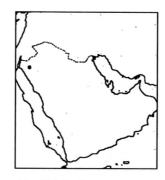

Map 380. A. albus

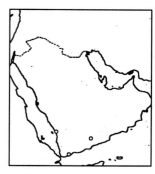

Map 381. A. tricolor

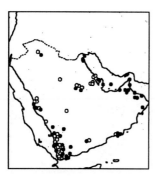

Map 382. A. graecizans subsp. graecizans

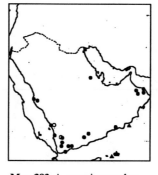

Map 383. A. graecizans subsp. silvestris ●
subsp. thellungianus ▲

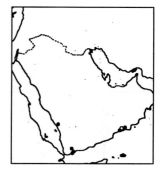

Map 384. A. sparganiocephalus

Distribution Maps

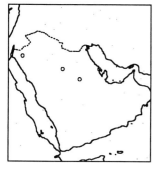

Map 385. A. lividus

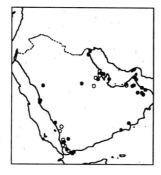

Map 386. A. viridis

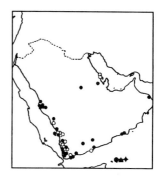

Map 387. Digera muricata subsp. muricata subsp. trivervis ●
var. trinervis ▲
var. patentipilosa ✦

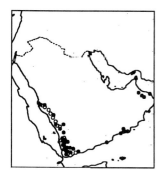

Map 388. Pupalia lappacea

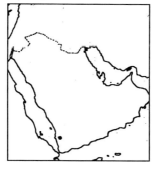

Map 389. P. grandiflora

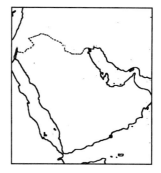

Map 390. P. robecchii

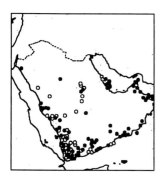

Map 391. Aerva javanica var. javanica

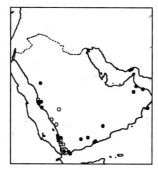

Map 392. Aerva javanica var. bovei

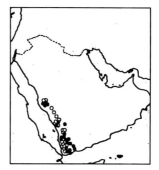

Map 393. A. lanata

Distribution Maps

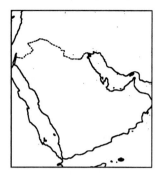
Map 394. A. revoluta

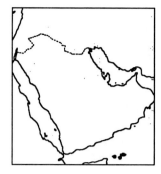
Map 395. A. microphylla

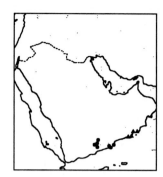
Map 396. A. artemisioides subsp. artemisiodes ● subsp. batharitica ▲

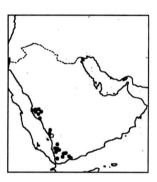
Map 397. Psilotrichum gnaphalobryum

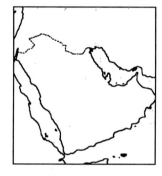
Map 398. P. sericeum

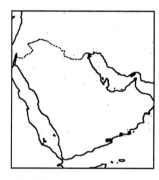
Map 399. P. virgatum

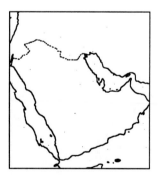
Map 400. P. aphyllum

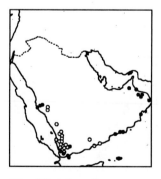
Map 401. Achyranthes aspera var. aspera

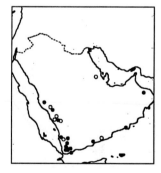
Map 402. A. aspera var. pubescens

Distribution Maps

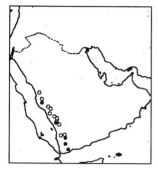

Map 403. A. aspera var. sicula

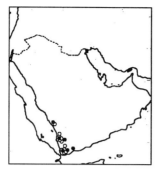

Map 404. Alternanthera pungens

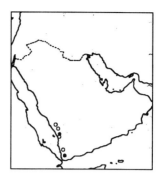

Map 405. A. sessilis

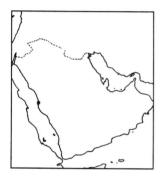

Map 406. A. tenella var. bettzickiana

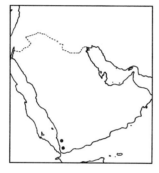

Map 407. Gomphrena globosa

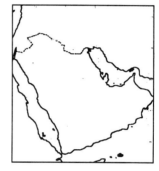

Map 408. G. celosioides

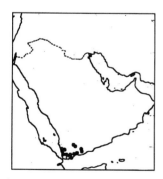

Map 409. Saltia papposa

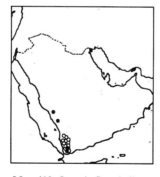

Map 410. Opuntia ficus-indica

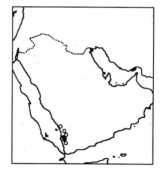

Map 411. O. dillenii

Distribution Maps

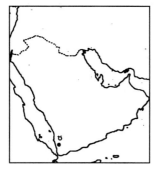 Map 412. Michelia champaca

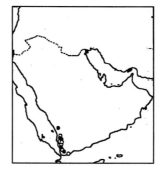 Map 413. Annona squamosa

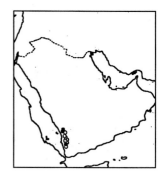 Map 414. Cassytha filiformis

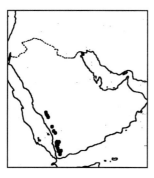 Map 415. Clematis hirsuta

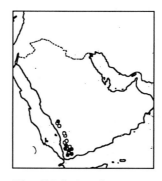 Map 416. C. simensis

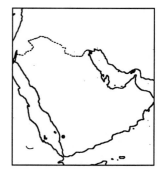 Map 417. C. longicauda

 Map 418. C. orientalis

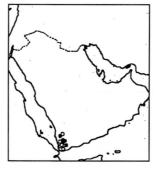 Map 419. Thalictrum minus

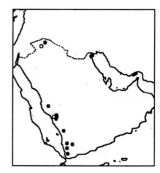 Map 420. Adonis dentata

Distribution Maps

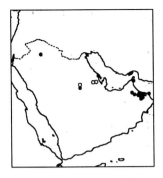

Map 421. Ranunculus muricatus

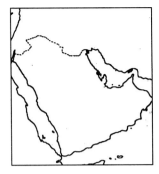

Map 422. R. cornutus

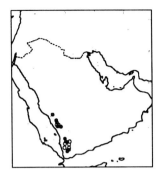

Map 423. R. multifidus

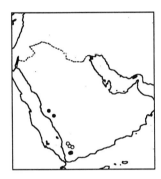

Map 424. R. rionii

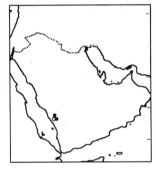

Map 425. R. sphaerospermus

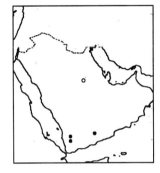

Map 426. Nigella sativa

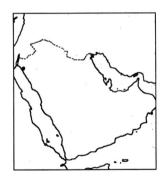

Map 427. Delphinium sheilae

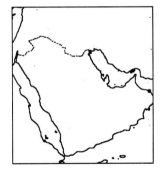

Map 428. D. penicillatum

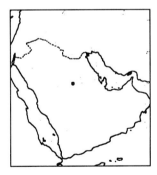

Map 429. D. orientale

Distribution Maps

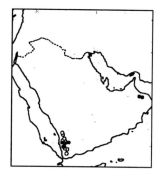
Map 430. Berberis holstii

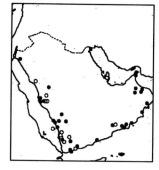
Map 431. Cocculus pendulus

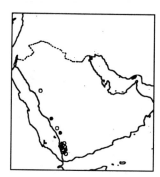
Map 432. C. hirsutus

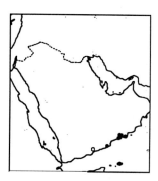
Map 433. C. balfourii

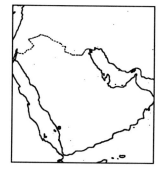
Map 434. Tinospora bakis

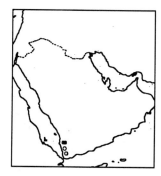
Map 435. Stephania abyssinica

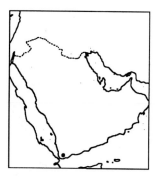
Map 436. Nymphaea nouchali var. caerulea

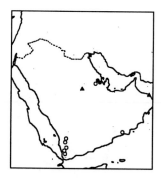
Map 437. Ceratophyllum demersum ● C. submersum ▲

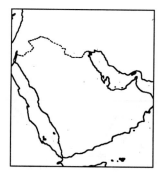
Map 438. Peperomia tetraphylla

Distribution Maps

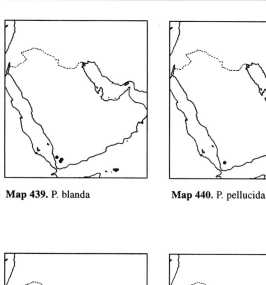

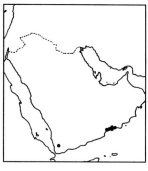

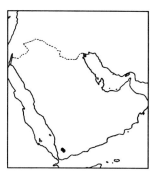

Map 439. P. blanda

Map 440. P. pellucida

Map 441. P. abyssinica

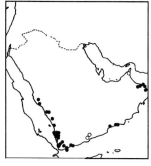

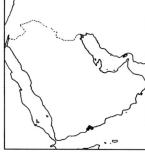

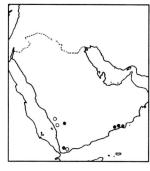

Map 442. Aristolochia bracteolata

Map 443. A. rigida

Map 444. Hydnora johannis

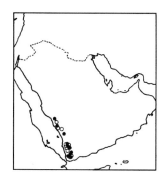

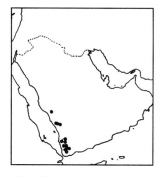

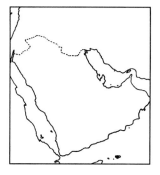

Map 445. Ochna inermis

Map 446. Hypericum revolutum

Map 447. H. balfourii

Distribution Maps

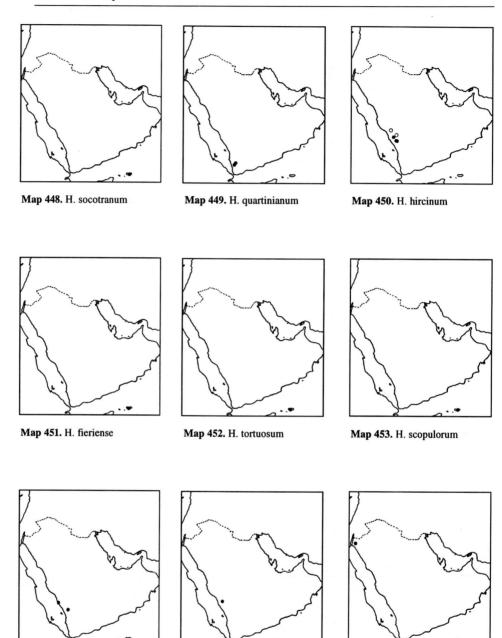

Distribution Maps

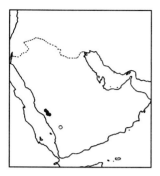
Map 457. H. perforatum

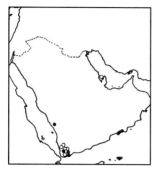
Map 458. Argemone mexicana

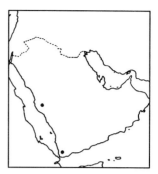
Map 459. A. ochroleuca

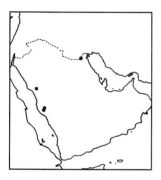
Map 460. Glaucium corniculatum

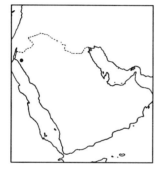
Map 461. G. arabicum

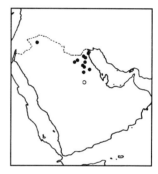
Map 462. Roemeria hybrida subsp. dodecandra

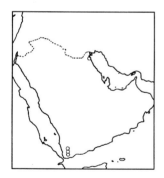
Map 463. Papaver somniferum

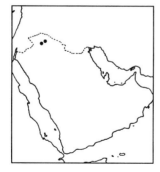
Map 464. P. glaucum

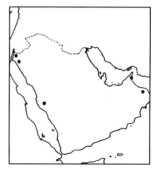
Map 465. P. decaisnei

Distribution Maps

 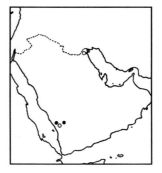 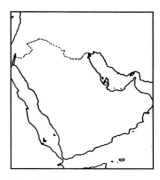

Map 466. P. macrostomum   **Map 467.** P. rhoeas   **Map 468.** P. umbonatum

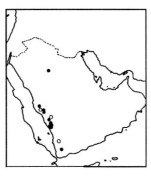 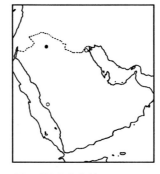 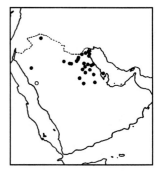

Map 469. P. dubium subsp. laevigatum   **Map 470.** P. hybridum   **Map 471.** Hypecoum pendulum

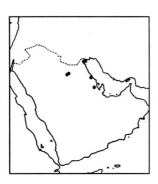 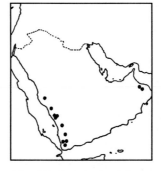 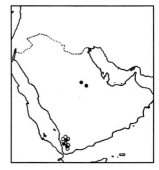

Map 472. H. geslinii   **Map 473.** Fumaria abyssinica   **Map 474.** F. parviflora

Distribution Maps

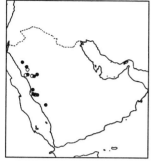

Map 475. Cleome chrysantha

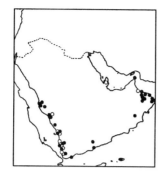

Map 476. C. scaposa

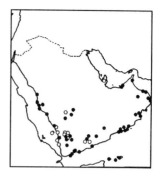

Map 477. C. brachycarpa

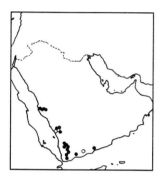

Map 478. C. ramosissima

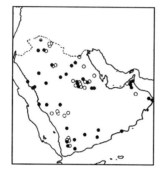

Map 479. C. amblyocarpa

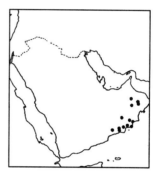

Map 480. C. brevipetiolata

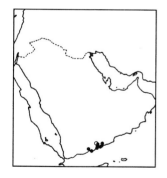

Map 481. C. macradenia

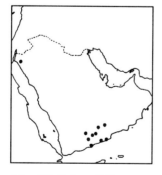

Map 482. C. droserifolia

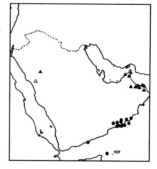

Map 483. C. austroarabica
subsp. austroarabica ●
subsp. muscatensis ▲

Distribution Maps

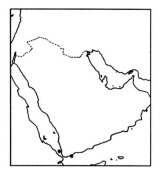

Map 484. C. pruinosa

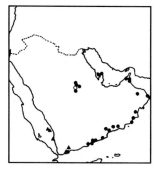

Map 485. C. noeana subsp. noeana ● subsp. brachystyla ▲

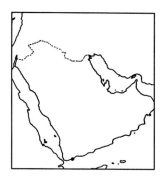

Map 486. C. polytricha

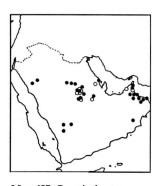

Map 487. C. rupicola

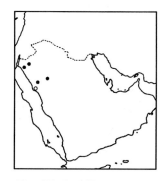

Map 488. C. arabica

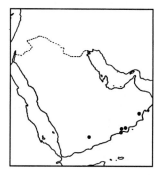

Map 489. C. albescens subsp. omanensis

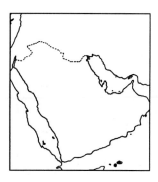

Map 490. C. socotrana

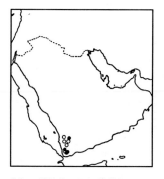

Map 491. C. angustifolia

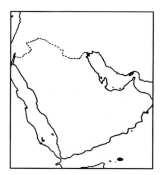

Map 492. C. tenella

Distribution Maps

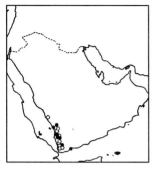
Map 493. C. viscosa

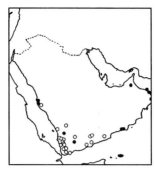
Map 494. C. gynandra

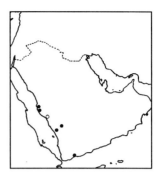
Map 495. C. hanburyana

Map 496. C. monophylla

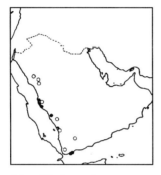
Map 497. C. paradoxa

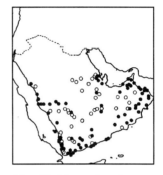
Map 498. Dipterygium glaucum

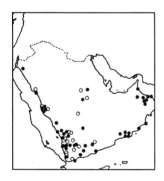
Map 499. Maerua crassifolia

Map 500. M. angolensis subsp. angolensis ● subsp. socotrana ▲

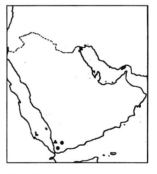
Map 501. M. triphylla var. calophylla type A ● var. johannis ▲

551

Distribution Maps

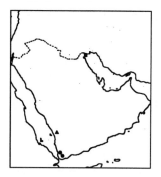
Map 502. M. triphylla var. calophylla type B ● type C ▲

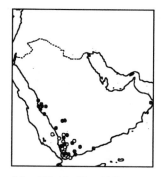
Map 503. M. oblongifolia

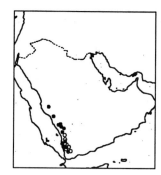
Map 504. Boscia angustifolia

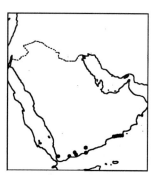
Map 505. B. arabica

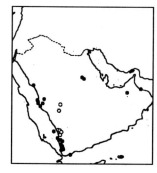
Map 506. Capparis decidua

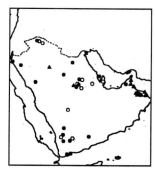
Map 507. C. spinosa var. spinosa ● var. mucronifolia ▲

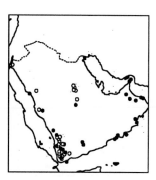
Map 508. C. cartilaginea

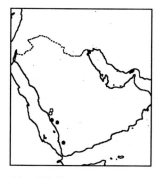
Map 509. C. tomentosa

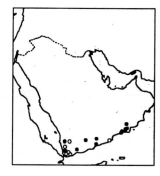
Map 510. Cadaba heterotricha

Distribution Maps

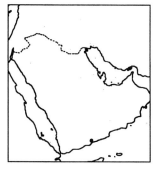 Map 511. C. mirabilis

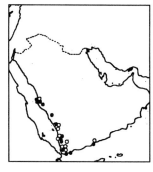 Map 512. C. glandulosa

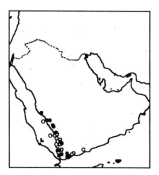 Map 513. C. rotundifolia

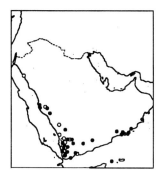 Map 514. C. farinosa

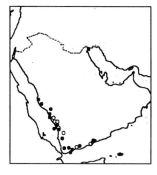 Map 515. C. longifolia

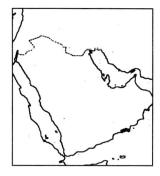 Map 516. C. baccarinii

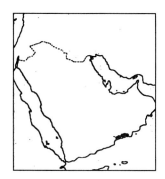 Map 517. Dhofaria macleishii

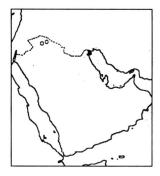 Map 518. Brassica deflexa

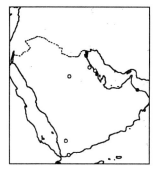 Map 519. B. juncea

Distribution Maps

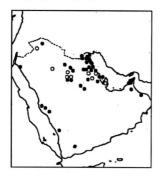

Map 520. B. tournefortii

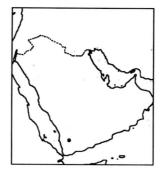

Map 521. B. oleracea

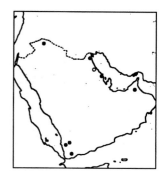

Map 522. B. rapa

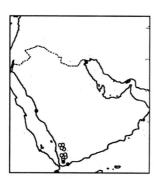

Map 523. B. napus

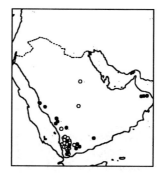

Map 524. Erucastrum arabicum

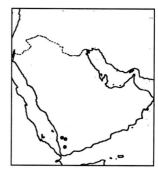

Map 525. E. meruense subsp. yemense

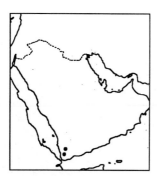

Map 526. E. woodiorum

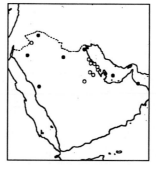

Map 527. Sinapis arvensis

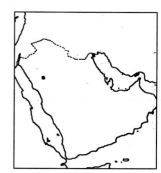

Map 528. S. alba

Distribution Maps

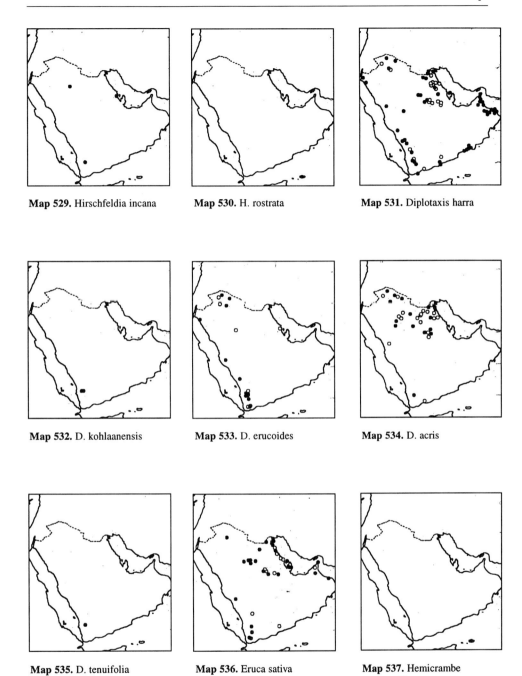

Map 529. Hirschfeldia incana

Map 530. H. rostrata

Map 531. Diplotaxis harra

Map 532. D. kohlaanensis

Map 533. D. erucoides

Map 534. D. acris

Map 535. D. tenuifolia

Map 536. Eruca sativa

Map 537. Hemicrambe townsendii

Distribution Maps

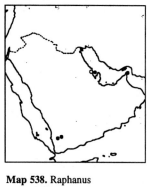
Map 538. Raphanus raphanistrum

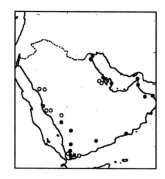
Map 539. R. sativus

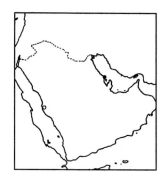
Map 540. Enarthrocarpus lyratus

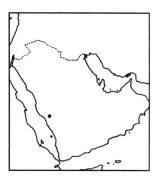
Map 541. Crambe aff. orientalis

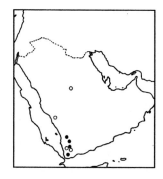
Map 542. Rapistrum rugosum

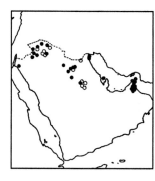
Map 543. Erucaria hispanica

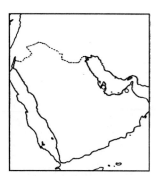
Map 544. E. crassifolia

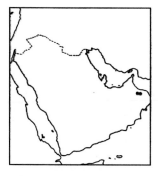
Map 545. E. sp. A

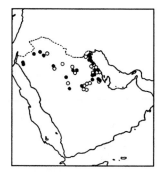
Map 546. Cakile arabica

Distribution Maps

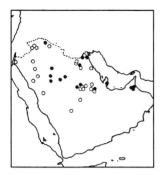

Map 547. Zilla spinosa

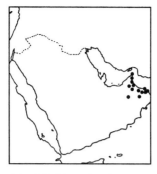

Map 548. Physorrhynchus chamaerapistrum

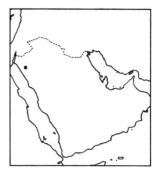

Map 549. Dolichorhynchus arabicus

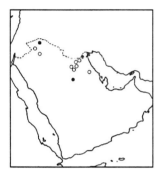

Map 550. Carrichtera annua

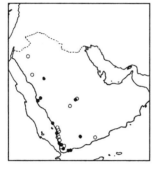

Map 551. Schouwia purpurea

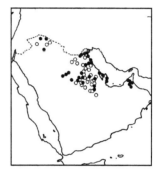

Map 552. Savignya parviflora

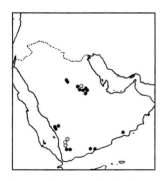

Map 553. Moricandia sinaica

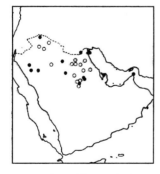

Map 554. Lepidium aucheri

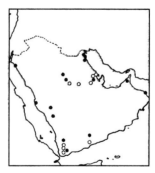

Map 555. L. sativum

Distribution Maps

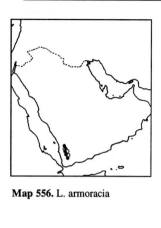
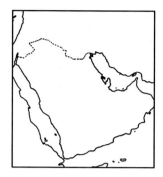
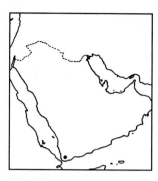

Map 556. L. armoracia **Map 557.** L. virginicum **Map 558.** L. bonariense

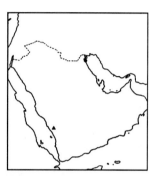
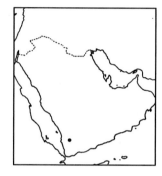
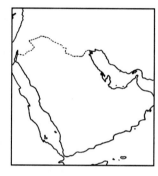

Map 559. L. ruderale ●
L. africanum ▲   **Map 560.** L. latifolium   **Map 561.** Coronopus squamatus

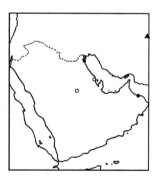
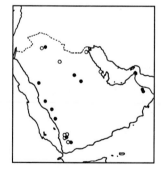
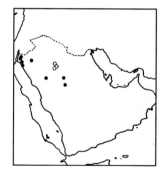

Map 562. C. didymus   **Map 563.** Cardaria draba   **Map 564.** Isatis lusitanica

Distribution Maps

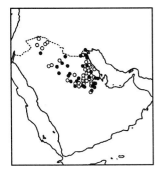

Map 565. Horwoodia dicksoniae

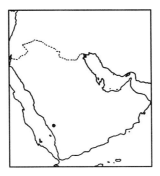

Map 566. Iberis umbellata

Map 567. Biscutella didyma

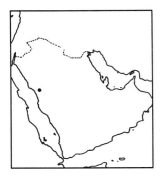

Map 568. Thlaspi perfoliatum

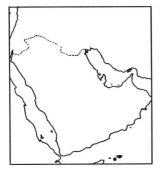

Map 569. Lachnocapsa spathulata

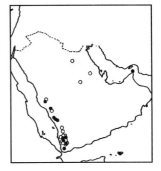

Map 570. Capsella bursa-pastoris

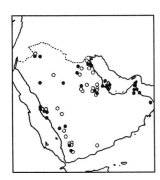

Map 571. Anastatica hierochuntica

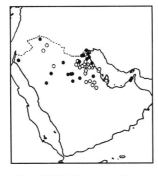

Map 572. Schimpera arabica

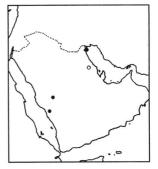

Map 573. Neslia apiculata

Distribution Maps

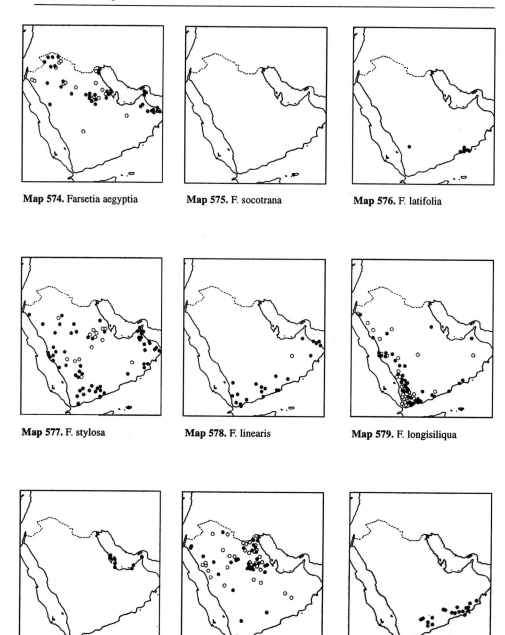

Map 574. Farsetia aegyptia

Map 575. F. socotrana

Map 576. F. latifolia

Map 577. F. stylosa

Map 578. F. linearis

Map 579. F. longisiliqua

Map 580. F. heliophila

Map 581. F. burtonae

Map 582. F. dhofarica

Distribution Maps

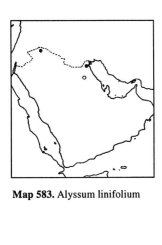
Map 583. Alyssum linifolium

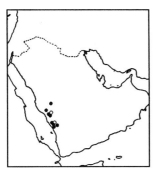
Map 584. A. desertorum

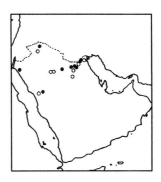
Map 585. A. homalocarpum

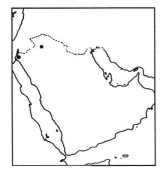
Map 586. A. marginatum

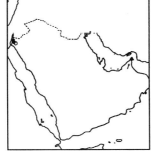
Map 587. A. subspinosum ●
Erophila verna ▲

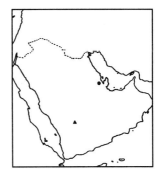
Map 588. Lobularia maritima ●
Lobularia sp. A ▲

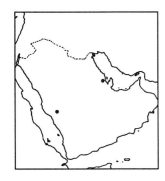
Map 589. L. libyca

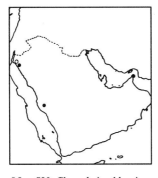
Map 590. Clypeola jonthlaspi

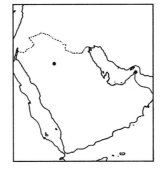
Map 591. C. aspera

Distribution Maps

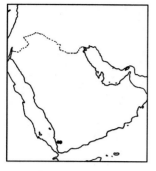
Map 592. Cardamine africana

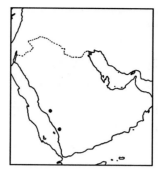
Map 593. C. hirsuta

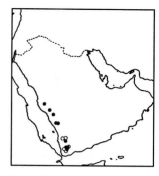
Map 594. Arabis alpina

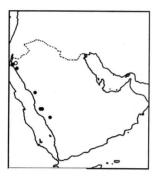
Map 595. Arabis nova

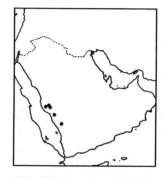
Map 596. Nasturtium officinale

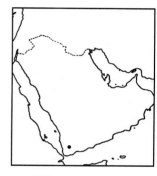
Map 597. N. microphyllum

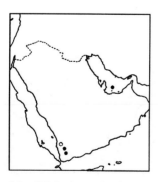
Map 598. Rorippa palustris

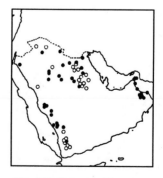
Map 599. Notoceras bicorne

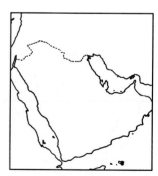
Map 600. Diceratella incana

Distribution Maps

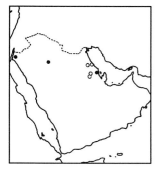
Map 601. Matthiola arabica

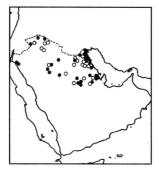
Map 602. M. longipetala

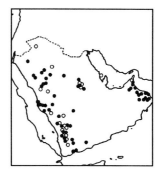
Map 603. Morettia parviflora

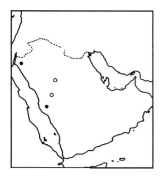
Map 604. M. canescens

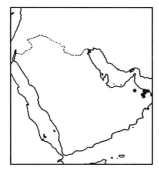
Map 605. M. philaeana

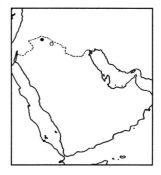
Map 606. Chorispora purpurascens

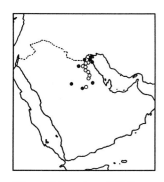
Map 607. Leptaleum filifolium

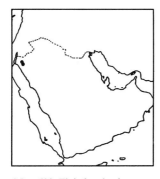
Map 608. Eigia longistyla

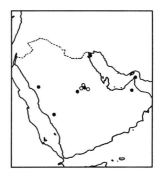
Map 609. Malcolmia africana

Distribution Maps

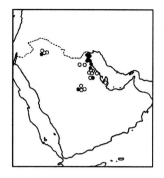

Map 610. M. grandiflora

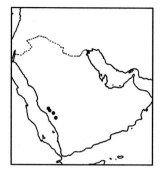

Map 611. M. chia

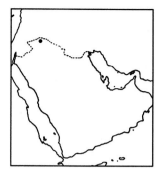

Map 612. M. crenulata

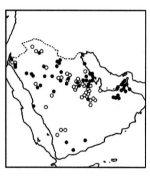

Map 613. Eremobium aegyptiacum

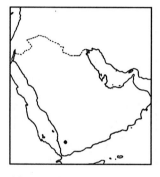

Map 614. Sterigmostemum sulphureum

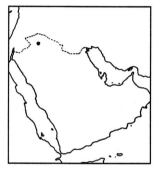

Map 615. Goldbachia laevigata

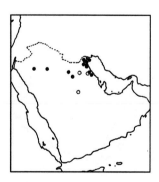

Map 616. Maresia pygmaea

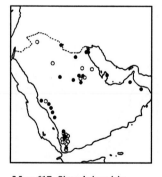

Map 617. Sisymbrium irio

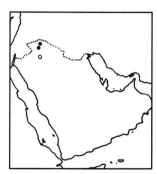

Map 618. S. loeseli

Distribution Maps

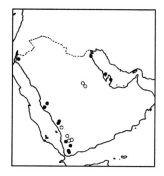

Map 619. S. orientale

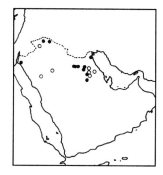

Map 620. S. septulatum

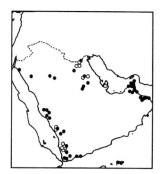

Map 621. S. erysimoides

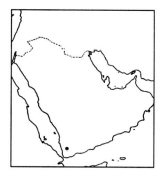

Map 622. S. officinale

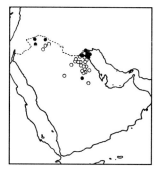

Map 623. Neotorularia torulosa

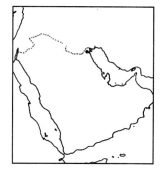

Map 624. Descurainia sophia

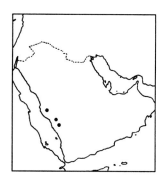

Map 625. Arabidopsis thaliana

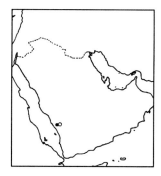

Map 626. A. kneuckeri

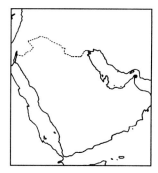

Map 627. A. pumila

Distribution Maps

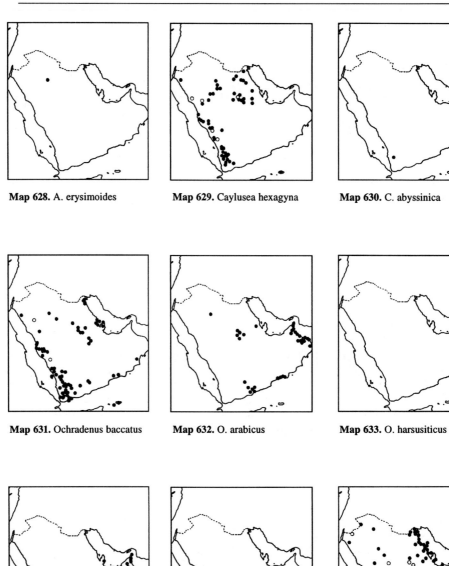

Map 628. A. erysimoides

Map 629. Caylusea hexagyna

Map 630. C. abyssinica

Map 631. Ochradenus baccatus

Map 632. O. arabicus

Map 633. O. harsusiticus

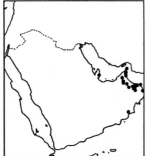

Map 634. O. aucheri

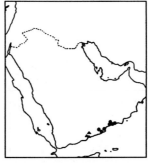

Map 635. O. spartioides ▲
O. gifrii ●

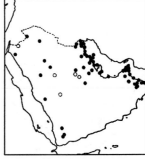

Map 636. Oligomeris linifolia

Distribution Maps

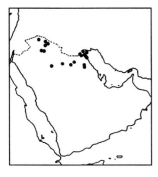

Map 637. Reseda alba subsp. decursiva

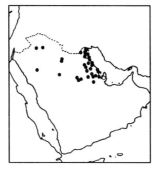

Map 638. R. arabica

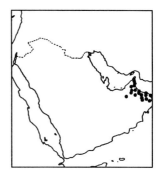

Map 639. R. aucheri var. bracteata

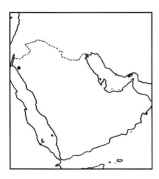

Map 640. R. pentagyna

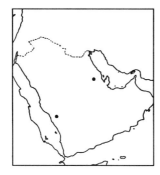

Map 641. R. lutea

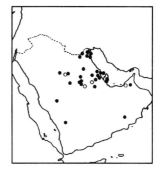

Map 642. R. muricata

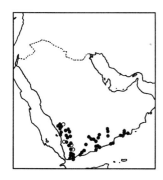

Map 643. R. sphenocleoides

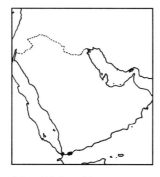

Map 644. R. amblyocarpa

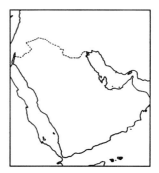

Map 645. R. viridis

Distribution Maps

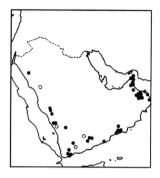
Map 646. Moringa peregrinaa

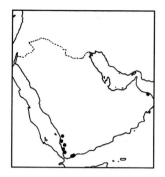
Map 647. M. oleifera

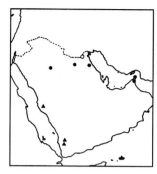
Map 648. Crassula alata subsp. alata ● subsp. pharnaceoides ▲

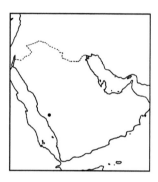
Map 649. C. tillaea

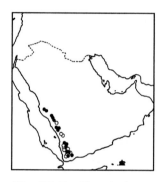
Map 650. C. schimperi subsp. schimperi ● subsp. phyturus ▲

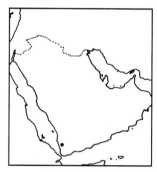
Map 651. C. alsinoides

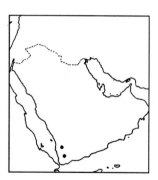
Map 652. C. granvikii

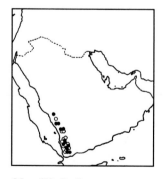
Map 653. C. alba

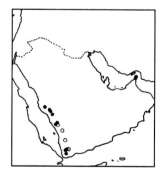
Map 654. Sedum hispanicum

Distribution Maps

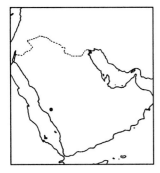
Map 655. S. caespitosum

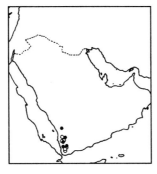
Map 656. Aeonium leucoblepharum

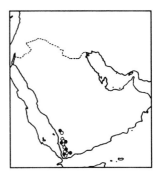
Map 657. Kalanchoe lanceolata

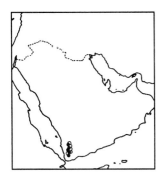
Map 658. K. yemensis

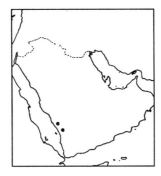
Map 659. K. laciniata

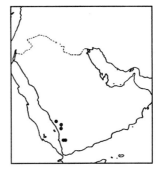
Map 660. K. citrina

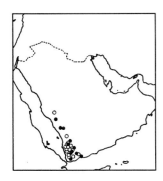
Map 661. K. alternans

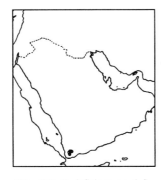
Map 662. K. deficiens var. deficiens

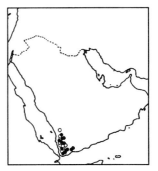
Map 663. K. deficiens var. glabra

Distribution Maps

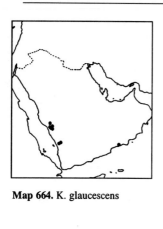 Map 664. K. glaucescens

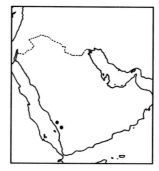 Map 665. K. crenata

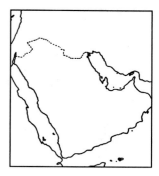 Map 666. K. rotundifolia

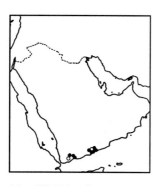 Map 667. K. bentii

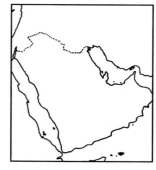 Map 668. K. robusta

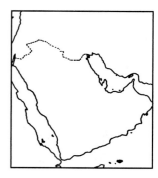 Map 669. K. farinacea

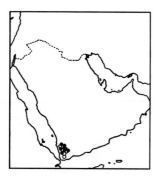 Map 670. Cotyledon barbeyi

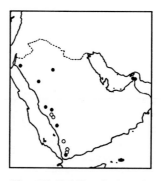 Map 671. Umbilicus horizontalis

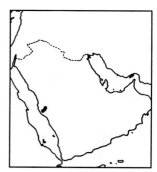 Map 672. U. rupestris

Distribution Maps

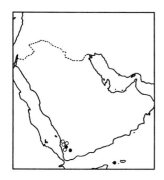

Map 673. Pittosporum viridiflorum

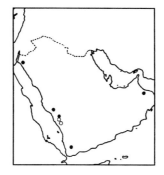

Map 674. Cotoneaster nummularia

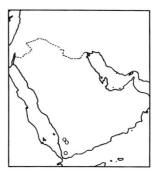

Map 675. Cydonia oblonga

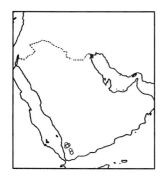

Map 676. Pyrus communis

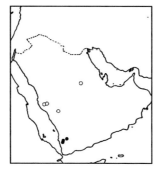

Map 677. Malus sylvestris

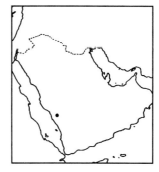

Map 678. Crataegus sinaica

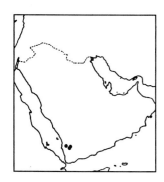

Map 679. Rubus caesius

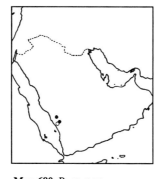

Map 680. R. sp. nov

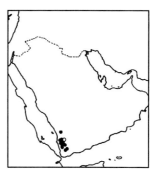

Map 681. R. arabicus

Distribution Maps

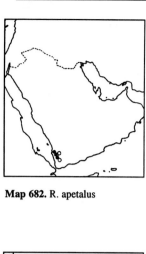

Map 682. R. apetalus

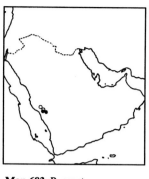

Map 683. R. sanctus

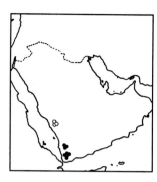

Map 684. Potentilla dentata

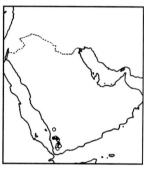

Map 685. P. reptans

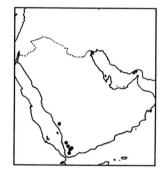

Map 686. Alchemilla cryptantha

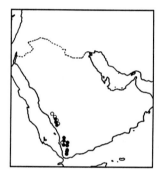

Map 687. Rosa abyssinica

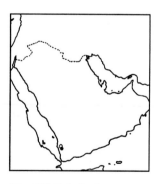

Map 688. R. barbeyi

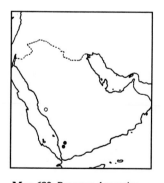

Map 689. Prunus x domestica

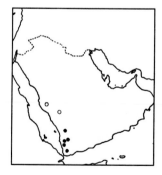

Map 690. P. armeniaca

Distribution Maps

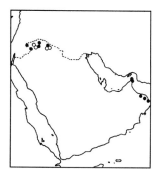

Map 691. P. arabica

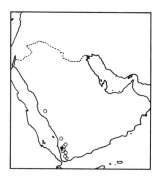

Map 692. P. persica

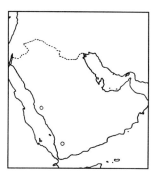

Map 693. P. dulcis

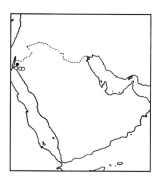

Map 694. P. korshinskyi

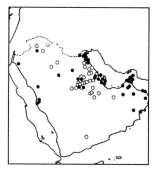

Map 695. Neurada procumbens

INDEX

Acacia abyssinica 20
  asak 18, 19, 20
  ehrenbergiana 18, 19, 20, 21
  etbaica 18, 19, 20, 21
  gerrardii 15, 20
  hamulosa 18, 19
  mellifera 19, 20
  nilotica 20
  oerfota 19, 20
  origena 19, 20, 26
  senegal 21
  tortilis 15, 18, 19, 20, 21
Acanthophyllum C.A. Meyer 177, 232
  bracteatum Boiss. 232
Achyranthes L. 284, 300
  aspera L. 301
    var. *argentea* (Lam.) Boiss. 301
    var. aspera 301
    var. pubescens (Moq.) C.C. Townsend 301, *303*
    var. sicula L. 301
  capitata Forsskal 301
  cordata Hochst. & Steud. 299
  decumbens Forsskal 285
  paniculata Forsskal 285
  papposa Forsskal 305
  polystachia Forsskal 285
  villosa Forsskal 296
Acokanthera schimperi 19
Acridocarpus orientalis 21
Acrostichaceae 54
Acrostichum L. 36, 54
  aureum L. 54, *56*
  dichotomum Forsskal 45
  filare Forsskal 62
Actiniopteridaceae 43
Actiniopteris Link 35, 43
  australis Schwartz 45
  dichotoma Balfour 45
  radiata (Swartz) Link 43, *44*, 45
  semiflabellata Pichi-Sermolli 43, *44*, 45
Adenium obesum 19, 20, 21
  venenata 20
Adiantaceae 45
Adiantum L. 35, 36, 51
  aethiopicum Balfour 53
  balfourii Baker 51, *52*, 53
  capillus-veneris L. 51, *52*, 53
  caudatum Schwartz 51
  incisum Forsskal 51, *52*
  philippense L. 51, *52*, 53
  poirettii Wikstr. 51, *52*, 53
  thalictroides Willd. ex Schlechtend. 53
Adina microcephala 20
Adonis L. 310, 314
  dentata Del. *313*, 314
Aellenia Ulbr. 277
  lancifolia (Boiss.) Ulbr. 279

postii (Eig) Aellen 265
subaphylla Collenette 279
Aeluropus lagopoides 17, 22
Aeonium Webb & Berth. 461, 468
  chrysanthum (Britten) Berger 468
  leucoblepharum Webb ex A. Rich. *465*, 468
Aerva Forsskal 284, 295
  artemisioides Vierh. & Schwartz 295, 298
    subsp. artemisioides *297*, 298
    subsp. batharitica A.G. Miller *297*, 298
  javanica (Burm.f.) Schultes 295, *297*
  lanata (L.) A.L. Juss. 295, 296, *297*
  microphylla Moq. 295, *297*, 298
  persica (Burm.f.) Merrill 295
  revoluta Balf.f. 295, 296, *297*
  tomentosa Forsskal 295
Agathophora (Fenzl) Bunge 234, 263
  algeriensis Botsch. 265
  alopecuroides (Del.) Fenzl ex Bunge 17, *254*, 263
    var. alopecuroides 265
    var. papillosa (Maire) Boulos 265
  galalensis Botsch. 265
  iraqensis Botsch. 265
  postii (Eig) Botsch. 265
Agriophyllum M. Bieb. 234, 249
  minus Fischer & C. Meyer ex Ledeb. 249, *254*
  montasirii El-Gazzar 249
Aizoaceae 155
Aizoon L. 156, 165
  canariense L. *163*, 165
  hispanicum L. 165, 166
Alchemilla L. 480, 488
  cryptantha Steud. ex A. Rich. 488, *489*
Allophylus rhoidiphyllus 22
Allosorus melanolepis Decne 54
Aloe sabaea 19
Alsine sect. *Minuartia* Forssk. 203
Alternanthera Forsskal 284, 302
  pungens Kunth 302, *303*
  repens (L.) Link 302
  sessilis (L.) DC. 302, *303*
  tenella Colla var. bettzickiana Veldk. 302, 304
Alyssum L. 382, 423
  anamense Velen. 426
  damascenum Boiss. & Gaill. 424, 426
  desertorum Stapf 424, *425*

  homalocarpum (Fischer & C. Meyer) Boiss. 424, *425*
  linifolium Stephan ex Willd. 424, *425*
  marginatum Steud. ex Boiss. 424, 426
  singarense Boiss. & Hausskn. 424, *425*, 426
  subspinosum T. Dudley 424, 426
Amaranthaceae 283
Amaranthus L. 284, 287
  albus L. 287, 289, *290*
  angustifolius Lam. subsp. *aschersonianus* Thell. 291
    subsp. *graecizans* (L.) Thell. 291
  blitum Balfour 291
  blitum L. 292
  caudatus L. 287, 288
  chlorostachys Willd. 288
  cruentus L. 289
  dubius Thell. 288, 289, *290*
  gangeticus L. 292
  gracilis Desf. 292
  graecizans L. 287, *290*, 291
    var. graecizans 291
    var. silvestris (Villars) Heukels 291, 292
    var. thellungianus (Nevski) Gusev 291, 292
  hybridus L. 288, *290*
    var. cruentus (L.) Thell. 288, 289
    var. hybridus 288
  hypochondriacus L. 288
  lanatus A.L. Juss. var *rotundifolia* Moq. 296
    var *viridis* Moq. 296
  lividus L. 287, 288, *290*, 292
  mangostanus L. 291
  oleraceus L. 292
  paniculatus L. 289
  polygamus Blatter 291
  retroflexus L. 293
  sparganiocephalus Thell. 287, *290*, 292
  spinosus L. 287, 289, *290*
  tricolor L. 287, 288, *290*, 291
  tristis L. 291
  viridus L. 288, *290*, 292
Amygdalus arabica Oliv. 491
  communis L. 492
  dulcis Miller 492
  korshinskyi (Hand.-Mazz.) Bornm. 492
  persica L. 491
Anabasis L. 233, 281
  alopecuroides (Del.) Moq. 263
  articulata (Forsskal) Moq. 281, 282

ehrenbergii Schweinf. ex Boiss. 281, 282
lachnantha Aellen & Rech.f. 17, 282
setifera Moq. *254*, 281, 282
Anagyris foetida 27
Anastatica L. 382, 417
hierochuntica L. 26, *414*, 417
Angiospermae 80
Anisotes trisulcus 19
Annona L. 308
squamosa L. 308
Annonaceae 307
Anogeissus dhofarica 21
Anogramma Link 35, 50
leptophylla (L.) Link *44*, 50
Anredera Jussieu 174
cordifolia (Tenore) van Steenis 174
Antiaris Lescheu 89, 90
challa (Schweinf.) Engl. 90
toxicaria Lescheu 90, *93*
Apluda mutica 22
Arabidopsis Heynh. 382, 383, 446
erysimoides Hedge & Kit Tan *444*, 446, 447
kneuckeri (Bornm.) O. Schulz *444*, 447
pumila (Stephan ex Willd.) N. Busch *444*, 446, 447
thaliana (L.) Heynh. *444*, 446, 447
Arabis L. 383, 429
alpina L. *425*, 430
caucasica Willd. 430
nova Villars *425*, 430
Arenaria L. 177, 204
deflexa Decne *205*, 206
filifolia Forsskal 203
foliacea Turrill *205*, 206, 207
glutinosa Bieb. 204
leptoclados (Reichenb.) Guss. 206
marina (L.) Roth. 201
rubra L. var. *marina* L. 201
serpyllifolia L. 206
serpyllifolia L. agg. 206
subsp. *leptoclados* (Reichenb.) Nyman 206
Argemone L. 339
mexicana L. 340, *343*
ochroleuca Sweet 340
Aristolochia L. 328
bracteata Retz. 328
bracteola Lam. 328, *329*
maurorum L. var. *latifolia* Blatter 328
rigida Duchartre 328, *329*
sempervirens Forsskal 328
Aristolochiaceae 327
Armeniaca vulgaris Lam. 491
Artemisia herba-alba 27
monosperma 16
Arthrocnemum Moq. 233, 253
glaucum Ung.-Sterb. 253

macrostachyum (Moric.) K. Koch 17, 253, *254*
Arthropteris J. Smith 37, 59
orientalis (J.F. Gmelin) Posthumus 59, *66*
Aspleniaceae 60
Asplenium L. 36, 37, 60
achilleifolium (Lam.) C. Chr. var. *bipinnatum* (Forsskal) C. Chr. 63
adiantum-nigrum L. 60, *61*, 62
aethiopicum (Burm. f.) Becherer 60, *61*, 62
ceterach 64
dalhousiae Hook. 64
filare (Forsskal) Alston 62
lanceolatum Forsskal 62
phillipsianum (Kümmerle) Bir, Fraser-Jenkins & Lovis 64
praemorsum Swartz 62
protensum Schrad. 60, 62
rutifolium (Berg.) Kunze 60, *61*, 63
var. *bipinnatum* (Forsskal) Schelpe 63
schweinfurthii Baker 60, *61*, 63
trichomanes L. 60, *61*
varians Wall. ex Hook. & Grev. *61*
subsp. fimbriatum (Kunze) Schelpe 60, 62
Asteriscus graveolens 26
Astragalus setiferus DC. 261
Atraphaxis L. 127, 142
spinosa L. *140*, 142
Atriplex L. 234, 243
alexandrina Boiss. 246
coriacea Forsskal 243, 244, *245*
crystallina Boiss. 246
dimorphostegia Karelin & Kir. 243, 247, *248*
farinosa Forsskal 243, *245*
subsp. farinosa 244
glauca L. 243, 246
griffithii Moq. subsp. stocksii (Boiss.) Boulos 243, 244, *245*
var. *stocksii* (Boiss.) Boiss. 244
halimus L. 243, 244
hastata Forsskal 244
inamoena Aellen 247
laciniata L. var. *turcomanica* Moq. 247
leucoclada Boiss. 243, 246
subsp. *turcomanica* (Moq.) Aellen 247
var. inamoena (Aellen) Zoh. 246-7, *248*
var. turcomanica (Moq.) Zoh. 246-7, *248*
ocymifolium Viv. 244
palaestina Boiss. 246
sokotranum Vierh. 246

stocksii Boiss. 244
stylosa Viv. 246
tatarica L. 243, 247
Avicennia marina 20, 22

Balanites aegyptiaca 20
Barbeya Schweinf. 88
oleoides Schweinf. 20, 88, *117*
Barbeyaceae 88
Barleria bispinosa 19
trispinosa 19
Basellaceae 174
Bassia All. 235, 249
arabica (Boiss.) Maire & Weiller 249, 251
eriophora (Schrader) Asch. 249, 250
hyssopifolia (Pallas) Kuntze 249, 250
indica (Wight) A.J. Scott 250, 251
joppensis Bornm. & Dinsm. 251
latifolia (Fresen.) Asch. & Schweinf. 250
muricata (L.) Asch. 249, 250, *258*
scoparia (L.) A.J. Scott 250, 251
Berberidaceae 318
Berberis L. 318
aristata Oliv. 318
forskaliana C.K. Schneider 318
holstii Engler *313*, 318
petitiana C.K. Schneider 318
Beta L. 234, 242
maritima L. 242
vulgaris L. subsp. maritima (L.) Arcang. 242
Bienertia Bunge ex Boiss. 235, 259
cycloptera Bunge ex Boiss. 17, 22, *258*, 259
Biscutella L. 383, 384, 415
didyma L. 414, 415
Boerhavia L. 143, 144
ascendens Willd. 145
boissieri Heimerl 152
coccinea Miller 145
diandra L. 145
dichotoma Vahl 151
diffusa L. 144, 145, *146*
var. *viscosa* (Lagasca & Rodriguez) Heimerl 145
elegans Choisy 144
subsp. elegans 145, *146*
subsp. stenophylla (Boiss.) A.G. Miller 145, *146*
var. *stenophylla* Boiss. 145
glutinosa Deflers 145
heimerlii Vierh. 152
helenae Schultes 153
pedunculosa A. Rich. 149
plumbaginea Cav. 151

Index

var. *dichotoma* (Vahl) Asch. & Schweinf. 151
var. *forskalei* Schweinf. 149
var. *glabrata* Boiss. 151
var. *socotrana* Heimerl 151
var. *viscosa* Boiss. 151
repens Balf.f. 145
repens L. 144, 145, *146*
var. *diffusa* (L.) Boiss. 145
var. *glabra* Choisy 145
var. *viscosa* Choisy 147
rubicunda Steud. 144
scandens Balf.f. 151
scandens Balfour 152
scandens Forsskal 151
simonyi Heimerl & Vierh. 151
stenocarpa Chiov. 149
verticillata Poiret 151
verticillata Schwartz 153
Boscia Lam. 350, 371
angustifolia A. Rich. *367*, 371
arabica Pestalozzi 21, *367*, 371
minimifolia Chiov. 372
senegalensis (Pers.) Lam. 372
Boswellia sacra 21
Boussingaultia cordifolia Tenore 174
Brassica L. 384, 387
arabica (Fischer & C. Meyer) Fiori 390
campestris L. 390
crassifolia Forsskal 402
deflexa Boiss. 387, *389*
juncea (L.) Czernj. & Cosson 387, 388, *389*
lasiocalycina (Boiss. & Hausskn.) Boiss. 387
napus L. 387, *389*, 390
oleracea L. 387, 388
rapa L. 387, 390
rostrata Balf. f. 393
tournefortii Gouan 387, 388, *389*
Buddleja polystachya 19, 20, 26
Bulliardia abyssinica A. Rich. 466
Buxus hildebrandtii 22

Cactaceae 305
Cadaba Forsskal 350, 375
baccarinii Chiov. 376, *377*, 378
farinosa Forsskal 21, 376, *377*, 378
glandulosa Forsskal 375, 376, *377*
heterotricha Stocks ex Hook. 375, 376, *377*
longifolia DC. 19, 376, *377*, 378
mirabilis Gilg 375, 376, *377*
rotundifolia Forsskal 375, *377*, 378
Cadia purpurea 19, 20
Caesalpinia erianthera 21
Caidbeja adhaerens Forsskal 115

Cakile Miller 385, 403
arabica Mandaville 401
arabica Velen. & Bornm. 403, *405*
Calligonum L. 127, 141
arabicum Soskov 142
comosum L'Hér. 16, 17, *140*, 141
crinitum Boiss. 141
subsp. arabicum (Soskov) Soskov 16, 17, 21, *140*, 142
polygonoides subsp. *comosum* (L'Hér.) Soskov 141
tetrapterum Jaub. & Spach *140*, 141, 142
Calotropis procera 20
Cannabaceae 102
Cannabis L. 104
sativa L. 104
Capparaceae 349
Capparis L. 350, 372
aegyptia Lam. 374
aphylla Roth 372
cartilaginea Decne 372, *373*, 374
decidua (Forsskal) Edgew. 20, 372
elliptica Hausskn. & Bornm. ex Bornm. var. *maskatensis* Hausskn. & Bornm. 374
galeata Fresen. 374
var. *montana* Schweinf. 374
inermis Forsskal 374
leucophylla DC. 374
mithradatica Forsskal 370
oblongifolia Forsskal 370
spinosa Balf. f. 374
spinosa Forsskal 374
spinosa L. 372
var. *aegyptia* (Lam.) Boiss. 374
var. *canescens* Cosson 374
var. *galeata* (Fresen.) Hook. f. 374
var. *mucronifolia* (Boiss.) Hedge & Lamond 374
var. *parviflora* (Boiss.) Boiss. 374
var. *spinosa* 374
tomentosa Lam. 372, 375
Capsella Medikus 381, 416
bursa-pastoris (L.) Medikus *414*, 416
Cardamine L. 386, 429
africana L. *425*, 429
hirsuta L. *425*, 429
Cardaria Desv. 384, 412
draba (L.) Desv. *409*, 412
Carissa edulis 20, 21
Caroxylon bottae (Jaub. & Spach) Moq. 277
imbricatum (Forsskal) Moq. 275
lancifolium Boiss. 279
salicornicum Moq. 266

Carphalea obovata 22
Carrichtera DC. 385, 404
annua (L.) DC. 406, *409*
vellae DC. 406
Caryophyllaceae 174
Cassytha L. 308
filiformis L. 309
Casuarina L. 80
equisetifolia L. 80
Casuarinaceae 80
Catha edulis 26
Caylusea A.St. Hil. 448
abyssinica (Fresen.) Fischer & C. Meyer 449
canescens A. St. Hil. 449
hexagyna (Forsskal) M.L. Green 449, *450*
jaberi Abedin 449
Cebatha Forsskal 319
villosa (Lam.) C. Christ. 320
Celosia L. 284
anthelminthica Asch. 284, 285
argentea L. 284, 285, *286*
caudata Vahl 285
polystachia (Forsskal) C.C. Townsend 285, *286*
populifolia Moq. 285
trigyna L. 285
var. *fasciculiflora* Fenzl ex Moq. 285
Celtis L. 86
africana Burm. f. 19, 86, *87*
integrifolia Lam. 86
kraussiana Bernh. 86
toka (Forsskal) Hepper & J.R.I. Wood 86, *87*
Centropodia fragilis 16
Cephalocroton socotranus 22
Cerastium L. 177, 207
dichotomum L. *205*, 207, 208
subsp. dichotomum 208
subsp. inflatum (Link.) Cullen 208
glomeratum Thuill. *205*, 207
inflatum Link 208
Ceratophyllaceae 323
Ceratophyllum L. 324
demersum L. 324, *325*
submersum L. 324, *325*
Ceratopteris Brongn. 35, 43
cornuta (P. Beauv.) Le Prieur 43, *44*
thalictroides (L.) Brongn. 43
Ceterach DC. 36, 63
dalhousiae (Hook.) C. Chr. *61*, 64
officinarum DC. *61*, 64
phillipsianum Kümmerle *61*, 64
Cheilanthes Swartz 36, 37, 45
catanensis (Cosent) H.P. Fuchs 48
concolor (Langsd. & Fisch.) R. & A. Tryon 50
coriacea Decne 45, 46, *47*
farinosa Balfour 48

farinosa (Forsskal) Kaulf. 45, 46, *47*
farinosa Schwartz 48
fragrans Swartz 48
involuta (Swartz) Schelpe & Anthony 50
marantae (L.) Dominin 46, *47*
pteridioides (Reichard) C. Chr. 45, 46, *47*, 48
quadripinnata (Forsskal) Kuhn 49
vellea (Aiton) F. Mueller 46, *47*, 48
viridis (Forsskal) Swartz 49
Cheiranthus linearis Forsskal 422
Chelidonium dodecandrum Forsskal 341
Chenolea arabica Boiss. 251
Chenoleoides arabica (Boiss.) Botsch. 251
Chenopodiaceae 233
Chenopodium L. 234, 235
aegyptiacum Hasselq. 256
album L. 236, *238*, 241
ambrosioides L. 236, 237, *238*
baryosmon Roemer & Schultes 275
botrys L. 236, *238*, 239
carinatum R. Br. 236, 237, *238*
fasciculosum Aellen 236
var. muraliforme Aellen *238*, 240
ficifolium Smith 236, *238*, 241
foetidum Schrader 237
glaucum L. 236, *238*, 240
maritimum L. 256
murale L. 236, *238*, 240
opulifolium Schrader ex W. Koch & Ziz 236, *238*, 241
procerum Hochst. ex Moq. 236, 237, *238*
schraderianum Schultes 236, 237, *238*
scoparia L. 251
triangulare Forsskal 242
vulvaria L. 236, *238*, 239
Choriptera Botsch. 280
semhahensis (Vierh.) Botsch. 280
Chorispora R. Br. ex DC. 385, 436
purpurascens (Banks & Sol.) Eig 436
syriaca Boiss. 436
Christella Léveillé 37, 65
dentata (Forsskal) Brownsey & Jermy 65, *66*
Clematis L. 309, 310
hirsuta Guillemin & Perr. 310, *311*
incisodentata A. Rich. 310
longicauda Steud. ex A. Rich. 310, *311*, 312
orientalis L. 310, *311*, 312
simensis Fresen. 310, *311*

wightiana Wallich 310
Cleome L. 350
albescens Franchet subsp. omanensis Chamberlain & Lamond 351, *353*, *356*, 362
amblyocarpa Barr. & Murb. 351, *353*, 354
angustifolia Forsskal 351, *353*, *356*, 362
arabica L. 351, *353*, 361
areysiana Deflers 364
austroarabica Chamberlain & Lamond 351, 357
subsp. austroarabica *353*, 357
subsp. muscatensis Chamberlain & Lamond *353*, 357
beckiana Rech. f. 360
brachyadenia Schwartz 355
brachycarpa Vahl ex DC. 351, 352, *353*, *356*
brachystyla Deflers 360
brevipetiolata Chamberlain & Lamond 351, *353*, 355, *358*
brevisiliqua Schultes f. 352
chrysantha Decne 351, 352, *353*, *356*
deflersii Blatter 364
digitata Forsskal 365
diversifolia Hochst. & Steud. 352
drepanocarpa O. Schwartz 360
droserifolia Del. 352, *353*, 355
glaucescens Mandaville 361
gynandra L. 350, *353*, 363
hanburyana Penzig 350, *353*, 364
macradenia Schweinf. 351, *353*, 355, *356*
monophylla L. 351, *353*, *356*, 364
noeana Boiss. 352, 359
subsp. brachystyla (Deflers) Chamberlain & Lamond *353*, 359, 360
subsp. noeana *353*, 359, 360
oxypetala Boiss. var. *micrantha* Boiss. 361
pallida Kotschy 365
papillosa Steud. 352
paradoxa R. Br. ex DC. 351, *353*, *356*, 364
pentaphylla L. 363
polytricha Franchet 352, *353*, *356*, 360
pruinosa T. Anders. 352, *353*, *356*, 359
ramosissima Webb ex Parl. 351, *353*, 354
rupicola Vicary 351, *353*, *356*, 361

scaposa DC. 351, 352, *353*, *356*
schweinfurthii Gilg 354
socotrana Balf. f. 351, *353*, 362
tenella L.f. 351, *353*, 363
trinervia Fresen. 361
viscosa L. 350, *353*, *356*, 363
Clypeola L. 382, 428
aspera (Grauer) Turrill *425*, 428
jonthlaspi L. *425*, 428
Cocculus DC. 319
balfourii Schweinf. ex Balfour 319, 320, *321*
cebatha DC. 320
hirsutus (L.) Theob. 319, 320
laeba DC. 320
pendulus (J. Forster) Diels 319, 320, *321*
villosus DC. 320
Combretum molle 20
Cometes L. 175, 176, 177
abyssinica (R.Br.) Wallich. 177, 178
subsp. *suffruticosa* Wagner & Vierhapper 178
surattensis L. 177, 178, *180*
Commicarpus Standley 143, 147
adenensis A.G. Miller 148, 153, *154*
ambiguus Meikle 148, 153, *154*
arabicus Meikle 148, 149, *150*
boissieri (Heimerl) Cuf. 148, 152, *154*
grandiflorus (A. Rich.) Standley 148, 149, *150*
heimerlii (Vierh.) Meikle 148, *150*, 152
helenae (Roemer & Schultes) Meikle 148, 153, *154*
mistus Thulin 148, *150*, 152
pedunculosus (A. Rich.) Cuf. 147, 149, *150*
plumbagineus (Cav.) Standley 148, *150*, 151
reniformis (Chiov.) Cuf. 147, 153
simonyi (Heimerl & Vierh.) Meikle 148, 149, *150*, 151
sinuatus Meikle 148, *150*, 151
squarrosus auctt. non (Heimerl) Standley 152
stellata Berhaut 153
stenocarpus (Chiov.) Cuf. 148, 149, *150*
verticillatus (Poiret) Standley 151
Commiphora abyssinica 20
foliacea 21
gileadensis 19, 20, 21
habessinica 21
kataf 19, 20
myrrha 19, 20
Consolida orientalis (J. Gay) Schröd. 318

577

Index

Convolvulus hystrix 21
Corbichonia Scop. 156, 159
   decumbens (Forsskal) Exell *158*, 159
Cordia abyssinica 20
Cornulaca Del. 234, 260
   amblyacantha Bunge 260, 261, *264*
   arabica Botsch. 261
   aucheri Moq. 260, *262*
   ehrenbergii Asch. 260, *262*, 263
   leucacantha Charif & Aellen 260
   monacantha Del. 16, 260, 261, *264*
   setifera (DC.) Moq. 260, 261, *262*
   tragcanthoides Moq. 261
Coronopus L. 384, 411
   didymus (L.) Smith *409*, 411, 412
   niloticus (Del.) Springer *409*, 411, 412
   squamatus (Forsskal) Asch. *409*, 411
Corrigiola repens Forsskal 192
Cosentinia vellea (Ait.) Tod. 48
Cotoneaster Medicus 480, 481
   nummularia Fischer & C. Meyer 21, 481, *482*
   racemiflorus auct. non (Desf.) Bosse 481
Cotyledon L. 463, 477
   alternans Vahl 472
   barbeyi Baker 477, *478*
   deficiens Forsskal 472
   lanceolata Forsskal 469
   nudicaulis Vahl 472
   orbiculata Forsskal 472
   rosulata Raadts 472
   umbilicus auct. non L. 477
Crambe Adans. 386, 400
   sp. A. 400
Crassula L. 463
   alata (Viv.) Berger 463, *465*
      subsp. alata 463, 464
      subsp. pharnaceoides (Fischer & C. Meyer) Wickens & Bywater 463, 464
   alba Forsskal 463, *465*, 466
   alsinoides (Hook. f.) Engl. 463, *465*, 466
   granvikii Mildbr. 463, *465*, 466
   hedbergii Wickens & Bywater 466
   pellucida L. subsp. *alsinoides* (Hook. f.) Tölken 466
   pentandra (Edgew.) Schönl. 464
   pharnaceoides Fischer & C. Meyer 464
   schimperi Fischer & C. Meyer 463, 464
      subsp. phyturus (Mildbr.) Ros. 464
      subsp. schimperi 464, *465*
   tillaea Lester-Garl. 463, 464
Crassulaceae 461
Crataegus L. 480, 484
   sinaica Boiss. 484
Cressa cretica 22
Croton confertus 21
   socotranus 21, 22
Cruciferae 380
Cryophytum nodiflorum (L.) L. Bolus 167
Cupressaceae 71
Cupressus L. 71, 74
   sempervirens L. 75
Cussonia holstii 26
Cyclosorus dentatus (Forsskal) Ching 65
Cydonia Miller 480, 481
   oblonga Miller 481
   vulgaris Pers. 481
Cymodocea ciliata 22
   rotunadata 22
Cyperus conglomeratus 16, 17, 21
Cystopteris Bernh. 37, 65
   fragilis (L.) Bernh. 44, 65

Darniella schweinfurthii (Solms-Laub.) Brullo 271
Debregeasia Gaudich. 105, 111
   bicolor (Roxb.) Wedd. 111
   saenab (Forsskal) Hepper & J.R.I. Wood 19, *110*, 111
   salicifolia (D. Don) Rendle 111
Delonix elata 20
Delosperma N.E.Br. 156, 166
   harazianum (Deflers) Poppend. & Ihlenf. 166, *168*
Delphinium L. 309, 317
   Subgenus Consolida (DC.) Hutch. 317
   Subgenus Delphinium 317
   ajacis L. 318
   orientale J. Gay 317, 318
   penicillatum Boiss. 317
   sheilae Kit Tan *313*, 317
Dendrosicyos socotranus 21, 22
Dennstaedtiaceae 58
Descurainia Webb & Berth. 382, 446
   sophia (L.) Webb & Berth. 446
Desmochaeta alternifolia (L.) DC. 293
Desmostachya bipinnata 20
Dhofaria A. Miller 350, 379
   macleishii A.G. Miller *373*, 379
Dianthus L. 177, 228
   Sect. Leiopetali Boiss. 229
   Sect. Plumaria (Opiz) Aschers & Graebn. 231
   Sect. Verruculosi Boiss. 228
   crinitus Sm. 228, *230*, 232
   cyri Fischer & C. Meyer 228, *230*
   deserti Kotschy 228, *230*, 231
   judaicus Boiss. 228, 229, *230*
   longiglumis Del. 228, *230*, 231
   pumilus Vahl 231
   sinaicus Boiss. 228, 232
   strictus Banks & Soland. 228, *230*
      subsp. sublaevis D.F. Chamb. 229
   uniflorus Forsskal 228, *230*, 231
Diceratella Boiss. 382, 433
   incana Balf. f. *432*, 433
Dichanthium foveaolatum 22
Dicotyledones 80
Digera Forsskal 284, 293
   alternifolia (L.) Aschers. 293
   arvensis Forsskal 293
   muricata (L.) Mart. 293
      subsp. muricata *290*, 293
      var. patentipilosa C.C. Townsend 293, 294
      subsp. trinervis C.C. Townsend 293, 294
      var. trinervis *290*, 294
Dionysia mira 27
Diplanthera uninervis 22
Diplotaxis DC. 385, 386, 393
   acris (Forsskal) Boiss. *394*, 395, 396
   erucoides (L.) DC. *394*, 395, 396
   harra (Forsskal) Boiss. *394*, 395
   kohlaanensis A.G. Miller & J. Nyberg *394*, 395
   tenuifolia (L.) DC. *394*, 395, 396
Dipterygium Decne 350, 365
   glaucum Decne 16, 17, 365, *373*
      var. *macrocarpa* Blatter 365
Dobera glabra 19, 20, 26
Dodonaea viscosa 19, 20, 21
Dolichorhynchus Hedge & Kit Tan 384, 404
   arabicus Hedge & Kit Tan 404, *405*
Dombeya torrida 26
Dorstenia L. 89, 91
   arabica Hemsley 92
   barnimiana Schweinf. 91, *93*
      var. *ophioglossoides* (Bureau) Engl. 91
   foetida (Forsskal) Schweinf. 91, 92, *93*
      var. foetida 92
      var. obovata (A. Rich.) Schweinf. & Engl. 92
   gigas Schweinf. ex Balf. f. 91, *93*, 94
   radiata Lam. 92
   socotrana A.G. Miller 91, 92
Doryopteris J.E. Smith 36, 50
   concolor (Langsd. & Fisch.) Kuhn *47*, 50

kirkii (Hook.) Alston 50
Dracaena cinnabari 22
 serrulata 20, 21
Droguetia Gaudich. 104, 105, 116
 debilis Rendle 116
 iners (Forsskal) Schweinf. 116, *117*
Dryopteridaceae 67
Dryopteris Adans. 37, 69
 crenata (Forsskal) O. Kuntze 67
 dentata (Forsskal) C. Chr. 65
 mauritiana Schwartz 65
 orientalis (J.F. Gmelin) C. Chr. 59
 rigida 69
 schimperiana (A. Br.) C. Chr. *66*, 69

Ebenus stellatus 21
Ecbolium viride 19
Echinopsilon hyssopifolium (Pallas) Moq. 250
 muricatus (L.) Moq. 250
Ehretia abyssinica 20
Eigia Soják 383, 436
 longistyla (Eig) Soják *437*, 438
Emex Neck. 128, 139
 australis Steinh. 139, 141, *142*
 spinosus (L.) Campderá 139, *140*
Enarthrocarpus Labill. 385, 399
 lyratus (Forsskal) DC. *398*, 399
Ephedra L. 76
 alata Decne 76, *77*
 alte C.A. Mey. 78
 alte Schwartz 79
 aphylla Forsskal 76, *77*, 78
 campylopoda C.A. Mey. 78
 ciliata C.A. Mey. 79
 foeminea Forsskal 76, *77*, 78
 foliata Boiss. ex. C.A. Mey. 76, *77*, 79
 fragilis Schwartz 79
 milleri Freitag & Maier-Stolte 76, *77*, 79
 pachyclada Boiss. 76, *77*, 79
  subsp. sinaica (H. Riedl) Freitag & Maier-Stolte *77*, 79
 peduncularis Boiss. 79
 sinaica H. Riedl 79
 transitoria H. Riedl 76, *77*, 78
Ephedraceae 75
Equisetaceae 41
Equisetum L. 34, 41
 ramosissimum Desf. *39*, 41
Eremobium Boiss. 382, 440
 aegyptiacum (Sprengel) Asch. & Schweinf. ex Boiss. 16, *437*, 440
 diffusum (Decne) Botsch. 440
 lineare (Del.) Botsch. 440
 nefudicum (Velen.) B.L. Burtt & Rech. f. 440

Erica arborea 26
Erophila DC. 427
 verna (L.) Bess. 427
Eruca Miller 385, 397
 lativalvis Boiss. 397
 sativa Miller 397, *398*
Erucaria Gaertner 385, 401
 aleppica Gaertn. 401
 crassifolia (Forsskal) Del. 401, 402, *405*
 hispanica (L.) Druce 401, *405*
 lineariloba Boiss. 401
 sp. A. 401, 402
 uncata (Boiss.) Asch. & Schweinf. 401, 402, *405*
Erucastrum C. Presl 386, 390
 arabicum Fischer & C. Meyer *389*, 390
 meruense Jonsell subsp. yemenense Jonsell 390, 391
 woodiorum Jonsell 390, 391
Euclea racemosa subsp. schimperi 20, 21, 26
Euonymus inermis Forsskal 331
Euphorbia arbuscula 21
 balsamifera 20
  subsp. adenensis 21
 cactus 21
 cuneata 19
 larica 21
 schimperiana 19
 smithii 21
 triaculeata 19
Euryops arabicus 19, 21

Fabrisinapis fruticosus C.C. Townsend 397
Farsetia L. 381, 382, 418
 aegyptia Turra 418, 419, *420*
 arabica Boulos 422
 burtoniae Oliver 418, *420*, 423
 depressa Kotschy 421
 dhofarica Jonsell & A.G. Miller 419, *420*, 423
 hamiltonii Royle 421
 hamiltonii Schwartz 422
 heliophila Bunge ex Cosson 419, *420*, 422
 latifolia Jonsell & A.G. Miller 419, *420*, 421
 linearis Decne ex Boiss. 419, *420*, 422
 longisiliqua Decne 419, *420*, 422
 ovalis Boiss. 419
 prostrata Balf. f. 419
 prostrata (Steud.) Hochst. 421
 ramosissima Fourn. 421
 socotrana B.L. Burtt 418, 419, *420*
 stylosa R. Br. 419, *420*, 421
 stylosa (Steud.) T. Anderson 422
Ficus L. 89, 94

ambiguum Forsskal 100
 amplissima J.E. Smith 95
 benghalensis L. 94
 benjamina L. 95
 capensis Thunb. 99
 carica L. 94, 95, 96, *97*
 carica Western 98
 challa Schweinf. 90
 chanas Forsskal 99
 cordata Thunb. 20
  subsp. salicifolia (Vahl) C.C. Berg 21, 96, 100, *101*
 deltoidea Jack. 94
 elastica Roxb. ex Hornem. 95
 exasperata Vahl 95, *97*, 99
 forskalaei Vahl 98
 geraniifolia Miq. 95
 glumosa Del. 96, 100, *103*
  var. *glaberrima* Martelli 100
 indica Forsskal 100
 ingens (Miq.) Miq. 96, 100, *101*
 ingentoides Hutch. 100
 johannis Boiss. 95, *97*, 98
 lutea Miller & Morris 100
 lyrata Warb. 94
 microcarpa L.f. 94
 morifolia Forsskal 98
 palmata Forsskal 95, 96, *97*
  subsp. palmata 98
  subsp. virgata (Roxb.) Browicz 98
 populifolia Vahl 20, 96, 102, *103*
 pseudosycomorus Decne 98
 religiosa Forsskal 102
 religiosa L. 95
 salicifolia Vahl 100
 serrata Forsskal 99
 socotrana Balf.f. 102
 sur Forsskal 95, 99, *101*
 sycomorus L. 20, 21, 95, 99, *101*
 taab Forsskal 102
 toka Forsskal 86
 vasta Forsskal 20, 21, 96, 102, *103*
Fleurya aestuans (L.) Gaudich. 106
Forsskaolea L. 105
 griersonii A.G. Miller *114*, 115, 116
 tenacissima L. *114*, 115
 viridis Ehrenb. *114*, 115
Fumaria L. 347
 abyssinica Hamm. 349
 parviflora Lam. *348*, 349
Fumariaceae 347

Girardinia Gaudich. 105, 108
 condensata (Steud.) Wedd. 108
 diversifolia (Link) I. Friis *107*, 108
Gisekia L. 156, 157

Index

pharnaceoides L. 157, *158*
Glaucium Miller 339, 340
  arabicum Fresen. 340, 341, *343*
  corniculatum (L.) J.H. Rudolph 340
Glinus L. 156, 159
  crystallinus Forsskal 165
  lotoides L. *158*, 159
  setiflorus Forsskal *158*, 159, 160
Goldbachia DC. 384, 441
  laevigata (M. Bieb.) DC. *437*, 441
Gomphrena L. 284, 304
  celosioides Mart. *303*, 304
  globosa L. *303*, 304
Grewia mollis 19, 20
  tembensis 19
  villosa 19
Guttiferae 331
Gymnocarpos Forssk. 175, 179
  argenteus Petruss. & Thulin 179, 181
  bracteatus (Balf.f.) Petruss. & Thulin 179, *180*, 182
  decandrus Forssk. 179, *180*, 181
  dhofarensis Petruss. & Thulin 179, 181
  fruticosus (Vahl) Pers. 181
  kuriensis (A.R. Smith) Petruss. & Thulin 179, *180*, 182
  mahranus Petruss. & Thulin 179, 183
  rotundifolius Petruss. & Thulin 179, 182
Gymnogramma cordata Balfour 64
Gymnospermae 71
Gynandropsis gynandra (L.) Briq. 363
  pentaphylla (L.) DC. 363
Gypsophila L. 177, 222
  antari Post & Beauv. 225
  arabica Barkoudah 225
  bellidifolia Boiss. 223, *226*
  capillaris (Forsskal) C. Chr. 223, 225, *226*
    subsp. capillaris 225
    subsp. confusa Zmarzty 225
  montana Balf.f. 223, 224, *226*
  obconica Barkoudah 225
  pilosa Huds. 223, 224, *226*
  porrigens (L.) Boiss. 224
  umbricola (J.R.I. Wood) R.A. Clement 223, 225, *226*
  viscosa J.A. Murray 223, 224, *226*
Gyroptera Botsch. 280
  cycloptera (Stapf) Botsch. 280

Halocharis Moq. 283
  sulphurea (Moq.) Moq. *258*, 283
Halocnemum M. Bieb. 233, 252
  strobilaceum (Pallas) M. Bieb. 17, 22, 252, *254*
Halogeton C. Meyer 263
  alopecuroides (Del.) Moq. 263
    var. *papillosa* Maire 265
  tetrandrus (Forsskal) Moq. 275
Halopeplis Bunge ex Ung.-Sternb. 235, 252
  perfoliata (Forsskal) Bunge ex Asch. 252, *258*
Halophila ovalis 22
  stipulacea 22
Halopyrum mucronatum 22
Halothamnus Jaub. & Spach 235, 277
  bottae Jaub. & Spach 277, *278*
  iraqensis Botsch. 277, *278*, 279
    var. *hispidulus* Botsch. 279
  lancifolius (Boiss.) Kothe-Heinrich 277, *278*, 279
Haloxylon Bunge 233, 266
  ammodendrum Collenette 266
  elegans (Bunge) Botsch. 266
  persicum Bunge 16, 266
  salicornicum (Moq.) Bunge ex Boiss. 17, 26, *254*, 266
Hammada salicornica (Moq.) Iljin 266
  scoparia Collenette 282
Haya Balf.f. 176, 190
  obovata Balf.f. 190, *191*
Helixanthera Lour. 121, 122
  thomsonii (Sprague) Danser 122, *123*
Hemicrambe Webb 386, 397
  townsendii Gómez Campo 397, *398*
Herniaria L. 176, 188
  cinerea DC. 189
  hemistemon J. Gay *186*, 188, 189
  hirsuta L. *186*, 189
  maskatensis Bornm. *186*, 189
Heteropogon contortus 22
Hirschfeldia Moench. 386, 392
  incana (L.) Lagr.-Fossat 393, *394*
  rostrata (Balf. f.) O. Schulz 393, *394*
Holosteum L. 176, 204
  glutinosum (Bieb.) Fisch. & C. Mey. 204, *205*
  umbellatum L. var. *glutinosum* (Bieb.) Gay 204
Homalodiscus aucheri Boiss. 453
Horwoodia Turrill 384, 413
  dicksoniae Turrill 413, *414*
Hydnora Thunb. 330
  abyssinica A. Br. ex Decne 330
  johannis Beccari *329*, 330
Hydnoraceae 328
Hypecoum L. 339, 346

  geslinii Coss. & Kral. 346, 347, *348*
  pendulum L. 346
Hyperanthera peregrina Forsskal 460
Hypericum L. 331
  annulatum Moris 332, 336, *337*
  balfourii N. Robson 332, 333, *334*
  collenettiae N. Robson 332, *337*, 338
  fieriense N. Robson 331, *334*, 336
  hircinum L. 27, 332, *334*, 335
  lanceolatum auctt. non Lam. 332
  lanceolatum Balf. f. 335
  mysorense Balf. f. 333
  perforatum L. 332, *337*, 339
  quartinianum A. Rich. 332, *334*, 335
  revolutum Vahl 20, 26, 332, *334*
  scopulorum Balf. f. 332, 336, *337*
  sinaicum Hochst. & Steud. ex Boiss. 332, *337*, 338
  socotranum Good 332, 333
    subsp. smithii N. Robson *334*, 335
    subsp. socotranum *334*, 335
  tortuosum Balf. f. 332, 336, *337*
Hypodematium Kunze 37, 67
  crenatum (Forsskal) Kuhn 66, 67

Iberis L. 383, 415
  umbellata L. 415
Illecebraceae, *see* Caryophyllaceae
Isatis L. 383, 413
  lusitanica L. 413, *414*

Jatropha unicostata 22
Juglans L. 83
  regia L. 83
Juniperus L. 71
  excelsa M. Bieb. subsp. polycarpos (K. Koch) Takhtajan 21, 72, *73*, 74
  macropoda Boiss. 74
  phoenicea L. 18, 27, 72
  procera Hochst. ex Endl. 19, 20, 72, *73*

Kalanchoe Adans. 463, 468
  abrupta Balf. f. 475
  aegyptiaca (Lam.) DC. 472
  alternans (Vahl) Pers. 469, 472, *476*
  bentii C.H. Wright subsp. bentii 468, 474, *476*
  brachycalyx A. Rich. var. *yemensis* Deflers 470

citrina Schweinf. 469, *471*, 472
crenata (Andrews) Haw. 469, *471*, 474
crenata J. Wood 472
deficiens (Forsskal) Asch. & Schweinf. 469, 472
  var. deficiens *471*, 473
  var. glabra Raadts 473
deflersii Gagnepain 474
farinacea Balf. f. 469, *475*, *476*
glaucescens Britten 469, *471*, 473
  subsp. *arabica* Cuf. 473
  var. *deficiens* (Forsskal) Senni 472
glaucescens Schweinf. 473
laciniata (L.) DC. 469, 470, *471*
laciniata Schwartz 468, 472, 473
lanceolata (Forsskal) Pers. 469, *471*
robusta Balf. f. 468, *475*, *476*
rotundifolia (Haw.) Haw. 469, 474, *476*
schweinfurthii Penzig 470
sp. A. 469, 473
teretifolia Deflers 474
yemensis (Deflers) Schweinf. 469, 470, *471*
Kleinia odora 20
Kochia eriophora Schrader 250
  hyssopifolia (Pallas) Roth 250
  indica Wight 251
  latifolia Fresen. 250
  muricata (L.) Schrader 250
  scoparia (L.) Schrader 251
    subsp. *indica* (Wight) Aellen 251
Kosaria foetida Forsskal 92

Lachnocapsa Balf. f. 381, 416
  spathulata Balf. f. *414*, 416
Laeba Forsskal 319
Lagenantha Chiov. 235, 280
  cycloptera (Stapf) M.G. Gilbert & Friis *258*, 280
Laportea Gaudich. 105, 106
  aestuans (L.) Chew 106, *107*
  interrupta (L.) Chew 106, *107*
Lauraceae 308
Lavandula dentata 19
Lepidium L. 384, 407
  africanum (Burm. f.) DC. 408, 411
  armoracia Fischer & C. Meyer 408, *409*
    subsp. *abyssinicum* A. Rich. Thell. 408
    subsp. *intermedium* A. Rich. Thell. var. *alpigenum* (A. Rich.) Thell. 410
  aucheri Boiss. 407, 408, *409*
  bonariense L. 408, *409*, 410

draba L. 412
  latifolium L. 408, *409*, 411
  ruderale L. 408, *409*, 410
  sativum L. 408, *409*
  schweinfurthii Thell. 410
  virginicum L. 408, *409*, 410
Leptadenia pyrotechnica 17, 20
Leptaleum DC. 382, 436
  filifolium (Willd.) DC. 436, 437
Limeum L. 156, 157
  arabicum Friedrich 16, 157, *158*
  humile Mandaville 157
  indicum Stocks ex T. Anderson 159
  obovatum Vicary 157, *158*
Limonium axillare 22
Lobularia Desv. 381, 427
  libyca (Viv.) Webb & Berth. *425*, 427, 428
  maritima (L.) Desv. *425*, 427
  sp. A. 427, 428
Lochia Balf.f. 179
  bracteata Balf.f. 182
    subsp. *abdulkuriana* Chaudhri 182
    subsp. *bracteata* forma *ciliata* Chaudhri 182
  kuriensis A.R. Smith 182
Loeflingia L. 176, 199
  hispanica L. 199, *200*
Lonchitis bipinnata Forsskal 63
Lonicera aucheri 21
Loranthaceae 121
Loranthus acaciae Zucc. 122
  arabicus Deflers 122
  curviflorus Benth. ex Olive 122
  doberae Schweinf. 124
  faurotii Franch. 122
  globiferus A. Rich. 125
  regularis auctt. arab. non Sprague 125
  rufescens auctt. arab. non DC. 125
  schimperi Hochst. ex A. Rich. 124
  thomsonii Sprague 122
Loxogramme (Blume) C. Presl. 35, 58
  lanceolata (Swartz) C. Presl. 58
Lunaria scabra Forsskal 419
Lycium shawii 17, 22
Lycopodium imbricatum Forsskal 38
  sanguinolentum Forsskal 40
  yemense Swartz 40

Maerua Forsskal 350, 366
  angolensis Balfour 368
  angolensis DC. 366, 368
    subsp. angolensis 368
    subsp. socotrana (Schweinf. ex Balf. f.) Kers var. socotrana *367*, 368

    var. *socotrana* Schweinf. ex Balf. f. 368
  arabica J.F. Gmel. 366
  crassifolia Forsskal 18-21, 366, *367*
  cylindricarpa Schwartz 370
  nervosa Blatter 370
  oblongifolia (Forsskal) A. Rich. 366, *367*, 370
  ovalifolia Blatter 370
  racemosa Vahl 371
  socotrana (Schweinf. ex Balf. f.) Gilg 368
  thomsoni T. Anders 366
  triphylla A. Rich. 366, 369
    var. calophylla (Gilg) De Wolf type A *367*, 369, 370
    type B *367*, 369, 370
    type C 369, 370
    var. johannis (Volkens & Gilg) De Wolf 369
  uniflora Vahl 366
  variifolia Schwartz 370
Maesa lanceolata 19
Magnoliaceae 307
Malcolmia R. Br. 382, 438
  aegyptiaca Sprengel 440
  africana (L.) R. Br. *437*, 438
  arabica Velen. 396
  behboudiana Rech. f. & Esfand. 439
  chia (L.) DC. 438, 439
  crenulata (DC.) Boiss. *437*, 438, 439
  grandiflora (Bunge) Kuntze *437*, 438, 439
  musilii Velen. 413
  nefudica Velen. 440
  pygmaea 441
  torulosa (Desf.) Boiss. 445
Malus Miller 480, 483
  communis Poir. 483
  sylvestris Miller 483
Maresia Pomel 383, 441
  nana (DC.) Battand. 441
  pulchella (Del.) O. Schulz 441
  pygmaea (Del.) O. Schulz *437*, 441
Marsilea L. 35, 42
  aegyptiaca Willd. *39*, 42
  coromandeliana Willd. *39*, 42
Marsileaceae 42
Matthiola R. Br. 382, 433
  arabica Boiss. *432*, 433, 434
  arabica Velen. 434
  bicornis (Smith) DC. 434
  humilis DC. 434
  livida (Del.) DC. 434
  longipetala (Vent.) DC. *432*, 433, 434
  oxyceras DC. 434
  prostrata Hochst. & Steud. ex Steud. 421
Maytenus dhofariensis 21
Menispermaceae 319
Mesembryanthemum L. 156

Index

cryptanthum Hook.f. 167
forsskalei Hochst. ex Boiss. 167, *168*
harazianum Deflers 166
nodiflorum L. 167, *168*
Michelia L. 307
champaca L. 307
Mimusops schimperi 20
Minuartia L. 176, 202
filifolia (Forsskal) Mattfeld *200*, 202, 203
hybrida (Vill.) Schischk. *200*, 203
meyeri (Boiss.) Bornm. *200*, 203, 204
picta (Sibth. & Sm.) Bornm. *200*, 202, 203
Mirabilis L. 143
jalapa L. 144, *146*
Mollugo L. 155, 160
cerviana (L.) Ser. *158*, 160
glinus A. Rich. 159
hirta Thunb. 159
nudicaulis Lam. 160
Moltkiopsis ciliata 16, 26
Monotheca buxifolia 21
Monsonia heliotropoides 16
Moraceae 89
Morettia DC. 383, 434
canescens Boiss. 435
parviflora Boiss. *432*, 435
philaeana (Del.) DC. *432*, 434, 435
Moricandia DC. 384, 407
nitens auctt. non (Viv.) E. Durand & Barratte 407
sinaica (Boiss.) Boiss. 407, *409*
Moringa Adans. 460
aptera Gaertn. 460
arabica Pers. 460
oleifera Lam. 460, 461, *462*
peregrina (Forsskal) Fiori 19, 21, 460, *462*
pterygosperma Gaertn. 461
Moringaceae 460
Morus L. 89
alba L. 90
var. *arabica* Bureau 90
arabica (Bureau) Koidzumi 90
nigra L. 90
Myrica L. 81
humilis Chamisso & Schlechtendal 81, *82*
salicifolia A. Rich. 81
Myricaceae 81
Myrsine africana 20

Nannorhops ritchieana 21
Nasturtium R. Br. 385, 430
microphyllum Boenn. 430, 431, *432*
officinale R. Br. 430, 431, *432*
Negripteris Pichi-Sermolli 36, 48

sciona Pichi-Sermolli *47*, 48
Neotorularia Hedge & Léonard 383, 386, 445
torulosa (Desf.) Hedge & Léonard 444, 445
Nephrodium crenatum (Forsskal) Bak. 67
molle Balfour 65
parasitica Vierhapper 65
pectinatum (Forsskal) Hier 59
Nephrolepis Schott 36, 59
cordifolia Balfour 59
undulata (Afzel. ex. Swartz) J. Sm. 59, *66*
Nerium mascatense 21
oleander 21
Neslia Desv. 381, 418
apiculata Fischer, C. Meyer & Avé-Lall. *414*, 418
Neurada L. 493
procumbens L. 16, 26, *489*, 493
Neuradaceae 493
Niebuhria oblongifolia (Forsskal) DC. 370
Nigella L. 309, 316
sativa L. 316
Nitraria retusa 22
Noaea Moq. 234, 281
mucronata (Forsskal) Asch. & Schweinf. *254*, 281
spinosissima (L.f.) Moq. 281
Notholaena vellea (Aiton) R. Br. 48
Nothosaerva Wight 284, 299
brachiata (L.) Wight 299
Notoceras R. Br. 382, 431
bicorne (Aiton f.) Amo *432*, 433
canariensis R. Br. 433
Nuxia congesta 19, 20, 26
Nyctaginaceae 143
Nymphaeaceae 323
Nymphaea L. 323
nouchali Burm. f. var. caerulea (Savigny) Verdc. 323

Obione coriacea (Forsskal) Moq. 244
leucoclada (Boiss.) Ulbr. 246
Ochna Schreb. 331
inermis (Forsskal) Schweinf. *87*, 331
parvifolia Vahl 331
Ochnaceae 330
Ochradenus Del. 448, 449
arabicus Chaudhary, Hillc. & A.G. Miller 451, *452*, 453
aucheri Boiss. 451, *452*, 453
baccatus Del. 17, 451, *452*
var. *monstruosa* Muell. Arg. 451
var. *scandens* Hochst. & Steud. ex Muell. Arg. 451
dewittii Abdallah 453

gifrii Thulin 451, *452*, 454
harusuticus A.G. Miller 451, *452*, 453
spartioides (Schwartz) Abdallah 451, 454
Odyssea mucronata 22
Olea europaea 20, 21, 26
subsp. africana 18, 19
Oleandraceae 59
Oligomeris Cambess. 448, 454
linifolia (Vahl) J.F. Macbr. 26, *450*, 455
Oncocalyx Tieghem 121, 124
doberae (Schweinf.) A.G. Miller & J. Nyberg 124
glabratus (Engl.) M.G. Gilbert 124
schimperi (Hochst. ex A. Rich.) M.G. Gilbert *123*, 124
Onychium Kaulfuss 36, 54
divaricatum (Poir.) Alston *44*, 54
melanolepis (Decne) Kunze 54
Ophioglossaceae 41
Ophioglossum L. 35, 41
aitchisonii (C.B. Clarke) D'Almeida 41
capense Swartz 41
polyphyllum A. Braun *39*, 41
reticulatum L. 41
Opophytum forskahlii (Hochst.) N.E.Br. 167
Opuntia Miller 306
dillenii (Ker Gawl.) Haw. 306
ficus-indica (L.) Miller 306
stricta (Haw.) Haw. var. *dillenii* (Ker Gawl) L. Benson 306
Orygia decumbens Forsskal 159
portulacifolia Forsskal 170
Osyris L. 118, 119
abyssinica A. Rich. 119
arborea Wall. 119
lanceolata Hochst. & Steud. 119
pendula Balf.f. 119
quadripartita Decne 119, *120*
sp. A. 119
Oxygonum Burchell 128, 133
atriplicifolium (Meissner) Martelli 134
sinuatum (Hochst. & Steud.) Dammer 134, *140*

Panicum turgidum 16-17, 20, 22
Papaver L. 339, 342
decaisnei Hochst. & Steud. ex Elkan 342, 344
dubium L. *343*
subsp. laevigatum (M. Bieb.) Kadereit 342, 345
glaucum Boiss. & Hausskn. 342, *343*, 344
hybridum L. 342, *343*, 346

macrostomum Boiss. & Huet ex Boiss. 342, *343*, 344
polytrichum Boiss. & Kotschy ex Boiss. 345
rhoeas L. 342, *343*, 345
somniferum L. 342, *343*, 344
syriacum Boiss. & Blanche ex Boiss. 345
umbonatum Boiss. 342, *343*, 345
Papaveraceae 339
Parietaria L. 105, 111
 alsinifolia Delile 112, *114*
 debilis G. Forster 112, *114*
 judaica L. 112, 113, *114*
 umbricola A.G. Miller 112, 113, *114*
Parkeriaceae 43
Paronychia (Tourn.) Miller 175, 184
 arabica (L.) DC. 184, 185, *186*
 chlorothrysa Murbeck 184, 185, *186*
 lenticulata Schwartz 185
 sclerocephala Decne. 187
 sinaica Fresen. 184, 185, *186*
Pedicellaria gynandra (L.) Schrank 363
Pellaea Link 37, 49
 concolor (Langsd. & Fisch.) Bak. 50
 involuta (Swartz) Bak. 49, 50
 quadripinnata (Forsskal) Prantl 47, 49
 viridis Balfour 50
 viridis (Forsskal) Prantl 47, 49
Peperomia Ruiz & Pavón 326
 abyssinica Miq. *325*, 326, 327
 arabica Decne ex Miq. 327
 blanda (Jacq.) Kunth 326, 327
  var. leptostachya (Hook. & Arn.) Düll. *325*, 327
 goudotii Balfour 327
 pellucida (L.) Kunth *325*, 326, 327
 reflexa (L.f.) A. Dietrich 326
 tetraphylla (J. Forster) Hook. f. & Arn. *325*, 326
Persicaria L. 128, 129
 amphibia (L.) S.F. Gray 130, *131*
 barbata (L.) Hara 130, *131*, 133
 decipiens (R.Br.) K.L. Wilson 130, 133
 dolichopoda (Ohki) Sasaki 133
 glabra (Willd.) M. Gomez 130, *131*, 132
 lapathifolia (L.) S.F. Gray 130, 132
 maculosa S.F. Gray 130, *131*, 133
 nepalensis (Meissner) Gross 130, *131*

salicifolia (Brouss. ex Willd.) Assenov 133
senegalensis (Meissner) Soják 130, *131*, 132
Petrorhagia (Ser.) Link 177, 227
 cretica (L.) Ball & Heywood 227, *230*
Pharnacium occultum Forsskal 157
 umbellatum Forsskal 160
Phoenix reclinata 20
Phragmanthera Tieghem 121, 125
 austroarabica A.G. Miller & J. Nyberg *123*, 125
Physorrhynchus Hook. f. 384, 385, 403
 chamaerapistrum (Boiss.) Boiss. 404, *405*
Pilea Lindley 104, 108
 tetraphylla (Steud.) Blume 109, *110*
Piperaceae 326
Pisonia L. 143, 155
 aculeata L. *146*, 155
Pistacia falcata 19
Pittosporaceae 479
Pittosporum Soland. 479
 abyssinicum Schwartz 479
 viridiflorum Sims *462*, 479
  subsp. *arabicum* Cufod. 479
Plantago boissieri 16
Pleopeltis Humb. & Bonpl. ex Willd. 35, 57
 macrocarpa (Bory ex Willd.) Kaulf. 44, 58
Plicosepalus Tieghem 121
 acaciae (Zucc.) Wiens & Pohl. 122, *123*
 curviflorus (Benth. ex Oliv.) Tieghem 122, *123*
 faurotii (Franchet) Tieghem 122
Pollichia Ait. 175, 178
 campestris Ait. 178, *180*
Polycarpaea Lam. 176, 190
 balfourii Briquet 192, 193, *194*
 caespitosa Balf.f. 192, *194*, 195
 corymbosa (L.) Lam. 190, 192
 divaricata Balf.f. 193
 fragilis Del. 192
 hassalensis Chamberlain *191*, 192, 196
 haufensis A.G. Miller 192, 196
 hayoides Chamberlain *191*, 192, 195
 jazirensis R.A. Clement *191*, 192, 194
 kuriensis Wagner 192, *194*, 196
 paulayana Wagner 192, *194*, 195

pulvinata M.G. Gilbert 192, 196
repens (Forsskal) Ashers. & Schweinf. 190, 192, *194*
robbairea (Kuntze) Greuter & Burdet 190, *194*, 197
spicata Wight ex Arn. 192, 193, *194*
 var. capillaris Balf. f. 193
 var. spicata 193
Polycarpon Loefl. ex L. 176, 198
 succulentum (Del.) J. Gay 198, 199, *200*
 tetraphyllum (L.) L. 198, *200*
Polygonaceae 127
Polygonum L. 128
 amphibium L. 130
 argyrocoleum Steud. ex Kunze 128, *131*
 aviculare L. 128, 129
 barbatum L. 133
 corrigioloides Jaub. & Spach. 128, 129
 glabrum Willd. 132
 lapathifolium L. 132
 nepalense Meissner 130
 palaestinum Zoh. 128, 129
 patulum Collenette 128
 persicaria L. 133
 salicifolium Brouss. ex Willd. 133
 senegalense Meissner 132
 serrulatum Lagasca 133
 setulosum Collenette 133
Polypodiaceae 57
Polypodium crenatum Forsskal 67
 dentatum Forsskal 65
 pectinatum Forsskal 59
Polystichum Roth 37, 68
 fuscopaleaceum Alston *66*, 68
 sp. A 68
Portulaca L. 169, 170
 cuneifolia Vahl 170
 foliosa Ker Gawl. 170, 171
 hareschata Forsskal 171
 imbricata Forsskal 171
 kermesina N.E. Br. 170, 173
 linifolia Forsskal 171
 oleracea L. 170, 171, *172*
 pilosa L. 170, 173
 quadrifida L. 170, 171, *172*
Portulacaceae 169
Potentilla L. 480, 487
 dentata Forsskal 487, *489*
 hispanica auctt. non Zimmet. 487
 pennsylvanica auct. non L. 487
 reptans L. 487, *489*
Pouzolzia Gaudich. 104, 105, 109
 arabica Deflers 109
 auriculata Wight 109
 hypoleuca Wedd. 109
 mixta Solms-Laub. 109, *110*

Index

parasitica (Forsskal)
   Schweinf. 109, *110*, 111
Prosopis cineraria 15, 16, 17, 21
Prunus L. 480, 490
  amygdalus Batsch. 492
  arabica (Oliv.) Meikle 27, *482*, 490, 491
  armeniaca L. 490, 491
  communis (L.) Arcang. 492
  dulcis (Mill.) D.A. Webb 491, 492
  korshynskii Hand.-Mazz. *482*, 491, 492
  persica (L.) Batsch. 490, 491
  x domestica L. 490, 491
Psilostachys sericea (Roxb.) Hook.f. 300
Psilotaceae 37
Psilotrichum Blume 284, 299
  aphyllum C.C. Townsend 299, 300
  cordatum Moq. 299
  gnaphalobryum (Hochst.) Schinz *286*, 299
  sericeum (Roxb.) Dalz. 299, 300
  virgatum C.C. Townsend 299, 300
Psilotum Swartz 34, 37
  nudum (L.) P. Beauv. 37, *39*
Psyllothamnus beevori Oliv. 183
Pteranthus Forsskal 175, 187
  dichotomus Forsskal *186*, 187
Pteridaceae 55
Pteridium Scop. 36, 58
  aquilinum (L.) Kuhn *56*, 58
Pteridophyta 33
Pteris L. 35, 36, 55
  catoptera Kunze 57
  cretica L. 55, *56*
  decursiva Forsskal 46
  dentata Forsskal 55, 57
  farinosa Forsskal 46
  longifolia Balfour 55
  longifolia Schwartz 55
  obliqua Forsskal 55
  quadriaurita Retz 55, *56*, 57
  quadripinnata Forsskal 49
  regularis Forsskal 57
  semiserrata Forsskal 55
  serrulata Forsskal 57
  subciliata Forsskal 55
  viridis Forsskal 49
  vittata L. 55, *56*
Pteropyrum Jaub. & Spach 127, 142
  scoparium Jaub. & Spach 21, *140*, 143
Pulicaria glutinosa 21
Pupalia A.L. Juss. 284, 294
  grandiflora Peter 294, 295
  lappacea (L.) A.L. Juss. *286*, 294
  robecchii Lopr. 294, 295
Pyrus L. 480, 483
  communis L. 483

cydonia L. 481
malus L. 483

Randonia spartioides Schwartz 454
Ranunculaceae 309
Ranunculus L. 310, 314
  Subgenus Batrachium (DC.) A. Gray 315
  Subgenus Ranunculus 314
  cornutus DC. 314, 315
  forskoehlii DC. 315
  multifidus Forsskal *311*, 314, 315
  muricatus L. *311*, 314
  rionii Lagger *311*, 314, 315
  sphaerospermus Boiss. & Blanche 314, 316
  trichophyllus Chaix var. *rionii* (Lagger) Rikli 315
Raphanus L. 385, 386, 399
  raphanistrum L. *398*, 399
  sativus L. *398*, 399
Rapistrum Crantz 386, 400
  rugosum (L.) All. *398*, 401
Reseda L. 448, 455
  alba L. 455, 456, *457*
    subsp. decursiva (Forsskal) Maire 456
  amblyocarpa Fresen. 456, *457*, 459
  arabica Boiss. 455, 456, *457*
  aucheri Boiss. 456, *457*
    var. bracteata (Boiss.) Abdallah & De Wit 458
  bracteata Boiss. 458
  decursiva Forsskal 456
  hexagyna Forsskal 449
  lutea L. 455, *457*, 458
  micrantha Schwartz 454
  muricata C. Presl 455, *457*, 458
  patzakiana Rech. f. 459
  pentagyna Abdallah & A.G. Miller 456, *457*, 458
  sphenocleoides Deflers 456, *457*, 459
  spinescens Schwartz 453
  stenostachya Boiss. 455, *457*, 459
  viridis Balf. f. 456, *457*, 460
Resedaceae 448
Rhanterium epapposum 17
Rhazya stricta 26
Rheum L. 128, 134
  palaestinum Feinbrun 134
Rhizophora mucronata 22
Rhus abyssinica 20
  retinorrhoea 19
  saeneb Forsskal 111
  somalensis 21
  thyrsiflora 22
  tripartita 27
Robbairea delileana Milne-Redhead 197
Rocama prostrata Forsskal 164

Roemeria Medic. 339, 341
  dodecandra (Forsskal) Stapf 341
  hybrida (L.) DC. 341
    subsp. dodecandra (Forsskal) E.A. Durande & Barratte 341, *348*
    subsp. *hybrida* Fl. Kuwait 341
Rokejeka capillaris Forsskal 225
Rorippa Scop. 385, 431
  microphylla (Boenn.) Hylander 431
  nasturtium-aquatica (L.) Hayek 431
  palustris (L.) Besser 431, *432*
Rosa L. 481, 488
  abyssinica Lindley 20, 26, 488, *489*
  barbeyi Boulenger 488, 490
  bottaiana Boulenger 488
  moschata Herrm. var. *abyssinica* (Lindley) Crépin 488
  schweinfurthii Boulenger 488
Rosaceae 480
Rubus L. 481, 484
  apetalus Poir. 484, 486
  arabicus (Deflers) Schweinf. *482*, 484, 485
  asirensis D.F. Chamb. 484, 485
  caesius L. 484, 485
  glandulosus Bell. var. *arabicus* Deflers 485
  petitianus A. Rich. 486
  sanctus Schreber *482*, 484, 486
Rumex L 128, 134
  conglomeratus Murr. 135, *137*, 138
  cyprius Murb. 135, *137*, 138
  dentatus L. 135, *137*, 139
  lacerus Balbis 136
  limoniastrum Jaub. & Spach. 135
  nepalensis Schwartz 138
  nervosus Vahl 135, *137*
  obtusifolius Schwartz 138
  persicarioides Forsskal 135
  pictus Forsskal 135, 136, *137*
  sp. A 135, 138
  steudelii Hochst. ex A. Rich. 135, *137*, 138
  vesicarius L. 135, 136, *137*
Ruppia maritima 22

Sageretia spiciflora 21
Salicaceae 83
Salicornia L. 233, 253
  cruciata Forsskal 252
  europaea L. 17, 253, *254*
  glauca Del. 253
  macrostachya Moric. 253
  perfoliata Forsskal 252
  strobilacea Pallas 252

Index

Salix L. 83
  acmophylla Boiss. 84
  excelsa S.G. Gmel. 84, 85
  mucronata Thunb. *82*, 84
  subserrata Willd. 84
Salsola L. 234, 235, 267
  Sect. *Sephragidanthus* Iljin 277
  aethiopica Botsch. 275
  alopecuroides Del. 263
  arabica Botsch. 268, 272
  articulata Forsskal 282
  baryosma Collenette 275
  baryosma (Roemer & Schultes) Dandy 275
  bottae (Jaub.& Spach) Boiss. 277
  chaudharyi Botsch. 276
  congesta N.E. Br. 275
  cyclophylla Baker 268, 272, *274*
  cycloptera Stapf 280
  delileana Botsch. 276
  drummondii Ulbr. 268, 271
  foetida Del. ex Sprengel 275
  forskalii Schweinf. 275
  hadramautica Baker 271
  hyssopifolia Pallas 250
  imbricata Collenette 275
  imbricata Forsskal 268, 275
  inermis Collenette 271
  inermis Forsskal 267, 269
  jordanicola Eig 267, 269, *270*
  kali L. 267, 269
  lachnantha (Botsch.) Botsch. 268, 276
  lancifolia (Boiss.) Boiss. 279
  leucophylla Baker 271
  longifolia Blatter 279
  mandavillei Botsch. 276
  mollis Desf. 257
  mucronata Forsskal 281
  muricata L. 250
  obpyrifolia Botsch. & Akhani 271
  omanensis Boulos 268, 272, *273*
  postii 265
  rosmarinus (Bunge ex Boiss.) Solms-Laub. 265
  rubescens Franchet 268, *270*, 271
  schweinfurthii Solms-Laub. 268, 271
  semhahensis Vierh. 280
  spinescens Moq. 268, *274*, 275
  tetrandra Forsskal 268, *274*, 275
  tomentosa (Moq.) Spach subsp. *lachnantha* Botsch. 276
  vermiculata L. subsp. *villosa* (Schultes) Eig 276
    var. *villosa* (Schultes) Moq. 276
  vermiculata Mandaville 276
  villosa Schultes 268, *270*, 276
  volkensii Asch. & Schweinf. 267, 269
Saltia R. Br. 284, 305
  papposa (Forsskal) Moq. *303*, 305
Salvadora persica 20
Sansevieria ehrenbergii 21
Santalaceae 118
Saponaria barbata Barkoudah 223
  montana (Balf.f.) Barkoudah 224
  umbricola J.R.I. Wood 225
  vaccaria L. 222
Savignyna DC. 384, 406
  aegyptiaca DC. 406
  parviflora (Del.) Webb 406, *409*
Schanginia C. Meyer 255
  aegyptiaca (Hasselq.) Aellen 256
  baccata (Forsskal ex J.F. Gmelin) Moq. 256
  hortensis (Forsskal ex J.F. Gmelin) Moq. 256
Schimpera Hochst. & Steud. 384, 417
  arabica Hochst. & Steud. *414*, 417
  persica Boiss. 417
Schouwia DC. 384, 406
  arabica DC. 406
  purpurea (Forsskal) Schweinf. 406, *409*
  schimperi Jaub. & Spach 406
  thebaica Webb 406
Scleranthus L. 176, 188
  orientalis Rössler *186*, 188
Sclerocephalus Boiss. 176, 187
  arabicus Boiss. *186*, 187
Sciophularia deserti 26
  hypericifolia 16
Sedum L. 463, 467
  caespitosum (Cav.) DC. 467, *478*
  hispanicum L. 467, *478*
  rubrum (L.) Thell. 467
Seidlitzia Bunge ex Boiss. 234, 265
  rosmarinus Bunge ex Boiss. 17, *258*, 265
Selaginella P. Beauv. 35, 38
  arabica Baker 40
  goudotana Spring *39*
    var. abyssinica (Spring) Bizzarri 40
  imbricata (Forsskal) Spring 38, *39*
  perpusilla Bak. *39*, 40
  yemensis (Swartz) Spring *39*, 40
Selaginellaceae 38
Sempervivum chrysanthum Britten 468
Sesuvium L. 156, 161
  sesuvioides (Fenzl) Verdc. 161, *163*
  verrucosum Raf. 161
Sevada Moq. 235, 279
  schimperi Moq. *258*, 279
Sideroxylon buxifolium 21
Silene L. 177, 209
  affinis Boiss. 216
  apetala Willd. 210, *212*, 214
  arabica Boiss. 210, 216
  arenosa C. Koch 211, 220
  asirensis D.F. Chamb. 210, 219
  austroiranica Rech. f. 211, 221
  bupleuroides L. subsp. *solenocalyx* (Boiss. & Huet) Melzh. 211
  burchellii Otth. 210, *212*, 214
  colorata Poiret subsp. olivieriana (Otth.) Rohrb. 210, 217
  conica L. 210, *212*, 213
  coniflora Otth. 210, *212*, 213
  conoidea L. 210, *212*, 213
  corylina D.F. Chamb. 210, *212*, 216
  crassipes Fenzl 210, 219
  engleri Pax 215
  flammolifolia Steud. ex A. Rich. var. canescens F.N. Williams 221
  gallica L. 210, *212*, 214
  hochstetteri Rohrb. *212*, 215
  hussonii Boiss. 210, 220
  inflata (Salisb.) Sm. 211
  leyseroides Boiss. 220
  linearis Decne 211, 221
  macrosolen Steud. ex A. Rich. 210, 211, *212*
  olivieriana Otth. 217
  peduncularis Boiss. 222
  schweinfurthii Rohrb. 215
  succulenta Forsskal 210, 220
  villosa Forsskal 210, 217
    form 'A' 217
    form 'B' 219
  vulgaris (Moench,) Garcke 209, 211, *212*
  yemensis Deflers 210, *212*, 215
Sinapis L. 387, 392
  alba L. 392, *394*
  arvensis L. 392, *394*
Sisymbrium L. 386, 442
  erysimoides Desf. 442, *444*, 445
  irio L. 442, *444*
  loeselii L. 442, 443
  officinale (L.) Scop. 442, *444*, 445
  orientale L. 442, 443, *444*
  pinnatifidum Forsskal 442
  septulatum DC. 442, 443, *444*
Sodada decidua Forsskal 372
Spergula L. 176, 199
  fallax (Lowe) Krause 199, *200*
Spergularia (Pers.) J. & C. Presl 176, 201

Index

bocconei (Scheele) Graebn. *200*, 201, 202
diandra (Guss.) Boiss. 201, 202
fallax Lowe 199
marina (L.) Bessler 201
marina (L.) Griseb. ·201
media (L.) Presl. 202
rubra (L.) J. & C. Presl. 202
salina J. & C. Presl. 201
Spermatophyta 71
Sphaerocoma T. Anderson 175, 183
aucheri Boiss. *180*, 183, 184
hookeri T. Anders. *180*, 183
Spinacia L. 234, 242
oleracea L. 243
Stellaria L. 177, 208
media (L.) Vill. *205*, 209
subsp. *pallida* (Dumort.) Aschers. & Graeb. 209
pallida (Dumort.) Murb. 209
Stephania Lour. 319, 322
abyssinica (Dillon & A. Rich.) Walp. *321*, 322
Sterculia africana 21
var. socotrana 22
Sterigmostemum M. Bieb. 382, 440
sulphureum (Banks & Sol.) Bornm. *437*, 440
Stigmatella longistyla Eig 438
Stipagrostis drarii 16
Strigosella africana (L.) Botsch. 438
grandiflora (Bunge) Botsch. 439
Suaeda Forsskal ex Scop. 235, 255
aegyptiaca (Hasselq.) Zoh. 26, 255, 256, *258*
baccata Forsskal ex J.F. Gmelin 256
fruticosa Forsskal ex J.F. Gmelin 157
hortensis Forsskal ex J.F. Gmelin 256
maris-mortui Post 256
maritima (L.) Dumort. 255, 256
mesopotamica Eig 257
mollis (Desf.) Del. 257
monodiana Maire 257
monoica Forsskal ex J.F. Gmelin 255, 256
moschata A.J. Scott 255, 257
paulayana Vierh. 257
prostrata Pallas 256
rosmarinus Ehrenb. ex Boiss. 265
schimperi (Moq.) Martelli 279
vermiculata Forsskal ex J.F. Gmelin 255, 257
var. *puberula* C.B. Clarke 279

volkensii C.B. Clarke 257
Subularia purpurea Forsskal 406
Sycomorus sur (Forsskal) Miq. 99

Talinum Adans. 169
cuneifolium Willd. 170
portulacifolium (Forsskal) Asch. ex Schweinf. 170, *172*
Tapinanthus Blume 121, 125
globiferus (A. Rich.) Tieghem *123*, 125
Tapinostemma arabicum (Deflers) Tieghem 122
Tarchonanthus camphoratus 19
Teclea nobilis 19, 26
Tectaria Cav. 37, 67
gemmifera (Fée) Alston *44*, 67
Telephium L. 156, 169
sphaerospermum Boiss. 169
Terminalia brownii 20
Tetragonia L. 156, 167
pentandra Balf.f. 167, *168*
Thalictrum L. 309, 312
minus L. 312, *313*
Thelypteridaceae 65
Themeda quadrivalvis 22
Thesium L. 118
humile Vahl 118, *120*
radicans Hochst. 118, 119, *120*
schweinfurthii Engl. 119
stuhlmannii Engl. 118, *120*
viride A.W. Hill 119
Thlaspi L. 383, 415
perfoliatum L. *414*, 415
Thuja L. 71, 75
orientalis L. 75
Tillaea alata Viv 463
alsinoides Hook. f. 466
pentandra Edgew. 464
pharnaceoides Steud. 464
Tinospora Miers. 322
bakis (A. Rich.) Miers *321*, 322
Torularia O. Schulz 445
torulosa (Desf.) O. Schulz 445
Traganum Del. 234, 259
nudatum Del. *254*, 259
undatum Collenette 263
Trema Lour. 88
guineensis (Schum.) Engl. var. hochstetteri (Buchinger) Engl. 88
hochstetteri (Buchinger) Engl. 88
orientalis (L.) Blume *87*, 88
Trianthema fruticosa Vahl 181
monogyna L. 164
pentandra L. 164
polysperma Oliver 161
sedifolia Vis. 164
Trianthemum L. 156, 162

crystallinum (Forsskal) Vahl 162, *163*
portulacastrum L. 162, 164
sheilae A.G. Miller 162
triquetrum Willd. 162, 164
Tribulus arabicus 21
Trichilia emetica 20

Ulmaceae 85
Umbilicus A. DC. 461, 477
horizontalis (Guss.) DC. 477, *478*
rupestris (salisb.) Dandy 477, *479*
Urochondra setulosa 22
Urtica L. 104, 105
divaricata Forsskal 106
iners Forsskal 116
muralis Vahl 111
palmata Forsskal 108
parasitica Forsskal 111
pilulifera L. 105, 106, *107*
urens L. 105
Urticaceae 104

Vaccaria Wolf 177, 222
hispanica (Miller) Rauschert 222, *226*
pyramidata L. 222
Valeriana scandens Forsskal 151
Velezia L. 177, 227
rigida L. 27, *226*, 227
Verea alternans (Vahl) Sprengel 472
lanceolata (Forsskal) Sprengel 469-70
Viola akhdarensis 27
Viscaceae 126
Viscum L. 126
cruciatum Sieb. ex Boiss. 126
schimperi Engl. *120*, 126, 127
triflorum DC. *120*, 126
Volutella aphylla Forsskal 309

Wiegmannia arabica (J.F. Gmel.) Hochst. & Steud. 366
Woodsiaceae 65

Xerotia Oliv. 176, 198
arabica Oliv. *194*, 198

Zaleya Burm. f. 156, 164
pentandra (L.) C. Jeffrey *163*, 164
Zilla Forsskal 383, 403
spinosa (L.) Prantl 403, *405*
Ziziphus spina-christi 15, 20, 21
Zygophyllum mandavillei 17
qatarense 17, 21